Fundamental Constants

Quantity	Symbol	Approximate Value	Current Best Value[†]
Speed of light in vacuum	c	3.00×10^8 m/s	2.99792458×10^8 m/s
Gravitational constant	G	6.67×10^{-11} N·m²/kg²	$6.67259(85) \times 10^{-11}$ N·m²/kg²
Avogadro's number	N_A	6.02×10^{23} mol⁻¹	$6.0221367(36) \times 10^{23}$ mol⁻¹
Gas constant	R	8.315 J/mol·K = 1.99 cal/mol·K = 0.082 atm·liter/mol·K	8.314510(70) J/mol·K
Boltzmann's constant	k	1.38×10^{-23} J/K	$1.380658(12) \times 10^{-23}$ J/K
Charge on electron	e	1.60×10^{-19} C	$1.60217733(49) \times 10^{-19}$ C
Stefan-Boltzmann constant	σ	5.67×10^{-8} W/m²·K⁴	$5.67051(19) \times 10^{-8}$ W/m²·K⁴
Permittivity of free space	$\epsilon_0 = (1/c^2\mu_0)$	8.85×10^{-12} C²/N·m²	$8.854187817... \times 10^{-12}$ C²/N·m²
Permeability of free space	μ_0	$4\pi \times 10^{-7}$ T·m/A	$1.2566370614... \times 10^{-6}$ T·m/A
Planck's constant	h	6.63×10^{-34} J·s	$6.6260755(40) \times 10^{-34}$ J·s
Electron rest mass	m_e	9.11×10^{-31} kg = 0.000549 u = 0.511 MeV/c^2	$9.1093897(54) \times 10^{-31}$ kg = $5.48579903(13) \times 10^{-4}$ u
Proton rest mass	m_p	1.6726×10^{-27} kg = 1.00728 u = 938.3 MeV/c^2	$1.6726231(10) \times 10^{-27}$ kg = 1.007276470(12) u
Neutron rest mass	m_n	1.6749×10^{-27} kg = 1.008665 u = 939.6 MeV/c^2	$1.6749286(10) \times 10^{-27}$ kg = 1.008664904(14) u
Atomic mass unit (1 u)		1.6605×10^{-27} kg = 931.5 MeV/c^2	$1.6605402(10) \times 10^{-27}$ kg = 931.49432(28) MeV/c^2

[†]Reviewed 1993 by B. N. Taylor, National Institute of Standards and Technology. Numbers in parentheses indicate one standard deviation experimental uncertainties in final digits. Values without parentheses are exact (i.e., defined quantities).

Other Useful Data

Joule equivalent (1 cal)		4.186 J
Absolute zero (0 K)		−273.15°C
Earth:	Mass	5.97×10^{24} kg
	Radius (mean)	6.38×10^3 km
Moon:	Mass	7.35×10^{22} kg
	Radius (mean)	1.74×10^3 km
Sun:	Mass	1.99×10^{30} kg
	Radius (mean)	6.96×10^5 km
Earth-sun distance (mean)		149.6×10^6 km
Earth-moon distance (mean)		384×10^3 km

The Greek Alphabet

Alpha	A	α	Nu	N	ν
Beta	B	β	Xi	Ξ	ξ
Gamma	Γ	γ	Omicron	O	o
Delta	Δ	δ	Pi	Π	π
Epsilon	E	ε	Rho	P	ρ
Zeta	Z	ζ	Sigma	Σ	σ
Eta	H	η	Tau	T	τ
Theta	Θ	θ	Upsilon	Y	υ
Iota	I	ι	Phi	Φ	ϕ, φ
Kappa	K	κ	Chi	X	χ
Lambda	Λ	λ	Psi	Ψ	ψ
Mu	M	μ	Omega	Ω	ω

Values of Some Numbers

π	= 3.1415927	$\sqrt{2}$	= 1.4142136	ln 2	= 0.6931472	$\log_{10}e$	= 0.4342945
e	= 2.7182818	$\sqrt{3}$	= 1.7320508	ln 10	= 2.3025851	1 rad	= 57.2957795°

Mathematical Signs and Symbols

$\propto$	is proportional to	$\leq$	is less than or equal to
$=$	is equal to	$\geq$	is greater than or equal to
$\approx$	is approximately equal to	Σ	sum of
$\neq$	is not equal to	$\bar{x}$	average value of x
$>$	is greater than	Δx	change in x
$\gg$	is much greater than	$\Delta x \rightarrow 0$	Δx approaches zero
$<$	is less than	$n!$	$n(n-1)(n-2)\ldots(1)$
$\ll$	is much less than		

Unit Conversions (Equivalents)

Length

1 in. = 2.54 cm
1 cm = 0.394 in.
1 ft = 30.5 cm
1 m = 39.37 in. = 3.28 ft
1 mi = 5280 ft = 1.61 km
1 km = 0.621 mi
1 nautical mile (U.S.) = 1.15 mi = 6076 ft = 1.852 km
1 fermi = 1 femtometer (fm) = 10^{-15} m
1 angstrom (Å) = 10^{-10} m
1 light-year (ly) = 9.46×10^{15} m
1 parsec = 3.26 ly = 3.09×10^{16} m

Volume

1 liter (L) = 1000 mL = 1000 cm^3 = 1.0×10^{-3} m^3 =
   1.057 quart (U.S.) = 54.6 in.^{3}
1 gallon (U.S.) = 4 qt (U.S.) = 231 in.^{3} = 3.78 L =
   0.83 gal (Imperial)
1 m^3 = 35.31 ft^3

Speed

1 mi/h = 1.47 ft/s = 1.609 km/h = 0.447 m/s
1 km/h = 0.278 m/s = 0.621 mi/h
1 ft/s = 0.305 m/s = 0.682 mi/h
1 m/s = 3.28 ft/s = 3.60 km/h
1 knot = 1.151 mi/h = 0.5144 m/s

Angle

1 radian (rad) = 57.30° = 57°18′
1° = 0.01745 rad
1 rev/min (rpm) = 0.1047 rad/s

Time

1 day = 8.64×10^4 s
1 year = 3.156×10^7 s

Mass

1 atomic mass unit (u) = 1.6605×10^{-27} kg
1 kg = 0.0685 slug
[1 kg has a weight of 2.20 lb where g = 9.81 m/s^2.]

Force

1 lb = 4.45 N
1 N = 10^5 dyne = 0.225 lb

Energy and Work

1 J = 10^7 ergs = 0.738 ft·lb
1 ft·lb = 1.36 J = 1.29×10^{-3} Btu = 3.24×10^{-4} kcal
1 kcal = 4.18×10^3 J = 3.97 Btu
1 eV = 1.602×10^{-19} J
1 kWh = 3.60×10^6 J = 860 kcal

Power

1 W = 1 J/s = 0.738 ft·lb/s = 3.42 Btu/h
1 hp = 550 ft·lb/s = 746 W

Pressure

1 atm = 1.013 bar = 1.013×10^5 N/m^2
     = 14.7 lb/in.^{2} = 760 torr
1 lb/in.^{2} = 6.90×10^3 N/m^2
1 Pa = 1 N/m^2 = 1.45×10^{-4} lb/in.^{2}

SI Derived Units and Their Abbreviations

Quantity	Unit	Abbreviation	In Terms of Base Units[†]
Force	newton	N	kg·m/s^2
Energy and work	joule	J	kg·m^2/s^2
Power	watt	W	kg·m^2/s^3
Pressure	pascal	Pa	kg/(m·s^2)
Frequency	hertz	Hz	s^{-1}
Electric charge	coulomb	C	A·s
Electric potential	volt	V	kg·m^2/(A·s^3)
Electric resistance	ohm	Ω	kg·m^2/(A^2·s^3)
Capacitance	farad	F	A^2·s^4/(kg·m^2)
Magnetic field	tesla	T	kg/(A·s^2)
Magnetic flux	weber	Wb	kg·m^2/(A·s^2)
Inductance	henry	H	kg·m^2/(s^2·A^2)

[†]kg = kilogram (mass), m = meter (length), s = second (time), A = ampere (electric current).

Metric (SI) Multipliers

Prefix	Abbreviation	Value
exa	E	10^{18}
peta	P	10^{15}
tera	T	10^{12}
giga	G	10^9
mega	M	10^6
kilo	k	10^3
hecto	h	10^2
deka	da	10^1
deci	d	10^{-1}
centi	c	10^{-2}
milli	m	10^{-3}
micro	μ	10^{-6}
nano	n	10^{-9}
pico	p	10^{-12}
femto	f	10^{-15}
atto	a	10^{-18}

PHYSICS

for

SCIENTISTS & ENGINEERS

Part 5

PHYSICS

for

SCIENTISTS & ENGINEERS

Third Edition

DOUGLAS C. GIANCOLI

PRENTICE HALL
Upper Saddle River, New Jersey 07458

Editor-in-Chief: Paul F. Corey
Production Editor: Susan Fisher
Executive Editor: Alison Reeves
Development Editor: David Chelton
Director of Marketing: John Tweedale
Senior Marketing Manager: Erik Fahlgren
Assistant Vice President of Production and Manufacturing: David W. Riccardi
Executive Managing Editor: Kathleen Schiaparelli
Manufacturing Manager: Trudy Pisciotti
Art Manager: Gus Vibal
Director of Creative Services: Paul Belfanti
Advertising and Promotions Manager: Elise Schneider
Editor in Chief of Development: Ray Mullaney
Project Manager: Elizabeth Kell
Photo Research: Mary Teresa Giancoli
Photo Research Administrator: Melinda Reo
Copy Editor: Jocelyn Phillips
Editorial Assistant: Marilyn Coco
Cover photo: Onne van der Wal/Young America
Composition: Emilcomp srl / Preparé Inc.

© 2000, 1989, 1984 by Douglas C. Giancoli
Published by Prentice Hall
Upper Saddle River, NJ 07458

Printed in the United States of America

10   9   8   7   6   5   4   3   2   1

ISBN   0-13-029098-X

Prentice-Hall International (UK) Limited, *London*
Prentice-Hall of Australia Pty. Limited, *Sydney*
Prentice-Hall Canada Inc., *Toronto*
Prentice-Hall Hispanoamericana, S.A., *Mexico City*
Prentice-Hall of India Private Limited, *New Delhi*
Prentice-Hall of Japan, Inc., *Tokyo*
Prentice-Hall (*Singapore*) Pte. Ltd.
Editora Prentice-Hall do Brasil, Ltda., *Rio de Janeiro*

CONTENTS

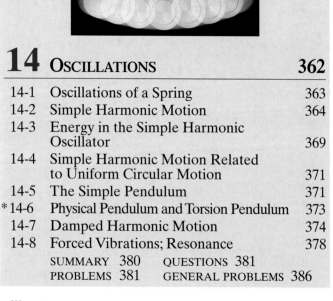

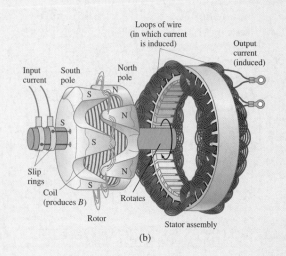

Loops of wire
(in which current
is induced)

Output
current
(induced)

Input
current

South
pole

North
pole

Slip
rings

Coil
(produces B)

Rotates

Rotor

Stator assembly

(b)

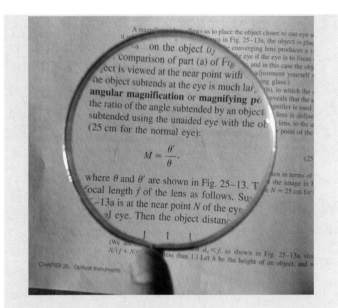

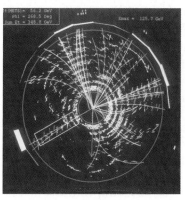

PREFACE

A Brand New Third Edition

It has been more than ten years since the second edition of this calculus-based introductory physics textbook was published. A lot has changed since then, not only in physics itself, but also in how physics is presented. Research in how students learn has provided textbook authors new opportunities to help students learn physics and learn it well.

This third edition comes in two versions. The standard version covers all of classical physics plus a chapter on special relativity and one on the early quantum theory. The extended version, with modern physics, contains a total of nine detailed chapters on modern physics, ending with astrophysics and cosmology. This book retains the original approach: in-depth physics, concrete and nondogmatic, readable.

This new third edition has many improvements in the physics and its applications. Before discussing those changes in detail, here is a list of some of the overall changes that will catch the eye immediately.

Full color throughout is not just cosmetic, although fine color photographs do help to attract the student readers. More important, full color diagrams allow the physics to be displayed with much greater clarity. We have not stopped at a 4-color process; this book has actually been printed in 5 pure colors (5 passes through the presses) to provide better variety and definition for illustrating vectors and other physics concepts such as rays and fields. I want to emphasize that color is used pedagogically to bring out the physics. For example, different types of vectors are given different colors—see the chart on page xxxi.

Many more diagrams, almost double the number in the previous edition, have all been done or redone carefully using full color; there are many more graphs and many more photographs throughout. See for example in optics where new photographs show lenses and the images they make.

Marginal notes have been added as an aid to students to (i) point out what is truly important, (ii) serve as a sort of outline, and (iii) help students find details about something referred to later that they may not remember so well. Besides such "normal" marginal notes, there are also marginal notes that point out brief *problem solving* hints, and others that point out interesting *applications*.

The great laws of physics are emphasized by giving them a marginal note all in capital letters and enclosed in a rectangle. The most important equations, especially those expressing the great laws, are further emphasized by a tan-colored screen behind them.

Chapter opening photographs have been chosen to illustrate aspects of each chapter. Each was chosen with an eye to writing a caption which could serve as a kind of summary of what is in that chapter, and sometimes offer a challenge. Some chapter-opening photos have vectors or other analysis superimposed on them.

Page layout: complete derivations. Serious attention has been paid to how each page was formatted, especially for page turns. Great effort has been made to keep important derivations and arguments on facing pages. Students then don't have to turn back to check. More important, readers repeatedly see before them, on two facing pages, an important slice of physics.

Two new kinds of Examples: Conceptual Examples and **Estimates.**

New Physics

The whole idea of a new edition is to improve, to bring in new material, and to delete material that is verbose and only makes the book longer or is perhaps too advanced and not so useful. Here is a brief summary of a few of the changes involving the physics iself. These lists are selections, not complete lists.

New discoveries:
- planets revolving around distant stars
- Hubble Space Telescope
- updates in particle physics and cosmology, such as inflation and the age of the universe

New physics topics added:
- new treatment of how to make estimates (Chapter 1), including new Estimating Examples throughout (in Chapter 1, estimating the volume of a lake, and the radius of the Earth)
- symmetry used much more, including for solving problems
- new Tables illustrating the great range of lengths, time intervals, masses, voltages
- gravitation as curvature of space, and black holes (Chapter 6)
- engine efficiency (Chapter 8 as well as Chapter 20)
- rolling with and without slipping, and other useful details of rotational motion (Chapter 10)
- forces in structures including trusses, bridges, arches, and domes (Chapter 12)
- square wave (Chapter 15)
- using the Maxwell distribution (Chapter 18)
- Otto cycle (Chapter 20)
- statistical calculation of entropy change in free expansion (Chapter 20)
- effects of dielectrics on capacitor connected and not (Chapter 24)
- grounding to avoid electric hazards (Chapter 25)
- three phase ac (Chapter 31)
- equal energy in **E** and **B** of EM wave (Chapter 32)
- radiation pressure, EM wave (Chapter 32)
- photos of lenses and mirrors with their images (Chapter 33)
- detailed outlines for ray tracing with mirrors and lenses (Chapters 33, 34)
- lens combinations (Chapter 34)
- new radiation standards (Chapter 43)
- Higgs boson, supersymmetry (Chapter 44)

Modern physics. A number of modern physics topics are discussed in the framework of classical physics. Here are some highlights:
- gravitation as curvature of space, and black holes (Chapter 6)
- planets revolving around distant stars (Chapter 6)
- kinetic energy at relativistic speeds (Chapter 7)
- nuclear collisions (Chapter 9)
- star collapse (Chapter 10)
- galaxy red shift, Doppler (Chapter 16)
- atoms, theory of (Chapters 17, 18, 21)
- atomic theory of thermal expansion (Chapter 17)
- mass of hydrogen atom (Chapter 17)
- atoms and molecules in gases (Chapters 17, 18)
- molecular speeds (Chapter 18)
- equipartition of energy; molar specific heats (Chapter 19)
- star size (Chapter 19)
- molecular dipoles (Chapters 21, 23)
- cathode ray tube (Chapters 23, 27)
- electrons in a wire (Chapter 25)
- superconductivity (Chapter 25)
- discovery and properties of the electron, e/m, oil drop experiment (Chapter 27)
- Hall effect (Chapter 27)

- magnetic moment of electrons (Chapter 27)
- mass spectrometer (Chapter 27)
- velocity selector (Chapter 27)
- electron spin in magnetic materials (Chapter 28)
- light and EM wave emission (Chapter 32)
- spectroscopy (Chapter 36)

Many other examples of modern physics are found as Problems, even in early chapters. Chapters 37 and 38 contain the modern physics topics of Special Relativity, and an introduction to Quantum Theory and Models of the Atom. The longer version of this text, "with Modern Physics," contains an additional seven chapters (for a total of nine) which present a detailed and extremely up-to-date treatment of modern physics: Quantum Mechanics of Atoms (Chapters 38 to 40); Molecules and Condensed Matter (Chapter 41); Nuclear Physics (Chapter 42 and 43); Elementary Particles (Chapter 44); and finally Astrophysics, General Relativity, and Cosmology (Chapter 45).

Revised physics and reorganizations. First of all, a major effort has been made to not throw everything at the students in the first few chapters. The basics have to be learned first; many aspects can come later, when the students are more prepared. Secondly, a great part of this book has been rewritten to make it clearer and more understandable to students. Clearer does not always mean simpler or easier. Sometimes making it "easier" actually makes it harder to understand. Often a little more detail, without being verbose, can make an explanation clearer. Here are a few of the changes, big and small:

- new graphs and diagrams to clarify velocity and acceleration; deceleration carefully treated.
- unit conversion now a new Section in Chapter 1, instead of interrupting kinematics.
- circular motion: Chapter 3 now gives only the basics, with more complicated treatment coming later: non-uniform circular motion in Chapter 5, angular variables in Chapter 10.
- Newton's second law now written throughout as $m\mathbf{a} = \Sigma\mathbf{F}$, to emphasize inclusion of all forces acting on a body.
- Newton's third law follows the second directly, with inertial reference frames placed earlier. New careful discussions to head off confusion when using Newton's third law.
- careful rewriting of chapters on Work and Energy, especially potential energy, conservative and nonconservative forces, and the conservation of energy.
- renewed emphasis that $\Sigma\tau = I\alpha$ is not always valid: only for an axis fixed in an inertial frame or if axis is through the CM (Chapters 10 and 11).
- rolling motion introduced early in Chapter 10, with more details later, including rolling with and without slipping.
- rotating frames of reference, and Coriolis, moved later, to Chapter 11, shortened, optional, but still including why an object does not fall straight down on Earth.
- fluids reduced to a single chapter (13); some topics and details dropped or greatly shortened.
- clearer details on how an object floats (Chapter 13).
- distinction between wave interference in space, and in time (beats) (Chapter 16).
- thermodynamics reduced to four chapters; the old chapters on Heat and on the First Law of Thermodynamics have been combined into one (19), with some topics shortened and a more rational sequence of topics achieved.
- heat transfer now follows the first law of thermodynamics (Chapter 19).
- electric potential carefully rewritten for accuracy (Chapter 23).
- CRT, computer monitors, TV, treated earlier (Chapter 23).
- use of Q_{encl} and I_{encl} for Gauss's and Ampère's laws, with subscripts meaning "enclosed".
- Ohm's law and definition of resistance carefully redone (Chapter 25).
- sources of magnetic field, Chapter 28, reorganized for ease of understanding, with some new material, and deletion of the advanced topic on magnetization vector.
- circuits with L, C, and/or R now introduced via Kirchhoff's loop rule, and clarified in other ways too (Chapters 30, 31).
- streamlined Maxwell's equations, with displacement current downplayed (Chapter 32).
- optics reduced to four chapters; polarization is now placed in the same chapter as diffraction.

New Pedagogy

All of the above mentioned revisions, rewritings, and reorganizations are intended to help students learn physics better. They were done in response to contemporary research in how students learn, as well as to kind and generous input from professors who have read, reviewed, or used the previous editions. This new edition also contains some new elements, especially an increased emphasis on conceptual development:

Conceptual Examples, typically 1 or 2 per chapter, sometimes more, are each a sort of brief Socratic question and answer. It is intended that students will be stimulated by the question to think, or reflect, and come up with a response—before reading the Response given. Here are a few:

- using symmetry (Chapters 1, 44, and elsewhere)
- ball moving upward: misconceptions (Chapter 2)
- reference frames and projectile motion: where does the apple land? (Chapter 3)
- what exerts the force that makes a car move? (Chapter 4)
- Newton's third law clarification: pulling a sled (Chapter 4)
- free-body diagram for a hockey puck (Chapter 4)
- advantage of a pulley (Chapter 4), and of a lever (Chapter 12)
- to push or to pull a sled (Chapter 5)
- which object rolls down a hill faster? (Chapter 10)
- moving the axis of a spinning wheel (Chapter 11)
- tragic collapse (Chapter 12)
- finger at top of a full straw (Chapter 13)
- suction cups on a spacecraft (Chapter 13)
- doubling amplitude of SHM (Chapter 14)
- do holes expand thermally? (Chapter 17)
- simple adiabatic process: stretching a rubber band (Chapter 19)
- charge inside a conductor's cavity (Chapter 22)
- how stretching a wire changes its resistance (Chapter 25)
- series or parallel (Chapter 26)
- bulb brightness (Chapter 26)
- spiral path in magnetic field (Ch. 27)
- practice with Lenz's law (Chapter 29)
- motor overload (Chapter 29)
- emf direction in inductor (Chapter 30)
- photo with reflection—is it upside down? (Chapter 33)
- reversible light rays (Chapter 33)
- how tall must a full-length mirror be? (Chapter 33)
- diffraction spreading (Chapter 36)

Estimating Examples, roughly 10% of all Examples, also a new feature of this edition, are intended to develop the skills for making order-of-magnitude estimates, even when the data are scarce, and even when you might never have guessed that any result was possible at all. See, for example, Section 1–6, Examples 1–5 to 1–8.

Problem Solving, with New and Improved Approaches

Learning how to approach and solve problems is a basic part of any physics course. It is a highly useful skill in itself, but is also important because the process helps bring understanding of the physics. Problem solving in this new edition has a significantly increased emphasis, including some new features.

Problem-solving boxes, about 20 of them, are new to this edition. They are more concentrated in the early chapters, but are found throughout the book. They each outline a step-by-step approach to solving problems in general, and/or specifically for the material being covered. The best students may find these separate "boxes" unnecessary (they can skip them), but many students will find it helpful to be reminded of the general approach and of steps they can take to get started; and, I think, they help to build confidence. The general problem solving box in Section 4–8 is placed there, after students have had some experience wrestling with problems, and so may be strongly motivated to read it with close attention. Section 4–8 can, of course, be covered earlier if desired.

Problem-solving Sections occur in many chapters, and are intended to provide extra drill in areas where solving problems is especially important or detailed.

Examples. This new edition has many more worked-out Examples, and they all now have titles for interest and for easy reference. There are even two new categories of Example: Conceptual, and Estimates, as described above. Regular Examples serve as "practice problems". Many new ones have been added, some of the old ones have been dropped, and many have been reworked to provide greater clarity and detail: more steps are spelled out, more of "why we do it this way", and more discussion of the reasoning and approach. In sum, the idea is "to think aloud with the students", leading them to develop insight. The total number of worked-out Examples is about 30% greater than in the previous edition, for an average of 12 to 15 per chapter. There is a significantly higher concentration of Examples in the early chapters, where drill is especially important for developing skills and a variety of approaches. The level of the worked-out Examples for most topics increases gradually, with the more complicated ones being on a par with the most difficult Problems at the end of each chapter, so that students can see how to approach complex problems. Many of the new Examples, and improvements to old ones, provide relevant applications to engineering, other related fields, and to everyday life.

Problems at the end of each chapter have been greatly increased in quality and quantity. There are over 30% more Problems than in the second edition. Many of the old ones have been replaced, or rewritten to make them clearer, and/or have had their numerical values changed. Each chapter contains a large group of Problems arranged by Section and graded according to difficulty: level I Problems are simple, designed to give students confidence; level II are "normal" Problems, providing more of a challenge and often the combination of two different concepts; level III are the most complex, typically combining different issues, and will challenge even superior students. The arrangement by Section number means only that those Problems depend on material up to and including that Section: earlier material may also be relied upon. The ranking of Problems by difficulty (I, II, III) is intended only as a guide.

General Problems. About 70% of Problems are ranked by level of difficulty (I, II, III) and arranged by Section. New to this edition are General Problems that are unranked and grouped together at the end of each chapter, and account for about 30% of all problems. The average total number of Problems per chapter is about 90. Answers to odd-numbered Problems are given at the back of the book.

Complete Physics Coverage, with Options

This book is intended to give students the opportunity to obtain a thorough background in all areas of basic physics. There is great flexibility in choice of topics so that instructors can choose which topics they cover and which they omit. Sections marked with an asterisk can be considered optional, as discussed more fully on p. xxv. Here I want to emphasize that topics not covered in class can still be read by serious students for their own enrichment, either immediately or later. Here is a partial list of physics topics, not the standard ones, but topics that might not usually be covered, and that represent how thorough this book is in its coverage of basic physics. Section numbers are given in parentheses.

- use of calculus; variable acceleration (2–8)
- nonuniform circular motion (5–4)
- velocity-dependent forces (5–5)
- gravitational versus inertial mass; principle of equivalence (6–8)
- gravitation as curvature of space; black holes (6–9)
- kinetic energy at very high speed (7–5)
- potential energy diagrams (8–9)
- systems of variable mass (9–10)
- rotational plus translational motion (10–11)
- using $\Sigma \tau_{CM} = I_{CM} \alpha_{CM}$ (10–11)
- derivation of $K = K_{CM} + K_{rot}$ (10–11)
- why does a rolling sphere slow down? (10–12)
- angular momentum and torque for a system (11–4)
- derivation of $d\mathbf{L}_{CM}/dt = \Sigma \boldsymbol{\tau}_{CM}$ (11–4)
- rotational imbalance (11–6)
- the spinning top (11–8)
- rotating reference frames; inertial forces (11–9)
- coriolis effect (11–10)
- trusses (12–7)
- flow in tubes: Poiseuille's equation (13–11)
- surface tension and capillarity (13–12)
- physical pendulum; torsion pendulum (14–6)
- damped harmonic motion: finding the solution (14–7)
- forced vibrations; equation of motion and its solution; Q-value (14–8)
- the wave equation (15–5)
- mathematical representation of waves; pressure wave derivation (16–2)
- intensity of sound related to amplitude (16–3)
- interference in space and in time (16–6)
- atomic theory of expansion (17–4)
- thermal stresses (17–5)
- ideal gas temperature scale (17–10)
- calculations using the Maxwell distribution of molecular speeds (18–2)
- real gases (18–3)
- vapor pressure and humidity (18–4)
- van der Waals equation of state (18–5)
- mean free path (18–6)
- diffusion (18–7)
- equipartition of energy (19–8)
- energy availability; heat death (20–8)
- statistical interpretation of entropy and the second law (20–9)
- thermodynamic temperature scale; absolute zero and the third law (20–10)
- electric dipoles (21–11, 23–6)
- experimental basis of Gauss's and Coulomb's laws (22–4)
- general relation between electric potential and electric field (23–2, 23–8)
- electric fields in dielectrics (24–5)
- molecular description of dielectrics (24–6)
- current density and drift velocity (25–8)
- superconductivity (25–9)
- RC circuits (26–4)
- use of voltmeters and ammeters; effects of meter resistance (26–5)
- transducers (26–6)
- magnetic dipole moment (27–5)
- Hall effect (27–8)
- operational definition of the ampere and coulomb (28–3)
- magnetic materials—ferromagnetism (28–7)
- electromagnets and solenoids (28–8)
- hysteresis (28–9)
- paramagnetism and diamagnetism (28–10)
- counter emf and torque; eddy currents (29–5)
- Faraday's law—general form (29–7)
- force due to changing **B** is nonconservative (29–7)
- LC circuits and EM oscillations (30–5)
- AC resonance; oscillators (31–6)
- impedance matching (31–7)
- three phase AC (31–8)
- changing electric fields produce magnetic fields (32–1)
- speed of light from Maxwell's equations (32–5)
- radiation pressure (32–8)
- fiber optics (33–7)
- lens combinations (34–3)
- aberrations of lenses and mirrors (34–10)
- coherence (35–4)
- intensity in double-slit pattern (35–5)
- luminous intensity (35–8)
- intensity for single-slit (36–2)
- diffraction for double–slit (36–3)
- limits of resolution, the λ limit (36–4, 36–5)
- resolution of the human eye and useful magnification (36–6)
- spectroscopy (36–8)
- peak widths and resolving power for a diffraction grating (36–9)
- x-rays and x-ray diffraction (36–10)
- scattering of light by the atmosphere (36–12)
- time–dependent Schrödinger equation (39–6)
- wave packets (39–7)
- tunneling through a barrier (39–9)
- free-electron theory of metals (41–6)
- semiconductor electronics (41–9)
- standard model, symmetry, QCD, GUT (44–9, 44–10)
- astrophysics, cosmology (Ch. 45)

New Applications

Relevant applications to everyday life, to engineering, and to other fields such as geology and medicine, provide students with motivation and offer the instructor the opportunity to show the relevance of physics. Applications are a good response to students who ask "Why study physics?" Many new applications have been added in this edition. Here are some highlights:

- airbags (Chapter 2)
- elevator and counterweight (Chapter 4)
- antilock brakes and skidding (Chapter 5)
- geosynchronous satellites (Chapter 6)
- hard drive and bit speed (Chapter 10)
- star collapse (Chapter 10)
- forces within trusses, bridges, arches, domes (Chapter 12)
- the Titanic (Chapter 12)
- Bernoulli's principle: wings, sailboats, TIA, plumbing traps and bypasses (Chapter 13)
- pumps (Chapter 13)
- car springs, shock absorbers, building dampers for earthquakes (Chapter 14)
- loudspeakers (Chapters 14, 16, 27)
- autofocusing cameras (Chapter 16)
- sonar (Chapter 16)
- ultrasound imaging (Chapter 16)
- thermal stresses (Chapter 17)
- R-values, thermal insulation (Ch. 19)
- engines (Chapter 20)
- heat pumps, refrigerators, AC; coefficient of performance (Chapter 20)
- thermal pollution (Chapter 20)
- electric shielding (Chapters 21, 28)
- photocopier (Chapter 21)
- superconducting cables (Chapter 25)
- jump starting a car (Chapter 26)
- aurora borealis (Chapter 27)
- solenoids and electromagnetics (Ch. 28)
- computer memory and digital information (Chapter 29)
- seismograph (Chapter 29)

- tape recording (Chapter 29)
- loudspeaker cross-over network (Ch. 31)
- antennas, for **E** or **B** (Chapter 32)
- TV and radio; AM and FM (Chapter 32)
- eye and corrective lenses (Chapter 34)
- mirages (Chapter 35)
- liquid crystal displays (Chapter 36)
- CAT scans, PET, MRI (Chapter 43)

Some old favorites retained (and improved):

- pressure gauges (Chapter 13)
- musical instruments (Chapter 16)
- humidity (Chapter 18)
- CRT, TV, computer monitors (Ch. 23, 27)
- electric hazards (Chapter 25)
- power in household circuits (Chapter 25)
- ammeters and voltmeters (Chapter 26)
- microphones (Chapters 26, 29)
- transducers (Chapter 26, and elsewhere)
- electric motors (Chapter 27)
- car alternator (Chapter 29)
- electric power transmission (Chapter 29)
- capacitors as filters (Chapter 31)
- impedance matching (Chapter 31)
- fiber optics (Chapter 33)
- cameras, telescopes, microscopes, other optical instruments (Chapter 34)
- lens coatings (Chapter 35)
- spectroscopy (Chapter 36)
- electron microscopes (Chapter 38)
- lasers, holography, CD players (Ch. 40)
- semiconductor electronics (Chapter 41)
- radioactivity (Chapters 42 and 43)

Deletions

Something had to go, or the book would have been too long. Lots of subjects were shortened—the detail simply isn't necessary at this level. Some topics were dropped entirely: polar coordinates; center-of-momentum reference frame; Reynolds number (now a Problem); object moving in a fluid and sedimentation; derivation of Poiseuille's equation; Stoke's equation; waveguide and transmission line analysis; electric polarization and electric displacement vectors; potentiometer (now a Problem); negative pressure; combinations of two harmonic motions; adiabatic character of sound waves; central forces.

Many topics have been shortened, often a lot, such as: velocity-dependent forces; variable acceleration; instantaneous axis; surface tension and capillarity; optics topics such as some aspects of light polarizarion. Many of the brief historical and philosophical issues have been shortened as well.

General Approach

This book offers an in-depth presentation of physics, and retains the basic approach of the earlier editions. Rather than using the common, dry, dogmatic approach of treating topics formally and abstractly first, and only later relating the material to the students' own experience, my approach is to recognize that physics is a description of reality and thus to start each topic with concrete observations and experiences that students can directly relate to. Then we move on to the generalizations and more formal treatment of the topic. Not only does this make the material more interesting and easier to understand, but it is closer to the way physics is actually practiced.

This new edition, even more than previous editions, aims to explain the physics in a readable and interesting manner that is accessible and clear. It aims to teach students by anticipating their needs and difficulties, but without oversimplifying. Physics is all about us. Indeed, it is the goal of this book to help students "see the world through eyes that know physics."

As mentioned above, this book includes of a wide range of Examples and applications from technology, engineering, architecture, earth sciences, the environment, biology, medicine, and daily life. Some applications serve only as examples of physical principles. Others are treated in depth. But applications do not dominate the text—this is, after all, a physics book. They have been carefully chosen and integrated into the text so as not to interfere with the development of the physics but rather to illuminate it. You won't find essay sidebars here. The applications are integrated right into the physics. To make it easy to spot the applications, a new *Physics Applied* marginal note is placed in the margin (except where diagrams in the margin prevent it).

It is assumed that students have started calculus or are taking it concurrently. Calculus is treated gently at first, usually in an optional Section so as not to burden students taking calculus concurrently. For example, using the integral in kinematics, Chapter 2, is an optional Section. But in Chapter 7, on work, the integral is discussed fully for all readers.

Throughout the text, *Système International* (SI) units are used. Other metric and British units are defined for informational purposes. Careful attention is paid to significant figures. When a certain value is given as, say, 3, with its units, it is meant to be 3, not assumed to be 3.0 or 3.00. When we mean 3.00 we write 3.00. It is important for students to be aware of the uncertainty in any measured value, and not to overestimate the precision of a numerical result.

Rather than start this physics book with a chapter on mathematics, I have instead incorporated many mathematical tools, such as vector addition and multiplication, directly in the text where first needed. In addition, the Appendices contain a review of many mathematical topics such as trigonometric identities, integrals, and the binomial (and other) expansions. One advanced topic is also given an Appendix: integrating to get the gravitational force due to a spherical mass distribution.

It is necessary, I feel, to pay careful attention to detail, especially when deriving an important result. I have aimed at including all steps in a derivation, and have tried to make clear which equations are general, and which are not, by explicitly stating the limitations of important equations in brackets next to the equation, such as

$$x = x_0 + v_0 t + \tfrac{1}{2} a t^2. \qquad\qquad \text{[constant acceleration]}$$

The more detailed introduction to Newton's laws and their use is of crucial pedagogic importance. The many new worked-out Examples include initially fairly simple ones that provide careful step-by-step analysis of how to proceed in solving dynamics problems. Each succeeding Example adds a new element or a new twist that introduces greater complexity. It is hoped that this strategy will enable even less-well-prepared students to acquire the tools for using Newton's laws correctly. If students don't surmount this crucial hurdle, the rest of physics may remain forever beyond their grasp.

Rotational motion is difficult for most students. As an example of attention to detail (although this is not really a "detail"), I have carefully distinguished the position vector (**r**) of a point and the perpendicular distance of that point from an axis, which is called R in this book (see Fig. 10–2). This distinction, which enters particularly in connection with torque, moment of inertia, and angular momentum, is often not made clear—it is a disservice to students to use **r** or r for both without distinguishing. Also, I have made clear that it is not always true that $\Sigma\tau = I\alpha$. It depends on the axis chosen (valid if axis is fixed in an inertial reference frame, or through the CM). To not tell this to students can get them into serious trouble. (See pp. 250, 283, 284.) I have treated rotational motion by starting with the simple instance of rotation about an axis (Chapter 10), including the concepts of angular momentum and rotational kinetic energy. Only in Chapter 11 is the more general case of rotation about a point dealt with, and this slightly more advanced material can be omitted if desired (except for Sections 11–1 and 11–2 on the vector product and the torque vector). The end of Chapter 10 has an optional subsection containing three slightly more advanced Examples, using $\Sigma\tau_{CM} = I_{CM}\alpha_{CM}$: car braking distribution, a falling yo-yo, and a sphere rolling with and without slipping.

Among other special treatments is Chapter 28, Sources of Magnetic Field: here, in one chapter, are discussed the magnetic field due to currents (including Ampère's law and the law of Biot-Savart) as well as magnetic materials, ferromagnetism, paramagnetism, and diamagnetism. This presentation is clearer, briefer, and more of a whole, and all the content is there.

Organization

The general outline of this new edition retains a traditional order of topics: mechanics (Chapters 1 to 12); fluids, vibrations, waves, and sound (Chapter 13 to 16); kinetic theory and thermodynamics (Chapters 17 to 20). In the two-volume version of this text, volume I ends here, after Chapter 20. The text continues with electricity and magnetism (Chapters 21 to 32), light (Chapters 33 to 36), and modern physics (Chapters 37 and 38 in the short version, Chapters 37 to 45 in the extended version "with Modern Physics"). Nearly all topics customarily taught in introductory physics courses are included. A number of topics from modern physics are included with the classical physics chapters as discussed earlier.

The tradition of beginning with mechanics is sensible, I believe, because it was developed first, historically, and because so much else in physics depends on it. Within mechanics, there are various ways to order topics, and this book allows for considerable flexibility. I prefer, for example, to cover statics after dynamics, partly because many students have trouble working with forces without motion. Besides, statics is a special case of dynamics—we study statics so that we can prevent structures from becoming dynamic (falling down)—and that sense of being at the limit of dynamics is intuitively helpful. Nonetheless statics (Chapter 12) can be covered earlier, if desired, before dynamics, after a brief introduction to vector addition. Another option is light, which I have placed after electricity and magnetism and EM waves. But light could be treated immediately after the chapters on waves (Chapters 15 and 16). Special relativity is Chapter 37, but could instead be treated along with mechanics—say, after Chapter 9.

Not every chapter need be given equal weight. Whereas Chapter 4 might require $1\frac{1}{2}$ to 2 weeks of coverage, Chapter 16 or 22 may need only $\frac{1}{2}$ week.

Some instructors may find that this book contains more material than can be covered completely in their courses. But the text offers great flexibility in choice of topics. Sections marked with a star (asterisk) are considered optional. These Sections contain slightly more advanced physics material, or material not usually covered in typical courses, and/or interesting applications. They contain no material needed in later chapters (except perhaps in later optional Sections). This does not imply that all nonstarred Sections must be covered: there still remains considerable flexibility in the choice of material. For a brief course, all optional material could be dropped as well as major parts of Chapters 11, 13, 16, 26, 30, 31, and 36 as well as selected parts of Chapters 9, 12, 19, 20, 32, 34, and the modern physics chapters. Topics not covered in class can be a valuable resource for later study; indeed, this text can serve as a useful reference for students for years because of its wide range of coverage.

Thanks

Some 60 physics professors provided input or direct feedback on every aspect of this textbook. The reviewers and contributors to this third edition are listed below. I owe each a debt of gratitude.

Ralph Alexander, University of Missouri at Rolla

Zaven Altounian, McGill University

Charles R. Bacon, Ferris State University

Bruce Birkett, University of California, Berkeley

Art Braundmeier, Southern Illinois University at Edwardsville

Wayne Carr, Stevens Institute of Technology

Edward Chang, University of Massachusetts, Amherst

Charles Chiu, University of Texas at Austin

Lucien Crimaldi, University of Mississippi

Robert Creel, University of Akron

Alexandra Cowley, Community College of Philadelphia

Timir Datta, University of South Carolina

Gary DeLeo, Lehigh University

John Dinardo, Drexel University

Paul Draper, University of Texas, Arlington

Alex Dzierba, Indiana University

William Fickinger, Case Western University

Jerome Finkelstein, San Jose State University

Donald Foster, Wichita State University

Gregory E. Frances, Montana State University

Lothar Frommhold, University of Texas at Austin

Thomas Furtak, Colorado School of Mines

Edward Gibson, California State University, Sacramento

Christopher Gould, University of Southern California

John Gruber, San Jose State University

Martin den Boer, Hunter College

Greg Hassold, General Motors Institute

Joseph Hemsky, Wright State University

Laurent Hodges, Iowa State University

Mark Holtz, Texas Tech University

James P. Jacobs, University of Montana

James Kettler, Ohio University Eastern Campus

Jean Krisch, University of Michigan

Mark Lindsay, University of Louisville

Eugene Livingston, University of Notre Dame

Bryan Long, Columbia State Community College

Daniel Mavlow, Princeton University

Pete Markowitz, Florida International University

John McCullen, University of Arizona, Tucson

Peter Nemeth, New York University

Hon-Kie Ng, Florida State University

Eugene Patroni, Georgia Institute of Technology

Robert Pelcovits, Brown University

William Pollard, Valdosta State University

Joseph Priest, Miami University

Carl Rotter, West Virginia University

Lawrence Rees, Brigham Young University

Peter Riley, University of Texas at Austin

Roy Rubins, University of Texas at Arlington

Mark Semon, Bates College

Robert Simpson, University of New Hampshire

Mano Singham, Case Western University

Harold Slusher, University of Texas at El Paso

Don Sparks, Los Angeles Pierce Community College

Michael Strauss, University of Oklahoma

Joseph Strecker, Wichita State University

William Sturrus, Youngstown State University

Arthur Swift, University of Massachusetts, Amherst

Leo Takahasi, The Pennsylvania State University

Edward Thomas, Georgia Institute of Technology

Som Tyagi, Drexel University

John Wahr, University of Colorado

Robert Webb, Texas A & M University

James Whitmore, The Pennsylvania State University

W. Steve Quon, Ventura College

I owe special thanks to Irv Miller, not only for many helpful physics discussions, but for having worked out all the Problems and managed the team that also worked out the Problems, each checking the other, and finally for producing the Solutions Manual and all the answers to the odd-numbered Problems at the end of this book. He was ably assisted by Zaven Altounian and Anand Batra.

I am particularly grateful to Robert Pelcovits and Peter Riley, as well as to Paul Draper and James Jacobs, who inspired many of the new Examples, Conceptual Examples, and Problems.

Crucial for rooting out errors, as well as providing excellent suggestions, were the perspicacious Edward Gibson and Michael Strauss, both of whom carefully checked all aspects of the physics in page proof.

Special thanks to Bruce Birkett for input of every kind, from illuminating discussions on pedagogy to a careful checking of details in many sections of this book. I wish also to thank Professors Howard Shugart, Joe Cerny, Roger Falcone and Buford Price for helpful discussions, and for hospitality at the University of California, Berkeley. Many thanks also to Prof. Tito Arecchi at the Istituto Nazionale di Ottica, Florence, Italy, and to the staff of the Institute and Museum for the History of Science, Florence, for their hospitality.

Finally, I wish to thank the superb editorial and production work provided by all those with whom I worked directly at Prentice Hall: Susan Fisher, Marilyn Coco, David Chelton, Kathleen Schiaparelli, Trudy Pisciotti, Gus Vibal, Mary Teresa Giancoli, and Jocelyn Phillips.

The biggest thanks of all goes to Paul Corey, whose constant encouragement and astute ability to get things done, provided the single strongest catalyst.

The final responsibility for all errors lies with me, of course. I welcome comments and corrections.

D.C.G.

AVAILABLE SUPPLEMENTS

For the Student

Student Study Guide and Solutions Manual
Douglas Brandt, Eastern Illinois University. (0-13-021475-2)
Contains chapter objectives, summaries with additional examples, self-study quizzes, key mathematical equations, and complete worked-out solutions to alternate odd problems in the text.

Doing Physics with Spreadsheets: A Workbook
Gordon Aubrecht, T. Kenneth Bolland, and Michael Ziegler, all of The Ohio State University.
(0-13-021474-4)
Designed to introduce students to the use of spreadsheets for solving simple and complex physics problems. Students are either provided with spreadsheets or must construct their own, then use the model to most closely approximate natural behavior. The amount of spreadsheet construction and the complexity of the spreadsheet increases as the student gains experience.

Science on the Internet: A Student's Guide, 1999
Andrew Stull and Carl Adler (0-13-021308-X)
The perfect tool to help students take advantage of the *Physics for Scientists and Engineers, Third Edition* Web page. This useful resource gives clear steps to access Prentice Hall's regularly updated physics resources, along with an overview of general World Wide Web navigation strategies. Available FREE for students when packaged with the text.

Prentice Hall/*New York Times* Themes of the Times — Physics
This unique newspaper supplement brings together a collection of the latest physics-related articles from the pages of *The New York Times*. Updated twice per year and available FREE to students when packaged with the text.

For the Instructor

Instructor's Solutions Manual
Irvin A. Miller, Drexel University.
Print version (0-13-021381-0); Electronic (CD-ROM) version (0-13-021481-7)
Contains detailed worked solutions to every problem in the text. Electronic versions are available in CD-ROM (dual platform for both Windows and Macintosh systems) for instructors with Microsoft Word or Word-compatible software.

Test Item File
Robert Pelcovits, Brown University; David Curott, University of North Alabama; and Edward Oberhofer, University of North Carolina at Charlotte (0-13-021482-5)
Contains over 2200 multiple choice questions, about 25% conceptual in nature. All are referenced to the corresponding Section in the text and ranked by difficulty.

Prentice Hall Custom Test Windows (0-13-021477-9); Macintosh (0-13-021476-0)
Based on the powerful testing technology developed by Engineering Software Associates, Inc. (ESA), Prentice Hall Custom Test includes all questions from the Test Item File and allows instructors to create and tailor exams to their own needs. With the Online Testing Program, exams can also be administered on line and data can then be automatically transferred for evaluation. A comprehensive desk reference guide is included along with online assistance.

Transparency Pack (0-13-021470-1)
Includes approximately 400 full color transparencies of images from the text.

Media Supplements

Physics for Scientists and Engineers Web Site www.prenhall.com/giancoli
A FREE innovative online resource that provides students with a wealth of activities and exercises for each text chapter. Features on the site include:

- Practice Questions, Destinations (links to related sites), NetSearch keywords and algorithmically generated numeric Practice Problems by Carl Adler of East Carolina University.
- Physlet Problems (Java-applet simulations) by Wolfgang Christian of Davidson College.
- Warmups and Puzzles essay questions and Applications from Gregor Novak and Andrew Gavrin at Indiana University-Purdue University, Indianapolis.
- Ranking Task Exercises edited by Tom O'Kuma of Lee College, Curtis Hieggelke of Joliet Junior College and David Maloney of Indiana University-Purdue University, Fort Wayne.

Using Prentice Hall CW '99 technology, the website grades and scores all objective questions, and results can be automatically e-mailed directly to the instructors if so desired. Instructors can also create customized syllabi online and link directly to activities on the Giancoli website.

Presentation Manager CD-ROM
Dual Platform (Windows/Macintosh; 0-13-214479-5)
This CD-ROM enables instructors to build custom sequences of Giancoli text images and Prentice Hall digital media for playback in lecture presentations. The CD-ROM contains all text illustrations, digitized segments from the Prentice Hall *Physics You Can See* videotape as well as additional lab and demonstration videos and animations from the Prentice Hall *Interactive Journey Through Physics* CD-ROM. Easy to navigate with Prentice Hall Presentation Manager software, instructors can preview, sequence, and play back images, as well as perform keyword searches, add lecture notes, and incorporate their own digital resources.

Physics You Can See *Video*
(0-205-12393-7)
Contains eleven two- to five-minute demonstrations of classical physics experiments. It includes segments such as "Coin and Feather" (acceleration due to gravity), "Monkey and Gun" (projectile motion), "Swivel Hips" (force pairs), and "Collapse a Can" (atmospheric pressure).

CAPA: A Computer-Assisted Personalized Approach to Assignments, Quizzes, and Exams
CAPA is an on-line homework system developed at Michigan State University that instructors can use to deliver problem sets with randomized variables for each student. The system gives students immediate feedback on their answers to problems, and records their participation and performance. Prentice Hall has arranged to have half of the even-numbered problems of Giancoli, *Physics for Scientists and Engineers, Third Edition*, coded for use with the CAPA system. For additional information about the CAPA system, please visit the web site at http://www.pa.msu.edu/educ/CAPA/.

WebAssign
WebAssign is a web-based homework delivery, collection, grading, and recording service developed and hosted by North Carolina State University. Prentice Hall will arrange for end-of-chapter problems from Giancoli, *Physics for Scientists and Engineers, Third Edition* to be coded for use with the *WebAssign* system for instructors who wish to take advantage of this service. For more information on the *WebAssign* system and its features, please visit http://webassign.net/info or e-mail webassign@ncsu.edu.

NOTES TO STUDENTS AND INSTRUCTORS ON THE FORMAT

1. Sections marked with a star (*) are considered optional. They can be omitted without interrupting the main flow of topics. No later material depends on them except possibly later starred sections. They may be fun to read though.

2. The customary conventions are used: symbols for quantities (such as m for mass) are italicized, whereas units (such as m for meter) are not italicized. Boldface (**F**) is used for vectors.

3. Few equations are valid in all situations. Where practical, the limitations of important equations are stated in square brackets next to the equation. The equations that represent the great laws of physics are displayed with a tan background, as are a few other equations that are so useful that they are indispensable.

4. The number of significant figures (see Section 1–3) should not be assumed to be greater than given: if a number is stated as (say) 6, with its units, it is meant to be 6 and not 6.0 or 6.00.

5. At the end of each chapter is a set of Questions that students should attempt to answer (to themselves at least). These are followed by Problems which are ranked as level I, II, or III, according to estimated difficulty, with level I Problems being easiest. These Problems are arranged by Section, but Problems for a given Section may depend on earlier material as well. There follows a group of General Problems, which are not arranged by Section nor ranked as to difficulty. Questions and Problems that relate to optional Sections are starred.

6. Being able to solve problems is a crucial part of learning physics, and provides a powerful means for understanding the concepts and principles. This book contains many aids to problem solving: (a) worked-out Examples and their solutions in the text, which are set off with a vertical blue line in the margin, and should be studied as an integral part of the text; (b) special "Problem-solving boxes" placed throughout the text to suggest ways to approach problem solving for a particular topic—but don't get the idea that every topic has its own "techniques," because the basics remain the same; (c) special problem-solving Sections (marked in blue in the Table of Contents); (d) "Problem solving" marginal notes (see point 8 below) which refer to hints for solving problems within the text; (e) some of the worked-out Examples are Estimation Examples, which show how rough or approximate results can be obtained even if the given data are sparse (see Section 1–6); and finally (f) the Problems themselves at the end of each chapter (point 5 above).

7. Conceptual Examples look like ordinary Examples but are conceptual rather than numerical. Each proposes a question or two, which hopefully starts you to think and come up with a response. Give yourself a little time to come up with your own response before reading the Response given.

8. Marginal notes: brief notes in the margin of almost every page are printed in blue and are of four types: (a) ordinary notes (the majority) that serve as a sort of outline of the text and can help you later locate important concepts and equations; (b) notes that refer to the great laws and principles of physics, and these are in capital letters and in a box for emphasis; (c) notes that refer to a problem-solving hint or technique treated in the text, and these say "Problem Solving"; (d) notes that refer to an application of physics, in the text or an Example, and these say "Physics Applied."

9. This book is printed in full color. But not simply to make it more attractive. The color is used above all in the Figures, to give them greater clarity for our analysis, and to provide easier learning of the physical principles involved. The Table on the next page is a summary of how color is used in the Figures, and shows which colors are used for the different kinds of vectors, for field lines, and for other symbols and objects. These colors are used consistently throughout the book.

10. Appendices include useful mathematical formulas (such as derivatives and integrals, trigonometric identities, areas and volumes, expansions), and a table of isotopes with atomic masses and other data. Tables of useful data are located inside the front and back covers.

USE OF COLOR

Vectors

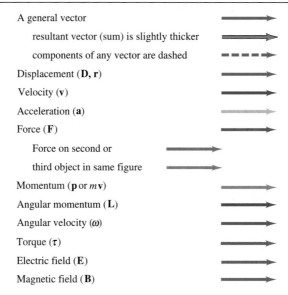

A general vector

    resultant vector (sum) is slightly thicker

    components of any vector are dashed

Displacement (**D, r**)

Velocity (**v**)

Acceleration (**a**)

Force (**F**)

    Force on second or

    third object in same figure

Momentum (**p** or m**v**)

Angular momentum (**L**)

Angular velocity (ω)

Torque (τ)

Electric field (**E**)

Magnetic field (**B**)

Electricity and magnetism

Electric field lines	
Equipotential lines	
Magnetic field lines	
Electric charge (+)	+ or ● +
Electric charge (−)	− or ● −

Electric circuit symbols

Wire	
Resistor	—⌁⌁⌁—
Capacitor	—‖—
Inductor	—⌒⌒⌒—
Battery	—‖—

Optics

Light rays	
Object	
Real image (dashed)	
Virtual image (dashed and paler)	

Other

Energy level (atom, etc.)	
Measurement lines	⊢—1.0 m—⊣
Path of a moving object	
Direction of motion or current	

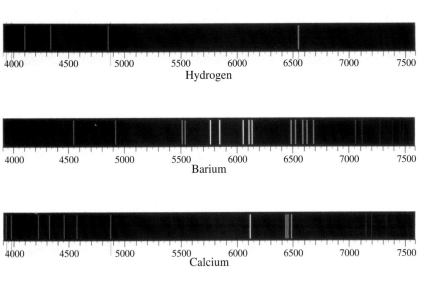

Hydrogen

Barium

Calcium

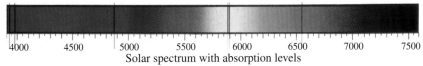

Solar spectrum with absorption levels

Atoms are tiny, and not observable using visible light. Trying to figure out the structure of atoms is a fascinating task. A major source of information about atoms comes from a study of the light emitted by the atoms of a pure material. Shown here are the line spectra (visible part) emitted by thin gases of hydrogen (top), barium, and calcium. At the bottom is the spectrum of the sun (visible light) in which are seen dark lines, indicating absorption at those wavelengths. What information do these spectra give us about atoms? We will see in this chapter that an entirely new theory is needed, the quantum theory.

CHAPTER 38

Early Quantum Theory and Models of the Atom

The second aspect of the revolution that shook the world of physics in the early part of the twentieth century (the first part was Einstein's theory of relativity) was the quantum theory. Unlike the special theory of relativity, the revolution of quantum theory required almost three decades to unfold, and many scientists contributed to its development. It began in 1900 with Planck's quantum hypothesis, and culminated in the mid-1920s with the theory of quantum mechanics of Schrödinger and Heisenberg which has been so effective in explaining the structure of matter.

38–1 | Planck's Quantum Hypothesis

One of the observations that was unexplained at the end of the nineteenth century was the spectrum of light emitted by hot objects. We saw in Chapter 19 that all objects emit radiation whose total intensity is proportional to the fourth power of the Kelvin temperature (T^4). At normal temperatures, we are not aware of this electromagnetic radiation because of its low intensity. At higher temperatures, there is sufficient infrared radiation that we can feel heat if we are close to the object. At still higher temperatures (on the order of 1000 K), objects actually glow, such as a red-hot electric stove burner or the element in a toaster. At temperatures above 2000 K, objects glow with a yellow or whitish color, such as white-hot iron and the filament of a lightbulb. As the temperature increases, the electromagnetic radiation emitted by bodies is most intense at higher and higher frequencies.

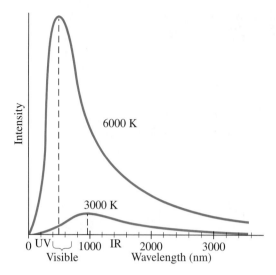

FIGURE 38–1 Spectrum of frequencies emitted by a blackbody at two different temperatures.

Blackbody radiation

The spectrum of light emitted by a hot dense object is shown in Fig. 38–1 for an idealized **blackbody**. A blackbody is a body that would absorb all the radiation falling on it (and so would appear black under reflection when illuminated from outside). The radiation such a blackbody would emit when hot and luminous, called **blackbody radiation** (though not necessarily black in color), is the easiest to deal with, and the radiation approximates that of many real objects. As can be seen, the spectrum contains a continuous range of frequencies. Such a continuous spectrum is emitted by any heated solid or liquid, and even by dense gases. The 6000-K curve in Fig. 38–1, corresponding to the temperature of the surface of the Sun, peaks in the visible part of the spectrum. For lower temperatures, the total radiation drops considerably and the peak occurs at longer wavelengths. Hence the blue end of the visible spectrum (and the UV) is relatively weaker. (This is why objects glow with a red color at around 1000 K.) It is found experimentally that the wavelength at the peak of the spectrum, λ_P, is related to the Kelvin temperature T by

$$\lambda_P T = 2.90 \times 10^{-3} \, \text{m} \cdot \text{K}. \tag{38–1}$$

This is known as **Wien's law**.

EXAMPLE 38–1 **The Sun's surface temperature.** Estimate the temperature of the surface of our Sun, given that the Sun emits light whose peak intensity occurs in the visible spectrum at around 500 nm.

SOLUTION Wien's law gives

$$T = \frac{2.90 \times 10^{-3} \, \text{m} \cdot \text{K}}{\lambda_P} = \frac{2.90 \times 10^{-3} \, \text{m} \cdot \text{K}}{500 \times 10^{-9} \, \text{m}} \approx 6000 \, \text{K}.$$

EXAMPLE 38–2 **Star color.** Suppose a star has a surface temperature of 32,500 K. What color would this star appear?

SOLUTION From Wien's law we have,

$$\lambda_P = \frac{2.90 \times 10^{-3} \, \text{m} \cdot \text{K}}{T} = \frac{2.90 \times 10^{-3} \, \text{m} \cdot \text{K}}{3.25 \times 10^4 \, \text{K}} = 89.2 \, \text{nm}.$$

The peak is in the UV range of the spectrum. In the visible region, the curve will be descending (see Fig. 38–1), so the shortest visible wavelengths will be strongest. Hence the star will appear bluish (or blue-white).

A major problem facing scientists in the 1890s was to explain blackbody radiation. Maxwell's electromagnetic theory had predicted that oscillating electric charges produce electromagnetic waves, and the radiation emitted by a hot object could be due to the oscillations of electric charges in the molecules of the material. Although this would explain where the radiation came from, it did not correctly predict the observed spectrum of emitted light. Two important theoretical curves based on classical ideas were those proposed by W. Wien (in 1896) and by Lord Rayleigh (in 1900). The latter was modified later by J. Jeans and since then has been known as the Rayleigh–Jeans theory. As experimental data came in, it became clear that neither Wien's nor the Rayleigh–Jeans formulations were in accord with experiment. Wien's was accurate at short wavelengths but deviated from experiment at longer wavelengths, whereas the reverse was true for the Rayleigh–Jeans theory (see Fig. 38–2).

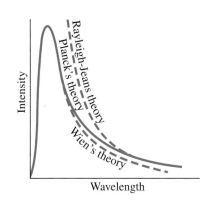

FIGURE 38–2 Comparison of the Wien and the Rayleigh–Jeans theories to that of Planck, which closely follows experiment.

The break came in late 1900 when Max Planck (1858–1947) proposed an empirical formula that nicely fit the data:

$$I(\lambda, T) = \frac{2\pi hc^2 \lambda^{-5}}{e^{hc/\lambda kT} - 1}.$$

$I(\lambda, T)$ is the radiation intensity as a function of wavelength λ at the temperature T; k is Boltzman's constant, c is the speed of light, and h is a new constant, now called **Planck's constant**. The value of h was estimated by Planck by fitting his formula for the blackbody radiation curve to experiment. The value accepted today is

$$h = 6.626 \times 10^{-34}\,\text{J·s}.$$

Planck then sought a theoretical basis for the formula and within two months found that he could obtain the formula by making a new and radical (though not so recognized at the time) assumption: that the energy distributed among the oscillations of atoms within molecules is not continuous but instead consists of a finite number of very small discrete amounts, each related to the frequency of oscillation by

$$E_{\text{min}} = hf.$$

Planck's assumption suggests that the energy of any molecular vibration could be only some whole number multiple of hf:

$$E = nhf, \qquad n = 1, 2, 3, \cdots. \qquad \textbf{(38–2)}$$

Planck's quantum hypothesis

This idea is often called **Planck's quantum hypothesis** ("quantum" means "discrete amount" as opposed to "continuous"), although little attention was brought to this point at the time. In fact, it appears that Planck considered it more as a mathematical device to get the "right answer" rather than as a discovery comparable to those of Newton. Planck himself continued to seek a classical explanation for the introduction of h. The recognition that this was an important and radical innovation did not come until later, after about 1905 when others, particularly Einstein, entered the field.

FIGURE 38–3 Ramp versus stair analogy. (a) On a ramp, a box can have continuous values of energy. (b) But on stairs, the box can have only discrete (quantized) values of energy.

The quantum hypothesis, Eq. 38–2, states that the energy of an oscillator can be $E = hf$, or $2hf$, or $3hf$, and so on, but there cannot be vibrations whose energy lies between these values. That is, energy would not be a continuous quantity as had been believed for centuries; rather it is **quantized**—it exists only in discrete amounts. The smallest amount of energy possible (hf) is called the **quantum of energy**. Recall from Chapter 14 that the energy of an oscillation is proportional to the amplitude squared. Thus another way of expressing the quantum hypothesis is that not just any amplitude of vibration is possible. The possible values for the amplitude are related to the frequency f.

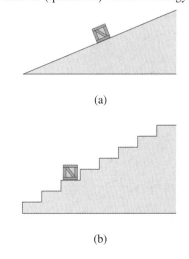

A simple analogy may help. A stringed instrument such as a violin or guitar can be played over a continuous range of frequencies by moving your finger along the string. A flute or piano, on the other hand, is "quantized" in the sense that only certain frequencies (notes) can be played. Or compare a ramp, on which a box can be placed at any height, to a flight of stairs on which the box can have only certain discrete amounts of potential energy, as shown in Fig. 38–3.

38-2 Photon Theory of Light and the Photoelectric Effect

In 1905, the same year that he introduced the special theory of relativity, Einstein made a bold extension of the quantum idea by proposing a new theory of light. Planck's work had suggested that the vibrational energy of molecules in a radiating object is quantized with energy $E = nhf$, where n is an integer. Einstein argued that therefore, when light is emitted by a molecular oscillator, the molecule's vibrational energy of nhf must decrease by an amount hf (or by $2hf$, etc.) to another integer times hf, namely $(n - 1)hf$. Then to conserve energy, the light ought to be emitted in packets or quanta, each with an energy

$$E = hf. \tag{38-3}$$

Again h is Planck's constant. Since all light ultimately comes from a radiating source, this suggests that perhaps *light is transmitted as tiny particles*, or **photons**, as they are now called, rather than as waves. This, too, was a radical departure from classical ideas. Einstein proposed a test of the quantum theory of light: quantitative measurements on the photoelectric effect.

Photoelectric effect

The **photoelectric effect** is the phenomenon that when light shines on a metal surface, electrons are emitted from the surface. (The photoelectric effect occurs in other materials, but is most easily observed with metals.) It can be observed using the apparatus shown in Fig. 38-4. A metal plate P and a smaller electrode C are placed inside an evacuated glass tube, called a **photocell**. The two electrodes are connected to an ammeter and a source of emf, as shown. When the photocell is in the dark, the ammeter reads zero. But when light of sufficiently high frequency illuminates the plate, the ammeter indicates a current flowing in the circuit. To explain completion of the circuit, we can imagine electrons, ejected by the impinging radiation, flowing across the tube from the plate to the "collector" C as shown in the diagram.

That electrons should be emitted when light shines on a metal is consistent with the electromagnetic (EM) wave theory of light: the electric field of an EM wave could exert a force on electrons in the metal and eject some of them. Einstein pointed out, however, that the wave theory and the photon theory of light give very different predictions on the details of the photoelectric effect. For example, one thing that can be measured with the apparatus of Fig. 38-4 is the maximum kinetic energy (K_{max}) of the emitted electrons. This can be done by using a variable voltage source and reversing the terminals so that electrode C is negative and P is positive. The electrons emitted from P will be repelled by the negative electrode, but if this reverse voltage is small enough, the fastest electrons will still reach C and there will be a current in the circuit. If the reversed voltage is increased, a point is reached where the current reaches zero—no electrons have sufficient kinetic energy to reach C. This is called the *stopping potential*, or *stopping voltage*, V_0, and from its measurement, K_{max} can be determined using conservation of energy:

$$K_{max} = eV_0.$$

Now let us examine the details of the photoelectric effect from the point of view of the wave theory versus Einstein's particle theory. First the wave theory, assuming monochromatic light. The two important properties of a light wave are its intensity and its frequency (or wavelength). When these two quantities are varied, the wave theory makes the following predictions:

Wave

theory

predictions

1. If the light intensity is increased, the number of electrons ejected and their maximum kinetic energy should be increased because the higher intensity means a greater electric field amplitude, and the greater electric field should eject electrons with higher speed.

2. The frequency of the light should not affect the kinetic energy of the ejected electrons. Only the intensity should affect K_{max}.

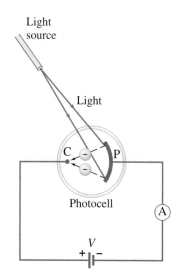

Light source

Light

C

P

Photocell

(A)

V

+ −

FIGURE 38-4 The photoelectric effect.

The photon theory makes completely different predictions. First we note that in a monochromatic beam, all photons have the same energy (= hf). Increasing the intensity of the light beam means increasing the number of photons in the beam, but does not affect the energy of each photon as long as the frequency is not changed. According to Einstein's theory, an electron is ejected from the metal by a collision with a single photon. In the process, all the photon energy is transferred to the electron and the photon ceases to exist. Since electrons are held in the metal by attractive forces, some minimum energy W_0 (called the **work function**, which is on the order of a few electron volts for most metals) is required just to get an electron out through the surface. If the frequency f of the incoming light is so low that hf is less than W_0, then the photons will not have enough energy to eject any electrons at all. If $hf > W_0$, then electrons will be ejected and energy will be conserved in the process. That is, the input energy (of the photon), hf, will equal the outgoing kinetic energy K of the electron plus the energy required to get it out of the metal, W:

$$hf = K + W. \qquad \textbf{(38–4a)}$$

The least tightly held electrons will be emitted with the most kinetic energy (K_{max}), in which case W in this equation becomes the work function W_0, and K becomes K_{max}:

$$hf = K_{max} + W_0. \qquad \textbf{(38–4b)}$$

Many electrons will require more energy than the bare minimum (W_0) to get out of the metal, and thus the kinetic energy of such electrons will be less than the maximum.

From these considerations, the photon theory makes the following predictions:

1. An increase in intensity of the light beam means more photons are incident, so more electrons will be ejected; but since the energy of each photon is not changed, the maximum kinetic energy of electrons is not changed by an increase in intensity.

2. If the frequency of the light is increased, the maximum kinetic energy of the electrons increases linearly, according to Eq. 38–4b. That is,

$$K_{max} = hf - W_0.$$

This relationship is plotted in Fig. 38–5.

3. If the frequency f is less than the "cutoff" frequency f_0, where $hf_0 = W_0$, no electrons will be ejected at all, no matter how great the intensity.

These predictions of the photon theory are clearly very different from the predictions of the wave theory. In 1913–1914, careful experiments were carried out by R. A. Millikan. The results were fully in agreement with Einstein's photon theory.

One other aspect of the photoelectric effect also confirmed the photon theory. If extremely low light intensity is used, the wave theory predicts a time delay before electron emission so that an electron can absorb enough energy to exceed the work function. The photon theory predicts no such delay—it only takes one photon (if its frequency is high enough) to eject an electron—and experiments showed no delay. This too confirmed Einstein's photon theory.

Photon

theory

predictions

FIGURE 38–5 Photoelectric effect: the maximum kinetic energy of ejected electrons increases linearly with the frequency of incident light. No electrons are emitted if $f < f_0$.

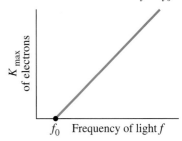

| EXAMPLE 38–3 | **Photon energy.** Calculate the energy of a photon of blue light, $\lambda = 450\ nm$.

SOLUTION Since $f = c/\lambda$, we have

$$E = hf = \frac{hc}{\lambda} = \frac{(6.63 \times 10^{-34}\ \mathrm{J \cdot s})(3.0 \times 10^8\ \mathrm{m/s})}{(4.5 \times 10^{-7}\ \mathrm{m})} = 4.4 \times 10^{-19}\ \mathrm{J},$$

or $(4.4 \times 10^{-19}\ \mathrm{J})/(1.6 \times 10^{-19}\ \mathrm{J/eV}) = 2.7\ \mathrm{eV}$.

EXAMPLE 38–4 ESTIMATE Photons from a lightbulb. Estimate how many visible light photons a 100-W lightbulb emits per second.

SOLUTION Let's assume an average wavelength in the middle of the visible spectrum, $\lambda \approx 500$ nm. The energy emitted in one second $(= 100 \text{ J})$ is $E = nhf$ when n is the number of photons emitted per second and $f = c/\lambda$. Hence

$$n = \frac{E}{hf} = \frac{E\lambda}{hc} = \frac{(100 \text{ J})(500 \times 10^{-9} \text{ m})}{(6.63 \times 10^{-34} \text{ J·s})(3.0 \times 10^{8} \text{ m/s})} = 2.5 \times 10^{20}.$$

This is an overestimate since much of the 100 J of electric energy input is transformed into heat rather than light. If the efficiency is between 1 percent and 10 percent, then the number of photons emitted is on the order of 10^{19}, still an enormous number.

EXAMPLE 38–5 Photoelectron speed and energy. What is the maximum kinetic energy and speed of an electron ejected from a sodium surface whose work function is $W_0 = 2.28$ eV when illuminated by light of wavelength: (a) 410 nm; (b) 550 nm?

SOLUTION (a) For $\lambda = 410$ nm,

$$hf = \frac{hc}{\lambda} = 4.85 \times 10^{-19} \text{ J} \quad \text{or} \quad 3.03 \text{ eV}.$$

From Eq. 38–4b, $K_{\text{max}} = 3.03 \text{ eV} - 2.28 \text{ eV} = 0.75 \text{ eV}$, or 1.2×10^{-19} J. Since $K = \frac{1}{2}mv^2$ where $m = 9.1 \times 10^{-31}$ kg,

$$v = \sqrt{\frac{2K}{m}} = 5.1 \times 10^{5} \text{ m/s}.$$

Notice that we used the nonrelativistic equation for kinetic energy. If v had turned out to be more than about $0.1c$, our calculation would have been inaccurate by more than a percent or so, and we would probably prefer to redo it using the relativistic form (Eq. 37–10).

(b) For $\lambda = 550$ nm, $hf = 3.61 \times 10^{-19}$ J $= 2.26$ eV. Since this photon energy is less than the work function, no electrons are ejected.

Applications of the Photoelectric Effect

The photoelectric effect, besides playing an important historical role in confirming the photon theory of light, also has many practical applications. Burglar alarms and automatic door openers often make use of the photocell circuit of Fig. 38–4. When a person interrupts the beam of light, the sudden drop in current in the circuit activates a switch—often a solenoid—which operates a bell or opens the door. UV or IR light is sometimes used in burglar alarms because of its invisibility. Many smoke detectors use the photoelectric effect to detect tiny amounts of smoke that interrupt the flow of light and so alter the electric current. Photographic light meters use this circuit as well. Photocells are used in many other devices, such as absorption spectrophotometers, to measure light intensity. One type of film sound track is a variably shaded narrow section at the side of the film. Light passing through the film is thus "modulated," and the output electrical signal of the photocell detector follows the frequencies on the sound track. See Fig. 38–6. For many applications today, the vacuum-tube photocell of Fig. 38–4 has been replaced by a semiconductor device known as a **photodiode**. In these semiconductors, the absorption of a photon liberates a bound electron, which changes the conductivity of the material, so the current through a photodiode is altered.

FIGURE 38–6 Optical sound track on movie film. In the projector, light from a small source (different from that for the picture) passes through the sound track on the moving film. The light and dark areas on the sound track vary the intensity of the transmitted light which reaches the photocell, whose current output is then a replica of the original sound. This output is amplified and sent to the loudspeakers. High-quality projectors can show movies containing several parallel sound tracks to go to different speakers around the theater.

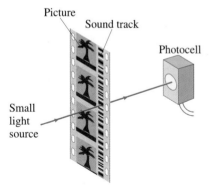

Picture
Sound track
Photocell
Small light source

EXAMPLE 38–6 Photosynthesis. In *photosynthesis*, the process by which pigments such as chlorophyll in plants capture the energy of sunlight to change CO_2 to useful carbohydrate, about nine photons are needed to trasform one molecule of CO_2 to carbohydrate and O_2. Assuming light of wavelength $\lambda = 670$ nm (chlorophyll absorbs most strongly in the range 650 nm to 700 nm), how efficient is the photosynthetic process? The reverse chemical reaction releases an energy of 4.9 eV/molecule of CO_2.

SOLUTION The energy of nine photons, each of energy $hf = hc/\lambda$ is $(9)(6.6 \times 10^{-34}\,\text{J}\cdot\text{s})(3.0 \times 10^{8}\,\text{m/s})/(6.7 \times 10^{-7}\,\text{m}) = 2.7 \times 10^{-18}\,\text{J}$ or 17 eV. Thus the process is $(4.9\,\text{eV}/17\,\text{eV}) = 29$ percent efficient.

38–3 | Photons and the Compton Effect

A number of other experiments were carried out in the early twentieth century which also supported the photon theory. One of these was the **Compton effect** (1923) named after its discoverer, A. H. Compton (1892–1962). Compton scattered short wavelength light (actually X-rays) from various materials. He found that the scattered light had a slightly longer wavelength than did the incident light, and therefore a slightly lower frequency indicating a loss of energy. This finding, he showed, could be explained on the basis of the photon theory as incident photons colliding with electrons of the material, Fig. 38–7. He applied the laws of conservation of energy and momentum to such collisions and found that the predicted energies of scattered photons were in accord with experimental results.

Let us look at the Compton effect in detail and seek to predict the shifted wavelength using the photon theory of light and Fig. 38–7. First we note that the photon is truly a relativistic particle—it travels at the speed of light. Thus we must use relativistic formulas for dealing with its mass, energy, and momentum. The momentum of any particle of rest mass m is given by $p = \gamma mv = mv/\sqrt{1 - v^2/c^2}$. Since $v = c$ for a photon, the denominator is zero. So the rest mass of a photon must also be zero, or its momentum would be infinite. Of course a photon is never at rest. The momentum of a photon, from Eq. 37–13 with $m = 0$, is given by $E^2 = p^2c^2$, or

$$p = \frac{E}{c}.$$

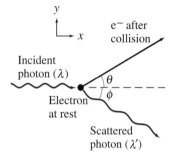

FIGURE 38–7 The Compton effect. A single photon of wavelength λ strikes an electron in some material, knocking it out of its atom. The scattered photon has less energy (since some is given to the electron) and hence has a longer wavelength λ'. Experiments found scattered X-rays of just the wavelengths predicted by conservation of energy and momentum using the photon model.

Since $E = hf$, the momentum of a photon is related to its wavelength by

$$p = \frac{hf}{c} = \frac{h}{\lambda}. \qquad (38\text{–}5)$$

If the incoming photon in Fig. 38–7 has wavelength λ, then its total energy and momentum are

$$E = hf = \frac{hc}{\lambda} \qquad \text{and} \qquad p = \frac{h}{\lambda}.$$

After the collision of Fig. 38–7, the photon scattered at the angle ϕ has a wavelength which we call λ'. Its energy and momentum are

$$E' = \frac{hc}{\lambda'} \qquad \text{and} \qquad p' = \frac{h}{\lambda'}.$$

The electron, assumed at rest before the collision but free to move when struck is scattered at an angle θ as shown in Fig. 38–7.

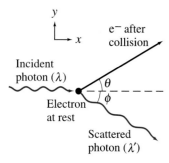

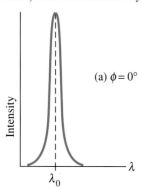

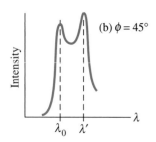

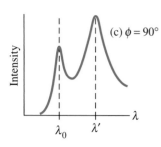

FIGURE 38–7 (repeated) The Compton effect.

FIGURE 38–8 Plots of intensity of radiation scattered from a target such as graphite (carbon), for three different angles. The values for λ' match Eq. 38–6. For (a) $\phi = 0°$, $\lambda' = \lambda_0$. In (b) and (c) a peak is found not only at λ' due to photons scattered from free electrons (or very nearly free) but also a peak at almost precisely λ_0. The latter is due to scattering from electrons very tightly bound to their atoms so the mass in Eq. 38–6 becomes very large (mass of the atom) and $\Delta\lambda$ becomes very small.

The electron's kinetic energy is (see Eq. 37–10):

$$K_e = \left(\frac{1}{\sqrt{1 - v^2/c^2}} - 1\right)m_e c^2$$

where m_e is the rest mass of the electron and v is its speed. The electron's momentum is

$$p_e = \frac{1}{\sqrt{1 - v^2/c^2}}\, m_e v.$$

We apply conservation of energy to the collision (see Fig. 38–7):

incoming photon → scattered photon + electron

$$\frac{hc}{\lambda} = \frac{hc}{\lambda'} + \left(\frac{1}{\sqrt{1 - v^2/c^2}} - 1\right)m_e c^2.$$

And we apply conservation of momentum to the x and y components of momentum:

$$\frac{h}{\lambda} = \frac{h}{\lambda'}\cos\phi + \frac{m_e v \cos\theta}{\sqrt{1 - v^2/c^2}}$$

$$0 = \frac{h}{\lambda'}\sin\phi - \frac{m_e v \sin\theta}{\sqrt{1 - v^2/c^2}}.$$

We can combine these three equations to eliminate v and θ (the algebra is left as an exercise—see Problem 25) and we obtain, as Compton did, an equation for the wavelength of the scattered photon in terms of its scattering angle ϕ:

$$\lambda' = \lambda + \frac{h}{m_e c}(1 - \cos\phi).$$

For $\phi = 0$, the wavelength is unchanged (there is no collision for this case of the photon passing straight through). At any other angle, λ' is longer than λ. The difference in wavelength,

$$\Delta\lambda = \lambda' - \lambda = \frac{h}{m_e c}(1 - \cos\phi), \tag{38–6}$$

is called the **Compton shift**. The quantity $h/m_e c$, which has the dimensions of length, is called the **Compton wavelength** λ_C of a free electron,

$$\lambda_C = \frac{h}{m_e c} = 2.43 \times 10^{-3}\,\text{nm} = 2.43\,\text{pm}. \qquad \text{[electron]}$$

Equation 38–6 predicts that λ' depends on the angle ϕ at which they are detected. Compton's measurements of 1923 were consistent with this formula, confirming the value of λ_C and the dependence of λ' on ϕ. See Fig. 38–8. The wave theory of light predicts no wavelength shift: an incoming electromagnetic wave of frequency f should set electrons into oscillation at the same frequency f, and such oscillating electrons should reemit EM waves of this same frequency f (Chapter 32), and would not change with the angle ϕ. Hence the Compton effect adds to the firm experimental foundation for the photon theory of light.

EXAMPLE 38–7 **X-ray scattering.** X-rays of wavelength 0.140 nm are scattered from a block of carbon. What will be the wavelengths of X-rays scattered at (a) 0°, (b) 90°, (c) 180°?

SOLUTION (a) For $\phi = 0°$, $\cos\phi = 1$, and Eq. 38–6 gives $\lambda' = \lambda = 0.140$ nm. This makes sense since for $\phi = 0°$, there really isn't any collision as the photon goes straight through without interacting.
(b) For $\phi = 90°$, $\cos\phi = 0$, so

$$\lambda' = \lambda + \frac{h}{m_e c} = 0.140 \text{ nm} + \frac{6.63 \times 10^{-34} \text{ J·s}}{(9.11 \times 10^{-31} \text{ kg})(3.00 \times 10^8 \text{ m/s})}$$
$$= 0.140 \text{ nm} + 2.4 \times 10^{-12} \text{ m} = 0.142 \text{ nm};$$

that is, the wavelength is longer by one Compton wavelength (= 0.0024 nm for an electron).
(c) For $\phi = 180°$, which means the photon is scattered backward, returning in the direction from which it came (a direct "head-on" collision), $\cos\phi = -1$, so

$$\lambda' = \lambda + 2\frac{h}{m_e c} = 0.140 \text{ nm} + 2(0.0024 \text{ nm}) = 0.145 \text{ nm}.$$

The Compton effect has been used to diagnose bone disease such as osteoporosis. Gamma rays, which are photons of even shorter wavelength than X-rays, coming from a radioactive source are scattered off bone material. The total intensity of the scattered radiation is proportional to the density of electrons, which is in turn proportional to the bone density. Changes in the density of bone material can indicate the onset of osteoporosis.

➡ **PHYSICS APPLIED**
Measuring bone density

38–4 | Photon Interactions; Pair Production

When a photon passes through matter, it interacts with the atoms and electrons. There are four important types of interactions that a photon can undergo:

Photon

interactions

1. The photon can be scattered from an electron (or a nucleus) and in the process lose some energy; this is the *Compton effect* (Fig. 38–7). But notice that the photon is not slowed down. It still travels with speed c, but its frequency will be lower.

2. The *photoelectric effect*: a photon may knock an electron out of an atom and in the process itself disappear.

3. The photon may knock an atomic electron to a higher energy state in the atom if its energy is not sufficient to knock the electron out altogether. In this process the photon also disappears, and all its energy is given to the atom. Such an atom is then said to be in an *excited state*, and we shall discuss this more later.

4. *Pair production*: A photon can actually create matter, such as the production of an electron and a positron, Fig. 38–9. (A positron has the same mass as an electron, but the opposite charge, +e.)

FIGURE 38–9 Pair production: a photon disappears and produces an electron and a positron.

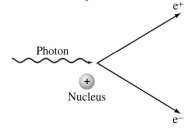

Pair production

This last process is called **pair production**, and the photon disappears in the process of creating the electron–positron pair. This is an example of rest mass being created from pure energy, and it occurs in accord with Einstein's equation $E = mc^2$. Notice that a photon cannot create an electron alone since electric charge would not then be conserved. The inverse of pair production also occurs: if an electron collides with a positron, the two **annihilate** each other and their energy, including their mass, appears as electromagnetic energy of photons. Because of this process positrons usually do not last long in nature.

EXAMPLE 38–8 **Pair production.** What is (*a*) the minimum energy of a photon that can produce an electron–positron pair? (*b*) What is this photon's wavelength?

SOLUTION (*a*) Because $E = mc^2$, and the mass created is equal to two electron masses, the photon must have energy

$$E = 2(9.11 \times 10^{-31}\,\text{kg})(3.0 \times 10^8\,\text{m/s})^2 = 1.64 \times 10^{-13}\,\text{J} = 1.02\,\text{MeV}.$$

A photon with less energy cannot undergo pair production.
(*b*) Since $E = hf = hc/\lambda$, the wavelength of a 1.02-MeV photon is

$$\lambda = \frac{hc}{E} = \frac{(6.6 \times 10^{-34}\,\text{J·s})(3.0 \times 10^8\,\text{m/s})}{(1.64 \times 10^{-13}\,\text{J})} = 1.2 \times 10^{-12}\,\text{m},$$

which is 0.0012 nm. Such photons are in the gamma-ray (or very short X-ray) region of the electromagnetic spectrum.

Pair production cannot occur in empty space, for energy and momentum could not simultaneously be conserved. In Example 38–8, for instance, energy is conserved, but only enough energy was provided to create the electron–positron pair at rest and thus with no momentum to carry away the initial momentum of the photon. Indeed, it can be shown that at any energy, an additional massive object, such as an atomic nucleus, must take part in the interaction to carry off some of the momentum.

38–5 Wave–Particle Duality; the Principle of Complementarity

The photoelectric effect, the Compton effect, and other experiments have placed the particle theory of light on a firm experimental basis. But what about the classic experiments of Young and others (Chapters 35 and 36) on interference and diffraction which showed that the wave theory of light also rests on a firm experimental basis?

We seem to be in a dilemma. Some experiments indicate that light behaves like a wave; others indicate that it behaves like a stream of particles. These two theories seem to be incompatible, but both have been shown to have validity. Physicists have finally come to the conclusion that this duality of light must be accepted as a fact of life. It is referred to as the **wave–particle duality**. Apparently, light is a more complex phenomenon than just a simple wave or a simple beam of particles.

To clarify the situation, the great Danish physicist Niels Bohr (1885–1962, Fig. 38–10) proposed his famous **principle of complementarity**. It states that to understand any given experiment, we must use either the wave or the photon theory, but not both. Yet we must be aware of both the wave and particle aspects of light if we are to have a full understanding of light. Therefore these two aspects of light complement one another.

It is not possible to "visualize" this duality. We cannot picture a combination of wave and particle. Instead, we must recognize that the two aspects of light are different "faces" that light shows to experimenters.

Part of the difficulty stems from how we think. Visual pictures (or models) in our minds are based on what we see in the everyday world. We apply the concepts of waves and particles to light because in the macroscopic world we see that energy is transferred from place to place by these two methods. We cannot see directly whether light is a wave or particle—so we do indirect experiments. To explain the experiments, we apply the models of waves or of particles to the nature of light. But these are abstractions of the human mind. When we try to conceive of what light really "is," we insist on a visual picture. Yet there is no reason why light should conform to these models (or visual images) taken from the macroscopic

FIGURE 38–10 Niels Bohr, walking with Enrico Fermi along the Appian Way outside Rome.

Wave–particle duality

Principle of complementarity

world. The "true" nature of light—if that means anything—is not possible to visualize. The best we can do is recognize that our knowledge is limited to the indirect experiments, and that in terms of everyday language and images, light reveals both wave and particle properties.

It is worth noting that Einstein's equation $E = hf$ itself links the particle and wave properties of a light beam. In this equation, E refers to the energy of a particle; and on the other side of the equation, we have the frequency f of the corresponding wave.

38–6 | Wave Nature of Matter

In 1923, Louis de Broglie (1892–1987) extended the idea of the wave–particle duality. He much appreciated the symmetry in nature, and argued that if light sometimes behaves like a wave and sometimes like a particle, then perhaps those things in nature thought to be particles—such as electrons and other material objects—might also have wave properties. De Broglie proposed that the wavelength of a material particle would be related to its momentum in the same way as for a photon, Eq. 38–5, $p = h/\lambda$. That is, for a particle having linear momentum p, the wavelength λ is given by

$$\lambda = \frac{h}{p}.$$

(38–7)   *de Broglie wavelength*

This is sometimes called the **de Broglie wavelength** of a particle.

EXAMPLE 38–9 **Wavelength of a ball.** Calculate the de Broglie wavelength of a 0.20-kg ball moving with a speed of 15 m/s. Discuss.

SOLUTION $\lambda = \dfrac{h}{p} = \dfrac{h}{mv} = \dfrac{(6.6 \times 10^{-34}\,\text{J}\cdot\text{s})}{(0.20\,\text{kg})(15\,\text{m/s})} = 2.2 \times 10^{-34}\,\text{m}.$

This is an incredibly small wavelength. Even if the speed were extremely small, say 10^{-4} m/s, the wavelength would be about 10^{-29} m. Indeed, the wavelength of any ordinary object is much too small to be measured and detected. The problem is that the properties of waves, such as interference and diffraction, are significant only when the size of objects or slits is not much larger than the wavelength. And there are no known objects or slits to diffract waves only 10^{-30} m long, so the wave properties of ordinary objects go undetected.

But tiny elementary particles, such as electrons, are another matter. Since the mass m appears in the denominator in Eq. 38–7, a very small mass should have a much larger wavelength.

EXAMPLE 38–10 **Wavelength of an electron.** Determine the wavelength of an electron that has been accelerated through a potential difference of 100 V.

SOLUTION We assume that the speed of the electron will be much less than c, so we use nonrelativistic mechanics. (If this assumption were to come out wrong, we would have to recalculate using relativistic formulas—see Sections 37–9 and 37–11.) The gain in kinetic energy will equal the loss in potential energy, so $\frac{1}{2}mv^2 = eV$ and

$$v = \sqrt{\frac{2\,eV}{m}} = \sqrt{\frac{(2)(1.6 \times 10^{-19}\,\text{C})(100\,\text{V})}{(9.1 \times 10^{-31}\,\text{kg})}} = 5.9 \times 10^6\,\text{m/s}.$$

Then

$$\lambda = \frac{h}{mv} = \frac{(6.6 \times 10^{-34}\,\text{J}\cdot\text{s})}{(9.1 \times 10^{-31}\,\text{kg})(5.9 \times 10^6\,\text{m/s})} = 1.2 \times 10^{-10}\,\text{m},$$

or 0.12 nm.

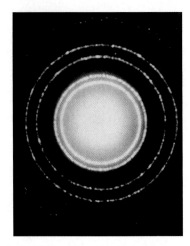

FIGURE 38-11 Diffraction pattern of electrons scattered from aluminum foil, as recorded on film.

From this Example, we see that electrons can have wavelengths on the order of 10^{-10} m. Although small, this wavelength can be detected: the spacing of atoms in a crystal is on the order of 10^{-10} m and the orderly array of atoms in a crystal could be used as a type of diffraction grating, as was done earlier for X-rays (see Section 36–10). C. J. Davisson and L. H. Germer performed the crucial experiment; they scattered electrons from the surface of a metal crystal and, in early 1927, observed that the electrons were scattered into a pattern of regular peaks. When they interpreted these peaks as a diffraction pattern, the wavelength of the diffracted electron wave was found to be just that predicted by de Broglie, Eq. 38–7. In the same year, G. P. Thomson (son of J. J. Thomson, who is credited with the discovery of the particle nature of electrons as we saw in Section 27–7), using a different experimental arrangement, also detected diffraction of electrons. (See Fig. 38–11. You can compare it to X-ray diffraction, Section 36–10.) Later experiments showed that protons, neutrons, and other particles also have wave properties.

Thus the wave–particle duality applies to material objects as well as to light. The principle of complementarity applies to matter as well. That is, we must be aware of both the particle and wave aspects in order to have an understanding of matter, including electrons. But again we must recognize that a visual picture of a "wave–particle" is not possible.

What is an electron?

We might ask ourselves: "What is an electron?" The early experiments of J. J. Thomson (Section 27–7) indicated a glow in a tube, and that glow moved when a magnetic field was applied. The results of these and other experiments were best interpreted as being caused by tiny negatively charged particles which we now call electrons. No one, however, has actually seen an electron directly. The drawings we sometimes make of electrons as tiny spheres with a negative charge on them are merely convenient pictures (now recognized to be inaccurate). Again we must rely on experimental results, some of which are best interpreted using the particle model and others using the wave model. These models are mere pictures that we use to extrapolate from the macroscopic world to the tiny microscopic world of the atom. And there is no reason to expect that these models somehow reflect the reality of an electron. We thus use a wave or a particle model (whichever works best in a situation) so that we can talk about what is happening. But we should not be led to believe that an electron *is* a wave or a particle. Instead we could say that an electron is the set of its properties that we can measure. Bertrand Russell said it well when he wrote that an electron is "a logical construction."

→ PHYSICS APPLIED

Electron diffraction

EXAMPLE 38-11 Electron diffraction. The wave nature of electrons is manifested in experiments where an electron beam interacts with the atoms on the surface of a solid. By studying the angular distribution of the diffracted electrons, one can indirectly measure the geometrical arrangement of atoms. Assume that the electrons strike perpendicular to the surface of a solid (see Fig. 38–12), and that their energy is low, $K = 100 \text{ eV}$, so that they interact only with the surface layer of atoms. If the smallest angle at which a diffraction maximum occurs is at 24°, what is the separation d between the atoms on the surface?

FIGURE 38-12 Exmple 38–11.

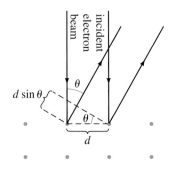

SOLUTION Treating the electrons as waves, we need to determine the condition where the difference in path traveled by the wave diffracted from adjacent atoms is an integer multiple of the de Broglie wavelength, so that constructive interference occurs. The path length difference is $d \sin \theta$; so for the smallest value of θ we must have

$$d \sin \theta = \lambda.$$

However, λ is related to the (non-relativistic) kinetic energy K by

$$K = \frac{p^2}{2m_e} = \frac{h^2}{2m_e \lambda^2}.$$

Thus

$$\lambda = \frac{h}{\sqrt{2m_e K}} = \frac{\left(6.63 \times 10^{-34}\,\text{J}\cdot\text{s}\right)}{\sqrt{2\left(9.11 \times 10^{-31}\,\text{kg}\right)\left(100\,\text{eV}\right)\left(1.6 \times 10^{-19}\,\text{J/eV}\right)}} = 0.123\,\text{nm}.$$

The surface inter-atomic spacing is

$$d = \frac{\lambda}{\sin \theta} = \frac{0.123\,\text{nm}}{\sin 24°} = 0.30\,\text{nm}.$$

*38–7 | Electron Microscopes

The idea that electrons have wave properties led to the development of the **electron microscope**, which can produce images of much greater magnification than a light microscope. Figures 38–13 and 38–14 are diagrams of two types, the **transmission electron microscope**, which produces a two-dimensional image, and the **scanning electron microscope**, which produces images with a three-dimensional quality. In each design, the objective and eyepiece lenses are actually magnetic fields that exert forces on the electrons to bring them to a focus. The fields are produced by carefully designed current-carrying coils of wire.

As discussed in Section 36–5, the best resolution of details on an object is on the order of the wavelength of the radiation used to view it. Electrons accelerated by voltages on the order of 10^5 V have wavelengths on the order of 0.004 nm. The maximum resolution obtainable would be on this order, but in practice, aberrations in the magnetic lenses limit the resolution in transmission electron microscopes to at best about 0.1 to 0.5 nm. This is still 10^3 times finer than that attainable with a light microscope, and corresponds to a useful magnification of about a million. Such magnifications are difficult to attain, and more common magnifications are 10^4 to 10^5. The maximum resolution attainable with a scanning electron microscope is somewhat less, about 5 to 10 nm at best.

FIGURE 38–13 Transmission electron microscope. The magnetic-field coils are designed to be "magnetic lenses," which bend the electron paths and bring them to a focus, as shown.

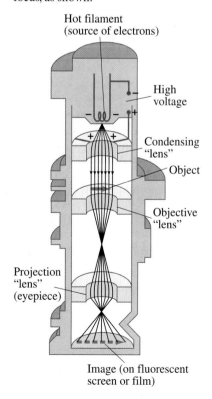

FIGURE 38–14 Scanning electron microscope. Scanning coils move an electron beam back and forth across the specimen. Secondary electrons produced when the beam strikes the specimen are collected and modulate the intensity of the beam in the CRT to produce a picture.

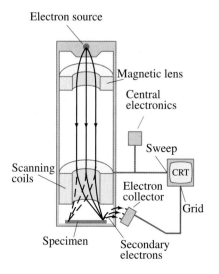

38–8 | Early Models of the Atom

The idea that matter is made up of atoms was accepted by most scientists by 1900. With the discovery of the electron in the 1890s, scientists began to think of the atom itself as having a structure and electrons as part of that structure. We now introduce our modern approach to the atom, and of the quantum theory with which it is intertwined.[†]

A typical model of the atom in the 1890s visualized the atom as a homogeneous sphere of positive charge inside of which there were tiny negatively charged electrons, a little like plums in a pudding, Fig. 38–15. J. J. Thomson, soon after his discovery of the electron in 1897, argued that the electrons in this model should be moving.

Around 1911, Ernest Rutherford (1871–1937) and his colleagues performed experiments whose results contradicted Thomson's model of the atom. In these experiments a beam of positively charged "alpha (α) particles" was directed at a thin sheet of metal foil such as gold, Fig. 38–16a. (These newly discovered α particles were emitted by certain radioactive materials and were soon shown to be doubly ionized helium atoms—that is, having a charge of $+2e$.) It was expected from Thomson's model that the alpha particles would not be deflected significantly since electrons are so much lighter than alpha particles, and the alpha particles should not have encountered any massive concentration of positive charge in that model to strongly repel them. The experimental results completely contradicted these predictions. It was found that most of the alpha particles passed through the foil unaffected, as if the foil were mostly empty space. And of those deflected, a few were deflected at very large angles—some even nearly back in the direction from which they had come. This could happen, Rutherford reasoned, only if the positively charged alpha particles were being repelled by a massive positive charge concentrated in a very small region of space (see Fig. 38–16b). He theorized that the atom must consist of a tiny but massive positively charged nucleus, containing over 99.9 percent of the mass of the atom, surrounded by electrons some distance away. The electrons would be moving in orbits about the nucleus—much as the planets move around the Sun—because if they were at rest, they would fall into the nucleus

[†] Some readers may say: "Tell us the facts as we know them today, and don't bother us with the historical background and its outmoded theories." Such an approach would ignore the creative aspect of science and thus give a false impression of how science develops. Moreover, it is not really possible to understand today's view of the atom without insight into the concepts that led to it

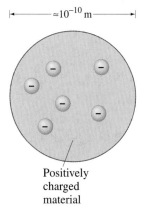

FIGURE 38–15 Plum-pudding model of the atom.

$\approx 10^{-10}$ m

Positively charged material

Rutherford's planetary model

FIGURE 38–16 (a) Experimental setup for Rutherford's experiment: α particles emitted by radon strike a metallic foil and some rebound backward; (b) backward rebound of α particles explained as the repulsion from a heavy positively charged nucleus.

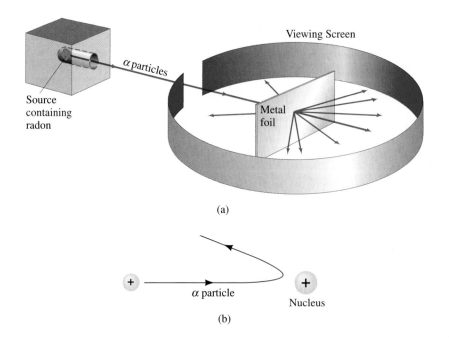

Viewing Screen

α particles

Source containing radon

Metal foil

(a)

$+$

α particle

$+$

Nucleus

(b)

due to electrical attraction, Fig. 38–17. Rutherford's experiments suggested that the nucleus must have a radius of about 10^{-15} to 10^{-14} m. From kinetic theory, and especially Einstein's analysis of Brownian movement (see Section 17–1), the radius of atoms was estimated to be about 10^{-10} m. Thus the electrons would seem to be at a distance from the nucleus of about 10,000 to 100,000 times the radius of the nucleus itself (if the nucleus were the size of a baseball, the atom would have the diameter of a big city several kilometers across). So an atom would be mostly empty space.

Rutherford's "planetary" model of the atom (also called the "nuclear model of the atom") was a major step toward how we view the atom today. It was not, however, a complete model and presented some major problems, as we shall see.

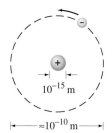

FIGURE 38–17 Rutherford's model of the atom, in which electrons orbit a tiny positive nucleus (not to scale). The atom is visualized as mostly empty space.

38–9 Atomic Spectra: Key to the Structure of the Atom

Earlier in this chapter we saw that heated solids (as well as liquids and dense gases) emit light with a continuous spectrum of wavelengths. This radiation is assumed to be due to oscillations of atoms and molecules, which are largely governed by the interaction of each atom or molecule with its neighbors.

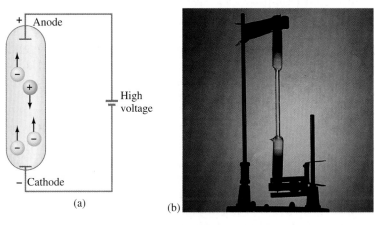

FIGURE 38–18 Gas-discharge tube: (a) is a diagram; (b) photo of actual discharge tube for hydrogen.

Rarefied gases can also be excited to emit light. This is done by intense heating, or more commonly by applying a high voltage to a "discharge tube" containing the gas at low pressure, Fig. 38–18. The radiation from excited gases had been observed early in the nineteenth century, and it was found that the spectrum was not continuous, but *discrete*. Since excited gases emit light of only certain wavelengths, when this light is analyzed through the slit of a spectroscope or spectrometer, a **line spectrum** is seen rather than a continuous spectrum. The line spectra emitted by a number of elements in the visible region are shown here in Fig. 38–19,

FIGURE 38–19 Emission spectra of the gases (a) atomic hydrogen, (b) helium, and (c) the *solar absorption* spectrum.

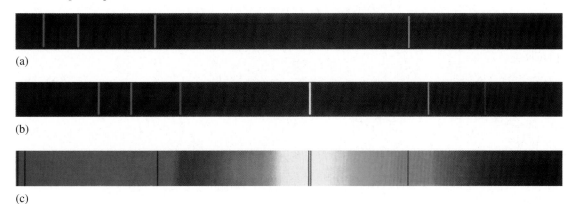

(a)

(b)

(c)

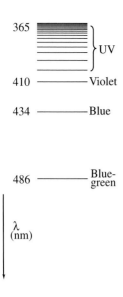

365 }UV

410 ——————— Violet

434 ——————— Blue

486 ——————— Blue-
green

λ
(nm)

656 ——————— Red

FIGURE 38–20 Balmer series of lines for hydrogen.

and also at the start of this chapter and in Chapter 35. The **emission spectrum** is characteristic of the material and can serve as a type of "fingerprint" for identification of the gas.

We also saw (Chapter 35) that if a continuous spectrum passes through a gas, dark lines are observed in the emerging spectrum, at wavelengths corresponding to lines normally emitted by the gas. This is called an **absorption spectrum** (Fig. 38–19c), and it became clear that gases can absorb light at the same frequencies at which they emit. Using film sensitive to ultraviolet and to infrared light, it was found that gases emit and absorb discrete frequencies in these regions as well as in the visible.

For our purposes here, the importance of the line spectra is that they are emitted (or absorbed) by gases at low pressure and low density. In such low density gases, the atoms are far apart on the average and hence the light emitted or absorbed is assumed to be by *individual atoms* rather than through interactions between atoms, as in a solid, liquid, or dense gas. Thus the line spectra serve as a key to the structure of the atom: any theory of atomic structure must be able to explain why atoms emit light only of discrete wavelengths, and it should be able to predict what these wavelengths are.

Hydrogen is the simplest atom—it has only one electron orbiting its nucleus. It also has the simplest spectrum. The spectrum of most atoms shows little apparent regularity. But the spacing between lines in the hydrogen spectrum decreases in a regular way, Fig. 38–20. Indeed, in 1885, J. J. Balmer (1825–1898) showed that the four visible lines in the hydrogen spectrum (with measured wavelengths 656 nm, 486 nm, 434 nm, and 410 nm) fit the following formula

$$\frac{1}{\lambda} = R\left(\frac{1}{2^2} - \frac{1}{n^2}\right), \qquad n = 3, 4, \cdots, \tag{38–8}$$

where n takes on the values 3, 4, 5, 6 for the four lines, and R, called the **Rydberg constant**, has the value $R = 1.0974 \times 10^7 \, \text{m}^{-1}$. Later it was found that this **Balmer series** of lines extended into the UV region, ending at $\lambda = 365$ nm, as shown in Fig. 38–20. Balmer's formula, Eq. 38–8, also worked for these lines with higher integer values of n. The lines near 365 nm became too close together to distinguish, but the limit of the series at 365 nm corresponds to $n = \infty$ (so $1/n^2 = 0$ in Eq. 38–8).

Later experiments on hydrogen showed that there were similar series of lines in the UV and IR regions, and each series had a pattern just like the Balmer series, but at different wavelengths, Fig. 38–21. Each of these series was found to fit a formula with the same form as Eq. 38–8 but with the $1/2^2$ replaced by $1/1^2$, $1/3^2$, $1/4^2$, and so on. For example, the so-called **Lyman series** contains lines with wavelengths from 91 nm to 122 nm (in the UV region) and fits the formula

$$\frac{1}{\lambda} = R\left(\frac{1}{1^2} - \frac{1}{n^2}\right), \qquad n = 2, 3, \cdots.$$

And the wavelengths of the **Paschen series** (in the IR region) fit

$$\frac{1}{\lambda} = R\left(\frac{1}{3^2} - \frac{1}{n^2}\right), \qquad n = 4, 5, \cdots.$$

FIGURE 38–21 Line spectrum of atomic hydrogen. Each series fits the formula $\frac{1}{\lambda} = R\left(\frac{1}{n'^2} - \frac{1}{n^2}\right)$ where $n' = 1$ for the Lyman series, $n' = 2$ for the Balmer series, $n' = 3$ for the Paschen series, and so on; n can take on all integer values from $n = n' + 1$ up to infinity. The only lines in the visible region of the electromagnetic spectrum are part of the Balmer series.

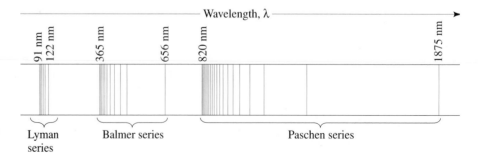

The Rutherford model, as it stood, was unable to explain why atoms emit line spectra. It had other difficulties as well. According to the Rutherford model, electrons orbit the nucleus, and since their paths are curved the electrons are accelerating. Hence they should give off light like any other accelerating electric charge (Chapter 32). Then, since energy is conserved, the electron's own energy must decrease to compensate. Hence electrons would be expected to spiral into the nucleus. As they spiraled inward, their frequency would increase in a short time and so too would the frequency of the light emitted. Thus the two main difficulties of the Rutherford model are these: (1) it predicts that light of a continuous range of frequencies will be emitted, whereas experiment shows line spectra; (2) it predicts that atoms are unstable—electrons should quickly spiral into the nucleus—but we know that atoms in general are stable, since the matter around us is stable.

Clearly Rutherford's model was not sufficient. Some sort of modification was needed, and it was Niels Bohr who provided it by adding an essential idea—the quantum hypothesis.

38–10 The Bohr Model

Bohr had studied in Rutherford's laboratory for several months in 1912 and was convinced that Rutherford's planetary model of the atom had validity. But in order to make it work, he felt that the newly developing quantum theory would somehow have to be incorporated in it. The work of Planck and Einstein had shown that in heated solids, the energy of oscillating electric charges must change discontinuously—from one discrete energy state to another, with the emission of a quantum of light. Perhaps, Bohr argued, the electrons in an atom also cannot lose energy continuously, but must do so in quantum "jumps." In working out his theory during the next year, Bohr postulated that electrons move about the nucleus in circular orbits, but that only certain orbits are allowed. He further postulated that an electron in each orbit would have a definite energy and would move in the orbit *without radiating energy* (even though this violated classical ideas since accelerating electric charges are supposed to emit EM waves; see Chapter 32). He thus called the possible orbits **stationary states**. Light is emitted, he hypothesized, only when an electron jumps from one stationary state to another of lower energy. When such a jump occurs, a single photon of light is emitted whose energy, since energy is conserved, is given by

Stationary state

$$hf = E_u - E_l, \tag{38–9}$$

where E_u refers to the energy of the upper state and E_l the energy of the lower state. See Fig. 38–22.

Bohr set out to determine what energies these orbits would have, since the spectrum of light emitted could then be predicted from Eq. 38–9. In the Balmer formula he had the key he was looking for. Bohr quickly found that his theory would be in accord with the Balmer formula if he assumed that the electron's angular momentum L is quantized and equal to an integer n times $h/2\pi$. As we saw in Chapter 10, angular momentum is given by $L = I\omega$, where I is the moment of inertia and ω is the angular velocity. For a single particle of mass m moving in a circle of radius r with speed v, $I = mr^2$ and $\omega = v/r$; hence, $L = I\omega = (mr^2)(v/r) = mvr$. Bohr's **quantum condition** is

$$L = mvr_n = n\frac{h}{2\pi}, \qquad n = 1, 2, 3, \cdots, \tag{38–10}$$

where n is an integer and r_n is the radius of the nth possible orbit. The allowed orbits are numbered 1, 2, 3, ..., according to the value of n, which is called the **quantum number** of the orbit.

FIGURE 38–22 An atom emits a photon (energy $= hf$) when its energy changes from E_u to a lower energy E_l.

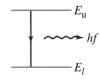

Angular momentum quantized

Quantum number, n

Equation 38–10 did not have a firm theoretical foundation. Bohr had searched for some "quantum condition," and such tries as $E = hf$ (where E represents the energy of the electron in an orbit) did not give results in accord with experiment. Bohr's reason for using Eq. 38–10 was simply that it worked; and we now look at how. In particular, let us determine what the Bohr theory predicts for the measurable wavelengths of emitted light.

An electron in a circular orbit of radius r_n (Fig. 38–23) would have a centripetal acceleration v^2/r_n produced by the electrical force of attraction between the negative electron and the positive nucleus. This force is given by Coulomb's law,

$$F = \frac{1}{4\pi\epsilon_0}\frac{(Ze)(e)}{r^2}.$$

The charge on the nucleus is $+Ze$, where Z is the number of positive charges[†] (i.e., protons). For the hydrogen atom, $Z = +1$.

In Newton's second law, $F = ma$, we substitute $a = v^2/r_n$ and Coulomb's law for F, and obtain

$$F = ma$$

$$\frac{1}{4\pi\epsilon_0}\frac{Ze^2}{r_n^2} = \frac{mv^2}{r_n}.$$

We solve this for r_n, and then substitute for v from Eq. 38–10 (which says $v = nh/2\pi m r_n$):

$$r_n = \frac{Ze^2}{4\pi\epsilon_0 mv^2} = \frac{Ze^2 4\pi^2 m r_n^2}{4\pi\epsilon_0 n^2 h^2}.$$

We solve for r_n (it appears on both sides, so we cancel one of them) and find

$$r_n = \frac{n^2 h^2 \epsilon_0}{\pi m Z e^2} = \frac{n^2}{Z}r_1 \qquad \textbf{(38–11)}$$

where

$$r_1 = \frac{h^2 \epsilon_0}{\pi m e^2}. \qquad \textbf{(38–12a)}$$

Equation 38–11 gives the radii of all possible orbits. The smallest orbit is for $n = 1$, and for hydrogen ($Z = 1$) has the value

$$r_1 = \frac{(6.626 \times 10^{-34}\,\text{J}\cdot\text{s})^2(8.85 \times 10^{-12}\,\text{C}^2/\text{N}\cdot\text{m}^2)}{(3.14)(9.11 \times 10^{-31}\,\text{kg})(1.602 \times 10^{-19}\,\text{C})^2}$$

or

$$r_1 = 0.529 \times 10^{-10}\,\text{m}. \qquad \textbf{(38–12b)}$$

The radius of the smallest orbit in hydrogen, r_1, is sometimes called the **Bohr radius**. From Eq. 38–11, we see that the radii of the larger orbits increase as n^2, so

$$r_2 = 4r_1 = 2.12 \times 10^{-10}\,\text{m},$$

$$r_3 = 9r_1 = 4.76 \times 10^{-10}\,\text{m},$$

$$\vdots$$

$$r_n = n^2 r_1.$$

The first four orbits are shown in Fig. 38–24. Notice that, according to Bohr's model, an electron can exist only in the orbits given by Eq. 38–11. There are no allowable orbits in between.

[†]We include Z in our derivation so that we can treat other single-electron ("hydrogenlike") atoms such as the ions He+ ($Z = 2$) and Li²⁺ ($Z = 3$). Helium in the neutral state has two electrons; if one electron is missing, the remaining He+ ion consists of one electron revolving around a nucleus of charge $+2e$. Similarly, doubly ionized lithium, Li²⁺, also has a single electron, and in this case $Z = 3$.

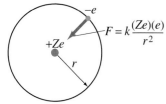
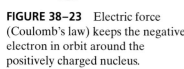

FIGURE 38–23 Electric force (Coulomb's law) keeps the negative electron in orbit around the positively charged nucleus.

Bohr radius

FIGURE 38–24 Possible orbits in the Bohr model of hydrogen; $r_1 = 0.529 \times 10^{-10}\,\text{m}$.

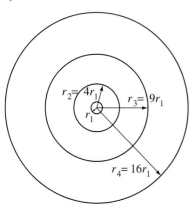

In each of its possible orbits, the electron would have a definite energy, as the following calculation shows. The total energy equals the sum of the kinetic and potential energies. The potential energy of the electron is given by $U = qV = -eV$, where V is the potential due to a point charge $+Ze$ as given by Eq. 23–5: $V = (1/4\pi\epsilon_0)(Q/r) = (1/4\pi\epsilon_0)(Ze/r)$. So

$$U = -eV = -\frac{1}{4\pi\epsilon_0}\frac{Ze^2}{r}.$$

The total energy E_n for an electron in the nth orbit of radius r_n is the sum of the kinetic and potential energies:

$$E_n = \tfrac{1}{2}mv^2 - \frac{1}{4\pi\epsilon_0}\frac{Ze^2}{r_n}.$$

When we substitute v from Eq. 38–10 and r_n from Eq. 38–11 into this equation, we obtain

$$E_n = -\left(\frac{Z^2e^4m}{8\epsilon_0^2 h^2}\right)\left(\frac{1}{n^2}\right) \qquad n = 1, 2, 3, \cdots \tag{38–13}$$

Energy levels

or

$$E_n = \frac{Z^2}{n^2}E_1 \qquad n = 1, 2, 3, \cdots$$

where E_1 is the lowest energy level ($n = 1$) for hydrogen ($Z = 1$). The value of E_1 is

$$E_1 = -\frac{me^4}{8\epsilon_0^2 h^2} = -2.17 \times 10^{-18}\,\text{J} = -13.6\,\text{eV},$$

Ground state of hydrogen

where we have converted joules to electron volts, as is customary in atomic physics. Since n^2 appears in the denominator of Eq. 38–13, the energies of the larger orbits in hydrogen ($Z = 1$) are given by

$$E_n = \frac{-13.6\,\text{eV}}{n^2}.$$

For example,

$$E_2 = \frac{-13.6\,\text{eV}}{4} = -3.40\,\text{eV},$$

$$E_3 = \frac{-13.6\,\text{eV}}{9} = -1.51\,\text{eV}.$$

Excited states (first two)

We see that not only are the orbit radii quantized, but from Eq. 38–13 so is the energy. The quantum number n that labels the orbit radii also labels the energy levels. The lowest **energy level** or **energy state** has energy E_1, and is called the **ground state**. The higher states, E_2, E_3, and so on, are called **excited states**.

Notice that although the energy for the larger orbits has a smaller numerical value, all the energies are less than zero. Thus, $-3.4\,\text{eV}$ is a higher energy than $-13.6\,\text{eV}$. Hence the orbit closest to the nucleus (r_1) has the lowest energy. The reason the energies have negative values has to do with the way we defined the zero for potential energy (U). For two point charges, $U = kq_1q_2/r$ corresponds to zero potential energy when the two charges are infinitely far apart. Thus, an electron that can just barely be free from the atom by reaching $r = \infty$ (or, at least, far from the nucleus) with zero kinetic energy will have $E = 0$, corresponding to $n = \infty$ in Eq. 38–13. If an electron is free and has kinetic energy, then $E > 0$. To remove an electron that is part of an atom requires an energy input (otherwise atoms would not be stable). Since $E \geq 0$ for a free electron, then an electron bound to an atom must have $E < 0$. That is, energy must be added to bring its energy up, from a negative value, to at least zero in order to free it.

The minimum energy required to remove an electron from the ground state of an atom is called the **binding energy** or **ionization energy**. The ionization energy for hydrogen has been measured to be 13.6 eV, and this corresponds precisely to removing an electron from the lowest state, $E_1 = -13.6$ eV, up to $E = 0$ where it can be free.

Binding energy
Ionization energy

It is useful to show the various possible energy values as horizontal lines on an energy-level diagram.[†] This is shown for hydrogen in Fig. 38–25. The electron in a hydrogen atom can be in any one of these levels according to Bohr theory. But it could never be in between, say at -9.0 eV. At room temperature, nearly all H atoms will be in the ground state ($n = 1$). At higher temperatures, or during an electric discharge when there are many collisions between free electrons and atoms, many atoms can be in excited states ($n > 1$). Once in an excited state, an atom's electron can jump down to a lower state, and give off a photon in the process. This is, according to the Bohr model, the origin of the emission spectra of excited gases. The vertical arrows in Fig. 38–25 represent the transitions or jumps that correspond to the various observed spectral lines. For example, an electron jumping from the level $n = 3$ to $n = 2$ would give rise to the 656-nm line in the Balmer series, and the jump from $n = 4$ to $n = 2$ would give rise to the 486-nm line (see Fig. 38–20). We can predict wavelengths of the spectral lines emitted by combining Eq. 38–9 with Eq. 38–13. Since $hf = hc/\lambda$, we have from Eq. 38–9

*Line spectra
emission
explained*

$$\frac{1}{\lambda} = \frac{hf}{hc} = \frac{1}{hc}\left(E_n - E_{n'}\right),$$

and then using Eq. 38–13,

$$\frac{1}{\lambda} = \frac{Z^2 e^4 m}{8\epsilon_0^2 h^3 c}\left(\frac{1}{(n')^2} - \frac{1}{(n)^2}\right), \tag{38–14}$$

where n refers to the upper state and n' to the lower state. This theoretical formula

[†]Note that above $E = 0$, an electron is free and can have any energy (E is not quantized). Thus there is a continuum of energy states above $E = 0$, as indicated in the energy-level diagram of Fig. 38–25.

FIGURE 38–25 Energy-level diagram for the hydrogen atom, showing the origin of spectral lines for the Lyman, Balmer, and Paschen series.

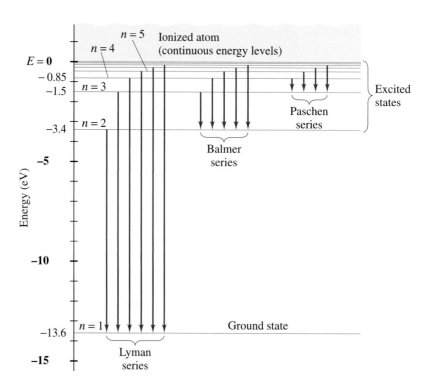

has the same form as the experimental Balmer formula, Eq. 38–8, with $n' = 2$. Thus we see that the Balmer series of lines corresponds to transitions or "jumps" that bring the electron down to the second energy level. Similarly, $n' = 1$ corresponds to the Lyman series and $n' = 3$ to the Paschen series (see Fig. 38–25). When the constant $(Z^2 e^4 m / 8\epsilon_0^2 h^3 c)$ in Eq. 38–14 is evaluated with $Z = 1$, it is found to have the measured value of the Rydberg constant, $R = 1.0974 \times 10^7 \, \text{m}^{-1}$ in Eq. 38–8, in accord with experiment (see Problem 47).

The great success of Bohr's theory is that it explains why atoms emit line spectra and accurately predicts, for hydrogen, the wavelengths of emitted light. The Bohr theory also explains absorption spectra: photons of just the right wavelength can knock an electron from one energy level to a higher one. To conserve energy, only photons that have just the right energy (and frequency) will be absorbed. This explains why a continuous spectrum of light entering a gas will emerge with dark (absorption) lines at frequencies that correspond to emission lines.

Absorption lines explained

The Bohr theory also ensures the stability of atoms. It establishes stability by fiat: the ground state is the lowest state for an electron and there is no lower energy level to which it can go and emit more energy. Finally, as we saw above, the Bohr theory accurately predicts the ionization energy of 13.6 eV for hydrogen. However, the Bohr theory was not so successful for other atoms, as we shall see.

EXAMPLE 38–12 **Wavelength of a Lyman line.** Use Fig. 38–25 to determine the wavelength of the first Lyman line, the transition from $n = 2$ to $n = 1$. In what region of the electromagnetic spectrum does this lie?

SOLUTION In this case, $hf = E_2 - E_1 = 13.6 \, \text{eV} - 3.4 \, \text{eV} = 10.2 \, \text{eV} = 1.63 \times 10^{-18} \, \text{J}$. Since $\lambda = c/f$, we have

$$\lambda = \frac{c}{f} = \frac{hc}{E_2 - E_1} = \frac{(6.63 \times 10^{-34} \, \text{J} \cdot \text{s})(3.00 \times 10^8 \, \text{m/s})}{1.63 \times 10^{-18} \, \text{J}} = 1.22 \times 10^{-7} \, \text{m},$$

or 122 nm, which is in the UV region. See Fig. 38–21.

EXAMPLE 38–13 **Wavelength of a Balmer line.** Determine the wavelength of light emitted when a hydrogen atom makes a transition from the $n = 6$ to the $n = 2$ energy level according to the Bohr model.

SOLUTION We can use Eq. 38–14 or its equivalent, Eq. 38–8, with $R = 1.097 \times 10^7 \, \text{m}^{-1}$. Thus

$$\frac{1}{\lambda} = (1.097 \times 10^7 \, \text{m}^{-1})\left(\frac{1}{4} - \frac{1}{36}\right) = 2.44 \times 10^6 \, \text{m}^{-1}.$$

So $\lambda = 1/(2.44 \times 10^6 \, \text{m}^{-1}) = 4.10 \times 10^{-7} \, \text{m}$ or 410 nm. This is the fourth line in the Balmer series, Fig. 38–20, and is violet in color.

EXAMPLE 38–14 **Absorption wavelength.** Use Fig. 38–25 to determine the maximum wavelength that hydrogen in its ground state can absorb. What would be the next smaller wavelength that would work?

SOLUTION Maximum λ corresponds to minimal energy, and thus the jump from the ground state to the first excited state (Fig. 38–25) for which the energy is 13.6 eV − 3.4 eV = 10.2 eV; the required wavelength, as we saw in Example 38–12, is 122 nm. The next possibility is to jump from the ground state to the second excited state, which requires 13.6 eV − 1.5 eV = 12.1 eV and corresponds to a wavelength

$$\lambda = \frac{c}{f} = \frac{hc}{E_3 - E_1}$$

$$= \frac{(6.63 \times 10^{-34} \, \text{J} \cdot \text{s})(3.00 \times 10^8 \, \text{m/s})}{(12.1 \, \text{eV})(1.60 \times 10^{-19} \, \text{J/eV})} = 103 \, \text{nm}.$$

EXAMPLE 38–15 Ionization energy. Use the Bohr model to determine the ionization energy of the He$^+$ ion, which has a single electron. Also calculate the minimum wavelength a photon must have to cause ionization.

SOLUTION We want to determine the minimum energy required to lift the electron from its ground state and to barely reach the free state at $E = 0$. The ground state energy of He$^+$ is given by Eq. 38–13 with $n = 1$ and $Z = 2$. Since all the symbols in Eq. 38–13 are the same as for the calculation for hydrogen, except that Z is 2 instead of 1, we see that E_1 will be $Z^2 = 2^2 = 4$ times the E_1 for hydrogen. That is,

$$E_1 = 4(-13.6 \text{ eV}) = -54.4 \text{ eV}.$$

Thus, to ionize the He$^+$ ion should require 54.4 eV, and this value agrees with experiment. The minimum wavelength photon that can cause ionization will have energy $hf = 54.4$ eV and wavelength $\lambda = c/f = hc/hf = (6.63 \times 10^{-34} \text{ J·s})(3.00 \times 10^8 \text{ m/s})/(54.4 \text{ eV})(1.60 \times 10^{-19} \text{ J/eV}) = 22.8$ nm. If the atom absorbed a photon of greater energy (wavelength shorter than 22.8 nm), the atom could still be ionized and the freed electron would have kinetic energy of its own.

In this last Example, we saw that E_1 for the He$^+$ ion is four times more negative than that for hydrogen. Indeed the energy-level diagram for He$^+$ looks just like that for hydrogen, Fig. 38–25, except that the numerical values for each energy level are four times larger. It is important to note, however, that we are talking here about the He$^+$ *ion*. Normal (neutral) helium has two electrons and its energy level diagram is entirely different.

CONCEPTUAL EXAMPLE 38–16 Hydrogen at 20°C. Estimate the average kinetic energy of hydrogen atoms (or molecules) at room temperature, and use the result to explain why nearly all H atoms are in the ground state at room temperature, and hence emit no light.

RESPONSE According to kinetic theory (Chapter 18), the average kinetic energy of atoms or molecules in a gas is given by Eq. 18–4,

$$\bar{K} = \tfrac{3}{2}kT,$$

where $k = 1.38 \times 10^{-23}$ J/K is Boltzmann's constant, and T is the kelvin (absolute) temperature. Room temperature is about $T = 300$ K, so

$$\bar{K} = \tfrac{3}{2}(1.38 \times 10^{-23} \text{ J/K})(300 \text{ K}) = 6.2 \times 10^{-21} \text{ J},$$

or, in electron volts:

$$\bar{K} = \frac{6.2 \times 10^{-21} \text{ J}}{1.6 \times 10^{-19} \text{ J/eV}} = 0.04 \text{ eV}.$$

The average kinetic energy is thus very small compared to the energy between the ground state and the next higher energy state (13.6 eV − 3.4 eV = 10.2 eV). Any atoms in excited states emit light and eventually fall to the ground state. Once in the ground state, collisions with other atoms can transfer energy of only 0.04 eV on the average. A small fraction of atoms can have much more energy (see Section 18–2 on the distribution of molecular speeds), but even kinetic energy that is 10 times the average is not nearly enough to excite atoms above the ground state. Thus, at room temperature, nearly all atoms are in the ground state. Atoms can be excited to upper states by very high temperatures, or by passing a current of high energy electrons through the gas, as in a discharge tube (Fig. 38–18).

We should note that Bohr made some radical assumptions that were at variance with classical ideas. He assumed that electrons in fixed orbits do not radiate light even though they are accelerating (moving in a circle), and he assumed that angular momentum is quantized. Furthermore, he was not able to say how an electron moved when it made a transition from one energy level to another. On the other hand, there is no real reason to expect that in the tiny world of the atom electrons would behave as ordinary-sized objects do. Nonetheless, he felt that where quantum theory overlaps with the macroscopic world, it should predict classical results. This is the **correspondence principle**, already mentioned in regard to relativity (Section 37–13). This principle does work for Bohr's theory of the hydrogen atom. The orbit sizes and energies are quite different for $n = 1$ and $n = 2$, say. But orbits with $n = 100{,}000{,}000$ and $100{,}000{,}001$ would be very close in size and energy (see Fig. 38–25). Indeed, jumps between such large orbits, which would approach macroscopic sizes, would be imperceptible. Such orbits would thus appear to be continuously spaced, which is what we expect in the everyday world.

Correspondence principle

Finally, be careful not to believe that the well-defined orbits of the Bohr model actually exist. The Bohr model is only a model, not reality. The idea of electron orbits was rejected a few years later, and today electrons are better thought of as forming "clouds."

38–11 | de Broglie's Hypothesis Applied to Atoms

Bohr's theory was largely of an *ad hoc* nature. Assumptions were made so that theory would agree with experiment. But Bohr could give no reason why the orbits were quantized, nor why there should be a stable ground state. Finally, ten years later, a reason was proposed by de Broglie.

We saw in Section 38–6 that in 1923, Louis de Broglie proposed that material particles, such as electrons, have a wave nature; and that this hypothesis was confirmed by experiment several years later.

One of de Broglie's original arguments in favor of the wave nature of electrons was that it provided an explanation for Bohr's theory of the hydrogen atom. According to de Broglie, a particle of mass m moving with a nonrelativistic speed v would have a wavelength (Eq. 38–7) of

$$\lambda = \frac{h}{mv}.$$

Each electron orbit in an atom, he proposed, is actually a standing wave. As we saw in Chapter 15, when a violin or guitar string is plucked, a vast number of wavelengths are excited. But only certain ones—those that have nodes at the ends—are sustained. These are the *resonant* modes of the string. Waves with other wavelengths interfere with themselves upon reflection and their amplitudes quickly drop to zero. With electrons moving in circles, according to Bohr's theory, de Broglie argued that the electron wave was a *circular* standing wave that closes on itself, Fig. 38–26. If the wavelength of a wave does not close on itself, as in Fig. 38–27, destructive interference takes place as the wave travels around the loop, and the wave quickly dies out. Thus, the only waves that persist are those for which the circumference of the circular orbit contains a whole number of wavelengths, Fig. 38–28. The circumference of a Bohr orbit of radius r_n is $2\pi r_n$, so we have

$$2\pi r_n = n\lambda, \qquad n = 1, 2, 3, \cdots.$$

When we substitute $\lambda = h/mv$, we get $2\pi r_n = nh/mv$, or

$$mvr_n = \frac{nh}{2\pi}.$$

This is just the *quantum condition* proposed by Bohr on an *ad hoc* basis, Eq. 38–10.

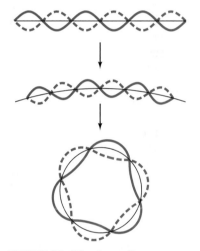

FIGURE 38–26 An ordinary standing wave compared to a circular standing wave.

FIGURE 38–27 When a wave does not close (and hence interferes destructively with itself), it rapidly dies out.

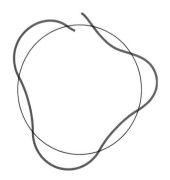

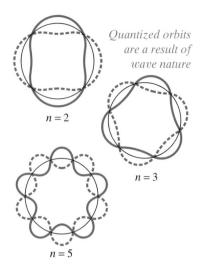

Quantized orbits are a result of wave nature

$n = 2$

$n = 3$

$n = 5$

FIGURE 38–28 Standing circular waves for two, three, and five wavelengths on the circumference; n, the number of wavelengths, is also the quantum number.

And it is from this equation that the discrete orbits and energy levels were derived. Thus we have an explanation for the quantized orbits and energy states in the Bohr model: they are due to the wave nature of the electron, and that only resonant "standing" waves can persist. This implies that the *wave–particle duality* is at the root of atomic structure.

It should be noted in viewing the circular electron waves of Fig. 38–28 that the electron is not to be thought of as following the oscillating wave pattern. In the Bohr model of hydrogen, the electron, considered as a particle, moves in a circle. The circular wave, on the other hand, represents the *amplitude* of the electron "matter wave," and in Fig. 38–28 the wave amplitude is shown superimposed on the circular path of the particle orbit for convenience.

Bohr's theory worked well for hydrogen and for one-electron ions. But it did not prove as successful for multielectron atoms. Bohr theory could not predict line spectra even for the next simplest atom, helium. It could not explain why some emission lines are brighter than others, nor why some lines are actually split into two or more closely spaced lines ("fine structure"). A new theory was needed and was indeed developed in the 1920s. This new and radical theory is called *quantum mechanics*, and finally solved the problem of atomic structure; but it gives us a very different view of the atom: the idea of electrons in well-defined orbits was replaced with the idea of electron "clouds." And this new theory of quantum mechanics has given us a wholly different view of the basic mechanisms underlying physical processes.

Summary

Quantum theory has its origins in **Planck's quantum hypothesis** that molecular oscillations are quantized: their energy E can only be integer (n) multiples of hf, where h is Planck's constant and f is the natural frequency of oscillation: $E = nhf$. This hypothesis explained the spectrum of radiation emitted by (black) bodies at high temperature.

Einstein proposed that for some experiments, light could be pictured as being emitted and absorbed as quanta (particles), which we now call **photons**, each with energy

$$E = hf.$$

He proposed the photoelectric effect as a test for the photon theory of light. In the **photoelectric effect**, the photon theory says that each incident photon can strike an electron in a material and eject it if it has sufficient energy. The maximum energy of ejected electrons is then linearly related to the frequency of the incident light. The photon theory is also supported by the **Compton effect** and the observation of electron–positron **pair production**.

The wave–particle duality refers to the idea that light and matter (such as electrons) have both wave and particle properties. The wavelength of a material object is given by

$$\lambda = \frac{h}{mv},$$

where mv is the momentum of the object. The **principle of complementarity** states that we must be aware of both the particle and wave properties of light and of matter for a complete understanding of them.

Early models of the atom include the plum-pudding model and Rutherford's planetary (or nuclear) model. Rutherford's model, which was created to explain the backscattering of alpha particles from thin metal foils, assumes that an atom consists of a tiny but massive positively charged nucleus surrounded (at a relatively great distance) by electrons.

To explain the line spectra emitted by atoms, as well as the stability of atoms, **Bohr theory** postulated (1) that electrons bound in an atom can only occupy orbits for which the angular momentum is quantized, which results in discrete values for the radius and energy; (2) that an electron in such a **stationary state** emits no radiation; (3) that, if an electron jumps to a lower state, it emits a photon whose energy equals the difference in energy between the two states; (4) that the angular momentum L of atomic electrons is quantized by the rule

$$L = \frac{nh}{2\pi},$$

where n is an integer called a **quantum number**. The $n = 1$ state in hydrogen is the **ground state**, which has an energy $E_1 = -13.6 \text{ eV}$; higher values of n correspond to **excited states** and their energies are

$$E_n = -\frac{13.6 \text{ eV}}{n^2}.$$

Atoms are excited to these higher states by collisions with other atoms or electrons or by absorption of a photon of just the right frequency.

De Broglie's hypothesis that electrons (and other matter) have a wavelength $\lambda = h/mv$ gave an explanation for Bohr's quantized orbits by bringing in the wave–particle duality: the orbits correspond to circular standing waves in which the circumference of the orbit equals a whole number of wavelengths.

Questions

1. What can be said about the relative temperature of whitish-yellow, reddish, and bluish stars?

2. If energy is radiated by all objects, why can we not see them in the dark? (See also Section 19–10.)

3. Does a lightbulb at a temperature of 2500 K produce as white a light as the Sun at 6000 K? Explain.

4. An ideal black body can be approximated by a small hole in an otherwise enclosed cavity. Explain. [*Hint*: The pupil of your eye is an approximate case.]

5. Darkrooms for developing black-and-white film are sometimes lit by a red bulb. Why red? Would such a bulb work in a darkroom for developing color photographs?

6. If the threshold wavelength in the photoelectric effect increases when the emitting metal is changed to a different metal, what can you say about the work functions of the two metals?

7. Explain why the existence of a cutoff frequency in the photoelectric effect more strongly favors a particle theory rather than a wave theory of light.

8. UV light causes sunburn, whereas visible light does not. Explain.

9. If an X-ray photon is scattered by an electron, does its wavelength change? If so, does it increase or decrease?

10. In both the photoelectric effect and in the Compton effect we have a photon colliding with an electron causing the electron to fly off. What then, is the difference between the two processes?

11. Show that the units of Planck's constant are those for angular momentum.

12. Consider a point source of light. How would the intensity of light vary with distance from the source according to (*a*) wave theory, (*b*) particle (photon) theory? Would this help to distinguish the two theories?

13. Explain how the photoelectric circuit of Fig. 38–4 could be used in (*a*) a burglar alarm, (*b*) a smoke detector, (*c*) a photographic light meter, and (*d*) a spectrophotometer (see Section 36–8).

14. Why do we say that light has wave properties? Why do we say that light has particle properties?

15. Why do we say that electrons have wave properties? Why do we say that electrons have particle properties?

16. What is the difference between a photon and an electron? Be specific: make a list.

17. If an electron and a proton travel at the same speed, which has the shorter wavelength?

18. In Rutherford's planetary model of the atom, what keeps the electrons from flying off into space?

19. Which of the following can emit a line spectrum: (*a*) gases, (*b*) liquids, (*c*) solids? Which can emit a continuous spectrum?

20. Why doesn't the O_2 gas in the air around us give off light?

21. How can you tell if there is oxygen near the surface of the Sun?

22. When a wide spectrum of light passes through hydrogen gas at room temperature, absorption lines are observed that correspond only to the Lyman series. Why don't we observe the other series?

23. Explain how the closely spaced energy levels for hydrogen near the top of Fig. 38–25 correspond to the closely spaced spectral lines at the top of Fig. 38–20.

24. Discuss the differences between Rutherford's and Bohr's theory of the atom.

25. Is it possible for the de Broglie wavelength of a "particle" to be greater than the dimensions of the particle? To be smaller? Is there any direct connection?

26. Approximately how large would *h* have to be so that quantum effects would be noticeable in the macroscopic world of everyday life?

27. Does it make sense that the potential energy of the electron in a hydrogen atom is negative and greater in magnitude than the kinetic energy, as given by the Bohr theory? What is the significance of this?

28. In a helium atom, which contains two electrons, do you think that on the average the electrons are closer to the nucleus or farther away than in a hydrogen atom? Why?

29. How can the spectrum of hydrogen contain so many lines when hydrogen contains only one electron?

30. The Lyman series is brighter than the Balmer series, because this series of transitions ends up in the most common state for hydrogen, the ground state. Why then was the Balmer series discovered first?

31. Use conservation of momentum to explain why photons emitted by hydrogen atoms have slightly less energy than that predicted by Eq. 38–9

Problems

Section 38–1

1. (I) How hot is a metal being welded if it radiates most strongly at 440 nm?

2. (I) (*a*) What is the temperature if the peak of a blackbody spectrum is at 25.0 nm? (*b*) What is the wavelength at the peak of a blackbody spectrum if the body is at a temperature of 2800 K?

3. (I) An HCl molecule vibrates with a natural frequency of 8.1×10^{13} Hz. What is the difference in energy (in joules and electron volts) between possible values of the oscillation energy?

4. (I) Estimate the peak wavelength for radiation from (*a*) ice at 0° C, (*b*) a floodlamp at 3300 K, (*c*) helium at 4 K, assuming blackbody emission. In what region of the EM spectrum is each?

5. (II) The steps of a flight of stairs are 20.0 cm high (vertically). If a 58.0-kg person stands with both feet on the same step, what is the gravitational potential energy of this person, relative to the ground, on (*a*) the first step, (*b*) the second step, (*c*) the third step, (*d*) the *n*th step? (*e*) What is the change in energy as the person descends from step 6 to step 2?

6. (II) Estimate the peak wavelength of light issuing from the pupil of the human eye (which approximates a blackbody) assuming normal body temperature.

7. (III) Planck's radiation law is given by:

$$I(\lambda, T) = \frac{2\pi h c^2 \lambda^{-5}}{e^{hc/\lambda kT} - 1}$$

where $I(\lambda, T)$ is the rate energy is radiated per unit surface area per unit wavelength interval at wavelength λ and Kelvin temperature T. (a) Show that Wien's displacement law follows from this relationship. (b) Determine the value of h from the experimental value of $\lambda_P T$ given in the text. (c) Derive the Stefan-Boltzmann law (the T^4 dependence of the rate at which energy is radiated—Eq. 19–15), by integrating Planck's formula over all wavelengths; that is, show that

$$\int I(\lambda, T) \, d\lambda \propto T^4.$$

Section 38–2

8. (I) What is the energy range in eV, of photons in the visible spectrum, of wavelength 400 nm to 750 nm?

9. (I) What is the energy of photons (in eV) emitted by an 88.5-MHz FM radio station?

10. (I) A typical gamma ray emitted from a nucleus during radioactive decay may have an energy of 300 keV. What is its wavelength? Would we expect significant diffraction of this type of light when it passes through an everyday opening, like a door?

11. (I) About 0.1 eV is required to break a "hydrogen bond" in a protein molecule. What are the minimum frequency and maximum wavelength of a photon that can accomplish this?

12. (I) What minimum frequency of light is needed to eject electrons from a metal whose work function is 4.3×10^{-19} J?

13. (II) What is the longest wavelength of light that will emit electrons from a metal whose work function is 3.10 eV?

14. (II) The work functions for sodium, cesium, copper and iron are 2.3, 2.1, 4.7 and 4.5 eV respectively. Which of these metals will not emit electrons when visible light shines on it?

15. (II) In a photoelectric-effect experiment it is observed that no current flows unless the wavelength is less than 570 nm. (a) What is the work function of this material? (b) What is the stopping voltage required if light of wavelength 400 nm is used?

16. (II) What is the maximum kinetic energy of electrons ejected from barium ($W_0 = 2.48$ eV) when illuminated by white light, $\lambda = 400$ to 750 nm?

17. (II) When UV light of wavelength 255 nm falls on a metal surface, the maximum kinetic energy of emitted electrons is 1.40 eV. What is the work function of the metal?

18. (II) The threshold wavelength for emission of electrons from a given surface is 350 nm. What will be the maximum kinetic energy of ejected electrons when the wavelength is changed to (a) 280 nm, (b) 380 nm?

19. (II) A certain type of film is sensitive only to light whose wavelength is less than 660 nm. What is the energy (eV and kcal/mol) needed for the chemical reaction to occur which causes the film to change?

20. (II) When 230-nm light falls on a metal, the current through a photoelectric circuit (Fig. 38–4) is brought to zero at a reverse voltage of 1.64 V. What is the work function of the metal?

Section 38–3

21. (II) The quantity h/mc, which has the dimensions of length, is called the *Compton wavelength*. Determine the Compton wavelength for (a) an electron, (b) a proton. (c) Show that if a photon has wavelength equal to the Compton wavelength of a particle, the photon's energy is equal to the rest energy of the particle.

22. (II) X-rays of wavelength $\lambda = 0.120$ nm are scattered from a carbon block. What is the Compton wavelength shift for photons detected at angles (relative to the incident beam) of (a) 45°, (b) 90°, (c) 180°?

23. (II) For each of the scattering angles in the previous Problem, determine (a) the fractional energy loss of the photon, and (b) the energy given to the scattered electron.

24. (II) (a) Show that the fractional energy loss of scattered photons in the Compton effect can be written as $\Delta\lambda/\lambda$. (b) Is this an exact expression? If not, specify what quantity limits the accuracy and what its value can be so this formula is accurate to 0.1 percent.

25. (III) In the Compton effect (see Fig. 38–7), use the relativistic equations for conservation of energy and of linear momentum to show that the Compton shift in wavelength is given by Eq. 38–6.

26. (III) In the Compton effect, a 0.100-nm photon strikes a free electron in a head-on collision and knocks it into the forward direction. The rebounding photon recoils directly backward. Use conservation of (relativistic) energy and momentum to determine (a) the kinetic energy of the electron, and (b) the wavelength of the recoiling photon. (Note: Use Eq. 38–5, but not Eq. 38–6.)

Section 38–4

27. (I) How much total kinetic energy will an electron–positron pair have if produced by a 2.84-MeV photon?

28. (II) What is the longest wavelength photon that could produce a proton–antiproton pair? (Each has a mass of 1.67×10^{-27} kg.)

29. (II) What is the minimum photon energy needed to produce a $\mu^+ - \mu^-$ pair? The mass of each μ (muon) is 207 times the mass of the electron. What is the wavelength of such a photon?

30. (II) A gamma-ray photon produces an electron–positron pair, each with a kinetic energy of 345 keV. What was the energy and wavelength of the photon?

Section 38–6

31. (I) Calculate the wavelength of a 0.21-kg ball traveling at 0.10 m/s.

32. (I) What is the wavelength of a neutron ($m = 1.67 \times 10^{-27}$ kg) traveling at 5.5×10^4 m/s?

33. (I) Through how many volts of potential difference must an electron be accelerated to achieve a wavelength of 0.28 nm?

34. (II) Calculate the de Broglie wavelength of an electron in your TV picture tube if it is accelerated by 30,000 V. Is it relativistic? How does its wavelength compare to the size of the "neck" of the tube, typically 5 cm? Do we have to worry about diffraction problems blurring our picture on the screen?

35. (II) What is the wavelength of an electron of energy (a) 10 eV, (b) 100 eV, (c) 1.0 keV?

36. (II) Show that if an electron and a proton have the same nonrelativistic kinetic energy, the proton has the shorter wavelength.

37. (II) Calculate the ratio of the kinetic energy of an electron to that of a proton if their wavelengths are equal. Assume that the speeds are non-relativistic.

38. (II) What is the wavelength of an O_2 molecule in the air at room temperature? [*Hint*: See Chapter 18.]

39. (III) A Cadillac with a mass of 2000 kg approaches a freeway underpass that is 10 m across. At what speed must the car be moving, in order for it to have a wavelength such that it might somehow "diffract" after passing through this single "slit"? How do these conditions compare to normal freeway speeds of 30 m/s?

* Section 38–7

* 40. (II) What voltage is needed to produce electron wavelengths of 0.10 nm? (Assume that the electrons are nonrelativistic.)

* 41. (II) Electrons are accelerated by 2250 V in an electron microscope. What is the maximum possible resolution?

Sections 38–9 and 38–10

42. (I) For the three hydrogen transitions indicated below, with n being the initial state and n' being the final state, is the transition an absorption or an emission? Which is higher, the initial state energy or the final state energy of the atom? Finally, which of these transitions involves the largest energy photon? (a) $n = 1$, $n' = 3$ (b) $n = 6$, $n' = 2$ (c) $n = 4$, $n' = 5$.

43. (I) How much energy is needed to ionize a hydrogen atom in the $n = 2$ state?

44. (I) The third longest wavelength in the Paschen series in hydrogen (Fig. 38–25) corresponds to what transition?

45. (I) Calculate the ionization energy of doubly ionized lithium, Li^{2+}, which has $Z = 3$.

46. (I) (a) Determine the wavelength of the second Balmer line ($n = 4$ to $n = 2$ transition) using Fig. 38–25. Determine likewise (b) the wavelength of the second Lyman line and (c) the wavelength of the third Balmer line.

47. (I) Evaluate the Rydberg constant R using Bohr theory (compare Eqs. 38–8 and 38–14) and show that its value is $R = 1.0974 \times 10^7 \, m^{-1}$.

48. (II) What is the longest wavelength light capable of ionizing a hydrogen atom in the ground state?

49. (II) What wavelength photon would be required to ionize a hydrogen atom in the ground state and give the ejected electron a kinetic energy of 10.0 eV?

50. (II) In the Sun, an ionized helium (He^+) atom makes a transition from the $n = 6$ state to the $n = 2$ state, emitting a photon. Can that photon be absorbed by hydrogen atoms present in the Sun? If so, between what energy states will the hydrogen atom jump?

51. (II) Construct the energy-level diagram for the He^+ ion (see Fig. 38–25).

52. (II) Construct the energy-level diagram for doubly ionized lithium, Li^{2+}.

53. (II) What is the potential energy and the kinetic energy of an electron in the ground state of the hydrogen atom?

54. (II) An excited hydrogen atom could, in principle, have a radius of 1.00 mm. What would be the value of n for a Bohr orbit of this size? What would its energy be?

55. (II) Is the use of nonrelativistic formulas justified in the Bohr atom? To check, calculate the electron's velocity, v, in terms of c for the ground state of hydrogen, and then calculate $\sqrt{1 - v^2/c^2}$.

56. (II) Suppose an electron was bound to a proton, as in the hydrogen atom, by the gravitational force rather than by the electric force. What would be the radius, and energy, of the first Bohr orbit?

57. (II) Show that the magnitude of the potential energy of an electron in any Bohr orbit of a hydrogen atom is twice the magnitude of its kinetic energy in that orbit.

58. (III) *Correspondence principle*: Show that for large values of n, the difference in radius Δr between two adjacent orbits (with quantum numbers n and $n - 1$) is given by

$$\Delta r = r_n - r_{n-1} \approx \frac{2r_n}{n},$$

so $\Delta r/r_n \to 0$ as $n \to \infty$, in accordance with the correspondence principle. [Note that we can check the correspondence principle by either considering large values of $n(n \to \infty)$ or by letting $h \to 0$. Are these equivalent?]

59. (III) (a) For very large values of n, show that when an electron jumps from the level n to the level $n - 1$, the frequency of the light emitted is equal to

$$f = \frac{v}{2\pi r_n}.$$

(b) Show that this is also just the frequency predicted by classical theory for an electron revolving in a circular orbit of radius r_n with speed v. (c) Explain why this is consistent with the correspondence principle.

General Problems

60. The Big Bang theory states that the beginning of the Universe was accompanied by a huge burst of photons. Those photons are still present today and make up the so called Cosmic Microwave Background Radiation. The Universe radiates like a blackbody with a temperature of about 2.7 K. Calculate the peak wavelength of this radiation.

61. At low temperatures, nearly all the atoms in hydrogen gas will be in the ground state. What minimum frequency photon is needed if the photoelectric effect is to be observed?

62. A beam of 85-eV electrons is scattered from a crystal, as in X-ray diffraction, and a first-order peak is observed at $\theta = 38°$. What is the spacing between planes in the diffracting crystal? (See Section 36–10).

63. Show that the energy E (in electron volts) of a photon whose wavelength is λ (meters) is given by
$$E = 1.24 \times 10^{-6}/\lambda.$$

64. A microwave oven produces electromagnetic radiation at $\lambda = 12.2$ cm and produces a power of 760 W. Calculate the number of microwave photons produced by the microwave oven each second.

65. Sunlight reaching the Earth's surface has an intensity of about $1000 \, \text{W/m}^2$. Estimate how many photons per square meter per second this represents. Take the average wavelength to be 550 nm.

66. A beam of red laser light ($\lambda = 633$ nm) hits a black wall and is fully absorbed. If this exerts a total force $F = 5.5$ nN on the wall, how many photons per second are hitting the wall?

67. If a 100-W light bulb emits 3.0 percent of the input energy as visible light (average wavelength 550 nm) uniformly in all directions, estimate how many photons per second of visible light will strike the pupil (4.0 mm diameter) of the eye of an observer 1.0 km away.

68. An electron and a positron collide head on, annihilate, and create two 0.90 MeV photons traveling in opposite directions. What were the initial kinetic energies of electron and positron?

69. In Compton's original experiment he saw X-rays scattered where the wavelength shifted from 0.0711 nm to 0.0735 nm. Through what angle did these X-rays scatter?

70. By what potential difference must (a) a proton $\left(m = 1.67 \times 10^{-27} \, \text{kg}\right)$, and (b) an electron $\left(m = 9.11 \times 10^{-31} \, \text{kg}\right)$, be accelerated to have a wavelength $\lambda = 5.0 \times 10^{-12}$ m?

71. In some of Rutherford's experiments (Fig. 38–16) the α particles $\left(\text{mass} = 6.64 \times 10^{-27} \, \text{kg}\right)$ had a kinetic energy of 4.8 MeV. How close could they get to a gold nucleus (charge = $+79e$)? Ignore the recoil motion of the nucleus.

72. By what fraction does the mass of an H atom decrease when it makes an $n = 3$ to $n = 1$ transition?

73. For what maximum kinetic energy is a collision between an electron and a hydrogen atom in its ground state definitely elastic?

74. Using Bohr theory, derive an equation for the angular velocity ω, and frequency f, of an electron in a hydrogen atom. Determine (a) for the ground state and (b) for the first excited state ($n = 2$).

75. Calculate the ratio of the gravitational to electric force for the electron in a hydrogen atom. Can the gravitational force be safely ignored?

76. A child's swing has a natural frequency of 0.75 Hz. (a) What is the separation between possible energy values (in joules)? (b) If the swing reaches a vertical height of 45 cm above its lowest point and has a mass of 20 kg (including the child), what is the value of the quantum number n? (c) What is the fractional change in energy between levels whose quantum numbers are n (as just calculated) and $n + 1$? Would quantization be measurable in this case?

77. Electrons accelerated by a potential difference of 12.3 V pass through a gas of hydrogen atoms at room temperature. What wavelengths of light will be emitted?

78. Atoms can be formed in which a muon (mass = 207 times the mass of an electron) replaces one of the electrons in an atom. Calculate, using Bohr theory, the energy of the photon emitted when a muon makes a transition from $n = 2$ to $n = 1$ in a muonic $^{208}_{82}\text{Pb}$ atom (lead whose nucleus has a mass 208 times the proton mass and charge $+82e$).

79. In a particular photoelectric experiment a stopping potential of 2.10 volts is measured when ultraviolet light having a wavelength of 290 nm is incident on the metal. Using the same setup, what will the new stopping potential be if blue light with a wavelength of 440 nm is used, instead of the ultraviolet light.

80. In an X-ray tube (see Fig. 36–25 and discussion in Section 36–10), the high voltage between filament and target is V. After being accelerated through this voltage, an electron strikes the target where it is decelerated (by positively charged nuclei) and in the process one or more X-ray photons are emitted. (a) Show that the photon of shortest wavelength will have
$$\lambda_0 = \frac{hc}{eV}.$$
(b) What is the shortest wavelength of X-ray emitted when accelerated electrons strike the face of a 30-kV television picture tube?

81. Show that the wavelength of a particle of mass m with kinetic energy K is given by the relativistic formula $\lambda = hc/\sqrt{K^2 + 2mc^2 K}$.

82. What is the kinetic energy and wavelength of a "thermal" neutron (one that is in equilibrium at room temperature–see Chapter 18)?

83. What is the theoretical limit of resolution for an electron microscope whose electrons are accelerated through 60 kV? (Relativistic formulas should be used.)

84. The intensity of the Sun's light in the vicinity of the Earth is about $1000 \, \text{W/m}^2$. Imagine a spacecraft with a mirrored square sail of dimension 1.0 km. Estimate how much thrust (in newtons) this craft will experience due to collisions with the Sun's photons. [Hint: Assume the photons bounce off the sail with no change in magnitude of their momentum.]

85. The human eye can respond to as little as 10^{-18} J of light energy. If this comes with a wavelength at the peak of visual sensitivity, 550 nm, how many photons lead to an observable flash?

86. Light with a wavelength of 300 nm strikes a metal whose work function is 2.2 eV. What is the shortest de Broglie wavelength for the electrons that are produced as photoelectrons?

Participants in the 1927 Solvay Conference. Albert Einstein, seated at front center, had difficulty accepting that nature could behave according to the rules of quantum mechanics, which we study in this chapter starting with the wave function and the uncertainty principle. We examine the Schrödinger equation and its solutions for some simple cases: free particles, the square well, and tunnelling through a barrier.

A. Piccard   E. Henriot   P. Ehrenfest   Ed. Herzen   Th. De Donder   E. Schrödinger   E. Verschaffelt   W. Pauli   W. Heisenberg   R. H. Fowler   L. Brillouin

P. Debye   M. Knudsen   W. L. Bragg   H. A. Kramers   P. A. M. Dirac   A. H. Compton   L. de Broglie   M. Born   N. Bohr

I. Langmuir   M. Planck   Mme Curie   H. A. Lorentz   A. Einstein   P. Langevin   Ch. E. Guye   C. T. R. Wilson   O. W. Richardson

CHAPTER 39

Quantum Mechanics

Bohr's model of the atom gave us a first (though rough) picture of what an atom is like. It proposed an explanation for why there should be emission and absorption of light by atoms at discrete wavelengths, as well as for the stability of atoms. The wavelengths of the line spectra and the ionization energy for hydrogen (and one-electron ions) that are predicted are in excellent agreement with experiment. But the Bohr theory had important limitations. It was not able to predict the line spectra for more complex atoms—not even for the neutral helium atom, which has only two electrons. Nor could it explain why emission lines, when viewed with great precision, consist of two or more very closely spaced lines (referred to as *fine structure*). The Bohr theory also did not explain why some spectral lines were brighter than others. And it could not explain the bonding of atoms in molecules or in solids and liquids.

From a theoretical point of view, too, the Bohr theory was not satisfactory. For it was a strange mixture of classical and quantum ideas. Moreover, the wave–particle duality was still not really resolved.

We mention these limitations of the Bohr theory not to disparage it—for it was a landmark in the history of science. Rather, we mention them to show why, in the early 1920s, it became increasingly evident that a new, more comprehensive theory was needed. It was not long in coming. Less than two years after de Broglie

FIGURE 39–1 Erwin Schrödinger with Lise Meitner (see Chapter 43).

FIGURE 39–2 Werner Heisenberg (center) on Lake Como with Wolfgang Pauli (right) and Enrico Fermi (left).

gave us his matter–wave hypothesis, Erwin Schrödinger (1887–1961; Fig. 39–1) and Werner Heisenberg (1901–1976; Fig. 39–2) independently developed a new comprehensive theory. Their separate approaches were quite different but were soon shown to be fully compatible. In this chapter we look at this new theory, quantum mechanics. In the following two chapters we will discuss the results of quantum mechanics applied to the hydrogen atom, to other atoms, and to molecules and solids.

39–1 Quantum Mechanics—A New Theory

The new theory, called **quantum mechanics**, unifies the wave–particle duality into a single consistent theory. As a theory, quantum mechanics has been extremely successful. It has successfully dealt with the spectra emitted by complex atoms, even the fine details. It explains the relative brightness of spectral lines and how atoms form molecules. It is also a much more general theory that covers all quantum phenomena from blackbody radiation to atoms and molecules. It has explained a wide range of natural phenomena and from its predictions many new practical devices have become possible. Indeed, it has been so successful that it is accepted today by nearly all physicists as the fundamental theory underlying physical processes.

Quantum mechanics deals mainly with the microscopic world of atoms and light. But in our everyday macroscopic world, we do perceive light and we accept that ordinary objects are made up of atoms. This new theory must therefore also account for the verified results of classical physics. That is, when it is applied to macroscopic phenomena, quantum mechanics must be able to produce the old classical laws. This, the **correspondence principle** (already mentioned in Section 37–13), is met fully by quantum mechanics. This doesn't mean we throw away classical theories such as Newton's laws. In the everyday world, the latter are far easier to apply and they give an accurate description. But when we deal with high speeds, close to the speed of light, we must use the theory of relativity; and when we deal with the tiny world of the atom, we use quantum mechanics.

Correspondence principle

Although we won't go into the detailed mathematics of quantum mechanics, we will discuss the main ideas and how they involve the wave and particle properties of matter to explain atomic structure and other applications.

39-2 The Wave Function and Its Interpretation; the Double-Slit Experiment

The important properties of any wave are its wavelength, frequency, and amplitude. For an electromagnetic wave, the wavelength determines whether the light is in the visible spectrum or not, and if so, what color it is. We also have seen that the wavelength (or frequency) is a measure of the energy of the corresponding photon ($E = hf$). The amplitude or displacement of an electromagnetic wave at any point is the strength of the electric (or magnetic) field at that point, and is related to the intensity of the wave (the brightness of the light).

For material particles such as electrons, quantum mechanics relates the wavelength to momentum according to de Broglie's formula, $\lambda = h/p$. But what corresponds to the amplitude or displacement of a matter wave? In quantum mechanics, this role is played by the **wave function**, which is given the symbol Ψ (the Greek capital letter psi, pronounced "sigh"). Thus Ψ represents the displacement as a function of time and position, of a new kind of field which we might call a "matter" field or a matter wave.

Wave function

To calculate the wave function Ψ in a given situation (say, for an electron in an atom) is one of the basic tasks of quantum mechanics. Indeed, the development of an equation to do so was Schrödinger's great contribution. The *Schrödinger wave equation*, as it is called, is considered to be the basic equation for the description of nonrelativistic material particles. We will discuss the Schrödinger equation and its solutions for some simple cases later in this chapter. Now let us ask: What is the meaning of the wave function Ψ for a particle? One way to interpret Ψ is simply as the displacement at any point in space and time of a "matter wave," so that it plays the role that E (the electric field) does for an electromagnetic wave. Another interpretation is possible, however, based on the wave–particle duality. To understand this, we make an analogy with light.

We saw in Chapter 15 that the intensity I of any wave is proportional to the square of the displacement amplitude. This holds true for light waves as well, and, as we saw in Chapter 32, the energy density in a light wave is proportional to E^2, where E is the electric field strength. From the *particle* point of view, the energy density of a light beam is proportional to the number of photons per unit volume, n. The more photons (of a given wavelength) per unit volume there are, the greater the energy density. Hence

$$n \propto E^2.$$

That is, the density of photons is proportional to the square of the electric field strength.

If the light beam is very weak, only a few photons will be involved. Indeed, it is possible to "build up" a photograph on film using very weak light so that the effect of individual photons can be seen. If we are dealing with only one photon, the relationship above ($n \propto E^2$) can be interpreted in a slightly different way. At any point the square of the electric field strength, E^2, is a measure of the *probability* that a photon will be found within a unit volume at that location. Said another way, if dV represents an infinitesimal volume element around that point, then $E^2 \, dV$ is proportional to the probability of finding a photon in the volume dV. At points where E^2 is large, there is a high probability of finding the photon; where E^2 is small, the probability of finding the photon is low.

We can interpret matter waves in the same way. The wave function Ψ may vary in magnitude from point to point in space and time. If Ψ describes a collection of many electrons, then $|\Psi|^2 \, dV$ will be proportional to the number of electrons expected to be found in the volume dV around any given point. When

$Probability \propto |\Psi|^2$

Light or
electrons

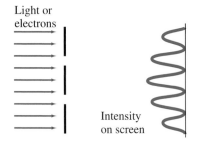

Intensity
on screen

FIGURE 39–3 Parallel beam, of light or electrons, falls on two slits whose sizes are comparable to the wavelength. An interference pattern is observed.

FIGURE 39–4 Young's double-slit experiment done with electrons—note that the pattern is not evident with only a few electrons (top photo), but with more and more electrons (second and third photos), the familiar double-slit interference pattern (Chapter 35) is seen.

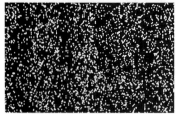

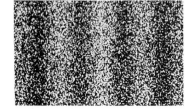

dealing with small numbers of electrons we can't make very exact predictions, so $|\Psi|^2$ takes on the character of a probability. If Ψ, which depends on time and position,[†] represents a single electron (say in an atom), then $|\Psi|^2$ is interpreted as follows: $|\Psi|^2 \, dV$ *at a certain point in space and time represents the probability of finding the electron within the volume* dV *about the given position at that time.* Thus, $|\Psi|^2$ is often referred to as the **probability density** or **probability distribution**, since it is the probability of finding the particle per unit volume.

To understand this better, we take as a thought experiment the familiar double-slit experiment, and consider it both for light and for electrons.

Consider two slits whose size and separation are on the order of the wavelength of whatever we direct at them, either light or electrons, Fig. 39–3. We know very well what would happen in this case for light, since this is just Young's double-slit experiment (Section 35–3): an interference pattern would be seen on the screen behind. If light were replaced by electrons with wavelength comparable to the slit size, they too would produce an interference pattern (recall Fig. 38–11). In the case of light, the pattern would be visible to the eye or could be recorded on film. For electrons, a fluorescent screen could be used (it glows where an electron strikes).

Now, if we reduced the flow of electrons (or photons) so that they passed one at a time through the slits, we would see a flash each time one struck the screen. At first, the flashes would seem random. Indeed, there is no way to predict just where any one electron would hit the screen. If we let the experiment run for a long time, however, and kept track of where each electron hit the screen, we would soon see a pattern emerging—namely the interference pattern predicted by the wave theory; see Fig. 39–4. Thus, although we could not predict where a given electron would strike the screen, we could predict probabilities. (The same can be said for photons.) The probability, as mentioned before, is proportional to $|\Psi|^2$. Where $|\Psi|^2$ is zero, we would get a minimum in the interference pattern. And where $|\Psi|^2$ is a maximum, we would get a peak in the interference pattern.

Since the interference pattern would occur even when electrons (or photons) passed through the slits one at a time, it is clear that the interference pattern would not arise from the interaction of one electron with another. It is as if the electron passed through both slits at the same time, interfering with itself. This is possible because, remember, an electron is not precisely a particle. It is as much a wave as it is a particle, and a wave could certainly travel through both slits at once. But what would happen if we covered one of the slits so we knew that the electrons passed through the other one, and a little later we covered the second slit so that electrons had to have passed through the first? The result would be that no interference pattern would be seen. We would see, instead, two bright areas (or diffraction patterns) on the screen behind the slits. This confirms our idea that if both slits are open, the screen shows an interference pattern as if each electron passed through both slits, like a wave. Yet each electron would make a tiny spot on the screen as if it were a particle.

The main point of this discussion is this: If we treat electrons (and other matter) as if they were waves, then Ψ represents the wave amplitude. If we treat them as particles, then we must treat them on a *probabilistic* basis. The square of the wave function, $|\Psi|^2$, gives the probability per unit volume of finding a given electron at a given point. We cannot predict—or even follow—the path of a single electron precisely through space and time.

[†]The wave function Ψ is generally a complex quantity (that is, it involves $i = \sqrt{-1}$) and hence is not directly observable. On the other hand, $|\Psi|^2$, the absolute value of Ψ squared, is always a real quantity and it is to $|\Psi|^2$ that we can give a physical interpretation.

39–3 | The Heisenberg Uncertainty Principle

Whenever a measurement is made, some uncertainty or error is always involved. For example, you cannot make an absolutely exact measurement of the length of a table. Even with a measuring stick that has markings 1 mm apart, there will be an inaccuracy of about $\frac{1}{2}$ mm or so. More precise instruments will produce more precise measurements. But there is always some uncertainty involved in a measurement no matter how good the measuring device. We expect that by using more precise instruments, the uncertainty in a measurement can be made indefinitely small.

But according to quantum mechanics, there is actually a limit to the accuracy of certain measurements. This limit is not a restriction on how well instruments can be made; rather, it is inherent in nature. It is the result of two factors: the wave–particle duality, and the unavoidable interaction between the thing observed and the observing instrument. Let us look at this in more detail.

Measurement uncertainty inherent in nature

To make a measurement on an object without somehow disturbing it, at least a little, is not possible. Consider trying to locate a Ping-pong ball in a completely dark room. You grope about trying to find its position; and just when you touch it with your finger, it bounces away. Whenever we measure the position of an object, whether it's a Ping-pong ball or an electron, we always touch it with something else that gives us the information about its position. To locate a lost Ping-pong ball in a dark room, you could probe about with your hand or a stick; or you could shine a light and detect the light reflecting off the ball. When you search with your hand or a stick, you find the ball's position when you touch it. But when you touch the ball you unavoidably bump it, and give it some momentum. Thus you won't know its *future* position. The same would be true, but to a much lesser extent, if you observe the Ping-pong ball using light. In order to "see" the ball, at least one photon must scatter from it, and the reflected photon must enter your eye or some other detector. When a photon strikes an ordinary-sized object, it does not appreciably alter the motion or position of the object. But when a photon strikes a very tiny object like an electron, it can transfer momentum to the object and thus greatly change the object's motion and position in an unpredictable way. The mere act of measuring the position of an object at one time makes our knowledge of its future position imprecise.

Now let us see where the wave–particle duality comes in. Imagine a thought experiment in which we are trying to measure the position of an object, say an electron, with photons, Fig. 39–5. (The arguments would be similar if we were using, instead, an electron microscope.) As we saw in Chapter 36, objects can be seen to an accuracy at best of about the wavelength of the radiation used. If we want an accurate position measurement, we must use a short wavelength. But a short wavelength corresponds to high frequency and large momentum ($p = h/\lambda$); and the more momentum the photons have, the more momentum they can give the object when they strike it. If photons of longer wavelength, and correspondingly smaller momentum are used, the object's motion when struck by the photons will not be affected as much. But the longer wavelength means lower resolution, so the object's position will be less accurately known. Thus the act of observing produces a significant uncertainty in either the *position* or the *momentum* of the electron. This is the essence of the *uncertainty principle* first enunciated by Heisenberg in 1927.

Quantitatively, we can make an approximate calculation of the magnitude of this effect. If we use light of wavelength λ, the position can be measured at best to an accuracy of about λ. That is, the uncertainty in the position measurement, Δx, is approximately

$$\Delta x \approx \lambda.$$

Suppose that the object can be detected by a single photon. The photon has a

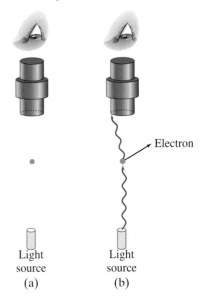

FIGURE 39–5 Thought experiment for observing an electron with a powerful light microscope. At least one photon must scatter from the electron (transferring some momentum to it) and enter the microscope.

Electron

Light source
(a)

Light source
(b)

momentum $p = h/\lambda$. When the photon strikes our object, it will give some or all of this momentum to the object, Fig. 39–5. Therefore, the final momentum of our object will be uncertain in the amount

$$\Delta p \approx \frac{h}{\lambda}$$

since we can't tell beforehand how much momentum will be transferred. The product of these uncertainties is

$$(\Delta x)(\Delta p) \approx h.$$

Of course, the uncertainties could be worse than this, depending on the apparatus and the number of photons needed for detection. In Heisenberg's more careful calculation, he found that at the very best

$$(\Delta x)(\Delta p_x) \gtrsim \frac{h}{2\pi}, \qquad \text{(39–1)}$$

where Δp_x is the uncertainty of the momentum in the x direction.[†] This is a mathematical statement of the **Heisenberg uncertainty principle**, or, as it is sometimes called, the **indeterminancy principle**. It tells us that we cannot measure both the position and momentum of an object precisely at the same time. The more accurately we try to measure the position, so that Δx is small, the greater will be the uncertainty in momentum, Δp_x. If we try to measure the momentum very precisely, then the uncertainty in the position becomes large. The uncertainty principle does not forbid individual exact measurements, however. For example, in principle we could measure the position of an object exactly. But then its momentum would be completely unknown. Thus, although we might know the position of the object exactly at one instant, we could have no idea at all where it would be a moment later.

Another useful form of the uncertainty principle relates energy and time, and we examine this as follows. The object to be detected has an uncertainty in position $\Delta x \approx \lambda$. Now the photon used to detect it travels with speed c, and it takes a time $\Delta t \approx \Delta x/c \approx \lambda/c$ to pass through the distance of uncertainty. Hence, the measured time when our object is at a given position is uncertain by about

$$\Delta t \approx \frac{\lambda}{c}.$$

Since the photon can transfer some or all of its energy ($= hf = hc/\lambda$) to our object, the uncertainty in energy of our object as a result is

$$\Delta E \approx \frac{hc}{\lambda}.$$

The product of these two uncertainties is

$$(\Delta E)(\Delta t) \approx h.$$

Heisenberg's more careful calculation gives

$$(\Delta E)(\Delta t) \gtrsim \frac{h}{2\pi}. \qquad \text{(39–2)}$$

This form of the uncertainty principle tells us that the energy of an object can be uncertain, or may even be nonconserved, by an amount ΔE for a time $\Delta t \approx h/(2\pi\, \Delta E)$.

[†]Note, however, that quantum mechanics does allow simultaneous precise measurements of p_x and y: that is $(\Delta y)(\Delta p_x) \gtrsim 0$.

The quantity $(h/2\pi)$ appears so often in quantum mechanics that for convenience it is given the symbol $\hbar$ ("h-bar"). That is

$$\hbar = \frac{h}{2\pi} = \frac{6.626 \times 10^{-34}\,\text{J}\cdot\text{s}}{2\pi} = 1.055 \times 10^{-34}\,\text{J}\cdot\text{s}.$$

By using this notation, Eqs. 39–1 and 39–2 for the uncertainty principle can be written

$$(\Delta x)(\Delta p_x) \gtrsim \hbar \qquad \text{and} \qquad (\Delta E)(\Delta t) \gtrsim \hbar.$$

We have been discussing the position and velocity of an electron as if it were a particle. But it isn't a particle. Indeed, we have the uncertainty principle because an electron—and matter in general—has wave as well as particle properties. What the uncertainty principle really tells us is that if we insist on thinking of the electron as a particle, then there are certain limitations on this simplified view—namely, that the position and velocity cannot both be known precisely at the same time; and that the energy can be uncertain (or nonconserved) in the amount ΔE for a time $\Delta t \approx \hbar/\Delta E$.

Because Planck's constant, h, is so small, the uncertainties expressed in the uncertainty principle are usually negligible on the macroscopic level. But at the level of the atom, the uncertainties are significant. Because we consider ordinary objects to be made up of atoms containing nuclei and electrons, the uncertainty principle is relevant to our understanding of all of nature. The uncertainty principle expresses, perhaps most clearly, the probabilistic nature of quantum mechanics. It thus is often used as a basis for philosophic discussion.

EXAMPLE 39–1 **Position uncertainty of electron.** An electron moves in a straight line with a constant speed $v = 1.10 \times 10^6\,\text{m/s}$ which has been measured to a precision of 0.10 percent. What is the maximum precision with which its position could be simultaneously measured?

SOLUTION The momentum of the electron is $p = mv = (9.11 \times 10^{-31}\,\text{kg}) \cdot (1.10 \times 10^6\,\text{m/s}) = 1.00 \times 10^{-24}\,\text{kg}\cdot\text{m/s}$. The uncertainty in the momentum is 0.10 percent of this, or $\Delta p = 1.0 \times 10^{-27}\,\text{kg}\cdot\text{m/s}$. From the uncertainty principle, the best simultaneous position measurement will have an uncertainty of

$$\Delta x = \frac{\hbar}{\Delta p} = \frac{1.06 \times 10^{-34}\,\text{J}\cdot\text{s}}{1.0 \times 10^{-27}\,\text{kg}\cdot\text{m/s}} = 1.1 \times 10^{-7}\,\text{m},$$

or 110 nm. This is about 1000 times the diameter of an atom.

EXAMPLE 39–2 **Position uncertainty of a baseball.** What is the uncertainty in position, imposed by the uncertainty principle, on a 150-g baseball thrown at $(93 \pm 2)\,\text{mph} = (42 \pm 1)\,\text{m/s}$?

SOLUTION The uncertainty in the momentum is

$$\Delta p = m\,\Delta v = (0.150\,\text{kg})(1\,\text{m/s}) = 0.15\,\text{kg}\cdot\text{m/s}.$$

Hence the uncertainty in a position measurement could be as small as

$$\Delta x = \frac{\hbar}{\Delta p} = \frac{1.06 \times 10^{-34}\,\text{J}\cdot\text{s}}{0.15\,\text{kg}\cdot\text{m/s}} = 7 \times 10^{-34}\,\text{m},$$

which is a distance incredibly smaller than any we could imagine observing or measuring. Indeed, the uncertainty principle sets no relevant limit on measurement for macroscopic objects.

EXAMPLE 39–3 **ESTIMATE** **J/ψ lifetime calculated.** The J/ψ meson, discovered in 1974, was measured to have an average mass of 3100 MeV/c^2 (note the use of energy units since $E = mc^2$) and an intrinsic width of 63 keV/c^2. By this we mean that the masses of different J/ψ mesons were actually measured to be slightly different from one another. This mass "width" is related to the very short lifetime of the J/ψ before it decays into other particles. Estimate its lifetime using the uncertainty principle.

SOLUTION The uncertainty of 63 keV/c^2 in the J/ψ's mass is an uncertainty in its rest energy, which in joules is

$$\Delta E = (63 \times 10^3 \, \text{eV})(1.60 \times 10^{-19} \, \text{J/eV}) = 1.01 \times 10^{-14} \, \text{J}.$$

Then we expect its lifetime τ ($= \Delta t$ here) to be

$$\tau \approx \frac{\hbar}{\Delta E} = \frac{1.06 \times 10^{-34} \, \text{J} \cdot \text{s}}{1.01 \times 10^{-14} \, \text{J}} \approx 1.0 \times 10^{-20} \, \text{s}.$$

Lifetimes this short are difficult to measure directly, and the assignment of very short lifetimes depends on this use of the uncertainty principle. (See Chapter 44.)

The uncertainty principle applies also for angular variables:

$$(\Delta L_z)(\Delta \phi) \gtrsim \hbar$$

where L is the component of angular momentum along a given axis (z) and ϕ is the angular position in a plane perpendicular to that axis.

39–4 Philosophic Implications; Probability versus Determinism

The classical Newtonian view of the world is a deterministic one (see Section 6–5). One of its basic ideas is that once the position and velocity of an object are known at a particular time, its future position can be predicted if the forces on it are known. For example, if a stone is thrown a number of times with the same initial velocity and angle, and the forces on it remain the same, the path of the projectile will always be the same. If the forces are known (gravity and air resistance, if any), the stone's path can be precisely predicted. This mechanistic view implies that the future unfolding of the universe, assumed to be made up of particulate bodies, is completely determined.

This classical deterministic view of the physical world has been radically altered by quantum mechanics. As we saw in the analysis of the double-slit experiment (Section 39–2), electrons all prepared in the same way will not all end up in the same place. According to quantum mechanics, certain probabilities exist that an electron will arrive at different points. This is very different from the classical view, in which the path of a particle is precisely predictable from the initial position and velocity and the forces exerted on it. According to quantum mechanics, the position and velocity of an object cannot even be known accurately at the same time. This is expressed in the uncertainty principle, and arises because basic entities, such as electrons, are not considered simply as particles: they have wave properties as well. Quantum mechanics allows us to calculate only the probability[†] that, say, an electron (when thought of as a particle) will be observed at various places. Quantum mechanics says there is some inherent unpredictability in nature.

[†]Note that these probabilities can be calculated precisely, just like exact predictions of probabilities at dice or playing cards, but unlike predictions of probabilities at sporting events or for natural or man-made disasters, which are only estimates.

Since matter is considered to be made up of atoms, even ordinary-sized objects are expected to be governed by probability, rather than by strict determinism. For example, quantum mechanics predicts a finite (but negligibly small) probability that when you throw a stone, its path will suddenly curve upward instead of following the downward-curved parabola of normal projectile motion. Quantum mechanics predicts with extremely high probability that ordinary objects will behave just as the classical laws of physics predict. But these predictions are considered probabilities, not certainties. The reason that macroscopic objects behave in accordance with classical laws with such high probability is due to the large number of molecules involved: when large numbers of objects are present in a statistical situation, deviations from the average (or most probable) approach zero. It is the average configuration of vast numbers of molecules that follows the so-called fixed laws of classical physics with such high probability, and gives rise to an apparent "determinism." Deviations from classical laws are observed when small numbers of molecules are dealt with. We can say, then, that although there are no precise deterministic laws in quantum mechanics, there are statistical laws based on probability.

It is important to note that there is a difference between the probability imposed by quantum mechanics and that used in the nineteenth century to understand thermodynamics and the behavior of gases in terms of molecules (Chapters 18 and 20). In thermodynamics, probability is used because there are far too many particles to keep track of. But the molecules are still assumed to move and interact in a deterministic way following Newton's laws. Probability in quantum mechanics is quite different; it is seen as *inherent* in nature, and not as a limitation on our abilities to calculate or to measure.

Although a few physicists have not given up the deterministic view of nature and have refused to accept quantum mechanics as a complete theory—one was Einstein—nonetheless, the vast majority of physicists do accept quantum mechanics and the probabilistic view of nature. This view, which as presented here is the generally accepted one, is called the **Copenhagen interpretation** of quantum mechanics in honor of Niels Bohr's home, since it was largely developed there through discussions between Bohr and other prominent physicists.

Copenhagen interpretation

Because electrons are not simply particles, they cannot be thought of as following particular paths in space and time. This suggests that a description of matter in space and time may not be completely correct. This deep and far-reaching conclusion has been a lively topic of discussion among philosophers. Perhaps the most important and influential philosopher of quantum mechanics was Bohr. He argued that a space–time description of actual atoms and electrons is not possible. Yet a description of experiments on atoms or electrons must be given in terms of space and time and other concepts familiar to ordinary experience, such as waves and particles. We must not let our *descriptions* of experiments lead us into believing that atoms or electrons themselves actually move in space and time as particles.

39–5 The Schrödinger Equation in One Dimension— Time-Independent Form

In order to describe physical systems quantitatively using quantum mechanics, we must have a means of determining the wave function Ψ mathematically. As mentioned in Section 39–2, the basic equation (in the nonrelativistic realm) for determining Ψ is the *Schrödinger equation*. We cannot, however, derive the Schrödinger equation from some higher principles, just as Newton's second law, for example, cannot be derived. The relation $\mathbf{F} = m\mathbf{a}$ was *invented* by Newton to describe how the motion of a body is related to the net applied force. As we saw early in this book, Newton's second law works exceptionally well. In the realm of classical physics it is the starting point for analytically solving a wide range of problems, and the solutions it yields are fully consistent with experiment. The validity of any fundamental equation resides in its agreement with experiment. The Schrödinger equation forms part of a new theory, and it too had to be *invented*—and then checked against experiment, a test that it passed splendidly.

The Schrödinger equation can be written in two forms: the time-dependent version and the time-independent version. We will mainly be interested in steady-state situations—that is, when there is no time dependence—and so we mainly deal with the time-independent version. (We briefly discuss the time-dependent version in the next, optional, section.) The time-independent version involves a wave function with only spatial dependence which we represent by lowercase psi, $\psi(x)$.

In classical mechanics, we solved problems using two approaches: via Newton's laws with the concept of force, and by using the energy concept with the conservation laws. The Schrödinger equation is based on the energy approach. Even though the Schrödinger equation cannot be derived, we can suggest what form it might take by using conservation of energy and considering a very simple case: that of a free particle on which no forces act, so that its potential energy U is constant. We assume that our particle moves along the x axis, and since no force acts on it, its momentum remains constant and its wavelength ($\lambda = h/p$) is fixed. To describe a wave for a free particle such as an electron, we expect that its wave function will satisfy a differential equation that is akin to (but not identical to) the classical wave equation. Let us see what we can infer about this equation. Consider a simple traveling wave of a single wavelength λ whose wave displacement, as we saw in Chapter 15 for mechanical waves and in Chapter 32 for electromagnetic waves, is given by $A \sin(kx - \omega t)$, or more generally as a superposition of sine and cosine: $A \sin(kx - \omega t) + B \cos(kx - \omega t)$. We are only interested in the spatial dependence, so we consider the wave at a specific moment, say $t = 0$. Thus we write as the wave function for our free particle

$$\psi(x) = A \sin kx + B \cos kx, \tag{39-3a}$$

where A and B are constants[†] and $k = 2\pi/\lambda$ (Eq. 15–11). For a particle of mass m and velocity v, the de Broglie wavelength is $\lambda = h/p$, where $p = mv$ is the particle's momentum. Hence

$$k = \frac{2\pi}{\lambda} = \frac{2\pi p}{h} = \frac{p}{\hbar}, \tag{39-3b}$$

One requirement for our wave equation, then, is that it have the wave function $\psi(x)$ as given by Eq. 39–3 as a solution for a free particle. A second requirement is that it be consistent with the conservation of energy, which we can express as

$$\frac{p^2}{2m} + U = E,$$

where E is the total energy, U is the potential energy, and (since we are considering the nonrelativistic realm) the kinetic energy K of our particle of mass m is $K = \frac{1}{2}mv^2 = p^2/2m$. Since $p = \hbar k$ (Eq. 39–3b), we can write the conservation of energy condition as

$$\frac{\hbar^2 k^2}{2m} + U = E. \tag{39-4}$$

Thus we are seeking a differential equation that satisfies conservation of energy

[†] In quantum mechanics, constants can be complex (i.e., with a real and/or imaginary part).

(Eq. 39–4) when $\psi(x)$ is its solution. Now, note that if we take two derivatives of our expression for $\psi(x)$, Eq. 39–3a, we get a factor $-k^2$ multiplied by $\psi(x)$:

$$\frac{d\psi(x)}{dx} = \frac{d}{dx}(A\sin kx + B\cos kx) = k(A\cos kx - B\sin kx)$$

$$\frac{d^2\psi(x)}{dx^2} = k\frac{d}{dx}(A\cos kx - B\sin kx) = -k^2(A\sin kx + B\cos kx) = -k^2\psi(x).$$

Can this last term be related to the k^2 term in Eq. 39–4? Indeed, if we multiply this last relation by $-\hbar^2/2m$, we obtain

$$-\frac{\hbar^2}{2m}\frac{d^2\psi(x)}{dx^2} = \frac{\hbar^2 k^2}{2m}\psi(x).$$

The right side is just the first term on the left in Eq. 39–4 multiplied by $\psi(x)$. If we multiply Eq. 39–4 through by $\psi(x)$, and make this substitution, we obtain

$$-\frac{\hbar^2}{2m}\frac{d^2\psi(x)}{dx^2} + U(x)\psi(x) = E\psi(x). \qquad \textbf{(39–5)}$$

SCHRÖDINGER EQUATION (time-independent form)

This is, in fact, the one-dimensional **time-independent Schrödinger equation**, where for generality we have written $U = U(x)$. It is the basis for solving problems in non-relativistic quantum mechanics. For a particle moving in three dimensions there would be additional derivatives with respect to y and z (see Chapter 40).

But note that we have by no means derived the Schrödinger equation. Although we have made a good argument in its favor, other arguments could also be made which might or might not lead to the same equation. The Schrödinger equation as written (Eq. 39–5) is useful and valid only because it has given results in accord with experiment for a wide range of situations.

There are some requirements we impose on any wave function that is a solution of the Schrödinger equation in order that it be physically meaningful. First, we insist that it be a continuous function; after all, if $|\psi|^2$ represents the probability of finding a particle at a certain point, we expect that probability to be continuous from point to point and not to take discontinuous jumps. Second, we want the wave function to be *normalized*. By this we mean that for a single particle, the probability of finding the particle at one point or another (i.e., the probabilities summed over all space) must be exactly 1 (or 100 percent). For a single particle, $|\psi|^2$ represents the probability of finding the particle in unit volume. Then

$$|\psi|^2\,dV$$

is the probability of finding the particle within a volume dV, where ψ is the value of the wave function in this infinitesimal volume dV. For the one-dimensional case, $dV = dx$. Then the sum of the probabilities over all space—that is, the probability of finding the particle at one point or another—becomes

$$\int_{\text{all space}} |\psi|^2\,dV = \int |\psi|^2\,dx = 1. \qquad \textbf{(39–6)}$$

Normalization

This is called the **normalization condition**, and the integral is taken over whatever region of space in which the particle has a chance of being found, which is often all of space, from $x = -\infty$ to $x = \infty$.

Schrödinger equation (time-dependent form)

The more general form of the Schrödinger equation, including time dependence, for a particle of mass m moving in one dimension, is

$$-\frac{\hbar^2}{2m}\frac{\partial^2\Psi(x,t)}{\partial x^2} + U(x)\Psi(x,t) = i\hbar\frac{\partial\Psi(x,t)}{\partial t}. \tag{39–7}$$

This is the **time-dependent Schrödinger equation**; here $U(x)$ is the potential energy of the particle as a function of position, and i is the imaginary number $i = \sqrt{-1}$. For a particle moving in three dimensions, there would be additional derivatives with respect to y and z, just as for the classical wave equation discussed in Section 15–5. Indeed, it is worth noting the similarity between the Schrödinger wave equation for zero potential energy ($U = 0$) and the classical wave equation: $\partial^2 D/\partial t^2 = v^2 \partial^2 D/\partial x^2$, where D is the wave displacement (equivalent of the wave function). In both equations there is the second derivative with respect to x; but in the Schrödinger equation there is only the first derivative with respect to time, whereas the classical wave equation has the second derivative for time.

As we pointed out in the preceding Section, we cannot derive the time-dependent Schrödinger equation. But we can show how the time-independent Schrödinger equation (Eq. 39–5) is obtained from it. For many problems in quantum mechanics, it is possible to write the wave function as a product of separate functions of space and time:

$$\Psi(x,t) = \psi(x)f(t).$$

Substituting this into the time-dependent Schrödinger equation (Eq. 39–7), we get:

$$-\frac{\hbar^2}{2m}f(t)\frac{d^2\psi(x)}{dx^2} + U(x)\psi(x)f(t) = i\hbar\psi(x)\frac{df(t)}{dt}.$$

We divide both sides of this equation by $\psi(x)f(t)$ and obtain an equation that involves only x on one side and only t on the other:

Separation of variables

$$-\frac{\hbar^2}{2m}\frac{1}{\psi(x)}\frac{d^2\psi(x)}{dx^2} + U(x) = i\hbar\frac{1}{f(t)}\frac{df(t)}{dt}.$$

This *separation of variables* is very convenient. Since the left side is a function only of x, and the right side is a function only of t, the equality can be valid for all values of x and all values of t only if each side is equal to a constant (the same constant, of course), which we call C:

$$-\frac{\hbar^2}{2m}\frac{1}{\psi(x)}\frac{d^2\psi(x)}{dx^2} + U(x) = C \tag{39–8a}$$

$$i\hbar\frac{1}{f(t)}\frac{df(t)}{dt} = C. \tag{39–8b}$$

We multiply the first of these (Eq. 39–8a) through by $\psi(x)$ and obtain

$$-\frac{\hbar^2}{2m}\frac{d^2\psi(x)}{dx^2} + U(x)\psi(x) = C\psi(x). \tag{39–8c}$$

This we recognize immediately as the time-independent Schrödinger equation, Eq. 39–5, where the constant C equals the total energy E. Thus we have obtained the time-independent form of Schrödinger's equation from the time-dependent form.

Equation 39–8b is easy to solve. Putting $C = E$, we rewrite Eq. 39–8b as

$$\frac{df(t)}{dt} = -i\frac{E}{\hbar}f(t)$$

(note that since $i^2 = -1$, $i = -1/i$), and then as

$$\frac{df(t)}{f(t)} = -i\frac{E}{\hbar}\,dt.$$

We integrate both sides to obtain

$$\ln f(t) = -i\frac{E}{\hbar}t$$

or

$$f(t) = e^{-i\left(\frac{E}{\hbar}\right)t}.$$

Thus the total wave function is

$$\Psi(x,t) = \psi(x)e^{-i\left(\frac{E}{\hbar}\right)t}, \tag{39–9}$$

where $\psi(x)$ satisfies Eq. 39–5. It is, in fact, the solution of the time-independent Schrödinger equation (Eq. 39–5) that is the major task of nonrelativistic quantum mechanics. Nonetheless, we should note that in general the wave function $\Psi(x, t)$ is a complex function since it involves $i = \sqrt{-1}$. It has both a real and an imaginary part.[†] Since $\Psi(x, t)$ is not purely real, it cannot itself be physically measurable. Rather it is only $|\Psi|^2$, which *is* real, that can be measured physically.

Note also that

$$\left|f(t)\right| = \left|e^{-i\left(\frac{E}{\hbar}\right)t}\right| = 1,$$

so $\left|f(t)\right|^2 = 1$. Hence the probability density in space does not depend on time:

$$\left|\Psi(x,t)\right|^2 = \left|\psi(x)\right|^2.$$

We thus will be interested only in the time-independent Schrödinger equation, Eq. 39–5, which we will now examine for a number of simple situations.

39–7 | Free Particles; Plane Waves and Wave Packets

A **free particle** is one that is not subject to any force, and we can therefore take its potential energy to be zero. (Although we dealt with the free particle in Section 39–5 in arguing for Schrödinger's equation, here we treat it directly using Schrödinger's equation as the basis.) Schrödinger's equation (Eq. 39–5) with $U(x) = 0$ becomes

$$-\frac{\hbar^2}{2m}\frac{d^2\psi(x)}{dx^2} = E\psi(x),$$

which can be written

$$\frac{d^2\psi}{dx^2} + \frac{2mE}{\hbar^2}\psi = 0.$$

This is a familiar equation that we encountered in Chapter 14 (Eq. 14–3) in connection with the simple harmonic oscillator. The solution to this equation, but with

[†] Recall that $e^{-i\theta} = \cos\theta - i\sin\theta$.

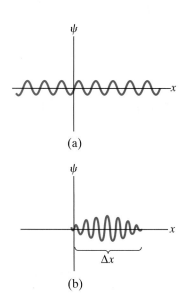

(a)

(b)

FIGURE 39–6 (a) A plane wave describing a free particle. (b) A wave packet of "width" Δx.

Wave packet

appropriate variable changes† for our case here, is

$$\psi = A \sin kx + B \cos kx, \qquad \textbf{(39–10)}$$

where

$$k = \sqrt{\frac{2mE}{\hbar^2}}. \qquad \textbf{(39–11a)}$$

Since $U = 0$, the total energy E of the particle is $E = \frac{1}{2}mv^2 = p^2/2m$ (where p is the momentum); thus

$$k = \frac{p}{\hbar} = \frac{h}{\lambda\hbar} = \frac{2\pi}{\lambda}. \qquad \textbf{(39–11b)}$$

So a free particle of momentum p and energy E can be represented by a plane wave that varies sinusoidally. If we are not interested in the phase, we can choose $B = 0$ in Eq. 39–10, and we show this sine wave in Fig. 39–6a.

Note in Fig. 39–6a that the sine wave will extend indefinitely‡ in the $+x$ and $-x$ directions. Thus, since $|\psi|^2$ represents the probability of finding the particle, the particle could be anywhere between $x = -\infty$ and $x = \infty$. This is fully consistent with the uncertainty principle (Section 39–3): the momentum of the particle was given and hence is known precisely ($p = \hbar k$), so the particle's position must be totally unpredictable. Mathematically, if $\Delta p = 0$, $\Delta x \gtrsim \hbar/\Delta p = \infty$.

To describe a particle whose position is well localized—that is, it is known to be within a small region of space—we can use the concept of a **wave packet**. Figure 39–6b shows an example of a wave packet whose width is about Δx as shown, meaning that the particle is most likely to be found within this region of space. A well-localized particle moving through space can thus be represented by a moving wave packet.

A wave packet can be represented mathematically as the sum of many plane waves (sine waves) of slightly different wavelengths. That this will work can be seen by looking carefully at Fig. 16–18. There we combined only two nearby frequencies (to explain why there are "beats") and found that the sum of two sine waves looked like a series of wave packets. If we add additional waves with other nearby frequencies, we can eliminate all but one of the packets and arrive at Fig. 39–6b. Thus a wave packet consists of waves of a *range* of wavelengths; hence it does not have a definite momentum $p(= h/\lambda)$, but rather, a range of momenta. This is, of course, consistent with the uncertainty principle: we have made Δx small, so the momentum cannot be precise; that is, Δp cannot be zero. Instead, our particle can be said to have a range of momenta, Δp, or to have an uncertainty in its momentum, Δp. It is not hard to show, even for this simple situation (see Problem 15), that $\Delta p \approx h/\Delta x$, in accordance with the uncertainty principle.

39–8 Particle in an Infinitely Deep Square Well Potential (a Rigid Box)

The Schrödinger equation can be solved analytically only for a few possible forms of the potential energy U. We consider two simple cases here which at first may not seem realistic, but have simple solutions that can be used as approximations to understand a variety of phenomena.

---

† In Eq. 14–3, t becomes x and ω becomes $k = \sqrt{2mE/\hbar^2}$. (Don't confuse this k with the spring constant k of Chapter 14.)

† Such an infinite wave makes problems for normalization since $\int_{-\infty}^{\infty}|\psi|^2\,dx = A^2\int_{-\infty}^{\infty}\sin^2 kx\,dx$ is infinite for any nonzero value for A. For practical purposes we can usually normalize the waves ($A \neq 0$) by assuming that the particle is in a large but finite region of space. The region can be chosen large enough so that momentum is still rather precisely fixed.

In our first case, we assume that a particle of mass m is confined to a one-dimensional box of width L whose walls are perfectly rigid. (This can serve as a very crude approximation for an electron in a metal, for example.) The particle is trapped in this box and collisions with the walls are perfectly elastic. The potential energy for this situation, which is commonly known as an **infinitely deep square well potential** or **rigid box**, is shown in Fig. 39–7. We can write the potential energy $U(x)$ as

$$U(x) = 0 \qquad 0 < x < L$$
$$U(x) = \infty \qquad x \le 0 \quad \text{and} \quad x \ge L.$$

For the region $0 < x < L$, where $U(x) = 0$, we already know the solution of the Schrödinger equation from our discussion in Section 39–7: it is just Eq. 39–10,

$$\psi(x) = A \sin kx + B \cos kx,$$

where (from Eq. 39–11a)

$$k = \sqrt{\frac{2mE}{\hbar^2}}.$$

Outside the well $U(x) = \infty$, so $\psi(x)$ must be zero. (If it weren't, the product $U\psi$ in the Schrödinger equation wouldn't be finite; besides, if $U = \infty$, we can't expect a particle of finite total energy to be in such a region.) So we are concerned only with the wave function within the well, and we must determine the constants A and B as well as any restrictions on the value of k (and hence on the energy E).

We have insisted that the wave function must be continuous. Hence, if $\psi = 0$ outside the well, it must be zero at $x = 0$ and at $x = L$:

$$\psi(0) = 0 \quad \text{and} \quad \psi(L) = 0.$$

FIGURE 39–7 A plot of potential energy U vs x for an infinitely deep square well potential.

These are the **boundary conditions** for this problem. At $x = 0$, $\sin kx = 0$ but $\cos kx = 1$, so

Boundary conditions

$$0 = \psi(0) = A \sin 0 + B \cos 0 = 0 + B.$$

Thus B must be zero. Our solution is reduced to

$$\psi(x) = A \sin kx.$$

Now we apply the other boundary condition, $\psi = 0$ at $x = L$:

$$0 = \psi(L) = A \sin kL.$$

We don't want $A = 0$ or we won't have a particle at all ($|\psi|^2 = 0$ everywhere). Therefore, we set

$$\sin kL = 0,$$

which, since the sine is zero for angles of $0, \pi, 2\pi, 3\pi \cdots$ radians, can happen only if $kL = 0, \pi, 2\pi, 3\pi, \cdots$. In other words,

$$kL = n\pi \qquad n = 1, 2, 3, \cdots, \tag{39–12}$$

where n is an integer. We eliminate the case $n = 0$ since this, too, produces $x = 0$ everywhere. Thus k, and hence E, cannot have just any value; rather, k is limited to values

$$k = \frac{n\pi}{L}.$$

Putting this expression in Eq. 39–11a (and substituting $h/2\pi$ for $\hbar$), we find that E can have only the values

$$E = n^2 \frac{h^2}{8mL^2}. \tag{39–13}$$

Energy levels, quantized (∞ potential well)

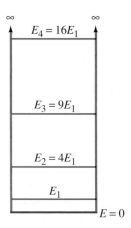

FIGURE 39–8 Possible energy levels for a particle in a box with perfectly rigid walls (infinite square well potential).

Zero-point energy

FIGURE 39–9 Wave functions corresponding to the quantum number n being 1, 2, 3 and 10 for a particle confined to a rigid box.

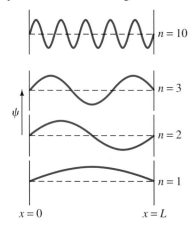

FIGURE 39–10 (below) The probability distribution for a particle in a rigid box for the states with $n = 1, 2, 3,$ and 10.

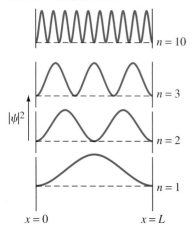

A particle trapped in a rigid box thus can only have certain *quantized energies*. The lowest energy (the ground state) has $n = 1$ and is given by

$$E_1 = \frac{h^2}{8mL^2}.$$

The next highest energy ($n = 2$) is

$$E_2 = 4E_1,$$

and for higher energies (see Fig. 39–8),

$$E_3 = 9E_1$$
$$\vdots$$
$$E_n = n^2E_1.$$

The integer n is called the **quantum number** of the state. That the lowest energy, E_1, is not zero means that the particle in the box can never be at rest. This is contrary to classical ideas, according to which a particle can have $E = 0$. E_1 is called the **zero-point energy**. One outcome of this result is that even at a temperature of absolute zero (0 K), quantum mechanics predicts that particles in a box would not be at rest but would have a zero-point energy.

We can also note that the energy E_1 and momentum $p_1 = \hbar k = \hbar\pi/L$ in the ground state are related inversely to the width of the box. The smaller the width L, the larger the momentum (and energy). This can be considered a direct result of the uncertainty principle (see Problem 21).

The wave function $\psi = A \sin kx$ for each of the quantum states is (since $k = n\pi/L$)

$$\psi_n = A \sin\left(\frac{n\pi}{L} x\right). \tag{39–14}$$

We can determine the constant A by imposing the normalization condition (Eq. 39–6):

$$1 = \int_{-\infty}^{\infty} \psi^2 \, dx = \int_0^L A^2 \sin^2\left(\frac{n\pi}{L} x\right) dx, \tag{39–15}$$

where the integral needs to be done only over the range $0 < x < L$ because outside these limits $\psi = 0$. The integral (see Example 39–5) is equal to $A^2L/2$, so we have

$$A = \sqrt{\frac{2}{L}} \quad \text{and} \quad \psi_n = \sqrt{\frac{2}{L}} \sin\left(\frac{n\pi}{L} x\right).$$

The amplitude A is the same for all the quantum numbers. Figure 39–9 shows the wave functions (Eq. 39–14) for $n = 1, 2, 3,$ and 10. They look just like standing waves on a string—see Fig. 15–27. This is not surprising since the wave function solutions, Eq. 39–14, are the same as for the standing waves on a string, and the condition $kL = n\pi$ is the same in the two cases.

Figure 39–10 shows the probability distribution, $|\psi|^2$, for the same states ($n = 1, 2, 3, 10$) for which ψ is shown in Fig. 39–9. We see immediately that the particle is more likely to be found in some places than in others. For example, in the ground state ($n = 1$), the electron is much more likely to be found near the center of the box than near the walls. This is clearly at variance with classical ideas, which predict a uniform probability density—the particle would be as likely to be found at one point in the box as at any other. The quantum-mechanical probability densities for higher states are even more complicated, with areas of low probability not only near the walls but also at regular intervals in between.

EXAMPLE 39–4 **Electron in an infinite potential well.** (*a*) Calculate the three lowest energy levels for an electron trapped in an infinitely deep square well potential of width $L = 1.00 \times 10^{-10}$ m (about the diameter of a hydrogen atom in its ground state). (*b*) If a photon were emitted when the electron jumps from the $n = 2$ state to the $n = 1$ state, what would its wavelength be?

SOLUTION (*a*) The ground state ($n = 1$) has energy

$$E_1 = \frac{h^2}{8mL^2} = \frac{(6.63 \times 10^{-34}\,\text{J·s})^2}{8(9.11 \times 10^{-31}\,\text{kg})(1.00 \times 10^{-10}\,\text{m})^2} = 6.03 \times 10^{-18}\,\text{J}.$$

In electron volts this is

$$E_1 = \frac{6.03 \times 10^{-18}\,\text{J}}{1.60 \times 10^{-19}\,\text{J/eV}} = 37.7\,\text{eV}.$$

Then

$$E_2 = (2)^2 E_1 = 151\,\text{eV}$$
$$E_3 = (3)^2 E_1 = 339\,\text{eV}.$$

(*b*) The energy difference is $E_2 - E_1 = 151\,\text{eV} - 38\,\text{eV} = 113\,\text{eV}$ or 1.81×10^{-17} J, and this would equal the energy of the emitted photon (energy conservation). Its wavelength would be

$$\lambda = \frac{c}{f} = \frac{hc}{E} = \frac{(6.63 \times 10^{-34}\,\text{J·s})(3.00 \times 10^8\,\text{m/s})}{1.81 \times 10^{-17}\,\text{J}} = 1.10 \times 10^{-8}\,\text{m}$$

or 11.0 nm, which is in the ultraviolet region of the spectrum.

EXAMPLE 39–5 **Calculating a normalization constant.** Show that the normalization constant A for all wave functions describing a particle in an infinite potential well of width L has a value of $A = \sqrt{2/L}$.

SOLUTION The wave functions are

$$\psi = A \sin \frac{n\pi x}{L}.$$

To normalize ψ, we must have (Eq. 39–15)

$$1 = \int_0^L |\psi|^2 \, dx = \int_0^L A^2 \sin^2 \frac{n\pi x}{L} \, dx.$$

We need integrate only from 0 to L since $\psi = 0$ for all other values of x. To evaluate this integral we let $\theta = n\pi x/L$ and use the trigonometric identity $\sin^2 \theta = \frac{1}{2}(1 - \cos 2\theta)$. Then, with $dx = L\,d\theta/n\pi$, we have

$$1 = A^2 \int_0^{n\pi} \sin^2 \theta \left(\frac{L}{n\pi}\right) d\theta = \frac{A^2 L}{2n\pi} \int_0^{n\pi} (1 - \cos 2\theta) \, d\theta$$

$$= \frac{A^2 L}{2n\pi} \left(\theta - \tfrac{1}{2}\sin 2\theta\right)\Big|_0^{n\pi}$$

$$= \frac{A^2 L}{2}.$$

Thus $A^2 = 2/L$ and

$$A = \sqrt{\frac{2}{L}}.$$

EXAMPLE 39-6 **ESTIMATE** **Confined bacterium.** A tiny bacterium with a mass of about 10^{-14} kg is confined between two rigid walls 0.1 mm apart. (a) Estimate its minimum speed. (b) If, instead, its speed is about 1 mm in 100 s, estimate the quantum number of its state.

SOLUTION (a) The minimum speed occurs in the ground state, $n = 1$; since $E = \frac{1}{2}mv^2$, we have

$$v = \sqrt{\frac{2E}{m}} = \sqrt{\frac{h^2}{4m^2L^2}} = \frac{h}{2mL} = \frac{6.6 \times 10^{-34}\,\text{J·s}}{2(10^{-14}\,\text{kg})(10^{-4}\,\text{m})} \approx 3 \times 10^{-16}\,\text{m/s}.$$

This is a speed so small that we could not measure it and the object would seem at rest, consistent with classical physics.

(b) Given $v = 10^{-3}\,\text{m}/100\,\text{s} = 10^{-5}\,\text{m/s}$, the kinetic energy of the bacterium is

$$E = \frac{1}{2}mv^2 = \frac{1}{2}\left(10^{-14}\,\text{kg}\right)\left(10^{-5}\,\text{m/s}\right)^2 = 0.5 \times 10^{-24}\,\text{J}.$$

From Eq. 39-13, the quantum number of this state is

$$n = \sqrt{E\left(\frac{8mL^2}{h^2}\right)} = \sqrt{\frac{(0.5 \times 10^{-24}\,\text{J})(8)(10^{-14}\,\text{kg})(10^{-4}\,\text{m})^2}{(6.6 \times 10^{-34}\,\text{J·s})^2}}$$
$$\approx \sqrt{1 \times 10^{21}} \approx 3 \times 10^{10}.$$

This number is so large that we could never distinguish between adjacent energy states (between $n = 3 \times 10^{10}$ and $3 \times 10^{10} + 1$). The energy states would appear to form a continuum. Thus, even though the energies involved here are small ($\ll 1\,\text{eV}$), we are still dealing with a macroscopic object (though visible only under a microscope) and the quantum result is not distinguishable from a classical one. This is in accordance with the correspondence principle.

* 39-9 Finite Potential Well

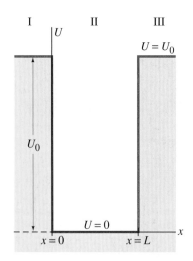

FIGURE 39-11 Potential energy U vs x for a finite one-dimensional square well.

Let us now look at a particle in a box whose walls are not perfectly rigid. That is, the potential energy outside the box or well is not infinite, but rises to some level U_0, as shown in Fig. 39-11. This is called a **finite potential well**. It can serve as an approximation for, say, a neutron in a nucleus. There are some significant new features that arise for the finite well as compared to the infinite well. We divide the well into three regions as shown in Fig. 39-11. In region II, inside the well, the Schrödinger equation is the same as before ($U = 0$), although the boundary conditions will be different. So we write the solution for region II as

$$\psi_{\text{II}} = A \sin kx + B \cos kx \qquad (0 < x < L)$$

but we don't immediately set $B = 0$ or assume that k is given by Eq. 39-12.

In regions I and III, the Schrödinger equation, now with $U(x) = U_0$, is

$$-\frac{\hbar^2}{2m}\frac{d^2\psi}{dx^2} + U_0\psi = E\psi.$$

We rewrite this as

$$\frac{d^2\psi}{dx^2} - \left[\frac{2m(U_0 - E)}{\hbar^2}\right]\psi = 0.$$

Let us assume that E is less than U_0, so the particle is "trapped" in the well (at least classically). There might be only one such **bound state**, or several, or even none, as we shall discuss later. We define the constant G by

$$G^2 = \frac{2m(U_0 - E)}{\hbar^2} \qquad\qquad \textbf{(39-16)}$$

and rewrite the Schrödinger equation as

$$\frac{d^2\psi}{dx^2} - G^2\psi = 0.$$

This equation has the general solution

$$\psi_{\text{I, III}} = Ce^{Gx} + De^{-Gx},$$

which can be confirmed by direct substitution, since

$$\frac{d^2}{dx^2}\left(e^{\pm Gx}\right) = G^2 e^{\pm Gx}.$$

In region I, x is always negative, so D must be zero (otherwise, $\psi \to \infty$ as $x \to -\infty$, giving an unacceptable result). Similarly in region III, where x is always positive, C must be zero. Hence

$$\psi_{\text{I}} = Ce^{Gx} \qquad (x < 0)$$
$$\psi_{\text{III}} = De^{-Gx} \qquad (x > L).$$

In regions I and III, the wave function decreases exponentially with distance from the well. The mathematical forms of the wave function inside and outside the well are different, but we insist that the wave function be continuous even at the two walls. We also insist that the slope of ψ, which is its first derivative, be continuous at the walls. Hence we have the boundary conditions:

$$\psi_{\text{I}} = \psi_{\text{II}} \quad \text{and} \quad \frac{d\psi_{\text{I}}}{dx} = \frac{d\psi_{\text{II}}}{dx} \quad \text{at } x = 0$$

$$\psi_{\text{II}} = \psi_{\text{III}} \quad \text{and} \quad \frac{d\psi_{\text{II}}}{dx} = \frac{d\psi_{\text{III}}}{dx} \quad \text{at } x = L.$$

At the left-hand wall $(x = 0)$ these boundary conditions become

$$Ce^0 = A\sin 0 + B\cos 0 \qquad \text{or} \qquad C = B$$

and

$$GCe^0 = kA\cos 0 - kB\sin 0 \qquad \text{or} \qquad GC = kA.$$

These are two of the relations that link the constants A, B, C, D and the energy E. We get two more relations from the boundary conditions at $x = L$, and a fifth relation from normalizing the wave functions over all space, $\int_{-\infty}^{\infty} |\psi|^2\, dx = 1$. These five relations allow us to solve for the five unknowns, including the energy E. We will not go through the detailed mathematics, but we will discuss some of the results.

Figure 39–12a shows the wave function ψ for the three lowest possible states, and Fig. 39–12b shows the probability distributions $|\psi|^2$. We see that the wave functions are smooth at the walls of the well. Within the well ψ has the form of a sinusoidal wave; for the ground state, there is less than a half wavelength. Compare this to the infinite well (Fig. 39–9), where the ground-state wave function is exactly a half wavelength: $\lambda = 2L$. For our finite well, $\lambda > 2L$. Thus for a finite well the momentum of a particle $(p = h/\lambda)$, and hence its ground-state energy, will be less than for an infinite well of the same width L.

Outside the finite well we see that the wave function drops off exponentially on either side of the walls. That ψ is not zero beyond the walls means that the particle can sometimes be found outside the well. This completely contradicts classical ideas. Outside the well, the potential energy of the particle is greater than its total energy: $U_0 > E$. This violates conservation of energy. But we clearly see in Fig. 39–12b that the particle can spend some time outside the well, where $U_0 > E$ (although the penetration into this classically forbidden region is generally not far since $|\psi|^2$ decreases exponentially with distance from either wall). The penetration

FIGURE 39–12 The wave functions (a), and probability distributions (b), for the three lowest possible states of a particle in a finite potential well. Each of the ψ and $|\psi|^2$ curves has been superposed on its energy level (dashed lines) for convenience.

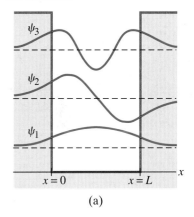

(a)

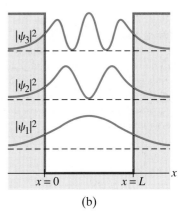

(b)

of a particle into a classically forbidden region is a very important result of quantum mechanics. But how can it be? How can we accept this nonconservation of energy? We can look to the uncertainty principle, in the form

$$\Delta E \, \Delta t \gtrsim \hbar.$$

It tells us that the energy is uncertain, and thus can even be nonconserved, by an amount ΔE for very short times $\Delta t \sim \hbar/\Delta E$.

Now let us consider the situation when the total energy E of the particle is greater than U_0. In this case the particle is a free particle and everywhere its wave function is sinusoidal, Fig. 39–13. Its wavelength is different outside the well than inside, as shown. Since $K = \frac{1}{2}mv^2 = p^2/2m$, the wavelength in region II is

$$\lambda = \frac{h}{p} = \frac{h}{\sqrt{2mE}} \qquad 0 < x < L,$$

whereas in regions I and III, where $p^2/2m = K = E - U_0$, the wavelength is

$$\lambda = \frac{h}{p} = \frac{h}{\sqrt{2m(E - U_0)}} \qquad x < 0 \quad \text{and} \quad x > L.$$

For $E > U_0$, any energy E is possible. But for $E < U_0$, as we saw above, the energy is quantized and only certain states are possible.

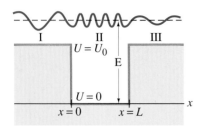

FIGURE 39–13 Particle of energy E traveling over a potential well whose depth U_0 is less than E (measured in the well).

Tunneling through a Barrier

We saw in Section 39–9 that according to quantum mechanics, a particle such as an electron can penetrate a barrier into a region forbidden by classical mechanics. There are a number of important applications of this phenomenon, particularly as applied to penetration of a thin barrier.

We consider a particle of mass m and energy E traveling to the right along the x axis in free space where the potential energy $U = 0$ so the energy is all kinetic energy $(E = K)$. The particle encounters a narrow potential barrier whose height U_0 (in energy units) is greater than E, and whose thickness is L (distance units); see Fig. 39–14a. Since $E < U_0$, we would expect from classical physics that the particle could not penetrate the barrier but would simply be reflected and would return in the opposite direction. Indeed, this is what happens for macroscopic objects. But quantum mechanics predicts a nonzero probability for finding the particle on the other side of the barrier. We can see how this can happen in part (b) of Fig. 39–14, which shows the wave function. The approaching particle has a sinusoidal wave function. Within the barrier the solution to the Schrödinger equation is a decaying exponential just as for the finite well of Section 39–9. However, before the exponential dies away to zero, the barrier ends (at $x = L$),

FIGURE 39–14 (a) A potential barrier of height U_0 and thickness L. (b) The wave function for a particle of energy $E\,(< U_0)$ that approaches from the left. The curve for ψ is superposed, for convenience, on the energy level line (dashed).

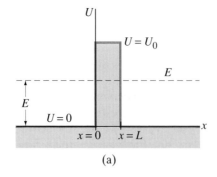
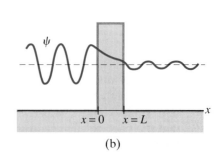

(a)   (b)

and for $x > L$ there is again a sinusoidal wave function, since $U = 0$ and $E = K > 0$. But it is a sine wave of greatly reduced amplitude. Nonetheless, because $|\psi|^2$ is nonzero beyond the barrier, we see that there is a nonzero probability that the particle penetrates the barrier. This process is called **tunneling** through the barrier, or **barrier penetration**. Although we cannot observe the particle within the barrier (it would violate conservation of energy), we can detect it after it has penetrated the barrier.

Quantitatively, we can describe the tunneling probability with a *transmission coefficient*, T, and a *reflection coefficient*, R. Suppose, for example, that $T = 0.03$ and $R = 0.97$; then if 100 particles struck the barrier, on the average 3 would tunnel through and 97 would be reflected. Note that $T + R = 1$, since an incident particle must either reflect or tunnel through. The transmission coefficient can be determined by writing the wave function for each of the three regions, just as we did for the finite well, and then applying the boundary conditions that ψ and $d\psi/dx$ must be continuous at the edges of the barrier ($x = 0$ and $x = L$). The calculation shows (see Problem 36) that if T is small ($\ll 1$), then

$$T \approx e^{-2GL}, \tag{39–17a}$$

where

$$G = \sqrt{\frac{2m(U_0 - E)}{\hbar^2}}. \tag{39–17b}$$

(This is the same G as in Section 39–9, Eq. 39–16.) We note that increasing the height of the barrier, U_0, or increasing its thickness, L, will drastically reduce T. Indeed, for macroscopic situations, T is extremely small, in accord with classical physics, which predicts no tunneling (again the correspondence principle).

EXAMPLE 39–7 **Barrier penetration.** A 50-eV electron approaches a square barrier 70 eV high and (*a*) 1.0 nm thick, (*b*) 0.10 nm thick. What is the probability that the electron will tunnel through?

SOLUTION (*a*) First we write the energy in SI units.

$$U_0 - E = (70\,\text{eV} - 50\,\text{eV})(1.6 \times 10^{-19}\,\text{J/eV}) = 3.2 \times 10^{-18}\,\text{J}.$$

Then, using Eq. 39–17, we have

$$2GL = 2\sqrt{\frac{2(9.11 \times 10^{-31}\,\text{kg})(3.2 \times 10^{-18}\,\text{J})}{(1.06 \times 10^{-34}\,\text{J·s})^2}}(1.0 \times 10^{-9}\,\text{m}) = 46$$

and

$$T = e^{-2GL} = e^{-46} = 1.1 \times 10^{-20},$$

which is extremely small.
(*b*) For $L = 0.10$ nm, $2GL = 4.6$ and

$$T = e^{-4.6} = 0.010.$$

Thus the electron has a 1 percent chance of penetrating a 0.1-nm-thick barrier, but only 1 chance in 10^{20} to penetrate a 1-nm barrier. By reducing the barrier thickness by a factor of 10, the probability of tunneling through was increased 10^{18} times! Clearly the transmission coefficient is extremely sensitive to the values of L, $U_0 - E$, and m.

Tunneling is a result of the wave properties of material particles, and also occurs for classical waves. For example, we saw in Section 33–7 that, when light traveling in glass strikes a glass–air boundary at an angle greater than the critical angle, the light is 100 percent totally reflected. We studied this phenomenon of

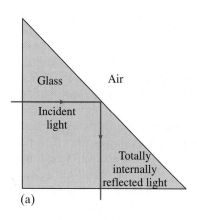

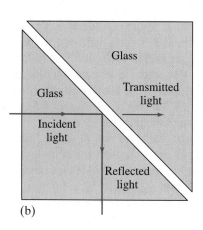

FIGURE 39–15 (a) Light traveling in glass strikes the interface with air at an angle greater than the critical angle, and is totally internally reflected. (b) A small amount of light tunnels through a narrow air gap between two pieces of glass.

(a)

(b)

FIGURE 39–16 Potential energy seen by an alpha particle (charge q) in presence of nucleus (charge Q), showing the wave function for tunneling out.

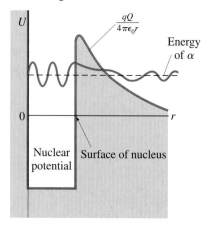

FIGURE 39–17 Probe tip of a scanning tunneling microscope moves up and down to maintain a constant tunneling current, producing an image of the surface.

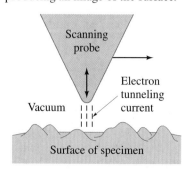

total internal reflection from the point of view of ray optics, and we show it here in Fig. 39–15a. The wave theory, however, predicts that waves actually penetrate the air for a few wavelengths—almost as if they "needed" to pass the interface to find out there is air beyond and hence need to be totally reflected. Indeed, if a second piece of glass is brought near the first as shown in Fig. 39–15b, a transmitted wave that has tunneled through the air gap can be experimentally observed. You can actually observe this for yourself by looking down into a glass of water at an angle such that light entering your eye has been totally internally reflected from the (outer) glass surface (it will look silvery). If you press a moistened fingertip against the glass, you can see the whorls of the ridges on your fingerprints, because at the ridges you have interfered with the total internal reflection at the outer surface of the glass. So you see light that has penetrated the gap and reflected off the ridges on your finger.

Applications of Tunneling

Tunneling thus occurs even for classical waves. What is new in quantum mechanics is that material particles have wave properties and hence can tunnel. Tunneling has provided the basis for a number of useful devices, as well as helped to explain a number of important phenomena, some of which we mention briefly now.

Some atomic nuclei undergo **radioactive decay** by the emission of an alpha (α) particle, which consists of two protons and two neutrons. Inside a radioactive nucleus, we can imagine that the protons and neutrons are moving about, and sometimes two of each come together and form this stable entity, the alpha particle. We will study alpha decay in more detail in Chapter 42, but for now we note that the potential energy diagram for the alpha particle inside this type of nucleus looks something like Fig. 39–16. The square well represents the attractive nuclear force that holds the nucleus together. To this is added the $1/r$ Coulomb potential energy of repulsion between the positive alpha particle and the remaining positively charged nucleus. The barrier that results is called the **Coulomb barrier**. The wave function for the tunneling particle shown must have energy greater than zero (or the barrier would be infinitely wide and tunneling could not occur), but less than the height of the barrier. If the alpha particle had energy higher than the barrier, it would always be free and the original nucleus wouldn't exist. Thus the barrier keeps the nucleus together, but occasionally a nucleus of this type can decay by the tunneling of an alpha particle. The probability of an α-particle escaping, and hence the "lifetime" of a nucleus, depends on the height and width of the barrier, and can take on a very wide range of values for only a limited change in barrier width as we saw in Ex. 39–7. Lifetimes of α-decaying radioactive nuclei range from less than 1 μs to 10^{10} yr.

A so-called **tunnel diode** is an electronic device made of two types of semiconductor carrying opposite-sign charge carriers, separated by a very thin neutral region. Current can tunnel through this thin barrier and can be controlled by the voltage applied to it, which affects the height of the barrier.

The **scanning tunneling electron microscope** (STM), developed in the 1980s, makes use of tunneling through a vacuum. A tiny probe, whose tip may be only one (or a few) atoms wide, is moved across the specimen to be examined in a series of linear passes, like those made by the electron beam in a TV tube or CRT. The tip, as it scans, remains very close to the surface of the specimen, about 1 nm above it, Fig. 39–17. A small voltage applied between the probe and the surface causes electrons to tunnel through the vacuum between them. This tunneling current is very sensitive to the gap width (see Example 39–7), so that a feedback mechanism can be used to raise and lower the probe to maintain a constant electron tunneling current. The probe's vertical motion, following the surface of the specimen, is then plotted as a function of position, producing a three-dimensional image of the surface. (See, for example, Fig. 39–18.) Surface features as fine as the size of an atom can be resolved: a resolution better than 0.1 nm laterally and 10^{-2} to 10^{-3} nm vertically. This kind of resolution was not available previously and has given a great impetus to the study of the surface structure of materials. The "topographic" image of a surface actually represents the distribution of electron charge (electron wave probability distributions).

The new **atomic force microscope** (AFM) is in many ways similar to an STM, but can be used on a wider range of sample materials. Instead of detecting an electric current, the AFM measures the force between a cantilevered tip and the sample, a force which depends strongly on the tip–sample separation at each point. The tip is moved as for the STM.

FIGURE 39–18 Image of cellular DNA, magnified about 2 million times, taken with a scanning tunneling microscope. Three turns of the DNA double helix can be seen in this false-color image.

Summary

In 1925, Schrödinger and Heisenberg separately worked out a new theory, **quantum mechanics**, which is now considered to be the basic theory at the atomic level. It is a statistical theory rather than a deterministic one.

An important aspect of quantum mechanics is the Heisenberg **uncertainty principle**. It results from the wave-particle duality and the unavoidable interaction between the observed object and the observer.

One form of the uncertainty principle states that the position x and momentum p_x of an object cannot both be measured precisely at the same time. The products of the uncertainties, $(\Delta x)(\Delta p_x)$, can be no less than $\hbar (= h/2\pi)$:

$$(\Delta p_x)(\Delta x) \gtrsim \hbar.$$

Another form states that the energy can be uncertain, or nonconserved, by an amount ΔE for a time Δt where

$$(\Delta E)(\Delta t) \gtrsim \hbar.$$

A particle such as an electron is represented by a **wave function** ψ. The square of the wave function, $|\psi|^2$, at any point in space represents the **probability** of finding the particle at that point. The wave function must be **normalized**, meaning that $\int |\psi|^2 \, dV$ over all space must equal 1, since the particle must be found at one place or another.

In nonrelativistic wave mechanics, ψ satisfies the **Schrödinger equation**:

$$-\frac{\hbar^2}{2m} \frac{d^2\psi}{dx^2} + U\psi = E\psi,$$

here in its one-dimensional time-independent form,

where U is the potential energy as a function of position and E is the total energy of the particle.

A **free particle** subject to no forces has a sinusoidal wave function $\psi = A \sin kx + B \cos kx$ with $k = p/\hbar$ and p is the particle's momentum. Such a wave of fixed momentum is spread out indefinitely in space as a plane wave.

A **wave packet**, localized in space, is a superposition of sinusoidal waves with a range of momenta.

For a particle confined to an **infinitely deep square well potential**, or **rigid box**, the Schrödinger equation gives the wave functions

$$\psi = A \sin kL,$$

where $A = \sqrt{2/L}$, as solutions inside the well. The energy is quantized,

$$E = \frac{\hbar^2 k^2}{2m} = \frac{n^2 h^2}{8mL^2},$$

where n is an integer.

In a **finite potential well**, the wave function extends into the classically forbidden region where the total energy is less than the potential energy. That this is possible is consistent with the uncertainty principle. The solutions to the Schrödinger equation in these areas are decaying exponentials.

Because quantum-mechanical particles can penetrate such classically forbidden areas, they can **tunnel** through thin barriers even though the potential energy in the barrier is greater than the total energy of the particle.

Questions

1. Compare a matter wave ψ to (a) a wave on a string, (b) an EM wave. Discuss similarities and differences.

2. Explain why Bohr's theory of the atom is not compatible with quantum mechanics, particularly the uncertainty principle.

3. Explain why it is that the more massive an object is, the easier it becomes to predict its future position.

4. In view of the uncertainty principle, why does a baseball seem to have a well-defined position and speed, whereas an electron does not?

5. Would it ever be possible to balance a very sharp needle precisely on its point? Explain.

6. When you check the pressure in a tire, doesn't some air inevitably escape? Is it possible to avoid this escape of air altogether? What is the relation to the uncertainty principle?

7. It has been said that the ground-state energy in the hydrogen atom can be precisely known but that the excited states have some uncertainty in their values (an "energy width"). Is this consistent with the uncertainty principle in its energy form? Explain.

8. If Planck's constant were much larger than it is, how would this affect our everyday life?

9. In what ways is Newtonian mechanics contradicted by quantum mechanics?

10. Discuss the connection between the zero-point energy for a particle in a box and the uncertainty principle.

11. The wave function for a particle in a box is zero at points within the box (except for $n = 1$). Does this mean that the probability of finding the particle at these points is zero? Does it mean that the particle cannot pass by these points? Explain.

12. What does the probability density look like for a particle in an infinite potential well for large values of n, say $n = 100$ or $n = 1000$? As n becomes very large, do your predictions approach classical predictions in accord with the correspondence principle?

13. For a particle in an infinite potential well the separation between energy states increases as n increases (see Eq. 39–13). But doesn't the correspondence principle require closer spacing between states as n increases so as to approach a classical (nonquantized) situation? Explain.

14. A particle is trapped in an infinite potential well. Describe what happens to the particle's ground-state energy and wave function as the potential walls become finite and get lower and lower until they finally reach zero ($U = 0$ everywhere).

15. A hydrogen atom and a helium atom, each with 4 eV of kinetic energy, approach a thin barrier 6 MeV high. Which has the greater probability of tunneling through?

Problems

Section 39–2

1. (II) The neutrons in a parallel beam, each having kinetic energy $\frac{1}{40}$ eV, are directed through two slits 1.0 mm apart. How far apart will the interference peaks be on a screen 1.0 m away?

2. (II) Bullets of mass 2.0 g are fired in parallel paths with speeds of 120 m/s through a hole 3.0 mm wide. How far from the hole must you be to detect a 1.0-cm spread in the beam?

Section 39–3

3. (I) A proton is traveling with a speed of $(4.825 \pm 0.012) \times 10^5$ m/s. With what maximum accuracy can its position be ascertained?

4. (I) An electron remains in an excited state of an atom for typically 10^{-8} s. What is the minimum uncertainty in the energy of the state (in eV)?

5. (I) If an electron's position can be measured to an accuracy of 1.6×10^{-8} m, how accurately can its velocity be known?

6. (II) A 12-g bullet leaves a rifle at a speed of 150 m/s. (a) What is the wavelength of this bullet? (b) If the position of the bullet is known to an accuracy of 0.55 cm (radius of the barrel), what is the minimum uncertainty in its vertical momentum? (c) If the accuracy of the bullet were determined only by the uncertainty principle (an unreasonable assumption), by how much might the bullet miss a pinpoint target 300 m away?

7. (II) An electron and a 150-g baseball are each traveling 75 m/s measured to an accuracy of 0.065 percent. Calculate and compare the uncertainty in position of each.

8. (II) Use the uncertainty principle to show that if an electron were present in the nucleus ($r \approx 10^{-15}$ m), its kinetic energy (use relativity) would be hundreds of MeV. (Since such electron energies are not observed, we conclude that electrons are not present in the nucleus.) [Hint: a particle can have energy as large as its uncertainty.]

9. (II) An electron in the $n = 2$ state of hydrogen remains there on the average about 10^{-8} s before jumping to the $n = 1$ state. (a) Estimate the uncertainty in the energy of the $n = 2$ state. (b) What fraction of the transition energy is this? (c) What is the wavelength, and width (in nm), of this line in the spectrum of hydrogen?

10. (II) How accurately can the position of a 2.50-keV electron be measured assuming its energy is known to 1.00 percent?

11. (III) In a double-slit experiment on electrons (or photons), suppose that we use indicators to determine which slit each electron went through (Section 39–2). These indicators must tell us the y coordinate to within $d/2$, where d is the distance between slits. Use the uncertainty principle to show that the interference pattern will be destroyed [Note: First show that the angle θ between maxima and minima of the interference pattern is given by $\frac{1}{2}\lambda/d$, Fig. 39–19.]

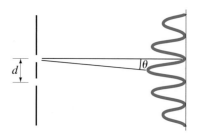

FIGURE 39–19
Problem 11.

*12. (III) (a) Show that $\Psi(x, t) = Ae^{i(kx-\omega t)}$ is a solution to the time-dependent Schrödinger equation for a free particle [$U(x) = U_0 = $ constant] but that $\Psi(x, t) = A\cos(kx - \omega t)$ and $\Psi(x, t) = A\sin(kx - \omega t)$ are not. (b) Show that the valid solution of part (a) satisfies conservation of energy if the de Broglie relations hold, $\lambda = h/p$, $\omega = E/\hbar$. That is, show that direct substitution in Eq. 39–7 gives

$$\hbar\omega = \frac{\hbar^2 k^2}{2m} + U_0.$$

Section 39–7

13. (I) Write the wave function for (a) a free electron and (b) a free proton, each having a constant velocity $v = 4.0 \times 10^5$ m/s.

14. (I) A free electron has a wave function $\psi(x) = A\sin(1.0 \times 10^{10}x)$, where x is given in meters. Determine the electron's (a) wavelength, (b) momentum, (c) speed, and (d) kinetic energy.

15. (II) Show that the uncertainty principle holds for a "wave packet" that is formed by two waves of similar wavelength λ_1 and λ_2. To do so, follow the argument leading up to Eq. 16–8, but use as the two waves $\psi_1 = A\sin k_1 x$ and $\psi_2 = A\sin k_2 x$. Then show that the width of each "wave packet" is $\Delta x = 2\pi/(k_1 - k_2) = 2\pi/\Delta k$ (from $t = 0.05$ s to $t = 0.15$ s in Fig. 16–18). Finally, show that $\Delta x \, \Delta p = h$ for this simple situation.

Section 39–8

16. (II) Show that for a particle in a perfectly rigid box, the wavelength of the wave function for any state is the de Broglie wavelength.

17. (II) What is the minimum speed of an electron trapped in a 0.10-nm-wide infinitely deep square well?

18. (II) An $n = 3$ to $n = 1$ transition for an electron trapped in a rigid box produces a 240-nm photon. What is the width of the box?

19. (II) An electron trapped in an infinitely deep square well has a ground-state energy $E = 8.0$ eV. (a) What is the longest wave-length photon this system can emit, and (b) what is the width of the well?

20. (II) The longest-wavelength line in the spectrum emitted by an electron trapped in an infinitely deep square well is 690 nm. What is the width of the well?

21. (II) For a particle in a box with rigid walls, determine whether our results for the ground state are consistent with the uncertainty principle by calculating the product $\Delta p \, \Delta x$. Take $\Delta x \approx L$, since the particle is known to be at least within the box. For Δp, note that although p is known ($= \hbar k$), the direction of $\mathbf{p}$ is not known, so the x component could vary from $-p$ to $+p$; hence take $\Delta p \approx 2p$.

22. (II) Write a formula for the positions of (a) the maxima and (b) the minima in $|\psi|^2$ for a particle in the nth state in an infinite square well.

23. (II) Determine the lowest four energy levels and wave functions for an electron trapped in an infinitely deep potential well of width 2.0 nm.

24. (II) An electron is trapped in a rigid box 0.50 nm wide. (a) Determine the energies and wave functions for the four lowest states. (b) Determine the wavelengths of photons emitted for all possible transitions between these states.

25. (II) Consider an atomic nucleus to be a rigid box of width 10^{-14} m. What would be the ground-state energy for (a) an electron, (b) a neutron, and (c) a proton?

26. (II) Suppose that the walls of an infinite square well are at $x = -L/2$ and $x = L/2$. What will be the wave functions for the four lowest energy levels? Sketch the wave functions and probability densities.

27. (II) An electron is trapped in an infinitely deep potential well of width L. Determine the probability of finding the electron within $\frac{1}{4}L$ of either wall if it is (a) in the ground state, (b) in the $n = 4$ state. [Hint: Evaluate $\int_0^{L/4}|\psi|^2\,dx + \int_{3L/4}^{L}|\psi|^2\,dx$.] (c) What is the classical prediction?

28. (II) An electron is trapped in a 1.00-nm-wide rigid box. Determine the probability of finding the electron within 0.10 nm of the center of the box (on either side of center) for (a) $n = 1$, (b) $n = 5$, and (c) $n = 20$. (d) Compare to the classical prediction.

*29. (II) Sketch the wave functions and the probability distributions for the $n = 4$ and $n = 5$ states for a particle trapped in a finite square well.

*30. (II) An electron with 100 eV of kinetic energy in free space passes over a finite potential well 50 eV deep that stretches from $x = 0$ to $x = 0.50$ nm. What is the electron's wavelength (a) in free space, (b) when over the well? (c) Draw a diagram showing the potential energy and total energy as a function of x, and on the diagram sketch a possible wave function.

*31. (II) An electron is trapped in a 0.10-nm-wide finite square well of height $U_0 = 1.0$ keV. Estimate at what distance outside the walls of the well the ground state wave function drops to 1 percent of its value at the walls.

*32. (II) Suppose that a particle of mass m is trapped in a finite potential well that has a rigid wall at $x = 0$ ($U = \infty$ for $x < 0$) and a finite wall of height $U = U_0$ at $x = L$, Fig. 39–20. (a) Sketch the wave functions for the lowest three states. (b) What is the form of the wave function in the ground state in the three regions $x < 0$, $0 < x < L$, $x > L$?

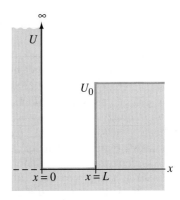

FIGURE 39–20 Problem 32.

33. (II) An electron approaches a potential barrier 10 eV high and 0.50 nm wide. If the electron has a 1.0 percent probability of tunneling through the barrier, what must its energy be?

34. (II) A 1.0-mA current of 1.0-MeV protons strikes a 2.0-MeV-high potential barrier 2.0×10^{-13} m thick. Estimate the transmitted current.

35. (II) For part (b) of Example 39–7, what effect will there be on the transmission coefficient if (a) the barrier height is raised 1 percent, (b) the barrier thickness is increased by 1 percent?

36. (III) Show that the transmission coefficient is given roughly by Eqs. 39–17 for a high or thick barrier, by calculating $|\psi(x = L)|^2/|\psi(0)|^2$. [Hint: Assume that ψ is a decaying exponential inside the barrier.]

37. (III) A uranium-238 nucleus $(Z = 92)$ lasts about 5×10^9 years before it decays by emission of an alpha particle $(Z = 2, M = 4M_{proton})$. (a) Assuming that the α particle is a point, and the nucleus is roughly 8 fm in radius, estimate the height of the Coulomb barrier (the peak in Fig. 39–16). (b) The alpha particle, when free, has kinetic energy ≈ 4 MeV. Estimate the width of the barrier. (c) Assuming that the square well has $U = 0$ inside (where $U = 0$ far from the nucleus), calculate the speed of the alpha particle and how often it hits the barrier, and from this (and Eq. 39–17) estimate its lifetime. [Hint: Replace the $1/r$ coulomb barrier with an "averaged" rectangular barrier (as in Fig. 39–14) of width equal to $\frac{1}{3}$ that calculated in (b).]

General Problems

38. If an electron's position can be measured to an accuracy of 2.0×10^{-8} m, how accurately can its velocity be known?

39. Estimate the lowest possible energy of a neutron contained in a typical nucleus of radius 1.0×10^{-15} m. [Hint: A particle can have an energy at least as large as its uncertainty.]

40. The Z^0 boson, discovered in 1985, is the mediator of the weak nuclear force, and it typically decays very quickly. Its average rest energy is 91.19 GeV, but its short lifetime shows up as an intrinsic width of 2.5 GeV. What is the lifetime of this particle?

41. What is the uncertainty in the mass of a muon $(m = 105.7 \text{ MeV}/c^2)$, specified in eV/c^2, given its average lifetime of 2.20 μs?

42. A free neutron $(m = 1.67 \times 10^{-27}$ kg) has a mean life of 900 s. What is the uncertainty in its mass (in kg)?

43. Use the uncertainty principle to estimate the position uncertainty for the electron in the ground state of the hydrogen atom. [Hint: Determine the momentum using the Bohr model of Section 38–10 and assume the momentum can be anywhere between this value and zero.] How does this compare to the Bohr radius?

44. A neutron is trapped in an infinitely deep potential well 2.0 fm in width. Determine (a) the four lowest possible energy states and (b) their wave functions. (c) What is the wavelength and energy of a photon emitted when the neutron makes a transition between the two lowest states? In what region of the EM spectrum does this photon lie? [Note: This is a rough model of an atomic nucleus.]

45. Estimate the kinetic energy and speed of an alpha particle $(q = 2e, m = 4$ proton masses) trapped in a nucleus 10^{-14} m wide. Assume an infinitely deep square well potential.

46. Simple Harmonic Oscillator. Suppose that an electron is trapped not in a square well, but one whose potential energy is that of a simple harmonic oscillator: $U(x) = \frac{1}{2}Cx^2$. That is, if the electron is displaced from $x = 0$, a restoring force $F = -Cx$ acts on it, where C is constant. (a) Sketch this potential energy. (b) Show that $x = Ae^{-Bx^2}$ is a solution to the Schrödinger equation and that the energy of this state is $E = \frac{1}{2}\hbar\omega$, where $\omega = 2\hbar B/m$, and A and B are constants. [Note: This is the ground state and this energy $\frac{1}{2}\hbar\omega$ is the zero-point energy for a harmonic oscillator. The energies of higher states are $E_n = (n + \frac{1}{2})\hbar\omega$, where n is an integer.]

47. A 10-gram pencil, 20 cm long, is balanced on its point. Classically, this is a configuration of (unstable) equilibrium, so the pencil could remain there forever if it were perfectly placed. A quantum mechanical analysis shows that the pencil must fall. (a) Why is this the case? (b) Estimate (within a factor of 2) how long it will take the pencil to hit the table if it is initially positioned as well as possible? [Hint: Use the uncertainty principle in its angular form to obtain an expression for the initial angle $\phi_0 \approx \Delta\phi$.]

48. The probability density for finding a particle at a particular location is $|\psi|^2$. The average value $\bar{x}$, the location of a particle averaged over many measurements, is given by

$$\bar{x} = \int_0^L x|\psi|^2 \, dx$$

for a particle in a rigid box. Calculate the average value of x for any value of n.

49. By how much does the tunneling current through the tip of an STM change if the tip rises 0.010 nm from some initial height above a sodium surface with a work function $W_0 = 2.28$ eV? [Hint: Let the work function (see Section 38–2) equal the energy needed to raise the electron to the top of the barrier.]

50. A small ball of mass 1.0×10^{-6} kg is dropped on a table from a height of 2.0 m. After each bounce the ball rises to 80 percent of its height before the bounce because of its inelastic collision with the table. Estimate how many bounces occur before the uncertainty principle plays a role in the problem. [Hint: Determine when the uncertainty in the ball's speed is comparable to its speed of impact on the table.]

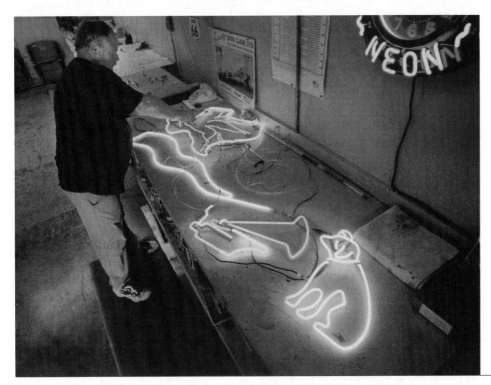

A neon tube is a thin glass tube, moldable into various shapes, filled with neon (or other) gas that glows with a particular color when a current at high voltage passes through it. Gas atoms, excited to upper energy levels, jump down to lower energy levels and emit light (photons) whose wavelengths (color) are characteristic of the type of gas.

In this chapter we study what quantum mechanics tells us about atoms, their wave functions and energy levels, including the effect of the exclusion principle. We also discuss interesting applications such as lasers and holography.

CHAPTER 40

Quantum Mechanics of Atoms

A t the beginning of Chapter 39 we discussed the limitations of the Bohr theory of atomic structure and why a new theory was needed. Although the Bohr theory had great success in predicting the wavelengths of light emitted and absorbed by the hydrogen atom, it could not do so for more complex atoms. Nor did it explain *fine structure*, the splitting of emission lines into two or more closely spaced lines. And, as a theory, it was an uneasy mixture of classical and quantum ideas.

Quantum mechanics came to the rescue in 1925 and 1926, and in this chapter we examine the quantum-mechanical theory of atomic structure, which is far more complete than the old Bohr theory.

40–1 Quantum-Mechanical View of Atoms

Although the Bohr model has been discarded as an accurate description of nature, nonetheless, quantum mechanics reaffirms certain aspects of the older theory, such as that electrons in an atom exist only in discrete states of definite energy, and that a photon of light is emitted (or absorbed) when an electron makes a transition from one state to another. But quantum mechanics is a much deeper theory, and has provided us with a very different view of the atom. According to quantum mechanics, electrons do not exist in well-defined circular orbits as in the Bohr theory. Rather, the electron (because of its wave nature) can be thought of as spread out in space as if it were a "cloud." The size and shape of the electron cloud can be calculated for a given state of an atom. For the ground state in the hydrogen atom,

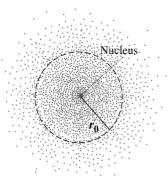

FIGURE 40–1 Electron cloud or "probability distribution" for the ground state of the hydrogen atom, as seen from afar. The circle represents the Bohr radius.

Probability distribution

the solution of the Schrödinger equation, as we will discuss in more detail in Section 40–3, gives

$$\psi(r) = \frac{1}{\sqrt{\pi r_0^3}} e^{-\frac{r}{r_0}}.$$

Here $\psi(r)$ is the wave function as a function of position, and it depends only on the radial distance r from the center, and not on angle θ or ϕ. (The constant r_0 has a value that happens to come out equal to the first Bohr radius.) Thus the electron cloud, whose density is $|\psi|^2$, for the ground state of hydrogen is spherically symmetric as shown in Fig. 40–1. The extent of the electron cloud at its higher densities roughly indicates the "size" of an atom, but just as a cloud may not have a distinct border, atoms do not have a precise boundary or a well-defined size. Not all electron clouds have a spherical shape, as we shall see later in this chapter. But note that $\psi(r)$, while becoming extremely small for large r (see the equation above), does not equal zero in any finite region. So quantum mechanics suggests that an atom is not mostly empty space. (Indeed, since $\psi \to 0$ only for $r \to \infty$, we might question the idea that there is any truly empty space in the universe.)

The electron cloud can be interpreted from either the particle or the wave viewpoint. Remember that by a particle we mean something that is localized in space—it has a definite position at any given instant. By contrast, a wave is spread out in space. The electron cloud, spread out in space as in Fig. 40–1, is a result of the wave nature of electrons. Electron clouds can also be interpreted as **probability distributions** for a particle. As we saw in Section 39–3, we cannot predict the path an electron will follow. After one measurement of its position we cannot predict exactly where it will be at a later time. We can only calculate the probability that it will be found at different points. If you were to make 500 different measurements of the position of an electron, considering it as a particle, the majority of the results would show the electron at points where the probability is high (dark area in Fig. 40–1). Only occasionally would the electron be found where the probability is low.

40–2 Hydrogen Atom: Schrödinger Equation and Quantum Numbers

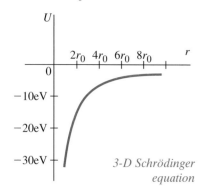

FIGURE 40–2 Potential energy $U(r)$ for the hydrogen atom. The radial distance r of the electron from the nucleus is given in terms of the Bohr radius r_0.

3-D Schrödinger equation

The hydrogen atom is the simplest of all atoms, consisting of a single electron of charge $-e$ moving around a central nucleus (a single proton) of charge $+e$. It is with hydrogen that a study of atomic structure must begin.

The Schrödinger equation (see Eq. 39–5) includes a term containing the potential energy. For the hydrogen (H) atom, the potential energy is due to the Coulomb force between electron and proton:

$$U = -\frac{1}{4\pi\epsilon_0}\frac{e^2}{r}$$

where r is the radial distance from the proton (situated at $r = 0$) to the electron. See Fig. 40–2. The (time-independent) Schrödinger equation, which must now be written in three dimensions, is then

$$-\frac{\hbar^2}{2m}\left(\frac{\partial^2\psi}{\partial x^2} + \frac{\partial^2\psi}{\partial y^2} + \frac{\partial^2\psi}{\partial z^2}\right) - \frac{1}{4\pi\epsilon_0}\frac{e^2}{r}\psi = E\psi, \qquad \text{(40–1)}$$

where $\partial^2\psi/\partial x^2$, $\partial^2\psi/\partial y^2$, and $\partial^2\psi/\partial z^2$ are partial derivatives with respect to x, y, and z. To solve the Schrödinger equation for the H atom, it is usual to write it in terms of spherical coordinates (r, θ, ϕ). We will not, however, actually go through the process of solving it. Instead, we look at the properties of the solutions, and (in the next Section) at the wave functions themselves.

Recall from Chapter 39 that the solutions of the Schrödinger equation in one dimension for the infinite square well were characterized by a single quantum number, which we called n, which arises from applying the boundary conditions. In the three-dimensional problem of the H atom, the solutions of the Schrödinger equation are characterized by three quantum numbers corresponding to boundary conditions applied in the three dimensions. However, four different quantum numbers are actually needed to specify each state in the H atom, the fourth coming from a relativistic treatment. We now discuss each of these quantum numbers. Much of our analysis here will also apply to more complex atoms, which we discuss starting in Section 40–4.

Quantum mechanics predicts the same energy levels (Fig. 38–25) for the H atom as does the Bohr theory. That is,

$$E_n = -\frac{13.6\,\text{eV}}{n^2} \qquad n = 1, 2, 3, \dots, \qquad \textbf{(40–2)}$$

where n is an integer called the **principal quantum number**. This is the same as the quantum number n that appeared in Bohr theory. It can have any integer value from 1 to ∞. The total energy of a state in the H atom depends on n, Eq. 40–2.

The **orbital quantum number**, l, is related to the magnitude of the orbital angular momentum of the electron; l can take on integer values from 0 to $(n - 1)$. For the ground state, $n = 1$, l can only be zero.[†] But for $n = 3$, say, l can be 0, 1, or 2. The actual magnitude of the orbital angular momentum L is related to the quantum number l by

$$L = \sqrt{l(l + 1)}\,\hbar. \qquad \textbf{(40–3)}$$

The value of l has almost no effect on the total energy in the hydrogen atom; only n does to any appreciable extent.[‡] But in atoms with two or more electrons, the energy does depend on l as well as n.

The **magnetic quantum number**, m_l, is related to the direction of the electron's orbital angular momentum, and it can take on integer values ranging from $-l$ to $+l$. For example, if $l = 2$, then m_l can be $-2, -1, 0, +1$, or $+2$. Since angular momentum is a vector, it is not surprising that both its magnitude and its direction would be quantized. For $l = 2$, the five different directions allowed can be represented by the diagram of Fig. 40–3. This limitation on the direction of $\mathbf{L}$ is often called **space quantization**. In quantum mechanics, the direction of the angular momentum is usually specified by giving its component along the z axis (this choice is arbitrary). Then L_z is related to m_l by the equation

$$L_z = m_l \hbar. \qquad \textbf{(40–4)}$$

The values of L_x and L_y are not definite, however (see Problem 65). The name for m_l derives not from theory (which relates it to L_z) but from experiment. It was found that when a gas discharge tube (Fig. 38–18) was placed in a magnetic field, the spectral lines were split into several very closely spaced lines. This splitting, known as the **Zeeman effect**, implies that the energy levels must be split (Fig. 40–4), and thus that the energy of a state depends not only on n but also on m_l when a magnetic field is applied—hence the name "magnetic quantum number." (Why the energy should depend on the direction of $\mathbf{L}$ can be seen from a semiclassical view of a moving electron as an electric current which interacts with the magnetic field—see Chapter 27, Section 27–5, and Section 40–7.)

Finally, there is the **spin quantum number**, m_s, which for an electron can have only two values, $m_s = +\frac{1}{2}$ and $m_s = -\frac{1}{2}$. The existence of this quantum number did not come out of Schrödinger's original theory, as did n, l, and m_l. Instead, a subsequent modification by P. A. M. Dirac (1902–1984) explained its presence as a relativistic effect. The first hint that m_s was needed, however, came from experiment. A careful study of the spectral lines of hydrogen showed that each actually consisted of two (or more) very closely spaced lines even in the absence of an external

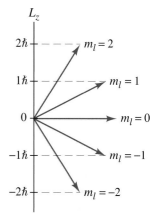

FIGURE 40–3 Quantization of angular momentum direction for $l = 2$.

FIGURE 40–4 When a magnetic field is applied, an $n = 3$, $l = 2$ energy level is split into five separate levels, corresponding to the five values of m_l (2, 1, 0, −1, −2). An $n = 2$, $l = 1$ level is split into three levels ($m_l = 1, 0, -1$). Transitions can occur between levels (not all transitions are shown), with photons of several slightly different frequencies being given off (the Zeeman effect).

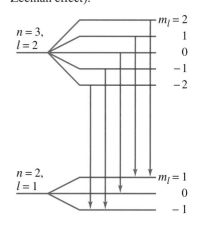

[†] Contrast this with the Bohr theory, which assigned $l = 1$ to the ground state (Eq. 38–10).

[‡] See discussion of *fine structure* starting at the bottom of this page.

magnetic field. It was at first hypothesized that this tiny splitting of energy levels, called **fine structure**, was due to angular momentum associated with a spinning of the electron. That is, the electron might spin on its axis as well as orbit the nucleus, just as the Earth spins on its axis as it orbits the Sun. The interaction between the tiny current of the spinning electron could then interact with the magnetic field due to the orbiting charge and cause the small observed splitting of energy levels. (So the energy depends slightly on m_l and m_s.) Today we consider this picture of a spinning electron as not legitimate. We cannot even view an electron as a localized object, much less a spinning one. What is important is that the electron can have two different states due to some intrinsic property that behaves as an angular momentum, and we still call this property "spin." The electron is said to have a spin quantum number $s = \frac{1}{2}$, which produces a spin angular momentum S given by

Spin angular momentum

$$S = \sqrt{s(s+1)}\,\hbar = \frac{\sqrt{3}}{2}\,\hbar.$$

(Compare Eq. 40–3.) This spin can have two different directions, $m_s = +\frac{1}{2}$ or $m_s = -\frac{1}{2}$, which are often said to be "spin up" and "spin down" (see Fig. 40–5). A state with spin down $\left(m_s = -\frac{1}{2}\right)$ has slightly lower energy than one with spin up. (Note that we include m_s, but not s, in our list of quantum numbers since s is the same for all electrons.)

The possible values of the four quantum numbers for an electron in the hydrogen atom are summarized in Table 40–1.

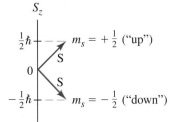

FIGURE 40–5 The spin angular momentum S can take on only two directions, $m_s = +\frac{1}{2}$ or $-\frac{1}{2}$, called "spin up" and "spin down."

TABLE 40–1 Quantum Numbers for an Electron

Name	Symbol	Possible Values
Principal	n	$1, 2, 3, \ldots, \infty$.
Orbital	l	For a given n: l can be $0, 1, 2, \ldots, n-1$.
Magnetic	m_l	For given n and l: m_l can be $l, l-1, \ldots, 0, \ldots, -l$.
Spin	m_s	For each set of n, l, and m_l: m_s can be $+\frac{1}{2}$ or $-\frac{1}{2}$.

CONCEPTUAL EXAMPLE 40–1 Possible states for $n = 3$. How many different states are possible for an electron whose principal quantum number is $n = 3$?

SOLUTION For $n = 3$, l can have the values $l = 2, 1, 0$. For $l = 2$, m_l can be $2, 1, 0, -1, -2$, which is five different possibilities. For each of these, m_s can be either up or down ($+\frac{1}{2}$ or $-\frac{1}{2}$), so for $l = 2$ there are $2 \times 5 = 10$ states. For $l = 1$, m_l can be $1, 0, -1$, and since m_s can be $+\frac{1}{2}$ or $-\frac{1}{2}$ for each of these, we have 6 more possible states. Finally, for $l = 0$, m_l can only be 0, and there are only 2 states corresponding to $m_s = +\frac{1}{2}$ and $-\frac{1}{2}$. The total number of states is $10 + 6 + 2 = 18$, as detailed in the following table:

n	l	m_l	m_s	n	l	m_l	m_s
3	2	2	$\frac{1}{2}$	3	2	-2	$-\frac{1}{2}$
3	2	2	$-\frac{1}{2}$	3	1	1	$\frac{1}{2}$
3	2	1	$\frac{1}{2}$	3	1	1	$-\frac{1}{2}$
3	2	1	$-\frac{1}{2}$	3	1	0	$\frac{1}{2}$
3	2	0	$\frac{1}{2}$	3	1	0	$-\frac{1}{2}$
3	2	0	$-\frac{1}{2}$	3	1	-1	$\frac{1}{2}$
3	2	-1	$\frac{1}{2}$	3	1	-1	$-\frac{1}{2}$
3	2	-1	$-\frac{1}{2}$	3	0	0	$\frac{1}{2}$
3	2	-2	$\frac{1}{2}$	3	0	0	$-\frac{1}{2}$

EXAMPLE 40–2 *E and L for n = 3.* Determine (*a*) the energy and (*b*) the orbital angular momentum for each of the states in Example 40–1.

SOLUTION (*a*) The energy of a state depends only on *n*, except for the very small corrections mentioned above, which we will ignore. Since *n* = 3 for all these states, they all have the same energy,

$$E_3 = -\frac{13.6 \text{ eV}}{(3)^2} = -1.51 \text{ eV}.$$

(*b*) For *l* = 0,

$$L = \sqrt{l(l+1)}\hbar = \sqrt{0(0+1)}\hbar = 0.$$

For *l* = 1,

$$L = \sqrt{1(1+1)}\hbar = \sqrt{2}\hbar$$
$$= 1.49 \times 10^{-34} \text{ J} \cdot \text{s}.$$

Atomic angular momenta are generally given as a multiple of $\hbar$ ($\sqrt{2}\hbar$ in this case), rather than in SI units. But note that for *l* = 1, *L* is on the order of 10^{-34} J·s. This means that macroscopic angular momenta will have such extremely high quantum numbers that the quantization of angular momentum will not be detectable: *L* will appear continuous, in accordance with the correspondence principle. Finally, for *l* = 2,

$$L = \sqrt{2(2+1)}\hbar = \sqrt{6}\hbar.$$

Another prediction of quantum mechanics is that when a photon is emitted or absorbed, transitions can occur only between states with values of *l* that differ by one unit:

$$\Delta l = \pm 1.$$

Selection rule

According to this **selection rule**, an electron in an *l* = 2 state can jump only to a state with *l* = 1 or *l* = 3. It cannot jump to a state with *l* = 2 or *l* = 0. A transition such as *l* = 2 to *l* = 0 is called a **forbidden transition**. Actually, such a transition is not absolutely forbidden and can occur, but only with very low probability compared to **allowed transitions**—those that satisfy the selection rule $\Delta l = \pm 1$. Since the orbital angular momentum of an H atom must change by one unit when it emits a photon, conservation of angular momentum tells us that the photon must carry off angular momentum. Indeed, experimental evidence of many sorts shows that the photon can be assigned a spin quantum number of 1.

"forbidden" transition

40–3 | Hydrogen Atom Wave Functions

The solution of the Schrödinger equation for the ground state of hydrogen—the state with lowest energy *E*—has an energy $E_1 = -13.6$ eV, as we have seen. The wave function for the ground state depends only on *r* and so is spherically symmetric. As already mentioned in Section 40–1, its form is

$$\psi_{100} = \frac{1}{\sqrt{\pi r_0^3}} e^{-\frac{r}{r_0}} \qquad \text{(40–5a)}$$

Ground state in hydrogen

where $r_0 = h^2 \epsilon_0 / \pi m e^2 = 0.0529$ nm is the Bohr radius (Section 38–10). The subscript 100 on ψ represents the quantum numbers *n*, *l*, m_l:

$$\psi_{nlm_l}.$$

For the ground state, *n* = 1, *l* = 0, m_l = 0, and there is only one wave function that serves for both $m_s = +\frac{1}{2}$ and $m_s = -\frac{1}{2}$ (the value of m_s does not affect the

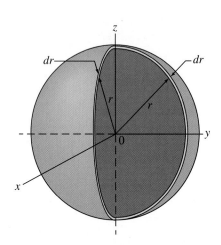

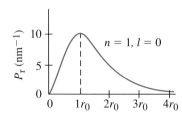

FIGURE 40–6 A spherical shell of thickness dr, inner radius r, and outer radius $r + dr$. Its volume is $dV = 4\pi r^2\, dr$.

Hydrogen ground state radial probability distribution

FIGURE 40–7 The radial probability distribution P_{r} for the ground state of hydrogen, $n = 1$, $l = 0$. The peak occurs at $r = r_0$, the Bohr radius.

spatial dependence of the wave function for any state, since spin is an *internal* or *intrinsic* property of the electron). The probability density for the ground state is

$$|\psi_{100}|^2 = \frac{1}{\pi r_0^3}\, e^{-\frac{2r}{r_0}} \qquad \textbf{(40–5b)}$$

which falls off exponentially with r. Note that ψ_{100}, as well as all other wave functions we discuss, has been normalized:

$$\int_{\text{all space}} |\psi_{100}|^2\, dV = 1.$$

The quantity $|\psi|^2\, dV$ gives the probability of finding the electron in a volume dV about a given point. It is often more useful to specify the **radial probability distribution**, P_{r}, which is defined so that $P_{\mathrm{r}}\, dr$ is the probability of finding the electron at a radial distance between r and $r + dr$ from the nucleus. That is, $P_{\mathrm{r}}\, dr$ specifies the probability of finding the electron within a thin shell of thickness dr of inner radius r and outer radius $r + dr$, regardless of direction (see Fig. 40–6). The volume of this shell is the product of its surface area, $4\pi r^2$, and its thickness, dr:

$$dV = 4\pi r^2\, dr.$$

Hence

$$|\psi|^2\, dV = |\psi|^2 4\pi r^2\, dr$$

and the radial probability distribution is

$$P_{\mathrm{r}} = 4\pi r^2 |\psi|^2. \qquad \textbf{(40–6)}$$

For the ground state of hydrogen, P_{r} becomes

$$P_{\mathrm{r}} = 4\frac{r^2}{r_0^3}\, e^{-\frac{2r}{r_0}} \qquad \textbf{(40–7)}$$

and is plotted in Fig. 40–7. The peak of the curve is the "most probable" value of r and occurs for $r = r_0$, the Bohr radius, which we now prove.

EXAMPLE 40–3 Most probable electron radius in hydrogen. Determine the most probable distance r from the nucleus at which to find the electron in the ground state of hydrogen.

SOLUTION The peak of the curve in Fig. 40–7 corresponds to the most probable value of r. At this point the curve has zero slope, so we take the derivative of Eq. 40–7 and set it equal to zero:

$$\frac{d}{dr}\left(4\frac{r^2}{r_0^3}\, e^{-\frac{2r}{r_0}}\right) = 0$$

$$\left(8\frac{r}{r_0^3} - \frac{8r^2}{r_0^4}\right)e^{-\frac{2r}{r_0}} = 0.$$

Since $e^{-\frac{2r}{r_0}}$ goes to zero only at $r = \infty$, it is the term in parentheses that must be zero:

$$8\frac{r}{r_0^3} - 8\frac{r^2}{r_0^4} = 0.$$

Therefore,

$$\frac{r}{r_0^3} = \frac{r^2}{r_0^4}$$

or

$$r = r_0.$$

The most probable radial distance of the electron from the nucleus according to quantum mechanics is at the Bohr radius, an interesting coincidence.

EXAMPLE 40–4 **Calculating probability.** Determine the probability of finding the electron in the ground state of hydrogen within two Bohr radii of the nucleus.

SOLUTION We need to integrate P_r from $r = 0$ out to $r = 2r_0$: that is, we want to find

$$P_r = \int_{r=0}^{2r_0} |\psi|^2 \, dV = \int 4 \frac{r^2}{r_0^3} e^{-\frac{2r}{r_0}} \, dr.$$

We first make the substitution

$$x = 2\frac{r}{r_0}$$

and then integrate by parts $\left(\int u \, dv = uv - \int v \, du\right)$ letting $u = x^2$ and $dv = e^{-x} \, dx$ (and note that $dx = 2dr/r_0$):

$$P_r = \frac{1}{2} \int_{x=0}^{4} x^2 e^{-x} \, dx = \frac{1}{2}\left[-x^2 e^{-x} + \int 2x e^{-x} \, dx \right]\Bigg|_0^4 .$$

The second term we also integrate by parts with $u = 2x$ and $dv = e^{-x} \, dx$:

$$P_r = \frac{1}{2}\left[-x^2 e^{-x} - 2x e^{-x} + 2\int e^{-x} \, dx \right]\Bigg|_0^4$$

$$= \left(-\frac{1}{2}x^2 - x - 1 \right)e^{-x}\Bigg|_0^4 .$$

We evaluate this at $x = 0$ and at $x = 2(2r_0)/r_0 = 4$:

$$P_r = (-8 - 4 - 1)e^{-4} + e^0 = 0.76$$

or 76 percent. Thus the electron would be found 76 percent of the time within 2 Bohr radii of the nucleus and 24 percent of the time farther away. Note that this result depends on our wave function being properly normalized, which it is, as is readily shown by letting $r \to \infty$ and integrating over all space: $\int_0^\infty |\psi|^2 \, dV = 1$.

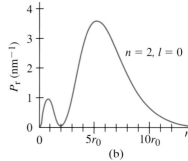

$|\psi_{200}|^2$

(a)

The first excited state in hydrogen has $n = 2$. For $l = 0$, the solution of the Schrödinger equation (Eq. 40–1) is a wave function that is again spherically symmetric:

$$\psi_{200} = \frac{1}{\sqrt{32\pi r_0^3}} \left(2 - \frac{r}{r_0} \right) e^{-\frac{r}{2r_0}}. \tag{40–8}$$

Figure 40–8a shows the probability distribution $|\psi_{200}|^2$ and Fig. 40–8b shows[†] a plot of the radial probability distribution

$$P_r = \frac{1}{8} \frac{r^2}{r_0^3} \left(2 - \frac{r}{r_0} \right)^2 e^{-\frac{r}{r_0}}.$$

There are two peaks in this curve; the second, at $r \approx 5r_0$, is higher and corresponds to the most probable value for r in the $n = 2$, $l = 0$ state. We see that the

FIGURE 40–8 (a) The probability distribution $|\psi_{200}|^2$ and (b) the radial probability distribution P_r for the $n = 2$, $l = 0$ state in hydrogen.

[†] Just as for a particle in a deep square well potential (see Figs. 39–9 and 39–10), the higher the energy, the more nodes there are in ψ and $|\psi|^2$ also for the H atom.

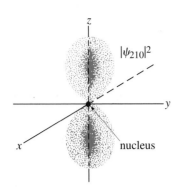

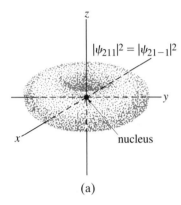

(a)

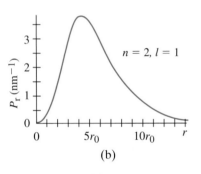

(b)

FIGURE 40–9 (a) The probability distribution for the three states with $n = 2$, $l = 1$. (b) Radial probability distribution for the sum of the three states with $n = 2$, $l = 1$, and $m_l = +1$, 0, or -1.

electron tends to be somewhat farther from the nucleus in the $n = 2$, $l = 0$ state than in the $n = 1$, $l = 0$ state.

For the state with $n = 2$, $l = 1$, there are three possible wave functions, corresponding to $m_l = +1$, 0, or -1:

$$\psi_{210} = \frac{z}{\sqrt{32\pi r_0^5}} e^{-\frac{r}{2r_0}}$$

$$\psi_{211} = \frac{x + iy}{\sqrt{64\pi r_0^5}} e^{-\frac{r}{2r_0}} \qquad \textbf{(40–9)}$$

$$\psi_{21-1} = \frac{x - iy}{\sqrt{64\pi r_0^5}} e^{-\frac{r}{2r_0}}.$$

where i is the imaginary number $i = \sqrt{-1}$. These wave functions are *not* spherically symmetric. The probability distributions, $|\psi|^2$, are shown in Fig. 40–9a, where we can see their directional orientation.

You may wonder how such non-spherically symmetric wave functions arise when the potential energy in the Schrödinger equation has spherical symmetry. Indeed, how could an electron select one of these states? In the absence of any external influence, such as a magnetic field in a particular direction, all three of these states are equally likely, and they all have the same energy. Thus an electron can be considered to spend one-third of its time in each of these states. The net effect, then, is the sum of these three wave functions squared:

$$\left|\psi_{210}\right|^2 + \left|\psi_{211}\right|^2 + \left|\psi_{21-1}\right|^2,$$

which is spherically symmetric, since $x^2 + y^2 + z^2 = r^2$. The radial probability distribution for this sum is shown in Fig. 40–9b.

Although the spatial distributions of the electron can be calculated for the various states, it is difficult to measure them experimentally. Indeed, most of the experimental information about atoms has come from a careful examination of the emission spectra under various conditions.

40–4 Complex Atoms; the Exclusion Principle

We have discussed the hydrogen atom in detail because it is the simplest to deal with. Now we briefly discuss more complex atoms, those that contain more than one electron, and whose energy levels can be determined experimentally from an analysis of the emission spectra. The energy levels are *not* the same as in the H atom, since the electrons interact with each other as well as with the nucleus. For atoms with more than one electron, the energy levels depend on both n and l.

Atomic number

The number of electrons in a neutral atom is called its **atomic number**, Z; Z is also the number of positive charges (protons) in the nucleus, and determines what kind of atom it is. That is, Z determines most of the properties that distinguish one atom from another.

Although modifications of the Bohr theory had been attempted in order to deal with complex atoms, the development of quantum mechanics in the years after 1925 proved far more successful. The mathematics becomes very difficult, however, since in multi-electron atoms, each electron is not only attracted to the nucleus but is repelled by the other electrons.

The simplest approach has been to treat each electron in an atom as occupying a particular state characterized by the quantum numbers n, l, m_l, and m_s. But to understand the possible arrangements of electrons in an atom, a new principle was needed. It was introduced by Wolfgang Pauli (1900–1958; Fig. 39–2) and is called the **Pauli exclusion principle**. It states:

No two electrons in an atom can occupy the same quantum state.

Pauli exclusion principle

Thus, no two electrons in an atom can have exactly the same set of the quantum numbers n, l, m_l, and m_s. The Pauli exclusion principle[†] forms the basis not only for understanding complex atoms, but also for understanding molecules and bonding, and other phenomena as well.

Let us now look at the structure of some of the simpler atoms when they are in the ground state. After hydrogen, the next simplest atom is *helium*, with two electrons. Both electrons can have $n = 1$, since one can have spin up $(m_s = +\frac{1}{2})$ and the other spin down $(m_s = -\frac{1}{2})$, thus satisfying the exclusion principle. Of course, since $n = 1$, l and m_l must be zero (Table 40–1). Thus the two electrons have the quantum numbers indicated in the table in the margin.

Lithium has three electrons, two of which can have $n = 1$. But the third cannot have $n = 1$ without violating the exclusion principle. Hence the third electron must have $n = 2$. Since it happens that the $n = 2$, $l = 0$ level has a lower energy than $n = 2$, $l = 1$, the electrons in the ground state have the quantum numbers indicated in the table in the margin. Of course, the quantum numbers of the third electron could also be, say, $(3, 1, -1, \frac{1}{2})$. But the atom in this case would be in an excited state since it would have greater energy. It would not be long before it jumped to the ground state with the emission of a photon. At room temperature, unless extra energy is supplied (as in a discharge tube), the vast majority of atoms are in the ground state because the average thermal energy $(K = \frac{3}{2}kT \approx 0.04\ \text{eV}$ at $T = 300\ \text{K})$ is much less than the energy needed to excite atoms (1 to 10 eV).

We can continue in this way to describe the quantum numbers of each electron in the ground state of larger and larger atoms. That for sodium, with its eleven electrons, is shown in the table in the margin.

Figure 40–10 shows a simple energy level diagram where occupied states are shown as up or down arrows $(m_s = +\frac{1}{2}$ or $-\frac{1}{2})$, and possible empty states are shown as small circles.

The ground-state configuration for all atoms is given in the **periodic table**, which is displayed inside the back cover of this book, and discussed in the next Section.

[†]The exclusion principle applies to identical particles whose spin quantum number is a half-integer $(\frac{1}{2}, \frac{3}{2}$, and so on), including electrons, protons, and neutrons; such particles are called **fermions** (after E. Fermi who derived a statistical theory describing them—see Section 41–6.). The exclusion principle does not apply to particles with integer spin quantum number (0, 1, 2, and so on), such as the photon and π meson, all of which are referred to as **bosons** (after S. N. Bose, who derived a statistical theory for them).

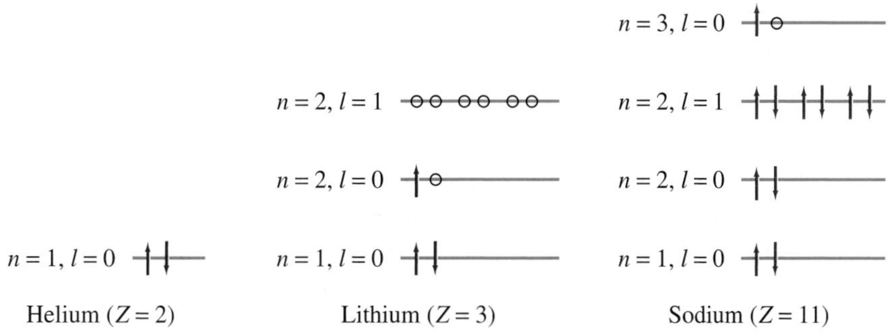

Helium, m $Z = 2$

n	l	m_l	m_s
1	0	0	$-\frac{1}{2}$
1	0	0	$\frac{1}{2}$

Lithium, $Z = 3$

n	l	m_l	m_s
1	0	0	$-\frac{1}{2}$
1	0	0	$\frac{1}{2}$
2	0	0	$-\frac{1}{2}$

Sodium, $Z = 11$

n	l	m_l	m_s
1	0	0	$-\frac{1}{2}$
1	0	0	$\frac{1}{2}$
2	0	0	$-\frac{1}{2}$
2	0	0	$\frac{1}{2}$
2	1	1	$-\frac{1}{2}$
2	1	1	$\frac{1}{2}$
2	1	0	$-\frac{1}{2}$
2	1	0	$\frac{1}{2}$
2	1	-1	$-\frac{1}{2}$
2	1	-1	$\frac{1}{2}$
3	0	0	$-\frac{1}{2}$

$n = 3, l = 0$

$n = 2, l = 1$

$n = 2, l = 0$

$n = 1, l = 0$

Helium ($Z = 2$)   Lithium ($Z = 3$)   Sodium ($Z = 11$)

FIGURE 40–10 Energy level diagram showing occupied states (arrows) and unoccupied states ($\circ$) for He, Li, and Na. Note that we have shown the $n = 2$, $l = 1$ level of Li even though it is empty.

TABLE 40–2 Values of *l*

Value of *l*	Letter Symbol	Maximum Number of Electrons in Subshell
0	s	2
1	p	6
2	d	10
3	f	14
4	g	18
5	h	22
⋮	⋮	⋮

TABLE 40–3
Electron Configuration of Some Elements

Z (No. of Electrons)	Element†	Ground State Configuration (outer electrons)
1	H	$1s^1$
2	He	$1s^2$
3	Li	$2s^1$
4	Be	$2s^2$
5	B	$2s^2 2p^1$
6	C	$2s^2 2p^2$
7	N	$2s^2 2p^3$
8	O	$2s^2 2p^4$
9	F	$2s^2 2p^5$
10	Ne	$2s^2 2p^6$
11	Na	$3s^1$
12	Mg	$3s^2$
13	Al	$3s^2 3p^1$
14	Si	$3s^2 3p^2$
15	P	$3s^2 3p^3$
16	S	$3s^2 3p^4$
17	Cl	$3s^2 3p^5$
18	Ar	$3s^2 3p^6$
19	K	$4s^1$
20	Ca	$4s^2$
21	Sc	$3d^1 4s^2$
22	Ti	$3d^2 4s^2$
23	V	$3d^3 4s^2$
24	Cr	$3d^5 4s^1$
25	Mn	$3d^5 4s^2$
26	Fe	$3d^6 4s^2$

†Names of elements can be found in Appendix D.

40–5 The Periodic Table of Elements

More than a century ago, Dmitri Mendeleev (1834–1907) arranged the then known elements into what we now call the **periodic table** of the elements. The atoms were arranged according to increasing mass, but also so that elements with similar chemical properties would fall in the same column. Today's version is shown inside the back cover. Each square contains the atomic number Z, the symbol for the element, and the atomic mass (in atomic mass units). Finally, in the lower left corner the configuration of the ground state of the atom is given. This requires some explanation. Electrons with the same value of n are referred to as being in the same **shell**. Electrons with $n = 1$ are in one shell (the K shell), those with $n = 2$ are in a second shell (the L shell), those with $n = 3$ are in the third (M) shell, and so on. Electrons with the same values of n and l are referred to as being in the same **subshell**. Letters are often used to specify the value of l as shown in Table 40–2. That is, $l = 0$ is the s subshell; $l = 1$ is the p subshell; $l = 2$ is the d subshell; beginning with $l = 3$, the letters follow the alphabet, f, g, h, i, and so on. (The first letters s, p, d, and f were originally abbreviations of "sharp," "principal," "diffuse," and "fundamental," experimental terms referring to the spectra.)

The Pauli exclusion principle limits the number of electrons possible in each shell and subshell. For any value of l, there are $2l + 1$ different m_l values (m_l can be any integer from 1 to l, from -1 to $-l$, or zero), and two different m_s values. There can be, therefore, at most $2(2l + 1)$ electrons in any l subshell. For example, for $l = 2$, five m_l values are possible $(2, 1, 0, -1, -2)$, and for each of these, m_s can be $+\frac{1}{2}$ or $-\frac{1}{2}$ for a total of $2(5) = 10$ states. Table 40–2 lists the maximum number of electrons that can occupy each subshell.

Since the energy levels depend almost entirely on the values of n and l, it is customary to specify the electron configuration simply by giving the n value and the appropriate letter for l, with the number of electrons in each subshell given as a superscript. The ground-state configuration of sodium, for example, is written as $1s^2 2s^2 2p^6 3s^1$. This is simplified in the periodic table by specifying the configuration only of the outermost electrons and any other nonfilled subshells (see Table 40–3 here, and the periodic table inside the back cover).

CONCEPTUAL EXAMPLE 40–5 Electron configurations. Which of the following electron configurations are possible, and which forbidden? (a) $1s^2 2s^2 2p^6 3s^3$; (b) $1s^2 2s^2 2p^6 3s^2 3p^5 4s^2$; (c) $1s^2 2s^2 2p^6 2d^1$.

RESPONSE (a) This is not allowed, because there are too many electrons in the s subshell in the M ($n = 3$) shell. The s subshell has $m_l = 0$, with two slots only: for "spin up" and "spin down" electrons. (b) This is allowed, but it is an excited state. One of the electrons in the 3p subshell has jumped up to the 4s subshell. Since there are 19 electrons, the element is potassium. (c) This is not allowed, because there is no d ($l = 2$) subshell in the $n = 2$ shell (Table 40–1). The outermost electron will have to be (at least) in the $n = 3$ shell.

The grouping of atoms in the periodic table is according to increasing atomic number, Z. There is also a strong regularity according to chemical properties. And although this is treated in chemistry textbooks, we discuss it here briefly because it is a result of quantum mechanics. See the periodic table on the inside back cover.

All the noble gases (in the last column of the periodic table) have completely filled shells or subshells. That is, their outermost subshell is completely full, and the electron distribution is spherically symmetric. With such full spherical symmetry, other electrons are not attracted nor are electrons readily lost (ionization energy is high). This is why the noble gases are nonreactive (more on this when we discuss molecules and bonding in Chapter 41). Column seven contains the **halogens**, which lack one electron from a filled shell. Because of the shapes of the orbits (see Section 41–1), an additional electron can be accepted from another atom, and hence these elements are quite reactive. They have a valence of −1, meaning that when an extra electron is acquired, the resulting ion has a net charge of −1e. At the left of the periodic table, column I contains the **alkali metals**, all of which have a single outer s electron. This electron spends most of its time outside the inner closed shells and subshells which shield it from most of the nuclear charge. Indeed, it is relatively far from the nucleus and is attracted to it by a net charge of only about +1e, because of the shielding effect of the other electrons. Hence this outer electron is easily removed and can spend much of its time around another atom, forming a molecule. This is why the alkali metals are highly reactive and have a valence of +1. The other columns of the periodic table can be treated similarly.

The presence of the transition elements in the center of the table, as well as the lanthanides (rare earths) and actinides below, is a result of incomplete inner shells. For the lowest Z elements, the subshells are filled in a simple order: first 1s, then 2s, followed by 2p, 3s, and 3p. You might expect that 3d ($n = 3$, $l = 2$) would be filled next, but it isn't. Instead, the 4s level actually has a slightly lower energy than the 3d (due to electrons interacting with each other), so it fills first (K and Ca). Only then does the 3d shell start to fill up, beginning with Sc. (The 4s and 3d levels are close, so some elements have only one 4s electron, such as Cr.) Most of the chemical properties of these **transition elements** are governed by the relatively loosely held 4s electrons and hence they usually have valences of +1 or +2. See also Table 40–3. A similar effect is responsible for the rare earths, which are shown at the bottom of the periodic table for convenience. All have very similar chemical properties, which are determined by their two outer 6s or 7s electrons, whereas the different numbers of electrons in the unfilled inner shells have little effect.

40–6 | X-Ray Spectra and Atomic Number

The line spectra of atoms in the visible, UV, and IR regions of the EM spectrum are mainly due to transitions between states of the outer electrons. Much of the charge of the nucleus is shielded from these electrons by the negative charge on the inner electrons. But the innermost electrons in the $n = 1$ shell "see" the full charge of the nucleus. Since the energy of a level is proportional to Z^2 (see Eq. 38–13), for an atom with $Z = 50$, we would expect wavelengths about $50^2 = 2500$ times shorter than those found in the Lyman series of hydrogen (around 100 nm), or 10^{-2} to 10^{-1} nm. Such short wavelengths lie in the X-ray region of the spectrum.

X-rays are produced when electrons accelerated by a high voltage strike the metal target inside the X-ray tube (Section 36–10). If we look at the spectrum of wavelengths emitted by an X-ray tube, we see that the spectrum consists of two parts: a continuous spectrum with a cutoff at some λ_0 which depends only on the voltage across the tube, and a series of peaks superimposed. A typical example is shown in Fig. 40–11. The smooth curve and the cutoff wavelength λ_0 move to the left as the voltage across the tube increases. The peaks (labeled K_α and K_β in Fig. 40–11), however, remain at the same wavelength when the voltage is changed, although they are located at different wavelengths when different target materials are used. This observation suggests that the peaks are characteristic of the material used. Indeed, we can explain them by imagining that the electrons accelerated by the high voltage of the tube can reach sufficient energies that when they collide with the atoms of the target, they can knock out one of the very tightly held inner electrons. Then we explain these **characteristic X-rays** (the peaks in Fig. 40–11) as photons emitted when an electron in an upper state drops down to fill the vacated lower state. The K lines result from transitions *into* the K shell ($n = 1$). The K_α line is a transition that originates from the $n = 2$ (L) shell, the K_β line from the $n = 3$ (M) shell. An L line is due to a transition into the L shell, and so on.

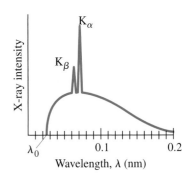

FIGURE 40–11 Spectrum of X-rays emitted from a molybdenum target in an X-ray tube operated at 50 kV.

Characteristic X-rays

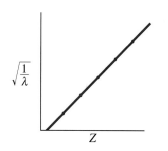

FIGURE 40–12 Plot of $\sqrt{1/\lambda}$ vs. Z for K_α X-ray lines.

Measurement of the characteristic X-ray spectra has allowed a determination of the inner energy levels of atoms. It has also allowed the determination of Z values for many atoms, since (as we have seen) the wavelength of the shortest X-rays emitted will be inversely proportional to Z^2. Actually, for an electron jumping from, say, the $n = 2$ to the $n = 1$ level, the wavelength is inversely proportional to $(Z - 1)^2$ because the nucleus is shielded by the one electron that still remains in the 1s level. In 1914, H. G. J. Moseley (1887–1915) found that a plot of $\sqrt{1/\lambda}$ versus Z produced a straight line, Fig. 40–12. The Z values of a number of elements were determined by fitting them to such a **Moseley plot**. The work of Moseley put the concept of atomic number on a firm experimental basis.

EXAMPLE 40–6 **X-ray wavelength.** Estimate the wavelength for an $n = 2$ to $n = 1$ transition in molybdenum ($Z = 42$). What is the energy of such a photon?

SOLUTION We use the Bohr formula, Eq. 38–14, with Z^2 replaced by $(Z - 1)^2 = (41)^2$. Or, more simply, we can use the result of Example 38–12 for the $n = 2$ to $n = 1$ transition in hydrogen ($Z = 1$). Since

$$\lambda \propto \frac{1}{(Z - 1)^2},$$

we will have

$$\lambda = \frac{\left(1.22 \times 10^{-7}\,\text{m}\right)}{(41)^2} = 0.073\,\text{nm}.$$

This is close to the measured value (Fig. 40–11) of 0.071 nm. Each of these photons would have energy (in eV) of:

$$E = hf = \frac{hc}{\lambda} = \frac{\left(6.63 \times 10^{-34}\,\text{J·s}\right)\left(3.00 \times 10^8\,\text{m/s}\right)}{\left(7.3 \times 10^{-11}\,\text{m}\right)\left(1.60 \times 10^{-19}\,\text{J/eV}\right)} = 17\,\text{keV}.$$

EXAMPLE 40–7 **Determining atomic number.** High-energy photons are used to bombard an unknown material. The strongest peak is found for X-rays emitted with an energy of 66 keV. Guess what the material is.

SOLUTION The strongest X-rays are generally for the K_α line (see Fig. 40–11) which occurs when photons knock out K shell electrons (the innermost orbit) and their place is taken by electrons from the L shell. We use the Bohr model, and assume the electrons "see" a nuclear charge of $Z - 1$ (screened by one electron). The hydrogen transition $n = 2$ to $n = 1$ would yield about 10.2 eV (see Fig. 38–25 or Example 38–12). Then since energy E is proportional to Z^2 (Eq. 38–13), or rather $(Z - 1)^2$ as we've just discussed, we can write

$$\frac{(Z - 1)^2}{1^2} = \frac{66 \times 10^3\,\text{eV}}{10.2\,\text{eV}} = 6.5 \times 10^3,$$

so $Z - 1 = \sqrt{6500} = 81$, and $Z = 82$, which makes it lead.

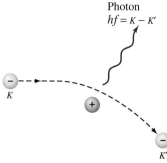

FIGURE 40–13 Bremsstrahlung photon produced by an electron decelerated by interaction with a target atom.

Now we briefly analyze the continuous part of an X-ray spectrum (Fig. 40–11) based on the photon theory of light. When electrons strike the target, they collide with atoms of the material and give up most of their energy as heat (about 99 percent, so X-ray tubes must be cooled). However, electrons can also give up energy by emitting a photon of light. An electron can be decelerated by interaction with atoms of the target (Fig. 40–13). But an accelerating charge can emit radiation (Chapter 32), and in this case it is called **bremsstrahlung** (German for "braking radiation"). Because energy is conserved, the energy of the emitted photon, hf, must equal the loss of kinetic energy K of the electron, $\Delta K = K - K'$, so

$$hf = \Delta K.$$

An electron may lose all or a part of its energy in such a collision. The continuous X-ray spectrum (Fig. 40–11) is explained as being due to such bremsstrahlung collisions in which varying amounts of energy are lost by the electrons. The shortest-wavelength X-ray (the highest frequency) must be due to an electron that gives up *all* its kinetic energy to produce one photon in a single collision. Since the initial kinetic energy of an electron is equal to the energy given it by the accelerating voltage, V, then $K = eV$. In a single collision in which the electron is brought to rest, we have

$$hf_0 = eV$$

or

$$\lambda_0 = \frac{hc}{eV}, \tag{40–10}$$

where $\lambda_0 = c/f_0$ is the cutoff wavelength, Fig. 40–11. This prediction for λ_0 corresponds precisely with that observed experimentally. This result is further evidence that X-rays are a form of electromagnetic radiation (light)[†] and that the photon theory of light is valid.

EXAMPLE 40–8 **Cutoff wavelength.** What is the shortest-wavelength X-ray photon emitted in an X-ray tube subjected to 50 kV?

SOLUTION From Eq. 40–10,

$$\lambda_0 = \frac{(6.6 \times 10^{-34}\,\text{J·s})(3.0 \times 10^8\,\text{m/s})}{(1.6 \times 10^{-19}\,\text{C})(5.0 \times 10^4\,\text{V})} = 2.5 \times 10^{-11}\,\text{m},$$

or 0.025 nm. This agrees well with experiment, Fig. 40–11.

---

* 40–7 Magnetic Dipole Moments; Total Angular Momentum

An electron orbiting the nucleus of an atom can be considered as a current loop, classically, and thus might be expected to have a **magnetic dipole moment** as discussed in Chapter 27. Indeed, in Example 27–9 we did a classical calculation of the magnetic dipole moment of the electron in the ground state of hydrogen based, essentially, on the Bohr model, and found it to give

$$\mu = IA = \tfrac{1}{2}evr.$$

Here v is the orbital velocity of the electron, and for a particle moving in a circle of radius r, its angular momentum is

$$L = mvr.$$

So we can write

$$\mu = \frac{1}{2}\frac{e}{m}L.$$

The direction of the angular momentum $\mathbf{L}$ is perpendicular to the plane of the current loop. So is the direction of the magnetic dipole moment vector $\boldsymbol{\mu}$, although in

---

[†] If X-rays were not photons but rather neutral particles with rest mass m_0, Eq. 40–10 would not hold.

the opposite direction since the electron's charge is negative. Hence we can write the vector equation

Magnetic dipole moment, H atom

$$\boldsymbol{\mu} = -\frac{1}{2}\frac{e}{m}\mathbf{L}.$$ (40–11)

This rough semiclassical derivation was based on the Bohr theory. The same result (Eq. 40–11), is obtained using quantum mechanics. Since $\mathbf{L}$ is quantized in quantum mechanics, the magnetic dipole moment, too, must be quantized.

As we saw in Section 27–5, a magnetic dipole moment in a magnetic field $\mathbf{B}$ experiences a torque, and the potential energy U of such a system depends on $\mathbf{B}$ and the orientation of $\boldsymbol{\mu}$ relative to $\mathbf{B}$ (Eq. 27–12):

$$U = -\boldsymbol{\mu} \cdot \mathbf{B}.$$

If the magnetic field $\mathbf{B}$ is in the z direction, then $U = -\mu_z B_z$ and from Eq. 40–4 ($L_z = m_l \hbar$) and Eq. 40–11, we have

$$\mu_z = -\frac{e\hbar}{2m}m_l.$$

(Be careful here not to confuse the electron mass m with the magnetic quantum number, m_l.) It is useful to define the quantity

Bohr magneton

$$\mu_{\mathrm{B}} = \frac{e\hbar}{2m}$$ (40–12)

which is called the **Bohr magneton** and has the value $\mu_{\mathrm{B}} = 9.27 \times 10^{-24}\,\mathrm{J/T}$ (joule/tesla). Then we can write

$$\mu_z = -\mu_{\mathrm{B}} m_l,$$ (40–13)

Zeeman effect

where m_l has integer values from 0 to $\pm l$ (see Table 40–1). An atom placed in a magnetic field would have its energy split into levels that differ by $\Delta U = \mu_{\mathrm{B}} B$; this is the *Zeeman effect*, and was shown in Fig. 40–4.

* Stern-Gerlach Experiment

The first evidence of this *space quantization* (Section 40–2) came in 1922 in a famous experiment known as the **Stern-Gerlach experiment**. Silver atoms (and later others) were heated in an oven from which they escaped as shown in Fig. 40–14. The atoms were made to pass through a collimator, which eliminated all but a narrow beam. The beam then passed into a *nonhomogeneous* magnetic field. The field was deliberately made nonhomogeneous so that it would exert a force on atomic magnetic moments: remember that the potential energy (in this case $-\boldsymbol{\mu} \cdot \mathbf{B}$) must change in space if there is to be a force ($F_x = -dU/dx$, etc., Section 8–2). If $\mathbf{B}$ has a gradient along the z axis, as in Fig. 40–14, then the force is along z:

$$F_z = -\frac{dU}{dz} = \mu_z \frac{dB_z}{dz}.$$

FIGURE 40–14 The Stern-Gerlach experiment, which is done inside a vacuum chamber.

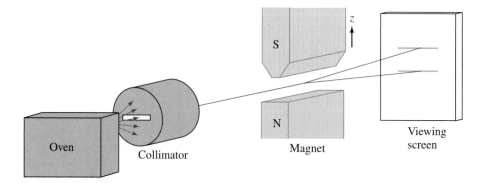

Thus the silver atoms would be deflected up or down depending on the value of μ for each atom. Classically, we would expect to see a continuous distribution on the viewing screen, since we would expect the atoms to have randomly oriented magnetic moments. But Stern and Gerlach saw instead two distinct lines for silver (and for other atoms sometimes more than two lines). These observations were the first evidence for space quantization, though not fully explained until a few years later. If the lines were due to orbital angular momentum, there should have been an odd number of them, corresponding to the possible values of m_l (since $\mu_z = -\mu_B m_l$). For $l = 0$, there is only one possibility, $m_l = 0$. For $l = 1$, m_l can be 1, 0, or -1, and we would expect three lines, and so on. Why there are only two lines was eventually explained by the concept of electron spin. With a spin of $\frac{1}{2}$, the electron spin can have only two orientations in space, as we saw in Fig. 40–5. Hence a magnetic dipole moment associated with spin would have only two positions. Thus, the two states for silver seen in the Stern-Gerlach experiment must be due to the spin of its one valence electron. Silver atoms must thus have zero orbital angular momentum but a total spin of $\frac{1}{2}$ due to this one valence electron (of its 47 electrons, the spins of the first 46 cancel). Two lines were found also for the H atom in its ground state, again due to the spin $\frac{1}{2}$ of its electron since the orbital angular momentum is zero.

The Stern-Gerlach deflection is proportional to the magnetic dipole moment, μ_z, and for a spin $\frac{1}{2}$ particle we expect $\mu_z = -\mu_B m_s = -(\frac{1}{2})(e\hbar/2m)$ as for the case of orbital angular momentum, Eq. 40–13. Instead, μ_z for spin was found to be about twice as large:

$$\mu_z = -g\mu_B m_s, \qquad\qquad\qquad (40\text{–}14)$$

Magnetic dipole moment for spin

where g, called the **g-factor** or **gyromagnetic ratio**, has been measured to be slightly larger than 2: $g = 2.0023\cdots$ for a free electron. This unexpected factor of (about) 2 clearly indicates that spin cannot be viewed as a classical angular momentum. It is a purely quantum-mechanical effect. Equation 40–14 is the same as Eq. 40–13 for orbital angular momentum with m_s replacing m_l. But for the orbital case, $g = 1$.

*Total Angular Momentum J

An atom can have both orbital and spin angular momenta. For example, in the 2p state of hydrogen $l = 1$ and $s = \frac{1}{2}$. In the 4d state, $l = 2$ and $s = \frac{1}{2}$. The **total angular momentum** is the vector sum of the orbital angular momentum **L** and the spin **S**:

$$\mathbf{J} = \mathbf{L} + \mathbf{S}.$$

Total angular momentum

According to quantum mechanics, the magnitude of the total angular momentum **J** is quantized:

$$J = \sqrt{j(j + 1)}\,\hbar. \qquad\qquad\qquad (40\text{–}15)$$

For the single electron in the H atom, quantum mechanics gives the result that j can be

$$j = l + s = l + \tfrac{1}{2}$$

or

$$j = l - s = l - \tfrac{1}{2}$$

but never less than zero, just as for l and s. For the 1s state, $l = 0$ and $j = \frac{1}{2}$ is the only possibility. For p states, say the 2p state, $l = 1$ and j can be either $\frac{3}{2}$ or $\frac{1}{2}$. The z component for j is quantized in the usual way:

$$m_j = j, j - 1, \cdots, -j.$$

For a 2p state with $j = \frac{1}{2}$, m_j can be $\frac{1}{2}$ or $-\frac{1}{2}$; for $j = \frac{3}{2}$, m_j can be $\frac{3}{2}, \frac{1}{2}, -\frac{1}{2}, -\frac{3}{2}$, for a total of four states. Note that the state of a single electron can be specified by giving n, l, m_l, m_s, or by giving n, j, l, m_j (only one of these descriptions at a time).

The interaction of magnetic fields with atoms, as in the Zeeman effect and the Stern-Gerlach experiment, involves the *total* angular momentum. Thus the Stern-Gerlach experiment on H atoms in the ground state shows two lines (for $m_j = +\frac{1}{2}$ and $-\frac{1}{2}$), but for the first excited state it shows four lines corresponding to the four possible m_j values $(\frac{3}{2}, \frac{1}{2}, -\frac{1}{2}, -\frac{3}{2})$.

* Spectroscopic Notation

We can specify the state of an atom, including the total angular momentum quantum number j, using the following **spectroscopic notation**. For a single electron state we can write

$$nL_j,$$

where the value of L is specified using the same letters as in Table 40–2, but in upper case:

$$L = 0 \quad 1 \quad 2 \quad 3 \quad 4 \cdots$$

$$\text{letter} = S \quad P \quad D \quad F \quad G \cdots.$$

So the $2P_{3/2}$ state has $n = 2$, $l = 1$, $j = \frac{3}{2}$, whereas $1S_{1/2}$ specifies the ground state in hydrogen.

* Fine Structure; Spin-Orbit Interaction

It is also a magnetic effect that produces the *fine structure* splitting mentioned in Section 40–2, which occurs in the absence of any external field. Instead, it is due to a magnetic field produced by the atom itself. We can see how it occurs by putting ourselves in the reference frame of the electron, in which case we see the nucleus revolving about us as a moving charge or electric current that produces a magnetic field, B_n. The electron has an intrinsic magnetic dipole moment μ_s (Eq. 40–14) and hence its energy will be altered by an amount (Eq. 27–12)

$$\Delta U = -\boldsymbol{\mu}_s \cdot \mathbf{B}_n.$$

Since μ_s takes on quantized values according to the values of m_s, the energy of a single electron state will split into two closely spaced energy levels (for $m_s = \frac{1}{2}$ and $-\frac{1}{2}$). This tiny splitting of energy levels produces a tiny splitting in spectral lines. For example, in the H atom, the $2P \to 1S$ transition is split into two lines corresponding to $2P_{1/2} \to 1S_{1/2}$ and $2P_{3/2} \to 1S_{1/2}$. The difference in energy between these two is only about 5×10^{-5} eV, which is very small compared to the $2P \to 1S$ transition energy of 13.6 eV $- 3.4$ eV $= 10.2$ eV.

The magnetic field $\mathbf{B}_n$ produced by the orbital motion is proportional to the orbital angular momentum $\mathbf{L}$, and since $\boldsymbol{\mu}_s$ is proportional to the spin $\mathbf{S}$, then $\Delta U = -\boldsymbol{\mu}_s \cdot \mathbf{B}_n$ can be written

Spin-orbit interaction

$$\Delta U \propto \mathbf{L} \cdot \mathbf{S}.$$

This interaction, which produces the fine structure, is thus called the **spin-orbit interaction**. Its magnitude is related to a dimensionless constant known as the **fine structure constant**,

Fine-structure constant

$$\alpha = \frac{e^2}{2\epsilon_0 hc} \approx \frac{1}{137},$$

which also appears elsewhere in atomic physics.

*40–8 Fluorescence and Phosphorescence

When an atom is excited from one energy state to a higher one by the absorption of a photon, it may return to the lower level in a series of two (or more) jumps if there is an energy level in between (Fig. 40–15). The photons emitted will consequently have lower energy and frequency than the absorbed photon. When the absorbed photon is in the UV and the emitted photons are in the visible region of the spectrum, this phenomenon is called **fluorescence** (Fig. 40–16).

The wavelength for which fluorescence will occur depends on the energy levels of the particular atoms. Because the frequencies are different for different substances, and because many substances fluoresce readily, fluorescence is a powerful tool for identification of compounds. It is also used for assaying—determining how much of a substance is present—and for following substances along a natural pathway as in plants and animals. For detection of a given compound, the stimulating light must be monochromatic, and solvents or other materials present must not fluoresce in the same region of the spectrum. Sometimes the observation of fluorescent light being emitted is sufficient to detect a compound. In other cases, spectrometers are used to measure the wavelengths and intensities of the emitted light.

Fluorescent lightbulbs work in a two-step process. The applied voltage accelerates electrons that strike atoms of the gas in the tube and cause them to be excited. When the excited atoms jump down to their normal levels, they emit UV photons which strike a fluorescent coating on the inside of the tube. The light we see is a result of this material fluorescing in response to the UV light striking it.

Materials such as those used for luminous watch dials are said to be **phosphorescent**. When an atom is raised to a normal excited state, it drops back down within about 10^{-8} s. In phosphorescent substances, atoms can be excited by photon absorption to energy levels, said to be **metastable**, which are states that last much longer—even a few seconds or longer. In a collection of such atoms, many of the atoms will descend to the lower state fairly soon, but many will remain in the excited state for over an hour. Hence light will be emitted even after long periods. When you put your watch dial close to a bright lamp, it excites many atoms to metastable states, and you can see the glow a long time after.

*40–9 Lasers

A laser is a device that can produce a very narrow intense beam of monochromatic coherent light. (By *coherent*, we mean that across any cross section of the beam, all parts have the same phase.) The emitted beam is a nearly perfect plane wave. An ordinary light source, on the other hand, emits light in all directions (so the intensity decreases rapidly with distance), and the emitted light is incoherent (the different parts of the beam are not in phase with each other). The excited atoms that emit the light in an ordinary lightbulb act independently, so each photon emitted can be considered as a short wave train that lasts about 10^{-8} s. These wave trains bear no phase relation to one another. Just the opposite is true of lasers.

The action of a laser is based on quantum theory. We have seen that a photon can be absorbed by an atom if (and only if) its energy hf corresponds to the energy difference between an occupied energy level of the atom and an available excited state, Fig. 40–17a. This is, in a sense, a resonant situation. If the atom is already in the excited state, it may of course jump spontaneously (i.e., no apparent stimulus) to the lower state with the emission of a photon. However, if a photon with this same energy strikes the excited atom, it can stimulate the atom to make the transition sooner to the lower state, Fig. 40–17b. This phenomenon is called **stimulated emission**, and it can be seen that not only do we still have the original photon, but also a second one of the same frequency as a result of the atom's transition. And these two photons are

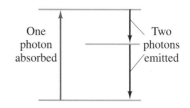
FIGURE 40–15 Fluorescence.

FIGURE 40–16 When UV light illuminates these various "fluorescent" rocks, they fluoresce in the visible region of the spectrum.

FIGURE 40–17 (a) Absorption of a photon. (b) Stimulated emission.

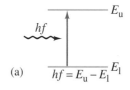

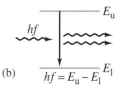

Stimulated emission

exactly *in phase*, and they are moving in the same direction. This is how coherent light is produced in a laser. Hence the name "laser," which is an acronym for **l**ight **a**mplification by **s**timulated **e**mission of **r**adiation.

The natural population of atoms in thermal equilibrium at any temperature T (in K) is given by the **Boltzmann distribution** (or **Boltzmann factor**):

$$N_n = Ce^{-\frac{E_n}{kT}}, \qquad \text{(40–16a)}$$

where N_n is the number of atoms in the state with energy E_n. For two states n and n', the ratio of the number of atoms in the two states is

$$\frac{N_n}{N_{n'}} = e^{-\left(\frac{E_n - E_{n'}}{kT}\right)} \qquad \text{(40–16b)}$$

Thus most atoms are in the ground state unless the temperature is very high. In the two-level system of Fig. 40–17, most atoms are normally in the lower state, so the majority of incident photons will be absorbed. In order to obtain the coherent light from stimulated emission, two conditions must be satisfied. First, atoms must be excited to the higher state, so that an **inverted population** is produced, one in which more atoms are in the upper state than in the lower one (Fig. 40–18). Then *emission* of photons will dominate over absorption. Hence the system will not be in thermal equilibrium. And second, the higher state must be a **metastable state**— a state in which the electrons remain longer than usual[†] so that the transition to the lower state occurs by stimulated emission rather than spontaneously. How these conditions are achieved for different lasers will be discussed shortly. For now, we assume that the atoms have been excited to an upper state. Figure 40–19 is a schematic diagram of a laser: the "lasing" material is placed in a long narrow tube at the ends of which are two mirrors, one of which is partially transparent (perhaps 1 or 2 percent). Some of the excited atoms drop down fairly soon after being excited. One of these is the atom shown on the far left in Fig. 40–19. If the emitted photon strikes another atom in the excited state, it stimulates this atom to emit a photon of the *same* frequency, moving in the *same* direction, and *in phase* with it. These two photons then move on to strike other atoms causing more stimulated emission. As the process continues, the number of photons multiplies. When the photons strike the end mirrors, most are reflected back, and as they move in the opposite direction, they continue to stimulate more atoms to emit photons. As the photons move back and forth between the mirrors, a small percentage passes through the partially transparent mirror at one end. These photons make up the narrow coherent external laser beam.

Inside the tube, some spontaneously emitted photons will be emitted at an angle to the axis, and these will merely go out the side of the tube and not affect the narrowness of the main beam. In a well-designed laser, the spreading of the beam is limited only by diffraction, so the angular spread is $\approx \lambda/a$ (see Eq. 36–1) where a is the diameter of the end mirror. The diffraction spreading can be incred-

[†] An excited atom may land in such a state and can jump to a lower state only by a so-called forbidden transition (discussed in Section 40–2), which is why its lifetime is longer than normal.

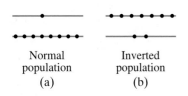

FIGURE 40–18 Two energy levels for a collection of atoms. Each dot represents the energy state of one atom. (a) A normal situation; (b) an inverted population.

FIGURE 40–19 Schematic laser diagram, showing excited atoms stimulated to emit light.

ibly small. The light energy, instead of spreading out in space as it does for an ordinary light source, is directed in a pencil-thin beam.

The excitation of the atoms in a laser can be done in several ways to produce the necessary inverted population. In a ruby laser, the lasing material is a ruby rod consisting of Al_2O_3 with a small percentage of aluminum (Al) atoms replaced by chromium (Cr) atoms. The Cr atoms are the ones involved in lasing. The atoms are excited by strong flashes of light of wavelength 550 nm, which corresponds to a photon energy of 2.2 eV. As shown in Fig. 40–20, the atoms are excited from state E_0 to state E_2. This process is called **optical pumping**. The atoms quickly decay either back to E_0 or to the intermediate state E_1, which is metastable with a lifetime of about 3×10^{-3} s (compared to 10^{-8} s for ordinary levels). With strong pumping action, more atoms can be forced into the E_1 state than are in the E_0 state. Thus we have the inverted population needed for lasing. As soon as a few atoms in the E_1 state jump down to E_0, they emit photons that produce stimulated emission of the other atoms and the lasing action begins. A ruby laser thus emits a beam whose photons have energy 1.8 eV and a wavelength of 694.3 nm (or "ruby-red" light).

In a helium–neon (He–Ne) laser, the lasing material is a gas, a mixture of about 15 percent He and 85 percent Ne. The atoms are excited by applying a high voltage to the tube so that an electric discharge takes place within the gas. In the process, some of the He atoms are raised to the metastable state E_1 shown in Fig. 40–21, which corresponds to a jump of 20.61 eV, almost exactly equal to an excited state in neon, 20.66 eV. The He atoms do not quickly return to the ground state by spontaneous emission, but instead often give their excess energy to a Ne atom when they collide—see Fig. 40–21. In such a collision, the He drops to the ground state and the Ne atom is excited to the state E_3' (the prime refers to neon states). The slight difference in energy (0.05 eV) is supplied by the kinetic energy of the moving molecules ($\frac{3}{2}kT \approx 0.04$ eV at $T = 300$ K). In this manner, the E_3' state in Ne—which is metastable—becomes more populated than the E_2' level. This inverted population between E_3' and E_2' is what is needed for lasing.

Other types of laser include: chemical lasers, in which the energy input comes from the chemical reaction of highly reactive gases; dye lasers, whose frequency is tunable; CO_2 gas lasers, capable of high power output in the infrared; rare-earth solid-state lasers such as the high-power Nd:Yag laser; and the *pn* junction laser in which the transitions occur between the bottom of the conduction band and the upper part of the valence band (Section 41–7).

The excitation of the atoms in a laser can be done continuously or in pulses. In a **pulsed laser**, the atoms are excited by periodic inputs of energy. The multiplication of photons continues until all the atoms have been stimulated to jump down

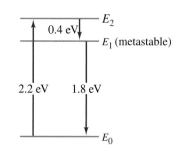

FIGURE 40–20 Energy levels of chromium in a ruby crystal. Photons of energy 2.2 eV "pump" atoms from E_0 to E_2, followed by decay to the metastable state E_1. Lasing action occurs by stimulated emission of photons in transition from E_1 to E_0.

He–Ne laser

Other lasers

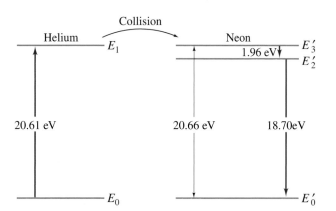

FIGURE 40–21 Energy levels for He and Ne. He is excited in the electric discharge to the E_1 state. This energy is transferred to the E_3' level of the Ne by collision. E_3' is metastable and decays to E_2' by stimulated emission, producing the output beam of the laser.

to the lower state, and the process is repeated with each input pulse. In a **continuous laser**, the energy input is continuous so that as atoms are stimulated to jump down to the lower level, they are soon excited back up to the upper level so that the output is a continuous laser beam. Any laser, of course, is not a source of energy. Energy must be put in, and the laser converts a part of this input energy into an intense narrow beam output.

The unique feature of light from a laser is, as mentioned before, that it is a coherent narrow beam of a single frequency (or several distinct frequencies). Because of this feature, the laser has found many applications. Lasers are a useful surgical tool. The narrow intense beam can be used to destroy tissue in a localized area, or to break up gallstones and kidney stones. Because of the heat produced, a laser beam can be used to "weld" broken tissue, such as a detached retina (Fig. 40–22). For some types of internal surgery, the laser beam can be carried by an optical fiber (Section 33–7) to the surgical point. An example is the removal of plaque clogging human arteries. Tiny organelles within a living cell have been destroyed using lasers by researchers studying how the absence of that organelle affects the behavior of the cell. Laser beams have been used to destroy cancerous and precancerous cells; at the same time, the heat seals off capillaries and lymph vessels, thus "cauterizing" the wound in the process to prevent spread of the disease. The intense heat produced in a small area by a laser beam is also used for welding and machining metals and for drilling tiny holes in hard materials. The beam of a laser is narrow in itself (typically, a few mm). But because the beam is coherent, monochromatic, and essentially parallel and narrow, lenses can be used to focus the light into incredibly small areas without the usual aberration problems. The limiting factor thus becomes diffraction, and the energy crossing unit area per unit time can be very large. The precise straightness of a laser beam is also useful to surveyors for lining up equipment precisely, especially in inaccessible places.

In everyday life, lasers are used as bar-code readers (at store checkout stands) and in compact disc (CD) players. The laser beam reflects off the stripes and spaces of a bar code, and off the tiny pits of a CD as shown in Fig. 40–23. The recorded information on a CD is a series of pits and spaces representing 0s and 1s (or "off" and "on") of a digitized code that is decoded electronically before being sent to the audio or video system. A bar-code reader is similar.

→ **PHYSICS APPLIED**

Medical and other uses of lasers

FIGURE 40–22 Laser being used in eye surgery.

→ **PHYSICS APPLIED**

CD players and bar codes

FIGURE 40–23 Reading a CD. The fine beam of a laser, focused even more finely with lenses, is directed at the undersurface of a rotating compact disc. The beam is reflected back from the areas between pits but reflects much less from pits. The reflected light is detected as shown, reflected by a half-reflecting mirror MS. The strong and weak reflections correspond to the 0s and 1s of the binary code representing the audio or video signal.

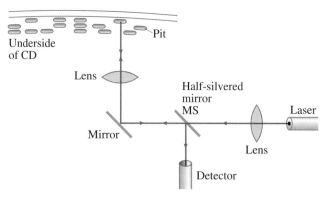

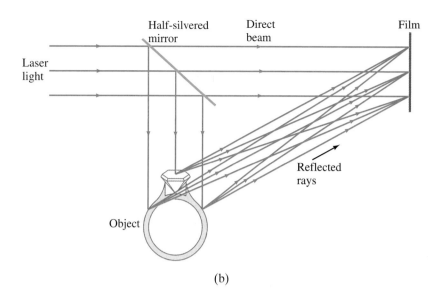

(a)   (b)

FIGURE 40–24 (a) Photo of a hologram of coins. (b) Making a hologram. Light reflected from various points on the object interferes (at the film) with light from the direct beam.

* 40–10 | Holography

One of the most interesting applications of laser light is the production of three-dimensional images called **holograms** (see Fig. 40–24). In an ordinary photograph, the film simply records the intensity of light reaching it at each point. When the photograph or transparency is viewed, light reflecting from it or passing through it gives us a two-dimensional picture. In holography, three-dimensional images are formed by interference, without lenses. When a laser hologram is made on film, a broadened laser beam is split into two parts by a half-silvered mirror, Fig. 40–24b. One part goes directly to the film; the rest passes to the object to be photographed, from which it is reflected to the film. Light from every point on the object reaches each point on the film, and the interference of the two beams allows the film to record both the intensity and relative phase of the light at each point. It is crucial that the light be coherent—that is, in phase at all points—which is why a laser is used. After the film is developed, it is placed again in a laser beam and a three-dimensional image of the object is created. You can walk around such an image and see it from different sides as if it were the original object. Yet, if you try to touch it with your hand, there will be nothing material there.

The details of how the image is formed are quite complicated. But we can get the basic idea by considering one single point on the object. In Fig. 40–25a the rays OA and OB have reflected from one point on our object. The rays CA and DB come directly from the source and interfere with OA and OB at points A and B on the film. A set of interference fringes is produced as shown in Fig. 40–25b. The spacing between the fringes changes from top to bottom as shown. Why this

FIGURE 40–25 Light from point O on the object interferes with light of the direct beam (rays CA and DB).

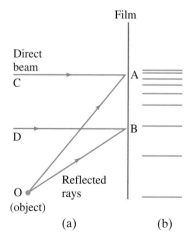

(a)   (b)

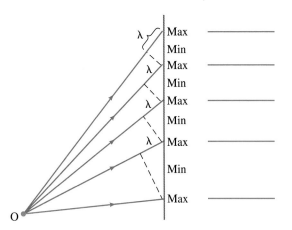

FIGURE 40–26 Each of the rays shown leaving point O is one wavelength shorter than the one above it. If the top ray is in phase with the direct beam (not shown), which has the same phase at all points on the screen, all the rays shown produce constructive interference. From this diagram it can be seen that the fringe spacing increases toward the bottom.

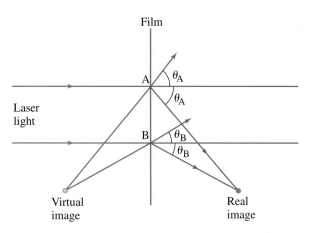

FIGURE 40–27 Reconstructing the image of one point on the object. Laser beam strikes film that is like a diffraction grating of variable spacing. Rays corresponding to the first diffraction maxima are shown emerging. The angle $\theta_A > \theta_B$ because the spacing at B is greater than at A ($\sin\theta = \lambda/d$). Hence real and virtual images of the point are reproduced as shown.

happens is explained in Fig. 40–26. Thus the hologram of a single point object would have the pattern shown in Fig. 40–25b. The film in this case looks like a diffraction grating with variable spacing. Hence, when coherent laser light is passed back through the developed film to reconstruct the image, the diffracted rays in the first order maxima occur at slightly different angles because the spacing changes. (Remember Eq. 36–13, $\sin\theta = \lambda/d$: where the spacing d is greater, the angle θ is less.) Hence, the rays diffracted upward (in first order) seem to diverge from a single point, Fig. 40–27. This is a virtual image of the original object, which can be seen with the eye. Rays diffracted in first order *downward* converge to make a real image, which can be seen and also photographed. (Note that the straight-through undiffracted rays are of no interest.) Of course real objects consist of many points, so a hologram will be a complex interference pattern which, when laser light is incident on it, will reproduce an image of the object. Each image point will be at the correct (three-dimensional) position with respect to other points, so the image accurately represents the original object. And it can be viewed from different angles as if viewing the original object. Holograms can be made in which a viewer can walk entirely around the image (360°) and see all sides of it.

White-light holograms   So-called **volume** or **white-light holograms** do not require a laser to see the image, but can be viewed with ordinary white light (preferably a nearby point source, such as the Sun or a clear bulb with a small bright filament). Such holograms must be made, however, with a laser. They are made not on thin film, but on a *thick* emulsion. The interference pattern, instead of being two-dimensional as for an ordinary hologram, is actually three-dimensional (hence the name "volume hologram"). The interference pattern in the film emulsion can be thought of as an array of bands or ribbons (consisting of the silver grains from the development process) where constructive interference occurred. This array, and the reconstruction of the image, can be compared to Bragg scattering of X-rays from the atoms in a crystal (see Section 36–10). White light can reconstruct the image because the Bragg condition (Eq. 36–20) selects out the appropriate single wavelength. If the hologram is originally produced by lasers emitting the three additive primary colors (red, green, and blue), the three-dimensional image can be seen in full color when viewed with white light.

Summary

In the quantum mechanical view of the atom, the electrons do not have well-defined orbits, but instead exist as a "cloud." Electron clouds can be interpreted as an electron wave spread out in space, or as a **probability distribution** for electrons considered as particles.

For the simplest atom, hydrogen, the Schrödinger equation contains the potential energy $U = -(1/4\pi\epsilon_0)(e^2/r)$. The solutions give the same values of energy as the old Bohr theory.

According to quantum mechanics, the state of an electron in an atom is specified by four **quantum numbers**: n, l, m_l, and m_s:

> the principal quantum number, n, can take on any integer value $(1, 2, 3, \cdots)$ and corresponds to the quantum number of the old Bohr theory;
>
> l can take on values from 0 up to $n - 1$;
>
> m_l can take on integer values from $-l$ to $+l$;
>
> m_s can be $+\frac{1}{2}$ or $-\frac{1}{2}$.

The energy levels in the hydrogen atom depend on n, whereas in other atoms they depend on n and l.

When an external magnetic field is applied, the spectral lines are split (the **Zeeman effect**), indicating that the energy depends also on m_l in this case.

Even in the absence of a magnetic field, precise measurements of spectral lines show a tiny splitting of the lines called **fine structure**, whose explanation is that the energy depends very slightly on m_l and m_s.

Transitions between states that obey the selection rule $\Delta l = \pm 1$ are far more probable than other so-called "forbidden" transitions.

The ground-state wave function in hydrogen has spherical symmetry, as do other $l = 0$ states. States with $l > 0$ have some directionality in space.

The probability density, $|\psi|^2$, and the *radial probability density*, $P_r = 4\pi r^2|\psi|^2$, are both useful to illustrate the spatial extent of the electron cloud.

The arrangement of electrons in multi-electron atoms is governed by the **Pauli exclusion principle**, which states that no two electrons can occupy the same quantum state—that is, they cannot have the same set of quantum numbers n, l, m_l, and m_s.

As a result, electrons in multi-electron atoms are grouped into **shells** (according to the value of n) and **subshells** (according to l).

Electron configurations are specified using the numerical values of n, and using letters for l: s, p, d, f, etc., for $l = 0, 1, 2, 3,$ and so on, plus a superscript for the number of electrons in that subshell. Thus, the ground state of hydrogen is $1s^1$, whereas that for oxygen is $1s^22s^22p^4$. The **periodic table** arranges the elements in horizontal rows according to increasing atomic number (number of electrons in the neutral atom). The shell structure gives rise to a periodicity in the properties of the elements, so that each vertical column can contain elements with similar chemical properties.

X-rays, which are a form of electromagnetic radiation of very short wavelength, are produced when high-speed electrons strike a target. The spectrum of X-rays so produced consists of two parts, a continuous spectrum produced when the electrons are decelerated by atoms of the target, and peaks representing photons emitted by atoms of the target after being excited by collision with the high-speed electrons. Measurement of these peaks allows determination of inner energy levels of atoms and determination of Z.

Questions

1. Discuss the differences between Bohr's view of the atom and the quantum-mechanical view.

2. The probability density $|\psi|^2$ is a maximum at the center of the H atom $(r = 0)$ for the ground state, whereas the radial probability density $P_r = 4\pi r^2|\psi|^2$ is zero at this point. Explain why.

3. Why do three quantum numbers come out of the Schrödinger theory (rather than, say, two or four)?

4. In Fig. 40–4, why do the upper and lower levels have different energy splittings in a magnetic field?

5. Which model of the hydrogen atom, the Bohr model or the quantum-mechanical model, predicts that the electron spends more time near the nucleus?

6. The size of atoms varies by only a factor of three or so from largest to smallest, yet the number of electrons varies from one to over 100. Why?

7. Excited hydrogen and excited helium atoms both radiate light as they jump down to the $n = 1$, $l = 0$, $m_l = 0$ state. Yet the two elements have very different emission spectra. Why?

8. The 589-nm yellow line in sodium is actually two very closely spaced lines. This splitting is due to an "internal" Zeeman effect. Can you explain this? [*Hint*: Put yourself in the reference frame of the electron.]

9. Which of the following electron configurations are forbidden? (*a*) $1s^22s^22p^43s^24p^2$; (*b*) $1s^22s^22p^83s^1$, (*c*) $1s^22s^22p^63s^23p^54s^24d^54f^1$.

10. Give the complete electron configuration for a uranium atom (careful scrutiny across the periodic table on the inside back cover will provide useful hints).

11. In what column of the periodic table would you expect to find the atom with each of the following configurations? (a) $1s^2 2s^2 2p^6 3s^2$; (b) $1s^2 2s^2 2p^6 3s^2 3p^6$; (c) $1s^2 2s^2 2p^6 3s^2 3p^6 4s^1$; (d) $1s^2 2s^2 2p^5$.

12. On what factors does the periodicity of the periodic table depend? Consider the exclusion principle, quantization of angular momentum, spin, and any others you can think of.

13. How would the periodic table look if there were no electron spin but otherwise quantum mechanics were valid? Consider the first 20 elements or so.

14. The ionization energy for neon ($Z = 10$) is 21.6 eV and that for sodium ($Z = 11$) is 5.1 eV. Explain the large difference.

15. Why do chlorine and iodine exhibit similar properties?

16. Explain why potassium and sodium exhibit similar properties.

17. Why are the chemical properties of the rare earths so similar?

18. Why do we not expect perfect agreement between measured values of X-ray line wavelengths and those calculated using Bohr theory, as in Example 40–6?

19. Why does the Bohr theory, which does not work at all well for normal transitions involving the outer electrons for He and more complex atoms, nevertheless predict reasonably well the atomic X-ray spectra for transitions deep inside the atom?

20. Why does the cutoff wavelength in Fig. 40–11 imply a photon nature for light?

21. How would you figure out which lines in an X-ray spectrum correspond to K_α, K_β, L, etc., transitions?

22. Why do the characteristic X-ray spectra vary in a systematic way with Z, whereas the visible spectra (Fig. 36–21) do not?

23. Why do we expect electron transitions deep within an atom to produce shorter wavelengths than transitions by outer electrons?

*24. Why is the direction of the magnetic dipole moment of an electron opposite to that of its orbital angular momentum?

*25. Why is a nonhomogeneous field used in the Stern-Gerlach experiment?

*26. Compare spontaneous emission to stimulated emission.

*27. Does the intensity of light from a laser fall off as the inverse square of the distance?

*28. How does laser light differ from ordinary light? How is it the same?

*29. Explain how a 0.0005-W laser beam, photographed at a distance, can seem much stronger than a 1000-W street lamp.

Problems

Section 40–2

1. (I) For $n = 6$, what values can l have?

2. (I) For $n = 5$, $l = 3$, what are the possible values of m_l and m_s?

3. (I) How many different states are possible for an electron whose principal quantum number is $n = 4$? Write down the quantum numbers for each state.

4. (I) If a hydrogen atom has $m_l = -3$, what are the possible values of n, l, m_s?

5. (I) A hydrogen atom is known to have $l = 4$. What are the possible values for n, m_l, and m_s?

6. (I) Calculate the magnitude of the angular momentum of an electron in the $n = 4$, $l = 2$ state of hydrogen.

7. (II) A hydrogen atom is in the 6g state. Determine (a) the principal quantum number, (b) the energy of the state, (c) the orbital angular momentum and its quantum number l, and (d) the possible values for the magnetic quantum number.

8. (II) (a) Show that the number of different states possible for a given value of l is equal to $2(2l + 1)$. (b) What is this number for $l = 0, 1, 2, 3, 4, 5$, and 6?

9. (II) Show that the number of different electron states possible for a given value of n is $2n^2$. (See Problem 8.)

10. (II) An excited H atom is in a 6d state. (a) Name all the states (n, l) to which the atom is "allowed" to jump with the emission of a photon. (b) How many different wavelengths are there (ignoring fine structure)?

Section 40–3

11. (I) Show that the ground-state wave function, Eq. 40–5, is normalized. [Hint: See Example 40–4.]

12. (II) For the ground state of hydrogen, what is the value of (a) ψ, (b) $|\psi|^2$, and (c) P_r at $r = r_0$?

13. (II) For the $n = 2$, $l = 0$ state of hydrogen, what is the value of (a) ψ, (b) $|\psi|^2$, and (c) P_r at $r = 5r_0$?

14. (II) Show that ψ_{200} as given by Eq. 40–8 is normalized.

15. (II) By what factor is it more likely to find the electron in the ground state of hydrogen at the Bohr radius (r_0) than at twice the Bohr radius $(2r_0)$?

16. (II) For the ground state of hydrogen, what is the probability of finding the electron within a spherical shell of inner radius $0.99r_0$ and outer radius $1.01r_0$?

17. (II) For the $n = 2$, $l = 0$ state of hydrogen, what is the probability of finding the electron within a spherical shell of inner radius $4.00r_0$ and outer radius $5.00r_0$?

18. (II) (a) Show that the probability of finding the electron in the ground state of hydrogen at less than one Bohr radius from the nucleus is 32 percent. (b) What is the probability of finding a 1s electron between $r = r_0$ and $r = 2r_0$?

19. (II) Determine the radius r of a sphere centered on the nucleus within which the probability of finding the electron for the ground state of hydrogen is (a) 50 percent, (b) 90 percent, (c) 99 percent.

20. (II) (a) Estimate the probability of finding an electron, in the ground state of hydrogen, within the nucleus assuming it to be a sphere of radius $r = 1.1$ fm. (b) What would be the probability if the electron were replaced with a muon, which is very similar to an electron (Chapter 44) except that its mass is 207 times greater?

21. (II) Determine the average radial probability distribution P_r for the $n = 2$, $l = 1$ state in hydrogen by calculating

$$P_r = 4\pi r^2 \left[\tfrac{1}{3} |\psi_{210}|^2 + \tfrac{1}{3} |\psi_{211}|^2 + \tfrac{1}{3} |\psi_{21-1}|^2 \right].$$

22. (II) Use the result of Problem 21 to show that the most probable distance r from the nucleus for an electron in the 2p state of hydrogen is $r = 4r_0$, which is just the second Bohr radius (Eq. 38–11, Fig. 38–24).

23. (II) Show that the mean value of r for an electron in the ground state of hydrogen is $\bar{r} = \tfrac{3}{2} r_0$, by calculating

$$\bar{r} = \int_{\text{all space}} r |\psi_{100}|^2 \, dV = \int_0^\infty r |\psi_{100}|^2 4\pi r^2 \, dr.$$

24. (III) Show that the probability of finding the electron within 1 Bohr radius of the nucleus in the hydrogen atom is (a) 3.4 percent for the $n = 2$, $l = 0$ state, and (b) 0.37 percent for the $n = 2$, $l = 1$ state.

25. (III) Show that ψ_{100} (Eq. 40–5a) satisfies the Schrödinger equation (Eq. 40–1) with the Coulomb potential, for energy $E = -me^4/8\epsilon_0^2 h^2$.

26. (III) For the $n = 2$, $l = 0$ state in hydrogen, what is the probability of finding the electron within the smaller peak in the radial probability distribution, Fig. 40–8b?

27. (III) The wave function for the $n = 3$, $l = 0$ state in hydrogen is

$$\psi_{300} = \frac{1}{\sqrt{27\pi r_0^3}} \left(1 - \frac{2r}{3r_0} + \frac{2r^2}{27r_0^2} \right) e^{-\frac{r}{3r_0}}.$$

(a) Determine the radial probability distribution P_r for this state, and (b) draw the curve for it on a graph. (c) Determine the most probable distance from the nucleus for an electron in this state.

Sections 40–4 and 40–5

28. (I) List the quantum numbers for each electron in the ground state of nitrogen ($Z = 7$).

29. (I) List the quantum numbers for each electron in the ground state of (a) carbon ($Z = 6$), (b) magnesium ($Z = 12$).

30. (I) How many electrons can be in the $n = 6$, $l = 3$ subshell?

31. (II) What is the full electron configuration for (a) selenium (Se), (b) gold (Au), (c) uranium (U)? [Hint: See the periodic table inside the back cover.]

32. (II) For each of the following atomic transitions, state whether the transition is allowed or forbidden, and why: (a) 4p → 3p; (b) 2p → 1s; (c) 3d → 2d; (d) 4d → 3s; (e) 4s → 2p.

33. (II) Using the Bohr formula for the radius of an electron orbit, estimate the average distance from the nucleus for an electron in the innermost ($n = 1$) orbit in uranium ($Z = 92$). Approximately how much energy would be required to remove this innermost electron?

34. (II) Estimate the binding energy of the third electron in lithium using Bohr theory. [Hint: This electron has $n = 2$ and "sees" a net charge of approximately $+1e$.] The measured value is 5.36 eV.

35. (II) Show that the total angular momentum is zero for a filled subshell.

36. (II) Let us apply the exclusion principle to an infinitely high square well (Section 39–8). Let there be five electrons confined to this rigid box whose width is L. Find the lowest energy state of this system, by placing the electrons in the lowest available levels, consistent with the Pauli exclusion principle.

Section 40–6

37. (I) What are the shortest-wavelength X-rays emitted by electrons striking the face of a 30-kV TV picture tube? What are the longest wavelengths?

38. (I) If the shortest-wavelength bremsstrahlung X-rays emitted from an X-ray tube have $\lambda = 0.029$ nm, what is the voltage across the tube?

39. (I) Show that the cutoff wavelength λ_0 is given by

$$\lambda_0 = \frac{1240 \text{ nm}}{V},$$

where V is the X-ray tube voltage in volts.

40. (II) Use the result of Example 40–6 to estimate the X-ray wavelength emitted when a cobalt atom ($Z = 27$) jumps from $n = 2$ to $n = 1$.

41. (II) Estimate the wavelength for an $n = 2$ to $n = 1$ transition in iron ($Z = 26$).

42. (II) Use Bohr theory to estimate the wavelength for an $n = 3$ to $n = 1$ transition in molybdenum. The measured value is 0.063 nm. Why do we not expect perfect agreement?

43. (II) A mixture of iron and an unknown material are bombarded with electrons. The wavelength of the K_α lines are 194 pm for iron and 229 pm for the unknown. What is the unknown material?

44. (II) Use conservation of energy and momentum to show that a moving electron cannot give off an X-ray photon unless there is a third body present, such as an atom or nucleus.

* Section 40–7

* 45. (I) Verify that the Bohr magneton has the value $\mu_B = 9.27 \times 10^{-24}$ J/T (see Eq. 40–12).

* 46. (II) Suppose that the splitting of energy levels shown in Fig. 40–4 was produced by a 2.0-T magnetic field. (a) What is the separation in energy between adjacent m_l levels for the same l? (b) How many different wavelengths will there be for 3d to 2p transitions, if m_l can change only by ± 1 or 0? (c) What is the wavelength for each of these transitions?

* 47. (II) In a Stern-Gerlach experiment, Ag atoms issued from the oven with an average speed of 700 m/s and passed through a magnetic field gradient $dB/dz = 1.5 \times 10^3$ T/m for a distance of 4.0 cm. (a) What is the separation of the two beams as they emerge from the magnet? (b) What would the separation be if the g-factor were 1 for electron spin?

*48. (II) (a) Write down the quantum numbers for each electron in the aluminum atom. (b) Which subshells are filled? (c) The last electron is in the 3p state; what are the possible values of the total angular momentum quantum number, j, for this electron? (d) Explain why the angular momentum of this last electron also represents the total angular momentum for the entire atom (ignoring any angular momentum of the nucleus). (e) How could you use a Stern-Gerlach experiment to determine which value of j the atom has?

*49. (II) What are the possible values of j for an electron in (a) the 3p, (b) the 4f, and (c) the 4d state of hydrogen? What is J in each case?

*50. (II) For an electron in a 5g state, what are all the possible values of j, m_j, J, and J_z?

*51. (II) The difference between the $2P_{3/2}$ and $2P_{1/2}$ energy levels in hydrogen is about 5×10^{-5} eV, due to the spin-orbit interaction. (a) Taking the electron's (orbital) magnetic moment to be 1 Bohr magneton, estimate the internal magnetic field due to the electron's orbital motion. (b) Estimate the internal magnetic field using a simple model of the nucleus revolving in a circle about the electron.

* Section 40–9

*52. (II) A laser used to weld detached retinas puts out 25-ms-long pulses of 640-nm light which average 0.65 W output during a pulse. How much energy can be deposited per pulse and how many photons does each pulse contain?

*53. (II) Estimate the angular spread of a laser beam due to diffraction if the beam emerges through a 3.0-mm-diameter mirror. Assume that $\lambda = 694$ nm. What would be the diameter of this beam if it struck a satellite 300 km above the Earth?

*54. (II) Suppose that the energy level system in Fig. 40–20 is not being pumped and is in thermal equilibrium. Determine the fraction of atoms in levels E_2 and E_1 relative to E_0 at $T = 300$ K.

*55. (II) To what temperature would the system in Fig. 40–20 have to be raised (see Problem 54) so that in thermal equilibrium the level E_2 would have half as many atoms as E_0? (Note that pumping mechanisms do not maintain thermal equilibrium.)

*56. (II) Show that a population inversion for two levels (as in a pumped laser) corresponds to a negative Kelvin temperature in the Boltzmann distribution. Explain why such a situation does not contradict the idea that negative Kelvin temperatures cannot be reached in the normal sense of temperature.

General Problems

57. The ionization (binding) energy of the outermost electron in boron is 8.26 eV. (a) Use the Bohr model to estimate the "effective charge," Z_{eff}, seen by this electron. (b) Estimate the average orbital radius.

58. Show that there can be 18 electrons in a "g" subshell.

59. What is the full electron configuration in the ground state for elements with Z equal to (a) 27, (b) 36, (c) 38? [Hint: See the periodic table inside the back cover.]

60. What are the largest and smallest possible values for the angular momentum L of an electron in the $n = 5$ shell?

61. Estimate (a) the quantum number l for the orbital angular momentum of the Earth about the Sun, and (b) the number of possible orientations for the plane of Earth's orbit.

62. Use the Bohr theory (especially Eq. 38–14) to show that the Moseley plot (Fig. 40–12) can be written

$$\sqrt{\frac{1}{\lambda}} = a(Z - b),$$

where $b \approx 1$, and evaluate a.

63. Determine the most probable distance from the nucleus of an electron in the $n = 2$, $l = 0$ state of hydrogen.

64. In the so-called *vector model* of the atom, space quantization of angular momentum (Fig. 40–3) is illustrated as shown in Fig. 40–28. The angular momentum vector of magnitude $L = \sqrt{l(l + 1)}\hbar$ is thought of as precessing around the z axis (like a spinning top or gyroscope) in such a way that the z component of angular momentum, $L_z = m_l\hbar$, also stays constant. Calculate the possible values for the angle θ between $\mathbf{L}$ and the z axis (a) for $l = 1$, (b) $l = 2$, and (c) $l = 3$. (d) Determine the minimum value of θ for $l = 100$ and $l = 10^6$. Is this consistent with the correspondence principle?

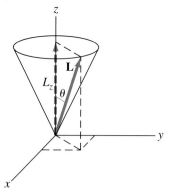

FIGURE 40–28 The vector model for orbital angular momentum. The orbital angular momentum vector $\mathbf{L}$ is imagined to precess about the z axis; L and L_z remain constant, but L_x and L_y continually change. Problems 64 and 65.

65. The vector model (Problem 64) gives some insight into the uncertainty principle for angular momentum, which is

$$\Delta L_z \, \Delta\phi \gtrsim \hbar$$

for the z component. Here ϕ is the angular position measured in the plane perpendicular to the z axis. Once m_l for an atom is known, L_z is known precisely, so $\Delta L_z = 0$. (a) What does this tell us about ϕ? (b) What can you say about L_x and L_y, which are *not* quantized (only L and L_z are)? (c) Show that although L_x and L_y are not quantized, nonetheless $(L_x^2 + L_y^2)^{1/2} = [l(l+1) - m_l^2]^{1/2}\hbar$ is.

* **66.** Show that the diffractive spread of a laser beam, $\approx \lambda/a$ as described in Section 40–9, is precisely what you might expect from the uncertainty principle. [*Hint*: Since the beam's width is constrained by the dimension of the aperture a, the component of the light's momentum perpendicular to the laser axis is uncertain.]

* **67.** Silver atoms (spin $= \frac{1}{2}$) are placed in a 1.0-T magnetic field which splits the ground state into two close levels. (a) What is the difference in energy between these two levels, and (b) what wavelength photon could cause a transition from the lower level to the upper one? (c) How would your answer differ if the atoms were hydrogen?

* **68.** Estimate the angular spread of a laser beam due to diffraction if the beam emerges through a 4.0-mm diameter mirror. Assume that $\lambda = 694$ nm. What would be the diameter of this beam if it struck (a) a satellite 1000 km above the Earth, or (b) the Moon?

* **69.** *Populations in the H atom.* Use the Boltzmann factor (Eq. 40–16) to estimate the fraction of H atoms in the $n = 2$ and $n = 3$ levels (relative to the ground state) for thermal equilibrium at (a) $T = 300$ K and (b) $T = 6000$ K. [Note: Since there are eight states with $n = 2$ and only two with $n = 1$, multiply your result for $n = 2$ by $\frac{8}{2} = 4$; do similarly for $n = 3$.] (c) Given 1.0 g of hydrogen, estimate the number of atoms in each state at $T = 6000$ K. (d) Estimate the number of $n = 3$ to $n = 1$ and $n = 2$ to $n = 1$ photons that will be emitted per second at $T = 6000$ K. Assume that the lifetime of each excited state is 10^{-8} s.

70. A very simple model of a "one-dimensional" metal consists of N electrons confined to a rigid box of width L. We neglect the Coulomb interaction between the electrons (this is called the "independent electron model" and does remarkably well in predicting many properties of real metals). The Fermi energy of a metal is the energy of the most energetic electron when the metal is in its ground state ($T = 0$ K). (a) Calculate the Fermi energy for this one-dimensional metal, taking into account the Pauli exclusion principle. You can assume for simplicity that N is even. (b) What is the smallest amount of energy that this metal can absorb? Consider the limit of your answer for large N. (The relatively small value you should obtain is indicative of how metals can readily conduct).

71. If the principal quantum number n were limited to the range from 1 to 6, how many elements would we find in nature?

72. It is possible for atoms to be excited into states with very high values of the principal quantum number. Electrons in these so-called *Rydberg states* have very small ionization energies and huge orbital radii. This makes them particularly sensitive to external perturbation, as would be the case if the atom were in an electric field. Consider the $n = 50$ state of the hydrogen atom. Determine the binding energy, the radius of the orbit, and the effective cross sectional area of this Rydberg state.

* **73.** In the analytical technique known as *electron spin resonance* (ESR), materials with atoms having one or more unpaired electrons are placed in a variable magnetic field plus a constant radio frequency (RF) electric field. The magnetic field splits the energy of the unpaired electrons according to the two possible spin orientations. If the splitting is the same as the photon energy which is available from the RF radiation, a resonant absorption takes place. This can be detected, and is a measure of the g-factor of the electrons in those atoms. The g-factor for molecules and solids can be different than for electrons in atoms, due to the chemical interactions in the material. Therefore, ESR provides a fingerprint of the system. In a particular ESR experiment, the RF radiation has a fixed wavelength of 2.0 cm. As the magnetic field is varied, a resonance is observed at 0.476 T. What is the g-factor for the unpaired electrons in the sample?

This Pentium chip is one of the leading processors used in computers today. It contains over 3 million transistors, not to mention diodes and other semiconductor electronic elements. Before discussing semiconductors and their applications later in this chapter, we begin with a study of how quantum theory describes how atoms bond together to form molecules, and how molecules behave. We then examine how atoms and molecules come together to form solids, with emphasis on metals and semi-conductors, and their use in electricity and electronics.

CHAPTER 41

Molecules and Solids

Since its development in the 1920s, quantum mechanics has had a profound influence on our lives, both intellectually and technologically. Even the way we view the world has changed, as we have seen in the last couple of chapters. In the present chapter, we will discuss how quantum mechanics has given us an understanding of the structure of molecules and matter in bulk, as well as a number of important applications including semiconductor devices. Our discussions will necessarily be qualitative for the most part.

41–1 | Bonding in Molecules

One of the great successes of quantum mechanics was to give scientists, at last, an understanding of the nature of chemical bonds. Since it is based in physics, and because this understanding is so important in many fields, we discuss it here.

By a molecule, we mean a group of two or more atoms that are strongly held together so as to function as a single unit. When atoms make such an attachment, we say that a chemical **bond** has been formed. There are two main types of strong chemical bond: covalent and ionic. Many bonds are actually intermediate between these two types.

Covalent Bonds

Covalent bond

To understand how **covalent bonds** are formed, we take the simplest case, the bond that holds two hydrogen atoms together to form the hydrogen molecule, H_2. The mechanism is basically the same for other covalent bonds. As two H atoms approach each other, the electron clouds begin to overlap, and the electrons from each atom can "orbit" both nuclei. (This is sometimes called "sharing" electrons.) If both electrons are in the ground state ($n = 1$) of their respective atoms, there are two possibilities: their spins can be parallel (both up or both down), in which case the total spin is $S = \frac{1}{2} + \frac{1}{2} = 1$; or their spins can be opposite ($m_s = +\frac{1}{2}$ for one, $m_s = -\frac{1}{2}$ for the other), so that the total spin $S = 0$. We shall now see that a bond is formed only for the $S = 0$ state, when the spins are opposite. First we consider the $S = 1$ state, for which the spins are the same. The two electrons cannot both be in the lowest energy state and be attached to the same atom, for then they would have identical quantum numbers in violation of the exclusion principle. The exclusion principle tells us that since no two electrons can occupy the same quantum state, if two electrons have the same quantum numbers, they must be different in some other way—namely, by being in different places in space (for example, attached to different atoms). When the two hydrogen atoms approach, the electrons will stay away from each other as shown by the probability distribution of Fig. 41–1. The positively charged nuclei then repel each other, and no bond is formed.

For the $S = 0$ state, on the other hand, the spins are opposite and the two electrons are consequently in different quantum states. Hence they can come close together spatially. In this case, the probability distribution looks like Fig. 41–2. As can be seen, the electrons spend much of their time between the two positively charged nuclei, and the latter are attracted to the negatively charged electron cloud between them. This attraction, which holds the two atoms together to form a molecule, constitutes a *covalent bond*.

The probability distributions of Figs. 41–1 and 41–2 can perhaps be better understood on the basis of waves. What the exclusion principle requires is that when the spins are the same, there is destructive interference of the electron wave functions in the region between the two atoms. But when the spins are opposite, constructive interference occurs in the region between the two atoms, resulting in a large amount of negative charge there. Thus a covalent bond can be said to be the result of constructive interference of the electron wave functions in the space between the two atoms, and of the electrostatic attraction of the two positive nuclei for the negative charge concentration between them.

Why a bond is formed can also be understood from the energy point of view. When the two H atoms approach close to one another, if the spins of their electrons are opposite, the electrons can occupy the same space, as discussed above. This means that each electron can now move about in the space of two atoms instead of in the volume of only one. Because each electron now occupies more space, it is less well localized. From the uncertainty principle, with Δx increased, the momentum, and hence the energy, can be less. Another way of understanding this is from the point of view of de Broglie waves (Section 38–11). Because each electron has a larger "orbit," its wavelength λ can be longer, so its momentum $p = h/\lambda$ (Eq. 38–7) can be less. With less momentum, each electron has less energy when the two atoms combine than when they are separate. That is, the molecule has less energy than the two separate atoms, and so is more stable. An energy input is required to break the H_2 molecule into two separate H atoms, so the H_2 molecule is a stable entity. This is what we mean by a *bond*. The energy required to break a bond is called the **bond energy**, the **binding energy**, or the **dissociation energy**. For the hydrogen molecule, H_2, the bond energy is 4.5 eV.

Ionic Bonds

An **ionic bond** is, in a sense, a special case of the covalent bond. Instead of the electrons being shared equally, they are shared unequally. For example, in sodium chloride (NaCl), the outer electron of the sodium spends nearly all its time around the chlorine (Fig. 41–3). The chlorine atom acquires a net negative charge as a result of the extra electron, whereas the sodium atom is left with a net positive charge. The electrostatic attraction between these two charged atoms holds them together. The resulting bond is called an *ionic bond* because it is created by the attraction between the two ions (Na^+ and Cl^-). But to understand the ionic bond,

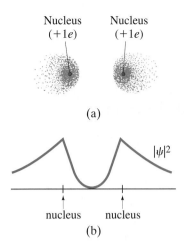

FIGURE 41–1 Electron probability distribution (electron cloud) for two H atoms when the spins are the same ($S = 1$): (a) electron cloud; (b) projection of $|\psi|^2$ along the line through the centers of the two atoms.

FIGURE 41–2 Electron probability distribution for two H atoms when the spins are opposite ($S = 0$): (a) electron cloud; (b) projection of $|\psi|^2$ along the line through the centers of the atoms. In this case a bond is formed because the positive nuclei are attracted to the concentration of negative charge between them. This is a hydrogen molecule, H_2.

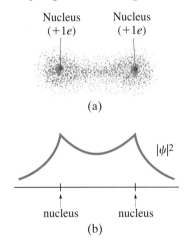

FIGURE 41–3 (below) Probability distribution $|\psi|^2$ for the last electron of Na in NaCl.

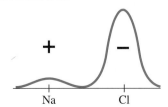

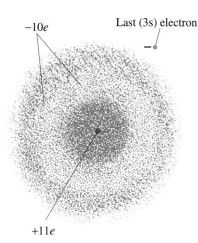

FIGURE 41–4 In a neutral sodium atom, the 10 inner electrons shield the nucleus, so the single outer electron is attracted by a net charge of +1e.

FIGURE 41–5 Neutral chlorine atom. The +17e of the nucleus is shielded by the 12 electrons in the inner shells and subshells. Four of the five 3p electrons are shown in doughnut-shaped clouds, and the fifth is in the (dashed-line) cloud concentrated about the vertical axis. An extra electron at x will be attracted by a net charge that can be as much as +5e.

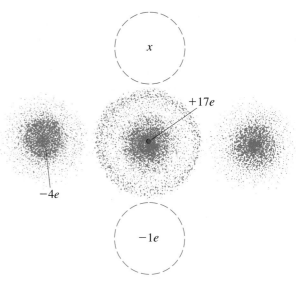

we must understand why the extra electron from the sodium spends so much of its time around the chlorine. After all, the chlorine is neutral; why should it attract another electron?

The answer lies in the probability distributions of the two neutral atoms. Sodium contains 11 electrons, 10 of which together form spherically symmetric closed shells (Fig. 41–4). The last electron spends most of its time beyond these closed shells. Because the closed shells have a total charge of −10e and the nucleus has charge +11e, the outermost electron in sodium "feels" a net attraction due to +1e. It is not held very strongly. On the other hand, 12 of chlorine's 17 electrons form closed shells, or subshells (corresponding to $1s^2 2s^2 2p^6 3s^2$). These 12 form a spherically symmetric shield around the nucleus. The other five electrons are in 3p states whose probability distributions are not spherically symmetric and have a form similar to those for the 2p states in hydrogen shown in Fig. 40–9a. Four of these 3p electrons can have "doughnut-shaped" distributions symmetric about the z axis, as shown in Fig. 41–5. The fifth can have a "barbell-shaped" distribution (as for $m_l = 0$ in Fig. 40–9a), which in Fig. 41–5 is shown only as a dashed outline because it is half empty. That is, the exclusion principle allows one more electron to be in this state (it will have spin opposite to that of the electron already there). If an extra electron—say from a Na atom—happens to be in the vicinity, it can be in this state, say at point x in Fig. 41–5. It could experience an attraction due to as much as +5e because the +17e of the chlorine nucleus is partly shielded at this point by the 12 inner electrons. Thus, the outer electron of a Na atom will be more strongly attracted by the +5e of the chlorine atom that it "sees" at certain points rather than by the +1e of its own atom. This, combined with the strong attraction between the two ions when the extra electron stays with the Cl⁻, produces the charge distribution of Fig. 41–3, and hence the ionic bond.

Partial Ionic Character of Covalent Bonds

A pure covalent bond in which the electrons are shared equally occurs mainly in symmetrical molecules such as H_2, O_2, and Cl_2. When the atoms involved are different from each other, it is usual to find that the shared electrons are more likely to be in the vicinity of one atom than the other. The extreme case is an ionic bond; in intermediate cases the covalent bond is said to have a *partial ionic character*. The molecules themselves are **polar**—that is, they have an electric dipole moment because one part (or parts) of the molecule has a net positive charge and other parts a net negative charge. An example is the water molecule, H_2O (Fig. 41–6). The shared electrons are more likely to be found around the oxygen atom than around the two hydrogens. The reason is similar to that discussed above in con-

FIGURE 41–6
The water molecule is polar.

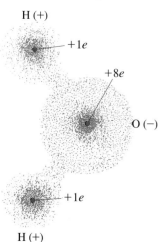

nection with ionic bonds. Oxygen has eight electrons $(1s^2 2s^2 2p^4)$, of which four form a spherically symmetric core and the other four could have, for example, a doughnut-shaped distribution. The barbell-shaped distribution on the z axis (like that shown dashed in Fig. 41–5) could be empty, so electrons from hydrogen atoms can be attracted by a net charge of $+4e$. They are also attracted by the H nuclei, so they partly orbit the H atoms as well as the O atom. The net effect is that there is a net positive charge on each H atom (less than $+1e$), because the electrons spend only part of their time there. And, there is a net negative charge on the O atom.

41–2 Potential-Energy Diagrams for Molecules

It is useful to analyze the interaction between two objects—say, between two atoms or molecules—with the use of a potential-energy diagram, a plot of the potential energy versus the separation distance.

For the simple case of two point charges, q_1 and q_2, the potential energy U is given by (see Chapter 23):

$$U = \frac{1}{4\pi\epsilon_0} \frac{q_1 q_2}{r},$$

where r is the distance between the charges, and the constant $(1/4\pi\epsilon_0)$ is equal to $9.0 \times 10^9 \,\mathrm{N\cdot m^2/C^2}$. If the two charges have the same sign, the potential energy is positive for all values of r, and a graph of U versus r in this case is shown in Fig. 41–7a. The force is repulsive (the charges have the same sign) and the curve rises as r decreases; this makes sense since work is done to bring the charges together, thereby increasing their potential energy. If, on the other hand, the two charges are of opposite sign, the potential energy is negative because the product $q_1 q_2$ is negative. The force is attractive in this case and the graph of $U(\propto -1/r)$ versus r looks like Fig. 41–7b. The potential energy becomes more *negative* as r decreases.

Now let us look at the potential energy diagram for the formation of a covalent bond, such as for the hydrogen molecule. The potential energy of one H atom in the presence of the other is plotted in Fig. 41–8. The potential energy U decreases as the atoms approach, because the electrons concentrate between the two nuclei (Fig. 41–2), so attraction occurs. However, at very short distances, the electrons would be "squeezed out"—there is no room for them between the two nuclei. Without the electrons between them, each nucleus would feel a repulsive force due to the other, so the curve rises as r decreases further. There is an optimum separation of the atoms, r_0 in Fig. 41–8, at which the energy is lowest. This is the point of greatest stability for the hydrogen molecule, and r_0 is the average separation of atoms in the H_2 molecule. The depth of this "well" is the *binding energy*,[†] as shown. This is how much energy must be put into the system to separate the two atoms to infinity, where the potential energy $U = 0$. For the H_2 molecule, the binding energy is about 4.5 eV, and $r_0 = 0.074\,\mathrm{nm}$. In molecules made

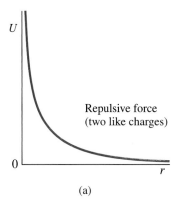

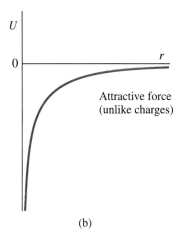

FIGURE 41–7 Potential energy U as a function of the separation for two point charges of (a) like sign and (b) opposite sign.

[†]The binding energy corresponds not quite to the bottom of the potential-energy curve, but to the lowest energy state, slightly above it, as shown in Fig. 41–8. Compare to a square well, Fig. 39–8.

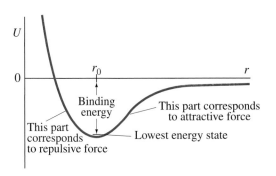

FIGURE 41–8 Potential-energy diagram for H_2 molecule; r is the separation of the two H atoms. The binding energy (the energy difference between $U = 0$ and the lowest energy state near the bottom of the well) is 4.5 eV, and $r_0 = 0.074\,\mathrm{nm}$.

of larger atoms, say, oxygen or nitrogen, repulsion also occurs at short distances, because the closed inner electron shells begin to overlap and the exclusion principle forbids their coming too close. The repulsive part of the curve rises even more steeply than $1/r$. A reasonable approximation to the potential energy, at least in the vicinity of r_0, is

$$U = -\frac{A}{r^m} + \frac{B}{r^n}, \tag{41–1}$$

where A and B are constants associated with the attractive and repulsive parts of the potential energy and m and n are small integers. For ionic and some covalent bonds, the attractive term can often be written with $m = 1$ (Coulomb potential).

For many bonds, the potential-energy curve has the shape shown in Fig. 41–9. There is still an optimum distance r_0 at which the molecule is stable. But when the atoms approach from a large distance, the force is initially repulsive rather than attractive. The atoms thus do not interact spontaneously. Instead, some additional energy must be injected into the system to get it over the "hump" (or barrier) in the potential-energy diagram. This required energy is called the **activation energy**.

The curve of Fig. 41–9 is much more common than that of Fig. 41–8. The activation energy often reflects a need to break other bonds, before the one under discussion can be made. For example, to make water from O_2 and H_2, the H_2 and O_2 molecules must first be broken into H and O atoms by an input of energy; this is what the activation energy represents. Then the H and O atoms can combine to form H_2O with the release of a great deal more energy than was put in initially. The initial activation energy can be provided by applying an electric spark to a mixture of H_2 and O_2, breaking a few of these molecules into H and O atoms. The resulting explosive release of energy when these atoms combine to form H_2O quickly provides the activation energy needed for further reactions, so additional H_2 and O_2 molecules are broken up and recombined to form H_2O.

The potential-energy diagrams for ionic bonds can have similar shapes. In NaCl, for example, the Na^+ and Cl^- ions attract each other at distances a bit larger than some r_0, but at shorter distances the overlapping of inner electron shells gives rise to repulsion. The two atoms thus are most stable at some intermediate separation, r_0, and there often is an activation energy.

FIGURE 41–9 Potential energy diagram for a bond requiring an activation energy.

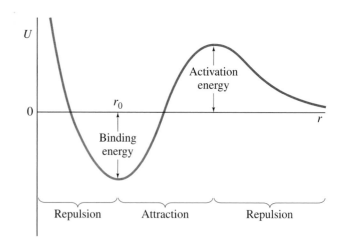

EXAMPLE 41–1 **ESTIMATE** **Sodium chloride bond.** A potential-energy diagram for the NaCl ionic bond is shown in Fig. 41–10, where we have set $U = 0$ for free Na and Cl neutral atoms (which are represented on the right in Fig. 41–10). Measurements show that 5.14 eV are required to remove an electron from a neutral Na atom to produce the Na^+ ion; and 3.61 eV of energy is released when an electron is "grabbed" by a Cl atom to form the Cl^- ion. Thus, forming Na^+ and Cl^- ions from neutral Na and Cl atoms requires 5.14 eV − 3.61 eV = 1.53 eV of energy, a form of activation energy. This is shown as the "bump" in Fig. 41–10. But note that the potential-energy diagram from here out to the right is not really a function of distance—it is drawn dashed to remind us that it only represents the energy difference between the ions and the neutral atoms (for which we have chosen $U = 0$). (a) Calculate the separation distance, r_1, at which the potential of the Na^+ and Cl^- ions drops to zero (measured value is $r_1 = 0.94$ nm). (b) Estimate the binding energy of the NaCl bond, which occurs at a separation $r_0 = 0.24$ nm. Ignore the repulsion of the overlapping electron shells that occurs at this distance (and causes the rise of the potential-energy curve for $r < r_0$, Fig. 41–10).

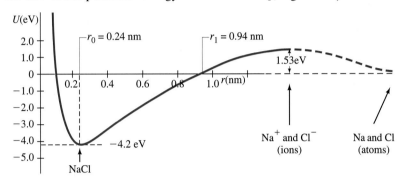

FIGURE 41–10 Potential-energy diagram for the NaCl bond. Beyond about $r = 1.2$ nm, the diagram is schematic only, and represents the energy difference between ions and neutral atoms. $U = 0$ is chosen for the two separated atoms Na and Cl (not for the ions). [For the two ions, Na^+ and Cl^-, the zero of potential energy at $r = \infty$, corresponds to $U \approx 1.53$ eV on this diagram.]

SOLUTION (a) The potential energy of two point charges is given by Coulombs' law:

$$U' = \frac{1}{4\pi\epsilon_0} \frac{q_1 q_2}{r},$$

where we distinguish U' from the U in Fig. 41–10. This formula works for our two ions if we set $U' = 0$ at $r = \infty$, which in the plot of Fig. 41–10 corresponds to $U = +1.53$ eV (Fig. 41–10 is drawn for $U = 0$ for the free *atoms*). The point r_1 in Fig. 41–10 corresponds to $U' = -1.53$ eV relative to the two free ions. Since $q_1 = +e$ and $q_2 = -e$, we have

$$r_1 = \frac{1}{4\pi\epsilon_0} \frac{q_1 q_2}{U'} = \frac{(9.0 \times 10^9\,\text{N}\cdot\text{m}^2/\text{C}^2)(-1.60 \times 10^{-19}\,\text{C})(+1.60 \times 10^{-19}\,\text{C})}{(-1.53\,\text{eV})(1.60 \times 10^{-19}\,\text{J/eV})}$$

$$= 0.94\,\text{nm},$$

which is just the measured value.

(b) At $r_0 = 0.24$ nm, the potential energy of the two ions (relative to $r = \infty$ for the two ions) is

$$U' = \frac{1}{4\pi\epsilon_0} \frac{q_1 q_2}{r}$$

$$= \frac{(9.0 \times 10^9\,\text{N}\cdot\text{m}^2/\text{C}^2)(-1.60 \times 10^{-19}\,\text{C})(+1.60 \times 10^{-19}\,\text{C})}{(0.24 \times 10^{-9}\,\text{m})(1.60 \times 10^{-19}\,\text{J/eV})} = -6.0\,\text{eV}.$$

Thus, we estimate that 6.0 eV of energy is given up when Na^+ and Cl^- ions form a NaCl bond. Put another way, it takes 6.0 eV to break the NaCl bond and form the Na^+ and Cl^- ions. To get the binding energy—the energy to separate the NaCl into Na and Cl atoms—we need to subtract out the 1.53 eV (the "bump" in Fig. 41–10) needed to ionize them:

binding energy = 6.0 eV − 1.53 eV = 4.5 eV.

The measured value (shown on Fig. 41–10) is 4.2 eV. The difference can be attributed to the energy associated with the repulsion of the electron shells at this distance.

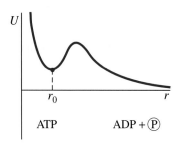

FIGURE 41–11 Potential energy diagram for the formation of ATP from ADP and phosphate (Ⓟ).

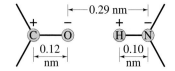

FIGURE 41–12 The C^+—O^- and H^+—N^- dipoles attract each other. (These dipoles may be part of larger molecules. See Fig. 41–13.)

Sometimes the potential energy of a bond looks like that of Fig. 41–11. In this case, the energy of the bonded molecule, at a separation r_0, is greater than when there is no bond ($r = \infty$). That is, an energy *input* is required to make the bond (hence the binding energy is negative), and there is energy release when the bond is broken. Such a bond is stable only because there is the barrier of the activation energy. This type of bond is important in living cells, for it is in such bonds that energy can be stored efficiently in certain molecules, particularly ATP (adenosine triphosphate). The bond that connects the last phosphate group (designated Ⓟ in Fig. 41–11) to the rest of the molecule (ADP, meaning adenosine diphosphate, since it contains only two phosphates) is of the form shown in Fig. 41–11. Energy is actually stored in this bond. When the bond is broken $\left(\text{ATP} \rightarrow \text{ADP} + Ⓟ \right)$, energy is released and this energy can be used to make other chemical reactions "go."

41–3 Weak (van der Waals) Bonds

Once a bond between two atoms or ions is made, energy must normally be supplied to break the bond and separate the atoms. As mentioned in Section 41–1, this energy is called the *bond energy* or *binding energy*. The binding energy for covalent and ionic bonds is typically 2 to 5 eV. These bonds, which hold atoms together to form molecules, are often called **strong bonds** to distinguish them from so-called "weak bonds." The term **weak bond**, as we use it here, refers to an attachment between molecules due to simple electrostatic attraction—such as *between* polar molecules (not *within* a polar molecule, which is a strong bond). The strength of the attachment is much less than for the strong bonds. Binding energies are typically in the range 0.04 to 0.3 eV—hence their name "weak bonds."

Weak bonds are generally the result of attraction between dipoles. For example, Fig. 41–12 shows two molecules, which have permanent dipole moments, attracting one another. Besides such **dipole–dipole bonds**, there can also be **dipole–induced dipole bonds**, in which a polar molecule with a permanent dipole moment can induce a dipole moment in an otherwise electrically balanced (nonpolar) molecule, just as a single charge can induce a separation of charge in a nearby object (see Fig. 21–7). There can even be an attraction between two nonpolar molecules. Even though a molecule may not have a permanent dipole moment on the average, we can think of its electrons as moving about so that at any instant there may be a separation of charge. Such transient dipoles can induce a dipole moment in a nearby molecule, creating a weak attraction. All these weak bonds are referred to as **van der Waals bonds**, and the forces involved **van der Waals forces**. The potential energy has the general shape shown in Fig. 41–8, with the attractive van der Waals potential energy varying as $1/r^6$.

When one of the atoms in a dipole–dipole bond is hydrogen, as in Fig. 41–12, it is called a **hydrogen bond**. A hydrogen bond is generally quite strong for a so-called "weak bond." This is because the hydrogen atom is the smallest atom and thus can be approached more closely. A hydrogen bond also has a partial "covalent" character: that is, electrons between the two dipoles may be shared to a small extent, thus making a stronger, more lasting, bond.

Weak bonds are important in liquids and solids when strong bonds are absent (see Section 41–5). They are also very important for understanding the activities of cells, such as the double-helix shape of DNA (Fig. 41–13), and DNA replication. The average kinetic energy of molecules in a cell is around $\frac{3}{2}kT \approx 0.04 \text{ eV}$, about the magnitude of weak bonds. This means that a weak bond can readily be broken just by a molecular collision. Hence weak bonds are not very permanent—they are, instead, brief attachments. But because of this, they play an important role in the cell. On the other hand, strong bonds—those that hold molecules together—are almost never broken simply by molecular collision.

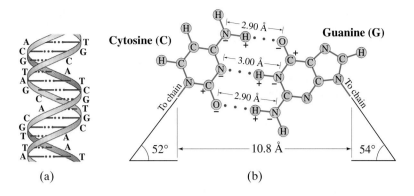

FIGURE 41–13 (a) Section of a DNA double helix. The red dots represent hydrogen bonds between the two chains. (b) "Close-up" view of helix: cytosine (C) and guanine (G) molecules on separate chains of a DNA double helix are held together by the hydrogen bonds (red dots) involving an H^+ on one molecule attracted to an N^- or C^+—O^- on the other molecule.

Thus they are relatively permanent. They can be broken by chemical action (the making of even stronger bonds), and this usually happens in the cell with the aid of an enzyme, which is a protein molecule.

EXAMPLE 41–2 **Nucleotide energy.** Calculate the interaction energy between the C$=$O dipole of thymine and the H—N dipole of adenine, assuming that the two dipoles are lined up as shown in Fig. 41–12. Dipole moment measurements (see Table 23–2) give $q_H = -q_N = 0.19e = 3.0 \times 10^{-20}$ C, and $q_C = -q_O = 0.42e = 6.7 \times 10^{-20}$ C.

SOLUTION The interaction energy will be equal to the potential energy of one dipole in the presence of the other, since this will be equal to the work needed to pull the two dipoles infinitely far apart. U will consist of four terms:

$$U = U_{CH} + U_{CN} + U_{OH} + U_{ON},$$

where U_{CH} means the potential energy of C in the presence of H, and similarly for the other terms. We do not have terms corresponding to C and O, or N and H, because the two dipoles are assumed to be stable entities (the bonds C$=$O and H—N remain intact). The potential energy for two point charges is $U_{12} = (1/4\pi\epsilon_0)(q_1 q_2/r)$, where r is the distance between them. So

$$U = \frac{1}{4\pi\epsilon_0}\left(\frac{q_C q_H}{r_{CH}} + \frac{q_C q_N}{r_{CN}} + \frac{q_O q_H}{r_{OH}} + \frac{q_O q_N}{r_{ON}}\right).$$

Using the distances shown in Fig. 41–12, we get:

$$U = (9.0 \times 10^9 \, \text{N} \cdot \text{m}^2/\text{C})\left(\frac{(6.7)(3.0)}{0.31} + \frac{(6.7)(-3.0)}{0.41} + \frac{(-6.7)(3.0)}{0.19}\right.$$
$$\left. + \frac{(-6.7)(-3.0)}{0.29}\right)\frac{(10^{-20} \, \text{C})^2}{(10^{-9} \, \text{m})}$$
$$= -1.82 \times 10^{-20} \, \text{J} = -0.11 \, \text{eV}.$$

The potential energy is negative, meaning 0.11 eV of work (or energy input) is required to separate the molecules. That is, the binding energy of this "weak" or hydrogen bond is 0.11 eV. This is only an estimate, of course, since other charges in the vicinity would have an influence too.

41–4 Molecular Spectra

When atoms combine to form molecules, the wave functions (and probability distributions) of the outer electrons overlap, and this interaction alters the energy levels. Nonetheless, molecules can undergo transitions between electron energy levels just as atoms do. For example, the H_2 molecule can absorb a photon of just the

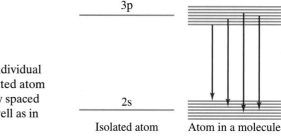

FIGURE 41–14 The individual energy levels of an isolated atom become bands of closely spaced levels in molecules, as well as in solids and liquids.

Isolated atom   Atom in a molecule

right frequency to excite one of its ground state electrons to an excited state. The excited electron can then return to the ground state, emitting a photon. The energy of photons emitted by molecules is of the same order of magnitude as for atoms, typically 1 to 10 eV.

Additional energy levels become possible for molecules (but not for atoms) because the molecule as a whole can rotate, and the atoms of the molecule can vibrate relative to each other. The energy levels for both rotational and vibrational levels are quantized, and are generally spaced much more closely (10^{-3} to 10^{-1} eV) than the electronic levels. Each atomic energy level thus becomes a set of closely spaced levels corresponding to the vibrational and rotational motions, Fig. 41–14. Transitions from one level to another appear as many very closely spaced lines. In fact, the lines are not always distinguishable, and these spectra are called **band spectra**. Each type of molecule has its own characteristic spectrum, which can be used for identification and for determination of structure.

Rotational Energy Levels in Molecules

Let us now look in more detail at rotational and vibrational states in molecules. We begin with rotation, considering here only diatomic molecules, although the analysis can be extended to polyatomic molecules. When a diatomic molecule rotates about an axis through the center of mass perpendicular to the line joining the two atoms as shown in Fig. 41–15, its kinetic energy of rotation (see Section 10–10) is

$$E_{\text{rot}} = \frac{1}{2} I \omega^2 = \frac{(I\omega)^2}{2I},$$

where $(I\omega)$ is the angular momentum (Section 10–9). Quantum mechanics predicts quantization of angular momentum just as in atoms (see Eq. 40–3):

$$I\omega = \sqrt{L(L + 1)}\,\hbar, \qquad L = 0, 1, 2, \cdots,$$

where L is an integer called the **rotational angular momentum quantum number**. Thus the rotational energy is quantized:

$$E_{\text{rot}} = \frac{(I\omega)^2}{2I} = L(L + 1)\frac{\hbar^2}{2I}, \qquad L = 0, 1, 2, \cdots. \tag{41–2}$$

Transitions between rotational energy levels are subject to the *selection rule* (see the end of Section 40–2):

Selection rule

$$\Delta L = \pm 1.$$

The energy of a photon emitted or absorbed for a transition between rotational states with angular momentum quantum number L and $L - 1$ will be

$$\Delta E_{\text{rot}} = E_L - E_{L-1} = \frac{\hbar^2}{2I} L(L + 1) - \frac{\hbar^2}{2I}(L - 1)(L)$$

$$= \frac{\hbar^2}{I} L. \qquad \begin{bmatrix} L \text{ is for the} \\ \text{upper state} \end{bmatrix} \tag{41–3}$$

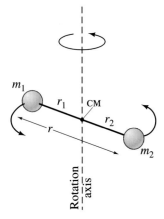

FIGURE 41–15 Diatomic molecule rotating about a vertical axis. The two atoms, of mass m_1 and m_2, are separated by a distance $r = r_1 + r_2$ where r_1 and r_2 are their respective distances from the center of mass.

We see that the transition energy increases directly with L. Figure 41–16 shows some of the allowed rotational energy levels and transitions. Measured absorption lines fall in the microwave or far-infrared regions of the spectrum, and their frequencies are generally $2, 3, 4, \cdots$ times higher than the lowest one, as predicted by Eq. 41–3.

The moment of inertia of the molecule in Fig. 41–15 rotating about its center of mass (Section 10–6) is

$$I = m_1 r_1^2 + m_2 r_2^2,$$

where r_1 and r_2 are the distances of each atom from their common center of mass. We show in Example 41–3 that I can be written

$$I = \frac{m_1 m_2}{m_1 + m_2} r^2 = \mu r^2, \qquad (41\text{–}4)$$

where $r = r_1 + r_2$ is the distance between the two atoms of the molecule and $\mu = m_1 m_2 / (m_1 + m_2)$ is called the **reduced mass**. If $m_1 = m_2$, then $\mu = \frac{1}{2}m_1 = \frac{1}{2}m_2$.

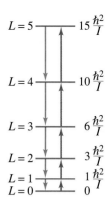

$L = 5$ ——— $15\frac{\hbar^2}{I}$

$L = 4$ ——— $10\frac{\hbar^2}{I}$

$L = 3$ ——— $6\frac{\hbar^2}{I}$

$L = 2$ ——— $3\frac{\hbar^2}{I}$

$L = 1$ ——— $1\frac{\hbar^2}{I}$

$L = 0$ ——— 0

FIGURE 41–16 Rotational energy levels and allowed transitions (emission and absorption) for a diatomic molecule. Upward-pointing arrows represent absorption of a photon, and downward arrows represent emission.

EXAMPLE 41–3 **Reduced mass.** Show that the moment of inertia of a diatomic molecule rotating about its center of mass can be written

$$I = \mu r^2,$$

where

$$\mu = \frac{m_1 m_2}{m_1 + m_2}$$

is the reduced mass, Eq. 41–4, and r is the distance between the atoms.

SOLUTION The moment of inertia of a single particle of mass m a distance r from the rotation axis is $I = mr^2$. Hence for our diatomic molecule (Fig. 41–15)

$$I = m_1 r_1^2 + m_2 r_2^2.$$

Now $r = r_1 + r_2$ and $m_1 r_1 = m_2 r_2$ because the axis of rotation passes through the center of mass. Hence

$$r_1 = r - r_2 = r - \frac{m_1}{m_2} r_1,$$

so

$$r_1 = \frac{r}{1 + \dfrac{m_1}{m_2}} = \frac{m_2 r}{m_1 + m_2}.$$

Similarly,

$$r_2 = \frac{m_1 r}{m_1 + m_2}.$$

Then

$$I = m_1 \left(\frac{m_2 r}{m_1 + m_2} \right)^2 + m_2 \left(\frac{m_1 r}{m_1 + m_2} \right)^2 = \frac{m_1 m_2 (m_1 + m_2) r^2}{(m_1 + m_2)^2}$$

$$= \frac{m_1 m_2}{m_1 + m_2} r^2 = \mu r^2,$$

where

$$\mu = \frac{m_1 m_2}{m_1 + m_2},$$

which is what we wished to show.

EXAMPLE 41–4 **Rotational transition.** A rotational transition $L = 1$ to $L = 0$ for the molecule CO has a measured absorption wavelength $\lambda = 2.60\,\text{mm}$ (microwave region). Use this to calculate (a) the moment of inertia of the CO molecule, and (b) the CO bond length, r. (c) Calculate the wavelengths of the next three rotational transitions, and the energies of the photon emitted for each of these four transitions.

SOLUTION (a) From Eq. 41–3, we can write

$$\frac{\hbar^2}{I}L = \Delta E = hf = \frac{hc}{\lambda_1}.$$

With $L = 1$ (the upper state) in this case, we solve for I:

$$I = \frac{\hbar^2 L}{hc}\lambda_1 = \frac{h\lambda_1}{4\pi^2 c} = \frac{(6.63 \times 10^{-34}\,\text{J·s})(2.60 \times 10^{-3}\,\text{m})}{4\pi^2(3.00 \times 10^8\,\text{m/s})}$$

$$= 1.46 \times 10^{-46}\,\text{kg·m}^2.$$

(b) The masses of C and O are 12.0 and 16.0 u, respectively, where $1\,\text{u} = 1.66 \times 10^{-27}\,\text{kg}$. Thus the reduced mass is

$$\mu = \frac{m_1 m_2}{m_1 + m_2} = \frac{(12.0)(16.0)}{28.0}(1.66 \times 10^{-27}\,\text{kg}) = 1.14 \times 10^{-26}\,\text{kg}$$

or 6.86 u. Then, from Eq. 41–4, the bond length is

$$r = \sqrt{\frac{I}{\mu}} = \sqrt{\frac{1.46 \times 10^{-46}\,\text{kg·m}^2}{1.14 \times 10^{-26}\,\text{kg}}} = 1.13 \times 10^{-10}\,\text{m} = 0.113\,\text{nm}.$$

(c) From Eq. 41–3, $\Delta E \propto L$. Hence $\lambda = c/f = hc/\Delta E$ is proportional to $1/L$. Thus, for $L = 2$ to $L = 1$ transitions, $\lambda_2 = \frac{1}{2}\lambda_1 = 1.30\,\text{mm}$. For $L = 3$ to $L = 2$, $\lambda_3 = \frac{1}{3}\lambda_1 = 0.87\,\text{mm}$. And for $L = 4$ to $L = 3$, $\lambda_4 = 0.65\,\text{mm}$. All are close to measured values. The energies of the photons, $hf = hc/\lambda$, are respectively $4.8 \times 10^{-4}\,\text{eV}$, $9.5 \times 10^{-4}\,\text{eV}$, $1.4 \times 10^{-3}\,\text{eV}$, and $1.9 \times 10^{-3}\,\text{eV}$.

Vibrational Energy Levels in Molecules

The potential energy of the two atoms in a typical diatomic molecule has the shape shown in Fig. 41–8 or 41–9, and Fig. 41–17 again shows the potential energy for the H_2 molecule. We note that the potential energy, at least in the vicinity of the equilibrium separation r_0, closely resembles the potential energy of a harmonic oscillator, $U = \frac{1}{2}kx^2$, which is shown superposed in dashed lines. Thus, for small displacements from r_0, each atom experiences a restoring force approximately proportional to the displacement, and the molecule vibrates as a simple harmonic oscillator (SHO). The classical frequency of vibration is

Classical vibration frequency

$$f = \frac{1}{2\pi}\sqrt{\frac{k}{\mu}}, \qquad\qquad (41\text{–}5)$$

FIGURE 41–17 Potential energy for the H_2 molecule and for a simple harmonic oscillator ($U_{SHO} = \frac{1}{2}kx^2$, with $|x| = |r - r_0|$). The 0.50 eV energy height marked is for use in Example 41–5 to estimate k. [Note that $U_{SHO} = 0$ is not the same as $U = 0$ for the molecule.]

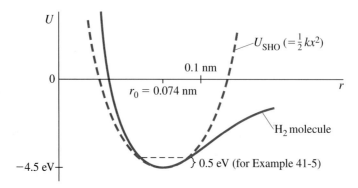

where k is the "stiffness constant" (as for a spring, Chapter 14) and instead of the mass m we must again use the reduced mass $\mu = m_1 m_2/(m_1 + m_2)$. (This is shown in Problem 59.) The Schrödinger equation for the SHO potential energy yields solutions for energy that are quantized according to

$$E_{\text{vib}} = \left(\nu + \tfrac{1}{2}\right)hf \qquad \nu = 0, 1, 2, \cdots, \qquad \textbf{(41–6)}$$

where f is given by Eq. 41–5 and ν is an integer called the **vibrational quantum number**. The lowest energy state $(\nu = 0)$ is not zero (as for rotation) but has $E = \tfrac{1}{2}hf$. This is called the **zero-point energy**.[†] Higher states have energy $\tfrac{3}{2}hf, \tfrac{5}{2}hf$, and so on, as shown in Fig. 41–18. Transitions are subject to the selection rule

$$\Delta\nu = \pm 1,$$

so allowed transitions occur only between adjacent states and all give off photons of energy

$$\Delta E_{\text{vib}} = hf. \qquad \textbf{(41–7)}$$

This is very close to experimental values for small ν, but for higher energies, the potential-energy curve (Fig. 41–17) begins to deviate from a perfect SHO curve, which affects the wavelengths and frequencies of the transitions. Typical transition energies are on the order of 10^{-1} eV, about 10 to 100 times larger than for rotational transitions, with wavelengths in the infrared region of the spectrum $(\approx 10^{-5}\,\text{m})$.[‡]

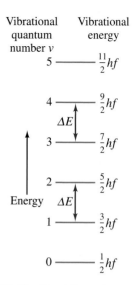

FIGURE 41–18 Allowed vibrational energies of a diatomic molecule, where f is the fundamental frequency of vibration, given by Eq. 41–5. Transitions are allowed only between adjacent levels $(\Delta\nu = \pm 1)$, which are equally spaced.

EXAMPLE 41–5 **ESTIMATE** **Wavelength for H_2.** (a) Use the curve of Fig. 41–17 to estimate the value of the stiffness constant k for the H_2 molecule, and then (b) estimate the fundamental wavelength for vibrational transitions.

SOLUTION (a) We choose arbitrarily an energy height of 0.50 eV which is indicated in Fig. 41–17. By measuring directly on the graph, we find that this energy corresponds to a vibration on either side of $r_0 = 0.074\,\text{nm}$ of about $x = \pm 0.017\,\text{nm}$. For SHO, $U_{\text{SHO}} = \tfrac{1}{2}kx^2$ and $U_{\text{SHO}} = 0$ at $x = 0$ $(r = r_0)$; then

$$k = \frac{2U_{\text{SHO}}}{x^2} \approx \frac{2(0.50\,\text{eV})(1.6 \times 10^{-19}\,\text{J/eV})}{(1.7 \times 10^{-11}\,\text{m})^2} \approx 550\,\text{N/m}.$$

This value of k would also be reasonable for a macroscopic spring. (b) The reduced mass is $\mu = m_1 m_2/(m_1 + m_2) = m_1/2 = \tfrac{1}{2}(1.0\,\text{u})(1.66 \times 10^{-27}\,\text{kg}) = 0.83 \times 10^{-27}\,\text{kg}$. Hence, using Eq. 41–5,

$$\lambda = \frac{c}{f} = 2\pi c\sqrt{\frac{\mu}{k}} = 2\pi(3.0 \times 10^8\,\text{m/s})\sqrt{\frac{0.83 \times 10^{-27}\,\text{kg}}{550\,\text{N/m}}}$$

$$= 2300\,\text{nm},$$

which is in the infrared region of the spectrum.

Experimentally, we do the inverse process: The wavelengths of vibrational transitions for a given molecule are measured, and from this the stiffness constant k can be calculated. The values of k calculated in this way are a measure of the strength of the molecular bond.

[†] Recall this phenomenon for a square well, Fig. 39–8.

[‡] Forbidden transitions with $\Delta\nu = 2$ are emitted somewhat more weakly, but their observation can be important in some cases, such as in astronomy.

EXAMPLE 41–6 **Vibrational energy levels in hydrogen.** Given that the hydrogen molecule emits infrared radiation of wavelength around 2300 nm, (*a*) what is the separation in energy between different vibrational levels, and (*b*) what is the lowest vibrational energy state?

SOLUTION (*a*)

$$\Delta E_{\text{vib}} = hf = \frac{hc}{\lambda} = \frac{(6.63 \times 10^{-34}\,\text{J}\cdot\text{s})(3.00 \times 10^8\,\text{m/s})}{(2300 \times 10^{-9}\,\text{m})(1.60 \times 10^{-19}\,\text{J/eV})} = 0.54\,\text{eV}.$$

(*b*) The lowest vibrational energy has $\nu = 0$ in Eq. 41–6: $E = \frac{1}{2}hf = 0.27\,\text{eV}$.

Rotational plus Vibrational Levels

When energy is imparted to a molecule, both the rotational and vibrational modes can be excited. Because rotational energies are an order of magnitude or so smaller than vibrational energies, which in turn are smaller than the electronic energy levels, we can represent the grouping of levels as shown in Fig. 41–19. Transitions between energy levels, with emission of a photon, are subject to the *selection rules*:

$$\Delta\nu = \pm 1$$

and

$$\Delta L = \pm 1.$$

Some allowed and forbidden (marked ✕) transitions are indicated in Fig. 41–19. Not all transitions and levels are shown, and the separation between vibrational levels, and (even more) between rotational levels, has been exaggerated. But we can clearly see the origin of the very closely spaced lines that give rise to the band spectra, as mentioned with reference to Fig. 41–14 earlier in this Section.

The spectra are quite complicated, so we consider briefly only transitions within the same electronic level, such as those at the top of Fig. 41–19. A transition from a state with quantum numbers ν and L, to one with quantum numbers $\nu + 1$ and $L \pm 1$ (see the selection rules above), will absorb[†] a photon of energy:

$$
\begin{aligned}
\Delta E &= \Delta E_{\text{vib}} + \Delta E_{\text{rot}} \\[4pt]
&= hf + (L+1)\frac{\hbar^2}{I} \qquad \begin{bmatrix} L \to L+1 \\ (\Delta L = +1) \end{bmatrix}, \quad L = 0, 1, 2, \cdots \\[4pt]
&= hf - L\frac{\hbar^2}{I} \qquad \begin{bmatrix} L \to L-1 \\ (\Delta L = -1) \end{bmatrix}, \quad L = 1, 2, 3, \cdots,
\end{aligned}
$$

(41–8)

where we have used Eqs. 41–3 and 41–7. Note that for $L \to L-1$ transitions, L cannot be zero since there is then no state with $L = -1$. Equations 41–8 predict an absorption spectrum like that shown schematically in Fig. 41–20, with transi-

[†]This is for absorption. For emission of a photon, the transition would be $\nu \to \nu - 1$, $L \to L \pm 1$.

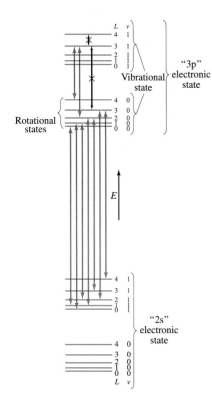

FIGURE 41–19 Combined electronic, vibrational, and rotational energy levels. Transitions marked with an ✕ are not allowed by the selection rules.

FIGURE 41–20 Expected spectrum for transitions between combined rotational and vibrational states.

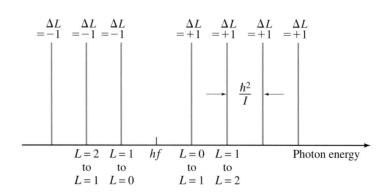

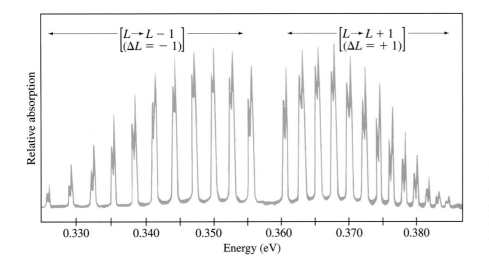

FIGURE 41–21 Absorption spectrum for HCl molecules. Lines on the left correspond to transitions where $L \rightarrow L - 1$; those on the right are for $L \rightarrow L + 1$. Each line has a double peak because chlorine has two isotopes of different mass and different moment of inertia.

tions $L \rightarrow L - 1$ on the left and $L \rightarrow L + 1$ on the right. Figure 41–21 shows the molecular absorption spectrum of HCl, which follows this pattern very well. (Each line in that spectrum is split into two because Cl consists of two isotopes of different mass; hence there are two kinds of HCl molecule with different moments of inertia I.)

EXAMPLE 41–7 **ESTIMATE** **The HCl molecule.** Estimate the moment of inertia of the HCl molecule using the absorption spectrum shown in Fig. 41–21. For the purposes of a rough estimate you can ignore the difference between the two isotopes.

SOLUTION The locations of the peaks in Fig. 41–21 should correspond to Eqs. 41–8. We don't know what value of L each peak corresponds to in Fig. 41–21, but we can estimate the energy difference between peaks to be about $\Delta E' = 0.0025$ eV. Then from Eqs. 41–8, the energy difference between two peaks is given by

$$\Delta E' = \Delta E_{L+1} - \Delta E_L$$
$$= \frac{\hbar^2}{I} \cdot$$

Then

$$I = \frac{\hbar^2}{\Delta E'} = \frac{\left(6.626 \times 10^{-34}\,\text{J} \cdot \text{s}/2\pi\right)^2}{\left(0.0025\,\text{eV}\right)\left(1.6 \times 10^{-19}\,\text{J/eV}\right)} = 2.8 \times 10^{-47}\,\text{kg} \cdot \text{m}^2.$$

If we use the results of Ex. 41–3, we can get a better idea of what this number means. We write $I = \mu r^2$ and calculate μ:

$$\mu = \frac{m_1 m_2}{m_1 + m_2} = \frac{(1.0\,\text{u})(35\,\text{u})}{36\,\text{u}}\left(1.66 \times 10^{-27}\,\text{kg/u}\right) = 1.6 \times 10^{-27}\,\text{kg}.$$

Then the bond length is given by (Eq. 41–4)

$$r = \left(\frac{I}{\mu}\right)^{\frac{1}{2}} = \left(\frac{2.8 \times 10^{-47}\,\text{kg} \cdot \text{m}^2}{1.6 \times 10^{-27}\,\text{kg}}\right)^{\frac{1}{2}} = 1.3 \times 10^{-10}\,\text{m},$$

which is the expected order of magnitude for a bond length.

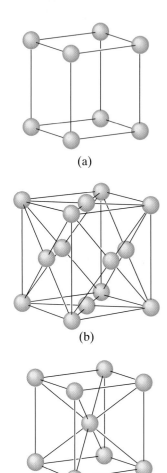

(a)

(b)

(c)

FIGURE 41–22 Arrangement of atoms in (a) a simple cubic crystal, (b) face-centered cubic crystal (note the atom at the center of each face) and (c) body-centered cubic crystal. Each shows the relationship of the bonds. Each of these "cells" is repeated in three dimensions to the edges of the macroscopic crystal.

FIGURE 41–23 Diagram of an NaCl crystal, showing the "packing" of atoms.

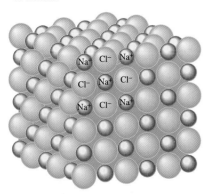

41–5 Bonding in Solids

Quantum mechanics has been a great tool for understanding the structure of solids. This active field of research today is called **solid-state physics**, or **condensed-matter physics** so as to include liquids as well. The rest of this chapter is devoted to this subject, and we begin with a brief look at the structure of solids and the bonds that hold them together.

Although some solid materials are *amorphous* in structure, in that the atoms and molecules show no long-range order, we will be interested here in the large class of *crystalline* substances whose atoms, ions, or molecules are generally believed to form an orderly array known as a *lattice*. Figure 41–22 shows three of the possible arrangements of atoms in a crystal: simple cubic, face-centered cubic, and body-centered cubic. The NaCl crystal is face-centered cubic (see Fig. 41–23), with one Na^+ ion or one Cl^- ion at each lattice point (i.e., considering Na and Cl separately).

The molecules of a solid are held together in a number of ways. The most common are by *covalent* bonding (such as between the carbon atoms of the diamond crystal) or *ionic* bonding (as in a NaCl crystal). Often the bonds are partially covalent and partially ionic. Our discussion of these bonds earlier in this chapter for molecules applies equally well here to solids.

Let us look for a moment at the NaCl crystal of Fig. 41–23. Each Na^+ ion feels an attractive Coulomb potential due to each of the six "nearest neighbor" Cl^- ions surrounding it. Note that one Na^+ does not "belong" exclusively to one Cl^-, so we must not think of ionic solids as consisting of individual molecules. Each Na^+ also feels a repulsive Coulomb potential due to other Na^+ ions, although this is weaker since the other Na^+ ions are farther away. Thus we expect a net attractive potential

$$U = -\alpha \frac{1}{4\pi\epsilon_0} \frac{e^2}{r}.$$

The factor α is called the *Madelung constant*. If each Na^+ were surrounded by only the six Cl^- ions, α would be 6, but the influence of all the other ions reduces it to a value $\alpha = 1.75$ for the NaCl crystal. The potential must also include a term representing the repulsive force when the wave functions of the inner shells and subshells overlap, and this has the form $U = B/r^m$, where m is a small integer. Thus

$$U = -\frac{\alpha}{4\pi\epsilon_0} \frac{e^2}{r} + \frac{B}{r^m}, \tag{41–9}$$

which has the same form as Eq. 41–1 for molecules (Section 41–2). It can be shown (Problem 19) that, at the equilibrium distance r_0,

$$U = U_0 = -\frac{\alpha}{4\pi\epsilon_0} \frac{e^2}{r_0} \left(1 - \frac{1}{m}\right).$$

This U_0 is known as the *ionic cohesive energy*; it is a sort of "binding energy"—the energy (per ion) needed to take the solid apart into separated ions.

A different type of bond occurs in metals. Metal atoms have relatively loosely held outer electrons. Present-day **metallic bond** theories propose that in a metallic solid, these outer electrons roam rather freely among all the metal atoms which, without their outer electrons, act like positive ions. The electrostatic attraction between the metal ions and this negative electron "gas" is what is believed, at least in part, to hold the solid together. The binding energy of metal bonds are typically 1 to 3 eV, somewhat weaker than ionic or covalent bonds (5 to 10 eV in solids). The "free electrons," according to this theory, are responsible for the high electrical and thermal conductivity of metals (see Sections 41–6 and 41–7). This theory also nicely accounts for the shininess of smooth metal surfaces: the electrons are free and can vibrate at any frequency, so when light of a range of frequencies falls on a metal, the electrons can vibrate in response and re-emit light of those same frequencies. Hence the reflected light will consist largely of the same frequencies as the incident light. Compare this to nonmetallic materials that have a distinct color—the atomic electrons exist only in certain energy states, and when white light falls on them, the atoms absorb at certain frequencies, and reflect other frequencies which make up the color we see.

The atoms or molecules of some materials, such as the noble gases, can form only **weak bonds** with each other. As we saw in Section 41–3, weak bonds have very low binding energies and would not be expected to hold atoms together as a liquid or solid at room temperature. The noble gases condense only at very low temperatures, where the atomic kinetic energy is small and the weak attraction can then hold the atoms together.

41–6 Free-Electron Theory of Metals

Let us look more closely at the free-electron theory of metals mentioned in the preceding Section. Let us imagine the electrons trapped within the metal as being in a potential well: inside the metal, the potential energy is zero, but at the edges of the metal there are high potential walls. Since very few electrons leave the metal at room temperature, we can imagine the walls as being infinitely high (as in Section 39–8). At higher temperatures, electrons do leave the metal (we know that thermionic emission occurs, Section 23–9), so we must recognize that the well is of finite depth. In this simple model, the electrons are trapped within the metal, but are free to move about inside the well whose size is macroscopic—the size of the piece of metal. The energy will be quantized, but the spacing between energy levels will be very tiny (see Eq. 39–13) since L is very large. Indeed, for a cube 1 cm on a side, the number of states with energy between, say, 5.0 and 5.5 eV, is on the order of 10^{22} (see Example 41–8).

To deal with such vast numbers of states, which are so closely spaced as to seem continuous, we need to use statistical methods. We define a quantity known as the **density of states**, $g(E)$, whose meaning is similar to the Maxwell distribution, Eq. 18–6 (see Section 18–2). That is, the quantity $g(E)\,dE$ represents the number of states per unit volume that have energy between E and $E + dE$. A careful calculation (see Problem 35), which must treat the potential well as three dimensional, shows that

$$g(E) = \frac{8\sqrt{2}\,\pi m^{\frac{3}{2}}}{h^3} E^{\frac{1}{2}}. \tag{41-10}$$

This is plotted in Fig. 41–24.

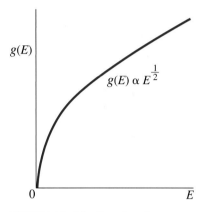

FIGURE 41–24 Density of states $g(E)$ as a function of energy E (Eq. 41–10).

Density of states

EXAMPLE 41–8 Electron states in copper. Estimate the number of states in the range 5.0 to 5.5 eV available to electrons in a 1.0-cm cube of copper metal.

SOLUTION Since $g(E)$ is the number of states per unit volume per unit energy interval, the number N of states is approximately (it is approximate because ΔE is not small)

$$N \approx g(E)V\,\Delta E,$$

where the volume $V = 1.0\,\text{cm}^3 = 1.0 \times 10^{-6}\,\text{m}^3$ and $\Delta E = 0.50\,\text{eV}$. We evaluate $g(E)$ at 5.25 eV, and find (Eq. 41–10):

$$N \approx g(E)V\,\Delta E = \frac{8\sqrt{2}\,\pi\left(9.1 \times 10^{-31}\,\text{kg}\right)^{\frac{3}{2}}}{\left(6.63 \times 10^{-34}\,\text{J·s}\right)^3}\,\sqrt{\left(5.25\,\text{eV}\right)\left(1.6 \times 10^{-19}\,\text{J/eV}\right)}$$

$$\times \left(1.0 \times 10^{-6}\,\text{m}^3\right)\left(0.50\,\text{eV}\right)\left(1.6 \times 10^{-19}\,\text{J/eV}\right)$$

$$\approx 8 \times 10^{21}$$

in $1.0\,\text{cm}^3$. Note that the type of metal did not enter the calculation.

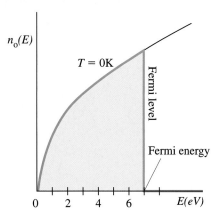

FIGURE 41–25 At $T = 0$ K, all states up to energy E_F, called the Fermi energy, are filled.

Fermi level

Fermi energy

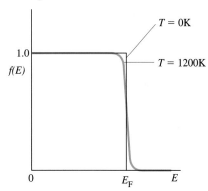

FIGURE 41–26 The Fermi–Dirac probability function for two temperatures, $T = 0$ K (black line) and $T = 1200$ K (blue curve). For $f(E) = 1$, a state with energy E is certainly occupied. For $f(E) = 0.5$, which occurs at $E = E_F$, the state with E_F has a 50% chance of being occupied.

Fermi factor

Equation 41–10 gives us the density of states. Now we must ask: How are the states available to an electron gas actually populated? Let us first consider the situation at absolute zero, $T = 0$ K. For a classical ideal gas, all the particles would be in the lowest state, with zero kinetic energy $\left(= \frac{3}{2}kT = 0\right)$. But the situation is vastly different for an electron gas because electrons obey the exclusion principle. Electrons do not obey classical statistics but rather a quantum statistics called **Fermi–Dirac statistics**[†] that takes into account the exclusion principle. All particles that have spin $\frac{1}{2}$ (or other half-integral spin: $\frac{3}{2}, \frac{5}{2}$, etc.), such as electrons, protons, and neutrons, obey Fermi–Dirac statistics and are referred to as **fermions**.[‡] The electron gas in a metal is often called a **Fermi gas**. According to the exclusion principle, no two electrons in the metal can have the same set of quantum numbers. Therefore, in each of the states of our potential well, there can be at most two electrons: one with spin up $\left(m_s = +\frac{1}{2}\right)$ and one with spin down $\left(m_s = -\frac{1}{2}\right)$. (This factor of 2 has already been included in Eq. 41–10.) Thus, at $T = 0$ K, the possible energy levels will be filled, two electrons each, up to a maximum level called the **Fermi level**. This is shown in Fig. 41–25, where the vertical axis is labeled $n_o(E)$ for "density of occupied states." The energy of the state at the Fermi level is called the **Fermi energy**, E_F. To determine E_F, we integrate Eq. 41–10 from $E = 0$ to $E = E_F$ (all states up to E_F are filled at $T = 0$ K):

$$\frac{N}{V} = \int_0^{E_F} g(E)\, dE, \tag{41–11}$$

where N/V is the number of conduction electrons per unit volume in the metal. Then, solving for E_F, the result (see Example 41–9) is

$$E_F = \frac{h^2}{8m}\left(\frac{3}{\pi}\frac{N}{V}\right)^{\frac{2}{3}}. \tag{41–12}$$

The average energy in this distribution (see Problem 31) is

$$\overline{E} = \tfrac{3}{5} E_F. \tag{41–13}$$

For copper, $E_F = 7.0$ eV (see Example 41–9) and $\overline{E} = 4.2$ eV. This is very much greater than the energy of thermal motion at room temperature $\left(\frac{3}{2}kT \approx 0.04 \text{ eV}\right)$. Clearly, all motion does not stop at absolute zero.

Thus, at $T = 0$, all states with energy below E_F are occupied, and all states above E_F are empty. What happens for $T > 0$? We expect that some (at least) of the electrons will increase in energy due to thermal motion. Classically, the distribution of occupied states would be given by the Boltzmann factor, $e^{-E/kT}$ (see Eqs. 40–16). But for our electron gas, a quantum-mechanical system obeying the exclusion principle, the probability of a given state of energy E being occupied is given by the **Fermi–Dirac probability function** (or **Fermi factor**):

$$f(E) = \frac{1}{e^{(E-E_F)/kT} + 1}, \tag{41–14}$$

where E_F is the Fermi energy. This function is plotted in Fig. 41–26 for two temperatures, $T = 0$ K and $T = 1200$ K (just below the melting point of copper). At $T = 0$ (or as T approaches zero) the factor $e^{(E-E_F)/kT}$ in Eq. 41–14 is zero if $E < E_F$ and is ∞ if $E > E_F$. Thus

$$f(E) = \begin{cases} 1 & E < E_F \\ 0 & E > E_F \end{cases} \quad \text{at} \quad T = 0.$$

This is what is plotted in black in Fig. 41–26 for $T = 0$ and is consistent with Fig. 41–25: all states up to the Fermi level are occupied [probability $f(E) = 1$]

[†] Developed independently by Enrico Fermi (Figs. 39–2, 39–10, and 42–7) in early 1926 and by P.A.M. Dirac a few months later.

[‡] Particles with integer spin (0, 1, 2, etc.), such as the photon, obey *Bose–Einstein* statistics and are called *bosons*.

and all states above are unoccupied. For $T = 1200\,\text{K}$, the Fermi factor changes only a little, as shown in Fig. 41–26 as the blue curve. Note that at any temperature T, when $E = E_\text{F}$, then Eq. 41–14 gives $f(E) = 0.50$, meaning the state at $E = E_\text{F}$ has a 50% chance of being occupied. To see how $f(E)$ affects the actual distribution of electrons in energy states, we must weight the density of possible states, $g(E)$, by the probability that those states will be occupied, $f(E)$. The product of these two functions then gives the **density of occupied states**,

$$n_\text{o}(E) = g(E)f(E) = \frac{8\sqrt{2}\,\pi m^{\frac{3}{2}}}{h^3}\frac{E^{\frac{1}{2}}}{e^{(E-E_\text{F})/kT} + 1}. \qquad \textbf{(41–15)}$$

Density of occupied states

Then $n_\text{o}(E)\,dE$ represents the number of electrons per unit volume with energy between E and $E + dE$ in thermal equilibrium at temperature T. This is plotted in Fig. 41–27 for $T = 1200\,\text{K}$, a temperature at which a metal is so hot it would glow. We see immediately that the distribution differs very little from that at $T = 0$. We see also that the changes that do occur are concentrated about the Fermi level. A few electrons from slightly below the Fermi level move to energy states slightly above it. The average energy of the electrons increases only very slightly when the temperature is increased from $T = 0\,\text{K}$ to $T = 1200\,\text{K}$. This is very different from the behavior of an ideal gas, for which kinetic energy increases directly with T. Nonetheless, this behavior is readily understood as follows. Energy of thermal motion at $T = 1200\,\text{K}$ is about $\frac{3}{2}kT \approx 0.1\,\text{eV}$. The Fermi level, on the other hand, is on the order of several eV: for copper it is $E_\text{F} \approx 7.0\,\text{eV}$. An electron at $T = 1200\,\text{K}$ may have $7\,\text{eV}$ of energy, but it can acquire at most only a few times $0.1\,\text{eV}$ of energy by a (thermal) collision with the lattice. Only electrons very near the Fermi level would find vacant states close enough to make such a transition. Essentially none of the electrons could increase in energy by, say, $3\,\text{eV}$, so electrons farther down in the electron gas are unaffected. Only electrons near the top of the energy distribution can be thermally excited to higher states. And their new energy is on the average only slightly higher than their old energy.

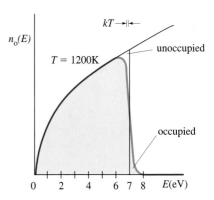

FIGURE 41–27 The density of occupied states for the electron gas in copper. The width kT represents thermal energy at $T = 1200\,\text{K}$.

EXAMPLE 41–9 **The Fermi level.** For the metal copper, determine (*a*) the Fermi energy, (*b*) the average energy of electrons, and (*c*) the speed of electrons at the Fermi level (this is called the *Fermi speed*).

SOLUTION (*a*) We combine Eqs. 41–10 and 41–11:

$$\frac{N}{V} = \frac{8\sqrt{2}\,\pi m^{\frac{3}{2}}}{h^3}\int_0^{E_\text{F}} E^{\frac{1}{2}}\,dE = \frac{8\sqrt{2}\,\pi m^{\frac{3}{2}}}{h^3}\frac{2}{3}E_\text{F}^{\frac{3}{2}}.$$

Solving for E_F, we obtain

$$E_\text{F} = \frac{h^2}{8m}\left(\frac{3}{\pi}\frac{N}{V}\right)^{\frac{2}{3}}$$

[note: $(\sqrt{2})^{\frac{2}{3}} = 2(\sqrt{2}/2^{\frac{3}{2}})^{\frac{2}{3}} = 2(\frac{1}{2})^{\frac{2}{3}}$], and this is Eq. 41–12. We calculated N/V, the number of conduction electrons per unit volume in copper, in Example 25–12 to be $N/V = 8.4 \times 10^{28}\,\text{m}^{-3}$. Hence for copper

$$E_\text{F} = \frac{(6.63 \times 10^{-34}\,\text{J}\cdot\text{s})^2}{8(9.1 \times 10^{-31}\,\text{kg})}\left[\frac{3(8.4 \times 10^{28}\,\text{m}^{-3})}{\pi}\right]^{\frac{2}{3}}\frac{1}{1.6 \times 10^{-19}\,\text{J/eV}} = 7.0\,\text{eV}.$$

(*b*) From Eq. 41–13, $\overline{E} = \frac{3}{5}E_\text{F} = 4.2\,\text{eV}$.
(*c*) In our model, we took $U = 0$ inside the metal, so $E = $ kinetic energy $= \frac{1}{2}mv^2$. Therefore, at the Fermi level, the Fermi speed is

$$v_\text{F} = \sqrt{\frac{2E_\text{F}}{m}} = \sqrt{\frac{2(7.0\,\text{eV})(1.6 \times 10^{-19}\,\text{J/eV})}{9.1 \times 10^{-31}\,\text{kg}}} = 1.6 \times 10^6\,\text{m/s},$$

a very high speed. The temperature of a classical gas would have to be extremely high to produce an average particle speed this large.

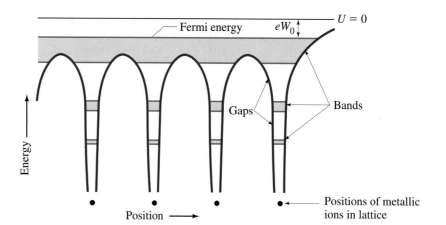

FIGURE 41–28 Potential energy for an electron in a metal crystal, with deep potential wells in the vicinity of each ion in the crystal lattice.

The simple model of the electron gas presented here provides good explanations for the electrical and thermal properties of conductors. But it does not explain why some materials are good conductors and others are good insulators. To provide an explanation, our model of a metal gas as a uniform potential well needs to be refined to include the effect of the lattice. Figure 41–28 shows a "periodic" potential that takes into account the attraction of electrons for each atomic ion in the lattice. Here we have taken $U = 0$ for an electron free of the metal; so within the metal, electron energies are less than zero (just as for molecules, or for the H atom, for example, in which the ground state has $E = -13.6\,\text{eV}$). The quantity eW_o represents the minimum energy to remove an electron from the metal, where W_o is the *work function* (see Section 38–2). The crucial outcome of putting a periodic potential (more easily approximated with narrow square wells) into the Schrödinger equation is that the allowed energy states are divided into *bands*, with energy gaps in between. Only electrons in the highest band, close to the Fermi level, are able to move about freely within the metal crystal. In the next Section we will see physically why there are bands and how they explain the properties of conductors, insulators, and semiconductors.

41–7 Band Theory of Solids

We saw in Section 41–1 that when two hydrogen atoms approach each other, the wave functions overlap, and the two 1s states (one for each atom) divide into two states of different energy. (As we saw, only one of these states, $S = 0$, has low enough energy to give a bound H_2 molecule.) Figure 41–29a shows this situation for 1s and 2s states for two atoms: as the atoms get closer, the 1s and 2s states split

FIGURE 41–29 The splitting of 1s and 2s atomic energy levels as (a) two atoms approach each other (the atomic separation decreases, moving toward the left); (b) the same for six atoms, and (c) for many atoms when they come together to form a solid.

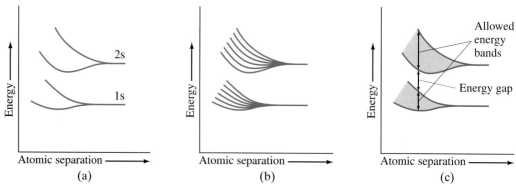

into two levels. If six atoms come together, as in Fig. 41–29b, each of the states splits into six levels. If a large number of atoms come together to form a solid, then each of the original atomic levels becomes a **band** as shown in Fig. 41–29c. The energy levels are so close together in each band that they seem essentially continuous. This is why the spectrum of heated solids (Section 38–1) appears continuous.

The crucial aspect of a good **conductor** is that the highest energy band containing electrons is only partially filled. Consider sodium, for example, whose energy bands are shown in Fig. 41–30. The 1s, 2s, and 2p bands are full (just as in a Na atom) and don't concern us. The 3s band, however, is only half full. To see why, recall that the exclusion principle stipulates that in an atom, only two electrons can be in the 3s state, one with spin up and one with spin down. These two states have slightly different energy. For a solid consisting of N atoms, the 3s band will contain $2N$ possible energy states. Now a sodium atom has a single 3s electron, so in a sample of sodium metal containing N atoms, there are N electrons in the 3s band, and N unoccupied states. When a potential difference is applied across the metal, electrons can respond by accelerating and increasing their energy, since there are plenty of unoccupied states of slightly higher energy available. Hence, a current flows readily and sodium is a good conductor. The characteristic of all good conductors is that the highest energy band is only partially filled, or two bands overlap so that unoccupied states are available. An example of the latter is magnesium, which has two 3s electrons, so its 3s band is filled. But the unfilled 3p band overlaps the 3s band in energy, so there are lots of available states for the electrons to move into. Thus magnesium, too, is a good conductor.

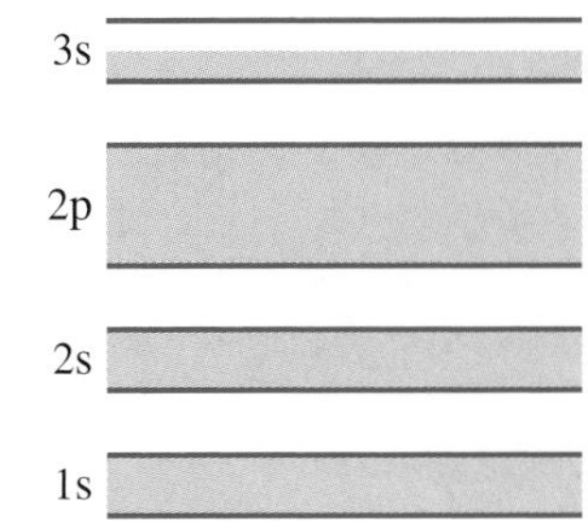

In a material that is a good **insulator**, on the other hand, the highest band containing electrons, called the **valence band**, is completely filled. The next higher energy band, called the **conduction band**, is separated from the valence band by a "forbidden" **energy gap**, (or **band gap**), E_g, of typically 5 to 10 eV. So at room temperature (300 K), where thermal energies (that is, average kinetic energy—see Chapter 18) are on the order of $\frac{3}{2}kT \approx 0.04\,\text{eV}$, almost no electrons can acquire the 5 eV needed to reach the conduction band. When a potential difference is applied across the material, no available states are accessible to the electrons, and no current flows. Hence, the material is a good insulator.

Figure 41–31 compares the relevant energy bands (a) for conductors, (b) for insulators, and also (c) for the important class of materials known as **semiconductors**. The bands for a pure (or **intrinsic**) semiconductor, such as silicon or germanium, are like those for an insulator, except that the unfilled conduction band is separated from the filled valence band by a much smaller energy gap, E_g, typically on the order of 1 eV. At room temperature, a few electrons can acquire enough thermal energy to reach the conduction band, and so a very small current may flow when a voltage is applied. At higher temperatures, more electrons have enough energy to jump the gap. This effect can often more than offset the effects of more frequent collisions due to increased disorder at higher temperature, so that the resistivity of semiconductors can *decrease* with increasing temperature (see Table 25–1). But this is not the whole story of semiconductor conduction. When a potential difference is applied to a semiconductor, the few electrons in the conduction band move toward the positive electrode. Electrons in the valence band try to do the same thing, and a few can because there are a small number of unoccupied

Conductors

3s

2p

2s

1s

FIGURE 41–30 Energy bands for sodium.

Insulators

Valence and conduction bands

Energy gap

Semiconductors (pure)

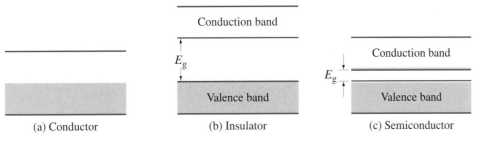

Conduction band

E_g

Valence band

(a) Conductor

(b) Insulator

Conduction band

E_g

Valence band

(c) Semiconductor

FIGURE 41–31 Energy bands for (a) a conductor, (b) an insulator, and (c) a semiconductor. Shading represents occupied states. Pale shading in part (c) represents electrons that can pass from the valence band to the conduction band due to thermal agitation at room temperature (exaggerated).

states which were left empty by the electrons reaching the conduction band. Such unfilled electron states are called **holes**. Each electron in the valence band that fills a hole in this way as it moves toward the positive electrode leaves behind a hole, so that the holes migrate toward the negative electrode. As the electrons tend to accumulate at one side of the material, the holes tend to accumulate on the opposite side. We will look at this phenomenon in more detail in the next Section.

EXAMPLE 41–10 **Calculating the energy gap.** It is found that the conductivity of a certain semiconductor increases when light of wavelength 345 nm or shorter strikes it, suggesting that electrons are being promoted from the valence band to the conduction band. What is the energy gap, E_g, for this semiconductor?

SOLUTION The longest wavelength, or lowest energy, photon to cause an increase in conductivity has $\lambda = 345$ nm, and it can transfer to an electron an energy

$$E_g = hf = \frac{hc}{\lambda} = \frac{(6.63 \times 10^{-34}\,\text{J}\cdot\text{s})(3.00 \times 10^8\,\text{m/s})}{(345 \times 10^{-9}\,\text{m})(1.60 \times 10^{-19}\,\text{J/eV})} = 3.6\,\text{eV}.$$

EXAMPLE 41–11 **ESTIMATE** **Free electrons in semiconductors and insulators.** Use the Fermi–Dirac probability function, Eq. 41–14, to estimate the order of magnitude of the numbers of free electrons in the conduction band of a solid containing 10^{21} atoms, assuming the solid is at room temperature ($T = 300$ K) and is (a) a semiconductor with $E_g \approx 1.1$ eV, (b) an insulator with $E_g \approx 5$ eV. Compare to a conductor.

SOLUTION At $T = 0$, all states above the Fermi energy E_F are empty, and all those below are full. So for semiconductors and insulators we can take E_F to be about midway between the valence and conduction bands, Fig. 41–32, and it does not change significantly as we go to room temperature. (a) For the semiconductor, the gap $E_g \approx 1.1$ eV, so $E - E_F \approx 0.55$ eV for the lowest states in the conduction band. Since at room temperature we have $kT \approx 0.026$ eV, then $(E - E_F)/kT \approx 0.55\,\text{eV}/0.026\,\text{eV} \approx 21$ and

$$f(E) = \frac{1}{e^{(E-E_F)/kT} + 1} \approx \frac{1}{e^{21}} \approx 10^{-9}.$$

Thus about 1 atom in 10^9 can contribute an electron to the conductivity. (b) For the insulator with $E - E_F \approx 5.0\,\text{eV} - \frac{1}{2}(5.0\,\text{eV}) = 2.5$ eV, we get

$$f(E) \approx \frac{1}{e^{2.5/0.026} + 1} \approx \frac{1}{e^{96}} \approx 10^{-42}.$$

Thus in an ordinary sample containing 10^{21} atoms, there would be no free electrons in an insulator $(10^{21} \times 10^{-42} = 10^{-21})$, about $10^{12}\,(10^{21} \times 10^{-9})$ free electrons in a semiconductor, and about 10^{21} free electrons in a good conductor.

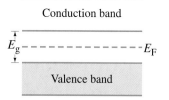

Conduction band

E_g ─ ─ ─ ─ ─ ─ ─ ─ ─ ─ E_F

Valence band

FIGURE 41–32 The Fermi energy is midway between the valence band and the conduction band.

CONCEPTUAL EXAMPLE 41–12 **Which is transparent?** The energy gap for silicon is 1.14 eV at room temperature while that of zinc sulfide (ZnS) is 3.6 eV. Which one of these is opaque and which is transparent to visible light?

RESPONSE Visible light photons span energies from roughly 1.8 eV to 3.2 eV ($E = hf = hc/\lambda$ where $\lambda = 400$ nm to 700 nm and $1\,\text{eV} = 1.6 \times 10^{-19}$ J). Light is absorbed by the electrons in a material. Silicon's gap is small enough to absorb these photons, thus bumping electrons well up into the conduction band, and so silicon is opaque. On the other hand, zinc sulfide's energy gap is too wide to absorb visible photons, and so the light can pass through the material; it can be transparent.

41–8 Semiconductors and Doping

The most commonly used semiconductors in modern electronics are silicon (Si) and germanium (Ge). An atom of silicon or germanium has four outer electrons that act to hold the atoms in the regular lattice structure of the crystal, shown schematically in Fig. 41–33a. Germanium and silicon acquire useful properties for use in electronics only when a tiny amount of impurity is introduced into the crystal structure (perhaps 1 part in 10^6 or 10^7). This is called **doping** the semiconductor. Two kinds of doped semiconductor can be made, depending on the type of impurity used. If the impurity is an element whose atoms have five outer electrons, such as arsenic, we have the situation shown in Fig. 41–33b, with the arsenic atoms holding positions in the crystal lattice where normally silicon atoms would be. Only four of arsenic's electrons fit into the bonding structure. The fifth does not fit in and can move relatively freely, somewhat like the electrons in a conductor. Because of this small number of extra electrons, a doped semiconductor becomes slightly conducting. The density of conduction electrons in an intrinsic (pure) semiconductor is about 1 per 10^9 atoms, as we saw in Example 41–11. With an impurity concentration of 1 in 10^6 or 10^7 when doped, the conductivity will be much higher and it can be controlled with great precision. An arsenic-doped silicon crystal is called an **n-type semiconductor** because *n*egative charges (electrons) carry the electric current.

Doped semiconductors

n-type

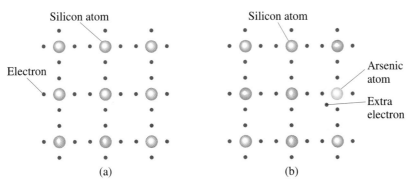

FIGURE 41–33 Two-dimensional representation of a silicon crystal. (a) Four (outer) electrons surround each silicon atom. (b) Silicon crystal doped with a few arsenic atoms: the extra electron doesn't fit into the crystal lattice and so is free to move about. This is an *n*-type semiconductor.

In a **p-type semiconductor**, a small percentage of semiconductor atoms are replaced by atoms with three outer electrons—such as gallium. As shown in Fig. 41–34a, there is a "hole" in the lattice structure near a gallium atom since it has only three outer electrons. Electrons from nearby silicon atoms can jump into this hole and fill it. But this leaves a hole where that electron had previously been, Fig. 41–34b. The vast majority of atoms are silicon, so holes are almost always next to a silicon atom. Since silicon atoms require four outer electrons to be neutral, this means that there is a net positive charge at the hole. Whenever an electron moves to fill a hole, the positive hole is then at the previous position of that electron. Another electron can then fill this hole, and the hole thus moves to a new location; and so on. This type of semiconductor is called *p*-type because it is the positive holes that seem to carry the electric current. Note, however, that both *p*-type and *n*-type semiconductors have *no net charge* on them.

p-type

Holes are positive

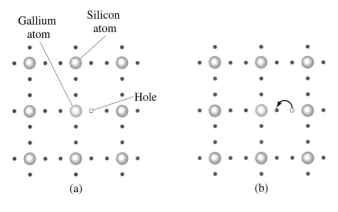

FIGURE 41–34 A *p*-type semiconductor, gallium-doped silicon. (a) Gallium has only three outer electrons, so there is an empty spot, or *hole*, in the structure. (b) Electrons from silicon atoms can jump into the hole and fill it. As a result, the hole moves to a new location (to the right in this figure), to where the electron used to be.

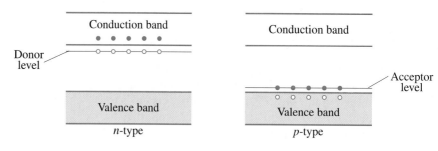

FIGURE 41-35 Impurity energy levels in doped semiconductors.

According to the band theory (Section 41–7), in a doped semiconductor, the impurity provides additional energy states between the bands, as shown in Fig. 41–35. In an *n*-type semiconductor, the impurity energy level lies just below the conduction band. Electrons in this energy level need only about 0.05 eV (in Si; even less in Ge) of energy to reach the conduction band; this is on the order of the thermal energy, $\frac{3}{2}kT$ (= 0.04 eV at 300 K), so transitions occur readily at room temperature. This energy level can thus supply electrons to the conduction band, so it is called a **donor** level. In *p*-type semiconductors, the impurity energy level is just above the valence band (Fig. 41–35). It is called an **acceptor** level because electrons from the valence band can easily jump into it. Positive holes are left behind in the valence band, and as other electrons move into these holes, the holes move about as discussed earlier.

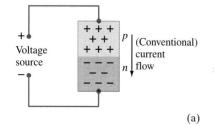

(a)

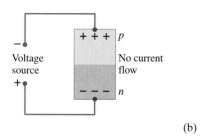

(b)

FIGURE 41-36 Schematic diagram showing how a semiconductor diode operates. Current flows when the voltage is connected in forward bias, as in (a), but not when connected in reverse bias, as in (b).

FIGURE 41-37 Current through a diode as a function of applied voltage.

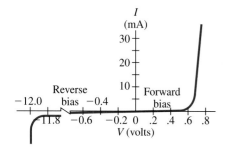

* ⎢41–9⎢ Semiconductor Diodes

Semiconductor diodes and transistors are essential components of modern electronic devices. The miniaturization achieved today allows many thousands of diodes, transistors, resistors, and so on, to be placed on a single *chip* only a millimeter on a side. We now discuss, briefly and qualitatively, the operation of diodes and transistors.

When an *n*-type semiconductor is joined to a *p*-type, a **p-n junction diode** is formed. Separately, the two semiconductors are electrically neutral. When joined, a few electrons near the junction diffuse from the *n*-type into the *p*-type semiconductor, where they fill a few of the holes. The *n*-type is left with a positive charge, and the *p*-type acquires a net negative charge. Thus a potential difference is established, with the *n* side positive relative to the *p* side, and this prevents further diffusion of electrons.

If a battery is connected to a diode with the positive terminal to the *p* side and the negative terminal to the *n* side as in Fig. 41–36a, the externally applied voltage opposes the internal potential difference and the diode is said to be **forward biased**. If the voltage is great enough (about 0.3 V for Ge, 0.6 V for Si at room temperature), a current will flow. The positive holes in the *p*-type semiconductor are repelled by the positive terminal of the battery and the electrons in the *n*-type are repelled by the negative terminal of the battery. The holes and electrons meet at the junction, and the electrons cross over and fill the holes. A current is flowing. Meanwhile, the positive terminal of the battery is continually pulling electrons off the *p* end, forming new holes, and electrons are being supplied by the negative terminal at the *n* end. Consequently, a large current flows through the diode.

When the diode is **reverse biased**, as in Fig. 41–36b, the holes in the *p* end are attracted to the battery's negative terminal and the electrons in the *n* end are attracted to the positive terminal. The current carriers do not meet near the junction and, ideally, no current flows.

A graph of current versus voltage for a typical diode is shown in Fig. 41–37. As can be seen, a real diode does allow a small amount of reverse current to flow due to thermal motion $(\frac{3}{2}kT)$ of electron pairs. For most practical purposes, this is negligible (at room temperature, a few μA in Ge, a few pA in Si; the reverse current increases rapidly with temperature, however, and may render a diode ineffective above 200°C).

EXAMPLE 41–13 **A diode.** The diode whose current–voltage characteristics are shown in Fig. 41–37 is connected in series with a 4.0-V battery and a resistor. If a current of 10 mA is to pass through the diode, what resistance must the resistor have?

SOLUTION In Fig. 41–37, we see that the voltage drop across the diode is about 0.7 V when the current is 10 mA. Therefore, the voltage drop across the resistor is 4.0 V − 0.7 V = 3.3 V, so $R = V/I = (3.3\ \text{V})/(1.0 \times 10^{-2}\ \text{A}) = 330\ \Omega$.

If the voltage across a diode connected in reverse bias is increased greatly, a point is reached where breakdown occurs. The electric field across the junction becomes so large that ionization of atoms results. The electrons thus pulled off their atoms contribute to a larger and larger current as breakdown continues. The voltage remains constant over a wide range of currents. This is shown on the far left in Fig. 41–37. This property of diodes can be used to accurately regulate a voltage supply. A diode designed for this purpose is called a **zener diode**. When placed across the output of an unregulated power supply, a zener diode can maintain the voltage at its own breakdown voltage as long as the supply voltage is always above this point. Zener diodes can be obtained corresponding to voltages of a few volts to hundreds of volts.

Since a *p-n* junction diode allows current to flow only in one direction (as long as the voltage is not too high), it can serve as a **rectifier**—to change ac into dc. A simple rectifier circuit is shown in Fig. 41–38a where the arrow inside the symbol for a diode indicates the direction in which a diode conducts conventional (+) current. The ac source applies a voltage across the diode alternately positive and negative. Only during half of each cycle will a current pass through the diode; so only then is there a current through the resistor R. Hence, a graph of the voltage V_{ab} across R as a function of time looks like Fig. 41–38b. This **half-wave rectification** is not exactly dc, but it is unidirectional. More useful is a **full-wave rectifier** circuit, which uses two diodes (or sometimes four) as shown in Fig. 41–39a. At any given instant, either one diode or the other will conduct current to the right. Therefore, the output across the load resistor R will be as shown in Fig. 41–39b. Actually this is the voltage if the capacitor C were not in the circuit. The capacitor tends to store charge, and thus helps to smooth out the current as shown in Fig. 41–39c.

Rectifier circuits are important because most line voltage is ac, and most electronic devices require a dc voltage for their operation. Hence, diodes are found in nearly all electronic devices including radio and TV sets, calculators, and computers.

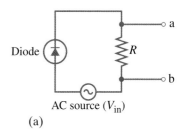

(a)

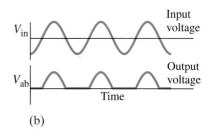

(b)

FIGURE 41–38 (a) A simple (half-wave) rectifier circuit using a semiconductor diode. (b) AC source input voltage, and output voltage across R, as functions of time.

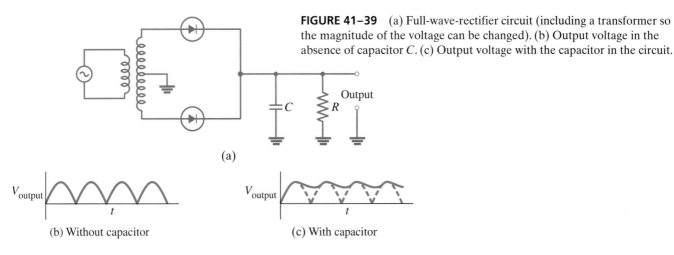

(a)

(b) Without capacitor

(c) With capacitor

FIGURE 41–39 (a) Full-wave-rectifier circuit (including a transformer so the magnitude of the voltage can be changed). (b) Output voltage in the absence of capacitor C. (c) Output voltage with the capacitor in the circuit.

LED

Another useful device is a **light-emitting diode** (LED). When a *p-n* junction is forward biased, a current begins to flow. Electrons cross from the *n* region into the *p* region and combine with holes, and a photon can be emitted with an energy approximately equal to the band gap, E_g (see Figs. 41–31c and 41–35). Often the energy, and hence the wavelength, is in the red region of the visible spectrum, producing the familiar LED displays on VCRs, CD players, car instrument panels, clocks, and so on. Infrared (i.e., nonvisible) LEDs are used in remote controls for TV, VCRs, and stereos.

Solar cells

Photodiodes and **solar cells** are *p-n* junctions used in the reverse way. Photons are absorbed, creating electron–hole pairs if the photon energy is greater than the band gap energy, E_g. The created electrons and holes produce a current that, when connected to an external circuit, becomes a source of emf and power.

A diode is called a **nonlinear device** because the current is not proportional to the voltage; that is, a graph of current versus voltage (Fig. 41–37) is not a straight line as it is for a resistor (which ideally *is* linear). Transistors are also *nonlinear* devices.

** 41–10* Transistors and Integrated Circuits

Transistors

A simple **junction transistor** consists of a crystal of one type of doped semiconductor sandwiched between two crystals of the opposite type. Both *pnp* and *npn* transistors are made, and they are shown schematically in Fig. 41–40a. The three semiconductors are given the names *collector, base,* and *emitter.* The symbols for *npn* and *pnp* transistors are shown in Fig. 41–40b. The arrow is always placed on the emitter and indicates the direction of (conventional) current flow in normal operation.

The operation of a transistor can be analyzed qualitatively—very briefly—as follows. Consider an *npn* transistor connected as shown in Fig. 41–41. A voltage V_{CE} is maintained between the collector and emitter by the battery $\mathscr{E}_C$. The voltage applied to the base is called the *base bias voltage*, V_{BE}. If V_{BE} is positive, conduction electrons in the emitter are attracted into the base. Since the base region is very thin (perhaps $1\ \mu$m), most of these electrons flow right across into the collector, which is maintained at a positive voltage. A large current, I_C, flows between collector and emitter and a much smaller current, I_B, through the base. A small variation in the base voltage due to an input signal causes a large change in the collector current and therefore a large change in the voltage drop across the output resistor R_C. Hence a transistor can *amplify* a small signal into a larger one.

Amplifiers

FIGURE 41–40 (a) Schematic diagram of *npn* and *pnp* transistors. (b) Symbols for *npn* and *pnp* transistors.

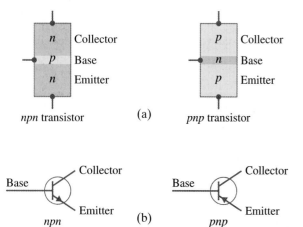

FIGURE 41–41 An *npn* transistor used as an amplifier.

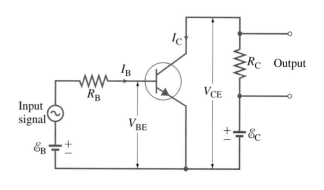

Normally a small ac signal is to be amplified, and when added to the base bias voltage causes the voltage and current at the collector to vary at the same rate but magnified. Thus, what is important for amplification is the *change* in collector current for a given input *change* in base current. The **current gain** is defined as the ratio

$$\beta_I = \frac{\text{output (collector) ac current}}{\text{input (base) ac current}} = \frac{i_C}{i_B}$$
Current gain

where i_B and i_C are the ac base current (input signal) and ac collector current (output). β_I is typically on the order of 10 to 100. Similarly, the **voltage gain** is

$$\beta_V = \frac{\text{output (collector) ac voltage}}{\text{input (base) ac voltage}}.$$
Voltage gain

Transistors are the basic elements in modern electronic **amplifiers** of all sorts.

A *pnp* transistor operates like an *npn*, except that holes move instead of electrons. The collector voltage is negative, and so is the base voltage in normal operation.

In Fig. 41–41 two batteries were shown: $\mathcal{E}_C$ supplied the collector voltage and $\mathcal{E}_B$ the base bias voltage. In practice, only one source is often used, and the base bias voltage can be obtained using a resistance voltage divider as in Fig. 41–42. Transistors can be connected in many other ways as well, and many new and innovative uses have been found for them.

Transistors were a great advance in miniaturization of electronic circuits. Although individual transistors are very small compared to the once used vacuum tubes, they are huge compared to **integrated circuits** or **chips** (see photo at start of this chapter). Tiny amounts of impurities can be placed at particular locations within a single silicon crystal. These can be arranged to form diodes, transistors, and resistors (undoped semiconductors). Capacitors and inductors can also be formed, although they are often connected separately. A tiny chip, only 1 mm on a side, may contain thousands of transistors and other circuit elements. Integrated circuits are now the heart of computers, television, calculators, cameras, and the electronic instruments that control aircraft, space vehicles, and automobiles. The "miniaturization" produced by integrated circuits not only allows extremely complicated circuits to be placed in a small space, but also has allowed a great increase in the speed of operation of, say, computers, because the distances the electronic signals travel are so tiny.

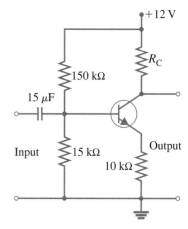

FIGURE 41–42 Typical transistor circuit involving an *npn* transistor. $\mathcal{E}_C = +12\,\text{V}$ and $\mathcal{E}_B$ is determined by the resistors.

Summary

Quantum mechanics explains the bonding together of atoms to form **molecules**. In a **covalent bond**, the electron clouds of two or more atoms overlap because of constructive interference between the electron waves. The positive nuclei are attracted to this concentration of negative charge between them, forming the bond. An **ionic bond** is an extreme case of a covalent bond in which one or more electrons from one atom spend much more time around the other atom than around their own. The atoms then act as oppositely charged ions that attract each other, forming the bond.

These **strong bonds** hold molecules together, and also hold atoms and molecules together in solids. Also important are **weak bonds** (or **van der Waals bonds**), which are generally dipole attractions between molecules.

When atoms combine to form molecules, the energy levels of the outer electrons are altered because they now interact with each other. Additional energy levels also become possible because the atoms can vibrate with respect to each other, and the molecule as a whole can rotate. The energy levels for both vibrational and rotational motion are quantized, and are very close together (typically, $10^{-1}\,\text{eV}$ to $10^{-3}\,\text{eV}$ apart). Each atomic energy level thus becomes a set of closely spaced levels corresponding to the vibrational and rotational motions. Transitions from one level to another appear as many very closely spaced lines. The resulting spectra are called **band spectra**.

The quantized rotational energy levels are given by

$$E_{\text{rot}} = L(L+1)\frac{\hbar^2}{2I}, \qquad L = 0, 1, 2, \cdots,$$

where I is the moment of inertia of the molecule.

The energy levels for vibrational motion are given by

$$E_{vib} = \left(\nu + \tfrac{1}{2}\right)hf, \qquad \nu = 0, 1, 2, \cdots,$$

where f is the classical natural frequency of vibration for the molecule. Transitions between energy levels are subject to the selection rules $\Delta L = \pm 1$ and $\Delta \nu = \pm 1$.

Some **solids** are bound together by covalent and ionic bonds, just as molecules are. In metals, the electrostatic force between free electrons and positive ions helps form the **metallic bond**.

In the free-electron theory of metals, electrons occupy the possible energy states according to the exclusion principle. At $T = 0\,K$, all possible states are filled up to a maximum energy level called the *Fermi energy*, E_F, the magnitude of which is typically a few eV. All states above E_F are vacant at $T = 0\,K$. At normal temperatures (300 K) the distribution of occupied states is only slightly altered and is given by the **Fermi–Dirac probability function**

$$f(E) = \frac{1}{e^{(E - E_F)/kT} + 1}.$$

In a crystalline solid, the possible energy states for electrons are arranged in **bands**. Within each band the levels are very close together, but between the bands there may be forbidden **energy gaps**. Good conductors are characterized by the highest occupied band (the **conduction band**) being only partially full, so there are many accessible states available to electrons to move about and accelerate when a voltage is applied. In a good insulator, the highest occupied energy band (the **valence band**) is completely full, and there is a large energy gap (5 to 10 eV) to the next highest band, the *conduction band*. At room temperature, molecular kinetic energy (thermal energy) available due to collisions is only about 0.04 eV, so almost no electrons can jump from the valence to the conduction band. In a **semiconductor**, the gap between valence and conduction bands is much smaller, on the order of 1 eV, so a few electrons can make the transition from the essentially full valence band to the nearly empty conduction band.

In a **doped** semiconductor, a small percentage of impurity atoms with five or three valence electrons replace a few of the normal silicon atoms with their four valence electrons. A five-electron impurity produces an **n-type** semiconductor with negative electrons as carriers of current. A three-electron impurity produces a **p-type** semiconductor in which positive **holes** carry the current. The energy level of impurity atoms lies slightly below the conduction band in an *n*-type semiconductor, and acts as a **donor** from which electrons readily pass into the conduction band. The energy level of impurity atoms in a *p*-type semiconductor lies slightly above the valence band and acts as an **acceptor** level, since electrons from the valence band easily reach it, leaving holes behind to act as charge carriers.

A semiconductor **diode** consists of a **p-n junction** and allows current to flow in one direction only; it can be used as a **rectifier** to change ac to dc. Common **transistors** consist of three semiconductor sections, either as **pnp** or **npn**. Transistors can amplify electrical signals and find many other uses. An integrated circuit consists of a tiny semiconductor crystal or "chip" on which many transistors, diodes, resistors, and other circuit elements have been constructed using careful placement of impurities.

Questions

1. What type of bond would you expect for (*a*) the N_2 molecule, (*b*) the HCl molecule, (*c*) Fe atoms in a solid?
2. Describe how the molecule $CaCl_2$ could be formed.
3. Does the H_2 molecule have a permanent dipole moment? Does O_2? Does H_2O? Explain.
4. Although the molecule H_3 is not stable, the ion H_3^+ is. Explain, using the Pauli exclusion principle.
5. The energy of a molecule can be divided into four categories. What are they?
6. If conduction electrons are free to roam about in a metal, why don't they leave the metal entirely?
7. A silicon semiconductor is doped with phosphorus. Will these atoms be donors or acceptors? What type of semiconductor will this be?
8. Explain why the resistivity of metals increases with temperature whereas the resistivity of semiconductors may decrease with increasing temperature.
9. Discuss the differences between an ideal gas and a Fermi electron gas.
10. Which aspects of Fig. 41–27 are peculiar to copper, and which are valid in general for other metals?

* 11. Can a diode be used to amplify a signal?
* 12. Figure 41–43 shows a "bridge-type" full-wave rectifier. Explain how the current is rectified and how current flows during each half cycle.

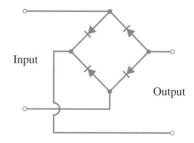

FIGURE 41–43 Question 12.

* 13. Compare the resistance of a *p-n* junction diode connected in forward bias to its resistance when connected in reverse bias.
* 14. Explain how a transistor could be used as a switch.

*15. If $\mathscr{E}_C$ were reversed in Fig. 41–41, how would the amplification be altered?

*16. Describe how a *pnp* transistor can operate as an amplifier.

*17. Do diodes and transistors obey Ohm's law?

*18. In a transistor, the base–emitter junction and the base–collector junction are essentially diodes. Are these junctions reverse-biased or forward-biased in the application shown in Fig. 41–41?

*19. What purpose does the capacitor in Fig. 41–42 serve?

Problems

Sections 41–1 to 41–3

1. (I) Estimate the binding energy of a KCl molecule by calculating the electrostatic potential energy when the K^+ and Cl^- ions are at their stable separation of 0.28 nm. Assume each has a charge of magnitude $1.0e$.

2. (I) Binding energies are often measured experimentally in kcal per mole, and then the binding energy in eV per molecule is calculated from that result. What is the conversion factor in going from kcal per mole to eV per molecule? What is the binding energy of KCl (= 4.43 eV) in kcal per mole?

3. (II) The measured binding energy of KCl is 4.43 eV. From the result of Problem 1, estimate the contribution to the binding energy of the repelling electron clouds at the equilibrium distance $r_0 = 0.28$ nm.

4. (II) Estimate the binding energy of the H_2 molecule, assuming the two H nuclei are 0.074 nm apart and the two electrons spend 33 percent of their time midway between them.

5. (III) Apply reasoning similar to that in the text for the $S = 0$ and $S = 1$ states in the formation of the H_2 molecule to show why the molecule He_2 is *not* formed. Show also why the He_2^+ molecular ion *is* formed (with a binding energy of 3.1 eV at $r_0 = 0.11$ nm).

Section 41–4

6. (I) Show that the quantity $\hbar^2/I$ has units of energy.

7. (I) What is the reduced mass of the molecules (a) NaCl; (b) N_2; (c) HCl?

8. (II) The so-called "characteristic rotational energy," $\hbar^2/2I$, for N_2 is 2.48×10^{-4} eV. Calculate the N_2 bond length.

9. (II) The fundamental vibration frequency for the CO molecule is 6.42×10^{13} Hz. Determine (a) the reduced mass, and (b) the effective value of the constant k. Compare to k for the H_2 molecule.

10. (II) Explain why there is no transition for $E = hf$ in Fig. 41–20 (and Fig. 41–21).

11. (II) (a) Calculate the characteristic rotational energy, $\hbar^2/2I$, for the O_2 molecule whose bond length is 0.121 nm. (b) What are the energy and wavelength of photons emitted in a $L = 2$ to $L = 1$ transition?

12. (II) The equilibrium separation of H atoms in the H_2 molecule is 0.074 nm (Fig. 41–8). Calculate the energies and wavelengths of photons for the rotational transitions (a) $L = 1$ to $L = 0$, (b) $L = 2$ to $L = 1$, and (c) $L = 3$ to $L = 2$.

13. (II) Calculate the bond length for the NaCl molecule given that three successive wavelengths for rotational transitions are 23.1 mm, 11.6 mm, and 7.71 mm.

14. (II) Derive Eqs. 41–8.

Section 41–5

15. (I) Estimate the ionic cohesive energy for NaCl taking $\alpha = 1.75$, $m = 8$, and $r_0 = 0.28$ nm.

16. (II) The spacing between "nearest neighbor" Na and Cl ions in a NaCl crystal is 0.24 nm. What is the spacing between two nearest neighbor Na ions?

17. (II) Common salt, NaCl, has a density of 2.165 g/cm³. The molecular weight of NaCl is 58.44. Estimate the distance between nearest neighbor Na and Cl ions. [*Hint*: Each ion can be considered to have one "cube" or "cell" of side s (our unknown) extending out from it.]

18. (II) Repeat the previous Problem for KCl whose density is 1.99 g/cm³.

19. (III) (a) Starting from Eq. 41–9, show that the ionic cohesive energy is given by $U_0 = -(\alpha e^2/4\pi\epsilon_0 r_0)(1 - 1/m)$. Determine U_0 for (b) NaI $(r_0 = 0.33$ nm$)$ and (c) MgO $(r_0 = 0.21$ nm$)$. Assume $m = 10$. (d) If you used $m = 8$ instead, how far off would your answers be?

20. (III) For a long one-dimensional chain of alternating positive and negative ions, show that the Madelung constant would be $\alpha = 2\ln 2$ [*Hint*: Use a series expansion for $\ln(1 + x)$.]

Section 41–6

21. (I) Estimate the number of states available to electrons in a 1.0-cm³ cube of copper between 6.90 and 7.00 eV.

22. (II) What, roughly, is the ratio of the density of molecules in an ideal gas at 300 K and 1 atm (say O_2) to the density of free electrons (assume one per atom) in a metal (copper) also at 300 K?

23. (II) Calculate the energy which has 90 percent occupancy probability for copper at (a) $T = 300$ K; (b) $T = 1200$ K.

24. (II) Calculate the energy which has 10 percent occupancy probability for copper at (a) $T = 300$ K; (b) $T = 1200$ K.

25. (II) What is the occupancy probability for a conduction electron in copper at $T = 300$ K for an energy $E = 1.010E_F$?

26. (II) Calculate the number of possible electron states in a 1.00-cm³ cube of silver between $0.99E_F$ and $E_F (= 5.48$ eV$)$.

27. (II) Calculate the Fermi energy and Fermi speed for sodium, which has a density of 0.97×10^3 kg/m³ and has one conduction electron per atom.

28. (II) The atoms in zinc metal $(\rho = 7.1 \times 10^3$ kg/m³$)$ have two free electrons. Calculate (a) the density of conduction electrons, (b) their Fermi energy, and (c) their Fermi speed.

29. (II) Given that the Fermi energy of aluminum is 11.63 eV, (a) calculate the density of free electrons using Eq. 41–12, and (b) estimate the valence of aluminum using this model and the known density $(2.70 \times 10^3 \, \text{kg/m}^3)$ and atomic weight (27.0) of aluminum.

30. (II) The neutrons in a neutron star (Chapter 45) can be treated as a Fermi gas with neutrons in place of the electrons in our model of an electron gas. Determine the Fermi energy for a neutron star of radius 10 km and mass twice that of our Sun. Assume that the star is made entirely of neutrons and is of uniform density.

31. (II) Show that the average energy of conduction electrons in a metal at $T = 0 \, \text{K}$ is $\bar{E} = \frac{3}{5} E_F$ (Eq. 41–13) by calculating

$$\bar{E} = \frac{\int E \, n_0(E) \, dE}{\int n_0(E) \, dE}.$$

32. (II) Show that the probability for the state at the Fermi energy being occupied is exactly $\frac{1}{2}$, independent of temperature.

33. (II) (a) For copper at room temperature $(T = 300 \, \text{K})$, calculate the Fermi factor, Eq. 41–14, for an electron with energy 0.10 eV above the Fermi energy. This represents the probability that this state is occupied. Is this reasonable? (b) What is the probability that a state 0.10 eV below the Fermi energy is occupied? (c) What is the probability that the state in part (b) is unoccupied?

34. (II) For a one-dimensional potential well, start with Eq. 39–13 and show that the number of states per unit energy interval for an electron gas is given by

$$g_L(E) = \sqrt{\frac{8mL^2}{h^2 E}}.$$

Remember that there can be two electrons (spin up and spin down) for each value of n. [Hint: Write the quantum number n in terms of E. Then $g_L(E) = 2 \, dn/dE$ where dn is the number of energy levels between E and $E + dE$.]

35. (III) Proceed as follows to derive the density of states, $g(E)$, the number of states per unit volume per unit energy interval, Eq. 41–10. Let the metal be a cube of side L. Extend the discussion of Section 39–8 for an infinite well to three dimensions, giving energy levels

$$E = \frac{h^2}{8mL^2} (n_1^2 + n_2^2 + n_3^2).$$

(Explain the meaning of n_1, n_2, n_3.) Each set of values for the quantum numbers n_1, n_2, n_3 corresponds to one state. Imagine a space where n_1, n_2, n_3 are the axes, and each state is represented by a point on a cubic lattice in this space, each separated by 1 unit along an axis. Consider the octant $n_1 > 0$, $n_2 > 0$, $n_3 > 0$. Show that the number of states N within a radius $R = (n_1^2 + n_2^2 + n_3^2)^{\frac{1}{2}}$ is $2(\frac{1}{8})(\frac{4}{3}\pi R^3)$. Then, to get Eq. 41–10, set $g(E) = (1/V)(dN/dE)$, where $V = L^3$ is the volume of the metal.

Section 41–7

36. (I) A semiconductor, bombarded with light of slowly increased frequency, begins to conduct when the wavelength of light is 640 nm; estimate the size of the energy gap E_g.

37. (I) Explain on the basis of energy bands why the sodium chloride crystal is a good insulator. [Hint: Consider the shells of Na^+ and Cl^- ions.]

38. (II) We saw that there are $2N$ possible electron states in the 3s band of Na, where N is the total number of atoms. How many possible electron states are there in the (a) 2s band, (b) 2p band, and (c) 3p band? (d) State a general formula for the total number of possible states in any given electron band.

39. (II) Calculate the longest-wavelength photon that can cause an electron in silicon $(E_g = 1.1 \, \text{eV})$ to jump from the valence band to the conduction band.

40. (II) The energy gap E_g in germanium is 0.72 eV. When used as a photon detector, roughly how many electrons can be made to jump from the valence to the conduction band by the passage of a 710-keV photon that loses all its energy in this fashion?

Section 41–8

41. (II) Suppose that a silicon semiconductor is doped with phosphorus so that one silicon atom in 10^6 is replaced by a phosphorus atom. Assuming that the "extra" electron in every phosphorus atom is donated to the conduction band, by what factor is the density of conduction electrons increased? The density of silicon is $2330 \, \text{kg/m}^3$, and the density of conduction electrons in pure silicon is about $10^{16} \, \text{m}^{-3}$ at room temperature.

* Section 41–9

* 42. (I) At what wavelength will an LED radiate if made from a material with an energy gap $E_g = 1.4 \, \text{eV}$?

* 43. (I) If an LED emits light of wavelength $\lambda = 650 \, \text{nm}$, what is the energy gap (in eV) between valence and conduction bands?

* 44. (II) A silicon diode, whose current–voltage characteristics are given in Fig. 41–37, is connected in series with a battery and a 760-Ω resistor. What battery voltage is needed to produce a 12-mA current?

* 45. (II) Suppose that the diode of Fig. 41–37 is connected in series to a 100-Ω resistor and a 2.0-V battery. What current flows in the circuit? [Hint: Draw a line on Fig. 41–37 representing the current in the resistor as a function of the voltage across the diode; the intersection of this line with the characteristic curve will give the answer.]

* 46. (II) Sketch the resistance as a function of current, for $V > 0$, for the diode shown in Fig. 41–37.

* 47. (II) An ac voltage of 120 V rms is to be rectified. Estimate very roughly the average current in the output resistor R (28 kΩ) for (a) a half-wave rectifier (Fig. 41–38), and (b) a full-wave rectifier (Fig. 41–39) without capacitor.

* 48. (III) A silicon diode passes significant current only if the forward-bias voltage exceeds about 0.6 V. Make a rough estimate of the average current in the output resistor R of (a) a half-wave rectifier (Fig. 41–38), and (b) a full-wave rectifier (Fig. 41–39) without a capacitor. Assume that $R = 150 \, \Omega$ in each case and that the ac voltage is 9.0 V rms in each case.

* 49. (III) A 120-V rms 60-Hz voltage is to be rectified with a full-wave rectifier as in Fig. 41–39, where $R = 18 \text{ k}\Omega$, and $C = 25 \mu\text{F}$. (a) Make a rough estimate of the average current. (b) What happens if $C = 0.10 \mu\text{F}$? [Hint: See Section 26–4.]

* Section 41–10

* 50. (II) Suppose that the current gain of the transistor in Fig. 41–41 is $\beta = i_C/i_B = 80$. If $R_C = 3.3 \text{ k}\Omega$, calculate the ac output voltage for a time-varying input current of $2.0 \mu\text{A}$.

* 51. (II) If the current gain of the transistor amplifier in Fig. 41–41 is $\beta = i_C/i_B = 100$, what value must R_C have if a 1.0-μA ac base current is to produce an ac output voltage of 0.40 V?

* 52. (II) A transistor, whose current gain $\beta = i_C/i_B = 70$, is connected as in Fig. 41–41 with $R_B = 3.2 \text{ k}\Omega$ and $R_C = 6.8 \text{ k}\Omega$. Calculate (a) the voltage gain, and (b) the power amplification.

* 53. (II) An amplifier has a voltage gain of 80 and a 15-kΩ load (output) resistance. What is the peak output current through the load resistor if the input voltage is an ac signal with a peak of 0.080 V?

General Problems

54. Estimate the binding energy of the H_2 molecule by calculating the difference in kinetic energy of the electrons between when they are in separate atoms and when they are in the molecule, using the uncertainty principle. Take Δx for the electrons in the separated atoms to be the radius of the first Bohr orbit, 0.053 nm, and for the molecule take Δx to be the separation of the nuclei, 0.074 nm.

55. The average translational kinetic energy of an atom or molecule is about $\bar{K} = \frac{3}{2}kT$ (see Chapter 18), where $k = 1.38 \times 10^{-23} \text{ J/K}$ is Boltzmann's constant. At what temperature T will $\bar{K}$ be on the order of the bond energy (and hence the bond able to be broken by thermal motion) for (a) a covalent bond (say H_2) of binding energy 4.5 eV, and (b) a "weak" hydrogen bond of binding energy 0.15 eV?

56. In the ionic salt KF, the separation distance between ions is about 0.27 nm. (a) Estimate the electrostatic potential energy between the ions assuming them to be point charges (magnitude $1e$). (b) It is known that F releases 4.07 eV of energy when it "grabs" an electron, and 4.34 eV is required to ionize K. Find the binding energy of KF relative to free K and F atoms, neglecting the energy of repulsion.

57. The fundamental vibration frequency for the HCl molecule is $8.66 \times 10^{13} \text{ Hz}$. Determine (a) the reduced mass, and (b) the effective value of the constant k. Compare to k for the H_2 molecule.

58. Imagine the two atoms of a diatomic molecule as if they were connected by a spring, Fig. 41–44. Show that the classical frequency of vibration is given by Eq. 41–5. [Hint: Let x_1 and x_2 be the displacements of each mass from initial equilibrium positions; then $m_1 d^2x_1/dt^2 = -kx$, and $m_2 d^2x_2/dt^2 = -kx$, where $x = x_1 + x_2$. Find another relationship between x_1 and x_2, assuming that the center of mass of the system stays at rest, and then show that $\mu d^2x/dt^2 = -kx$.]

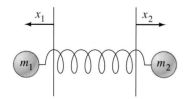

FIGURE 41–44 Problem 58.

59. Explain, using the Boltzmann factor (Eq. 40–16), why the heights of the peaks in Fig. 41–21 are different from one another. Explain also why the lines are not equally spaced. [Hint: Does the moment of inertia necessarily remain constant?]

60. Do we need to consider quantum effects for everyday rotating objects? Estimate the differences between rotational energy levels for a spinning baton compared to the energy of the baton. Assume the baton consists of a 30-cm-long bar with a mass of 200 g and two end masses (each points), each of mass 300 g, and that it rotates at 1.6 rev/s.

61. When solid argon melts at $-189°C$, its latent heat of fusion goes directly into breaking the bonds between the atoms. Solid argon is a weakly bound cubic lattice, with each atom connected to six neighbors, each bond having a binding energy of $3.9 \times 10^{-3} \text{ eV}$. What is the latent heat of fusion for argon, in J/kg? [Hint: Show that in a simple cubic lattice (Fig. 41–45), there are three times as many bonds as there are atoms, when the number of atoms is large.]

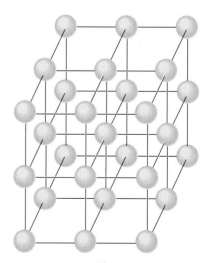

FIGURE 41–45 Problem 61.

62. A strip of silicon 1.5 cm wide and 1.0 mm thick is immersed in a magnetic field of strength 1.6 T perpendicular to the strip (Fig. 41–46). When a current of 0.20 mA is run through the strip, there is a resulting Hall effect voltage of 18 mV across the strip (Section 27–8). How many electrons per silicon atom are in the conduction band? The density of silicon is 2330 kg/m^3.

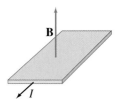

FIGURE 41–46 Problem 62.

63. Most of the Sun's radiation has wavelengths shorter than 1000 nm. For a solar cell to absorb all this, what energy gap ought the material have?

64. The energy gap between valence and conduction bands in germanium is 0.72 eV. What range of wavelengths can a photon have to excite an electron from the top of the valence band into the conduction band?

65. A TV remote control emits IR light. If the detector on the TV set is *not* to react to visible light, could it make use of silicon as a "window" with its energy gap $E_g = 1.14$ eV? What is the shortest-wavelength light that can strike silicon without causing electrons to jump from the valence band to the conduction band?

66. The *Fermi temperature* T_F is defined as that temperature at which the thermal energy kT (without the $\frac{3}{2}$) is equal to the Fermi energy: $kT_F = E_F$. (*a*) Determine the Fermi temperature for copper. (*b*) Show that for $T \gg T_F$, the Fermi factor (Eq. 41–14) approaches the Boltzmann factor. (Note: This last result is not very useful for understanding conductors. Why?)

67. For an arsenic donor atom in a doped silicon semiconductor, assume that the "extra" electron moves in a Bohr orbit about the arsenic ion. For this electron in the ground state, take into account the dielectric constant $K = 12$ of the Si lattice (which represents the weakening of the Coulomb force due to all the other atoms or ions in the lattice), and estimate (*a*) the binding energy, and (*b*) the orbit radius for this extra electron. [*Hint*: Substitute $\epsilon = K\epsilon_0$ in Coulomb's law.]

* 68. A full-wave rectifier (Fig. 41–39) uses two diodes to rectify a 75-V rms 60 Hz ac voltage. If $R = 8.8$ kΩ and $C = 30$ μF, what will be the approximate percent variation in the output voltage? The variation in output voltage (Fig. 41–39c) is called *ripple voltage*. [*Hint*: See Section 26–4 and assume the discharge of the capacitor is approximately linear.]

* 69. A zener diode voltage regulator is shown in Fig. 41–47. Suppose that $R = 1.80$ kΩ, and that the diode breakdown voltage is 130 V: the diode is rated at a maximum current of 110 mA. (*a*) If $R_{load} = 16.0$ kΩ, over what range of supply voltages will the circuit maintain the voltage at 130 V? (*b*) If the supply voltage is 200 V, over what range of load resistance will the voltage be regulated?

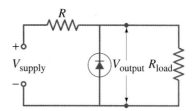

FIGURE 41–47 Problem 69.

This archeologist has unearthed the remains of a sea-turtle within an ancient man-made stone circle. Carbon dating of the remains can tell her when humans inhabited the site.

In this chapter we begin our discussion of nuclear physics including the properties of nuclei, the various forms of radioactivity, and how radioactive decay can be used in a variety of fields to determine the age of old objects, from bones and trees to rocks and other mineral substances, and obtain information on the history of the Earth.

Nuclear Physics and Radioactivity

In the early part of the twentieth century, Rutherford's experiments led to the idea that at the center of an atom there is a tiny but massive nucleus. At the same time that the quantum theory was being developed and scientists were attempting to understand the structure of the atom and its electrons, investigations into the nucleus itself had also begun. In this chapter and the next, we take a brief look at *nuclear physics*.

42–1 | Structure and Properties of the Nucleus

An important question to physicists in the early part of this century was whether the nucleus had a structure, and what that structure might be. It turns out that the nucleus is a complicated entity and is not fully understood even today. However, by the early 1930s, a model of the nucleus had been developed that is still useful.

According to this model, a nucleus is considered as an aggregate of two types of particles: protons and neutrons. (Of course, we must remember that these "particles" also have wave properties, but for ease of visualization and language, we often refer to them simply as "particles.") A **proton** is the nucleus of the simplest atom, hydrogen. It has a positive charge ($= +e = +1.60 \times 10^{-19}\,\text{C}$) and a mass

Proton

$$m_\text{p} = 1.67262 \times 10^{-27}\,\text{kg}.$$

The **neutron**, whose existence was ascertained only in 1932 by the Englishman James Chadwick (1891–1974), is electrically neutral ($q = 0$), as its name implies. Its mass, which is almost identical to that of the proton, is

Neutron

$$m_\text{n} = 1.67493 \times 10^{-27}\,\text{kg}.$$

Nucleons

These two constituents of a nucleus, neutrons and protons, are referred to collectively as **nucleons**.

Although the hydrogen nucleus consists of a single proton alone, the nuclei of all other elements consist of both neutrons and protons. The different types of nuclei are often referred to as **nuclides**. The number of protons in a nucleus (or nuclide) is called the **atomic number** and is designated by the symbol Z. The total number of nucleons, neutrons plus protons, is designated by the symbol A and is called the **atomic mass number**. This name is used since the mass of a nucleus is very closely A times the mass of one nucleon. A nuclide with 7 protons and 8 neutrons thus has $Z = 7$ and $A = 15$. The **neutron number** N is $N = A - Z$.

Z and A

To specify a given nuclide, we need give only A and Z. A special symbol is commonly used which takes the form

$$^A_Z\text{X},$$

where X is the chemical symbol for the element (see Appendix D, and the periodic table inside the back cover), A is the atomic mass number, and Z is the atomic number. For example, $^{15}_7\text{N}$ means a nitrogen nucleus containing 7 protons and 8 neutrons for a total of 15 nucleons. In a neutral atom, the number of electrons orbiting the nucleus is equal to the atomic number Z (since the charge on an electron has the same magnitude but opposite sign to that of a proton). The main properties of an atom, and how it interacts with other atoms, are largely determined by the number of electrons. Hence Z determines what kind of atom it is: carbon, oxygen, gold, or whatever. It is redundant to specify both the symbol of a nucleus and its atomic number Z as described above. If the nucleus is nitrogen, for example, we know immediately that $Z = 7$. The subscript Z is thus sometimes dropped and $^{15}_7\text{N}$ is then written simply ^{15}N; in words we say "nitrogen fifteen."

Isotopes

For a particular type of atom (say, carbon), nuclei are found to contain different numbers of neutrons, although they all have the same number of protons. For example, carbon nuclei always have 6 protons, but they may have 5, 6, 7, 8, 9, or 10 neutrons. Nuclei that contain the same number of protons but different numbers of neutrons are called **isotopes**. Thus, $^{11}_6\text{C}$, $^{12}_6\text{C}$, $^{13}_6\text{C}$, $^{14}_6\text{C}$, $^{15}_6\text{C}$, and $^{16}_6\text{C}$ are all isotopes of carbon. Of course, the isotopes of a given element are not all equally common. For example, 98.9 percent of naturally occurring carbon (on Earth) is the isotope $^{12}_6\text{C}$, and about 1.1 percent is $^{13}_6\text{C}$. These percentages are referred to as the **natural abundances**.[†] Many isotopes that do not occur naturally can be produced in the laboratory by means of nuclear reactions (more on this later). Indeed, all elements beyond uranium ($Z > 92$) do not occur naturally and are only produced artificially, as are many nuclides with $Z \leq 92$.

The approximate size of nuclei was determined originally by Rutherford from the scattering of charged particles. Of course, we cannot speak about a definite size for nuclei because of the wave–particle duality: their spatial extent must remain somewhat fuzzy. Nonetheless a rough "size" can be measured by scattering high-speed electrons off nuclei. It is found that nuclei have a roughly spherical

---

[†]The mass value for each element as given in the periodic table (inside back cover) is an average weighted according to the natural abundances of its isotopes.

shape with a radius that increases with A according to the approximate formula

$$r \approx (1.2 \times 10^{-15}\,\text{m})(A^{\frac{1}{3}}).$$

(42–1)   *Nuclear radii*

Since the volume of a sphere is $V = \frac{4}{3}\pi r^3$, we see that the volume of a nucleus is proportional to the number of nucleons, $V \propto A$. This is what we would expect if nucleons were like impenetrable billiard balls: if you double the number of balls, you double the total volume. Hence, all nuclei have nearly the same density, and it is enormous (see Problem 5).

EXAMPLE 42–1 ESTIMATE **Nuclear sizes.**   Estimate the diameter of the following nuclei: (a) $^1_1$H, (b) $^{40}_{20}$Ca, (c) $^{208}_{82}$Pb, (d) $^{235}_{92}$U.

SOLUTION   (a) For hydrogen, $A = 1$, Eq. 42–1 gives

$$d = \text{diameter} = 2r \approx 2.4 \times 10^{-15}\,\text{m}$$

since $A^{\frac{1}{3}} = 1^{\frac{1}{3}} = 1$.

(b) For calcium $d = 2r \approx (2.4 \times 10^{-15}\,\text{m})(40)^{\frac{1}{3}} = 8.2 \times 10^{-15}\,\text{m}$.

(c) For lead $d \approx (2.4 \times 10^{-15}\,\text{m})(208)^{\frac{1}{3}} = 14 \times 10^{-15}\,\text{m}$.

(d) For uranium $d \approx (2.4 \times 10^{-15}\,\text{m})(235)^{\frac{1}{3}} = 15 \times 10^{-15}\,\text{m}$.

The range of nuclear diameters is only from 2.4 fm to 15 fm.

The masses of nuclei can be determined, in one method, by measuring the radius of curvature of fast-moving nuclei in a magnetic field using a mass spectrometer, as discussed in Section 27–9. Indeed, as mentioned there, the existence of different isotopes of the same element was discovered using this device. Nuclear masses can be specified in **unified atomic mass units** (u). On this scale, a neutral $^{12}_6$C atom is given the precise value 12.000000 u. A neutron then has a measured mass of 1.008665 u, a proton 1.007276 u, and a neutral hydrogen atom, $^1_1$H (proton plus electron) 1.007825 u. The masses of many nuclides are given in Appendix D. It should be noted that the masses in this table, as is customary, are for the *neutral atom*, and not for a bare nucleus.

➡ **PROBLEM SOLVING**

Masses are for neutral atom

Masses are often specified using the electron-volt energy unit. This can be done because mass and energy are related, and the precise relationship is given by Einstein's equation $E = mc^2$ (Chapter 37). Since the mass of a proton is $1.67262 \times 10^{-27}\,\text{kg}$, or 1.007276 u, then

$$1.0000\,\text{u} = \left(\frac{1.0000\,\text{u}}{1.007276\,\text{u}}\right)(1.67262 \times 10^{-27}\,\text{kg})$$
$$= 1.66054 \times 10^{-27}\,\text{kg};$$

this is equivalent to an energy (see table inside front cover)

$$E = mc^2 = \frac{(1.66054 \times 10^{-27}\,\text{kg})(2.9979 \times 10^8\,\text{m/s})^2}{(1.6022 \times 10^{-19}\,\text{J/eV})}$$
$$= 931.5\,\text{MeV}.$$

Thus

$$1\,\text{u} = 1.6605 \times 10^{-27}\,\text{kg} = 931.5\,\text{MeV}/c^2.$$

Atomic mass unit

The rest masses of some of the basic particles are given in Table 42–1.

TABLE 42–1   Rest Masses in Kilograms, Unified Atomic Mass Units, and MeV/c^2

	Mass		
Object	kg	u	MeV/c^2
Electron	9.1094×10^{-31}	0.00054858	0.51100
Proton	1.67262×10^{-27}	1.007276	938.27
$^1_1$H atom	1.67353×10^{-27}	1.007825	938.78
Neutron	1.67493×10^{-27}	1.008665	939.57

Just as an electron has intrinsic spin and angular momentum quantum numbers, so too do nuclei and their constituents, the proton and neutron. Both the proton and the neutron are spin-$\frac{1}{2}$ particles. A nucleus, made up of protons and neutrons, has a **nuclear spin** quantum number, I, that can be either integer or half integer, depending on whether it is made up of an even or an odd number of nucleons. The *nuclear angular momentum* of a nucleus is given, as might be expected (see Section 40–2), by $\sqrt{I(I + 1)}\,\hbar$.

Nuclear magnetic moments are measured in terms of the **nuclear magneton**

$$\mu_\text{N} = \frac{e\hbar}{2m_\text{p}}, \tag{42-2}$$

which is defined by analogy with the Bohr magneton for electrons ($\mu_\text{B} = e\hbar/2m_\text{e}$, Section 40–7). Since μ_N contains the proton mass, m_p, instead of the electron mass, it is about 2000 times smaller. The electron magnetic moment is about 1 Bohr magneton, so we might expect the proton to have a magnetic moment μ_p of about $1\,\mu_\text{N}$. Instead, it is

$$\mu_\text{p} = 2.7928\mu_\text{N}.$$

There is no satisfactory explanation for this large factor. Also surprising is the fact that the neutron has a magnetic moment:

$$\mu_\text{n} = -1.9135\mu_\text{N}.$$

This suggests that, although the neutron carries no net charge, there may be some sort of electric current within the neutron. The minus sign for μ_n indicates that its magnetic moment is opposite to its spin.

Important applications based on nuclear spin are nuclear magnetic resonance (NMR) and magnetic resonance imaging (MRI). They are discussed in the next chapter (Section 43–10).

42–2 Binding Energy and Nuclear Forces

The total mass of a stable nucleus is always less than the sum of the masses of its separate protons and neutrons, as the following Example shows.

EXAMPLE 42–2 $^4_2$**He mass compared to its constituents.** Compare the mass of a $^4_2$He nucleus to that of its constituent nucleons.

SOLUTION The mass of a neutral $^4_2$He atom, from Appendix D, is 4.002603 u. The mass of two neutrons and two protons (including the two electrons) is

$$
\begin{aligned}
2m_\text{n} &= 2.017330\ \text{u} \\
2m(^1_1\text{H}) &= \underline{2.015650\ \text{u}} \\
&\ \ \ \ 4.032980\ \text{u}.
\end{aligned}
$$

We almost always deal with masses of neutral atoms—that is, nuclei with Z electrons—since this is how masses are measured. We must therefore be sure to balance out the electrons when we compare masses, which is why we used the mass of $^1_1$H in this Example rather than that of the proton alone.

➡ **PROBLEM SOLVING**

Keep track of electron masses

Thus the mass of $^4_2$He is measured to be $4.032980\ \text{u} - 4.002603\ \text{u} = 0.030377\ \text{u}$ less than the masses of its constituents. Where has this lost mass gone?

It has, in fact, gone into energy of another kind (such as radiation, or kinetic energy, for example). The mass (or energy) difference in the case of $^4_2$He, given in energy units, is $(0.030377\ \text{u})(931.5\ \text{MeV/u}) = 28.30\ \text{MeV}$. This difference is referred to as the **total binding energy** of the nucleus. The total binding energy represents the amount of energy that must be put into a nucleus in order to break it apart into its constituent protons and neutrons. If the mass of, say, a $^4_2$He nucleus were exactly equal to the mass of two neutrons plus two protons, the nucleus could

Binding energy

fall apart without any input of energy. To be stable, the mass of a nucleus *must* be less than that of its constituent nucleons, so that energy input *is* needed to break it apart. Note that the binding energy is not something a nucleus has—it is energy it "lacks" relative to the total mass of its separate constituents.

Nuclear binding energy can be compared to the binding energy of electrons in an atom. We saw in Chapter 38 that the binding energy of the one electron in the hydrogen atom, for example, is 13.6 eV. The mass of a $^1_1$H atom is less than that of a single proton plus a single electron by 13.6 eV. Compared to the total mass of the atom (939 MeV), this is incredibly small (1 part in 10^8), and for practical purposes the mass difference can be ignored. The binding energies of nuclei are on the order of 10^6 times greater than the binding energies of electrons in atoms.

The **average binding energy per nucleon** is defined as the total binding energy of a nucleus divided by A, the total number of nucleons. For $^4_2$He, it is 28.3 MeV/4 = 7.1 MeV. Figure 42–1 shows the average binding energy per nucleon as a function of A for stable nuclei. The curve rises as A increases and reaches a plateau at about 8.7 MeV per nucleon above about $A \approx 40$. Beyond about $A \approx 80$, the curve decreases slowly, indicating that larger nuclei are held together a little less tightly than those in the middle of the periodic table. (We will see later that these characteristics allow the release of nuclear energy in the processes of fission and fusion.)

Binding energy per nucleon

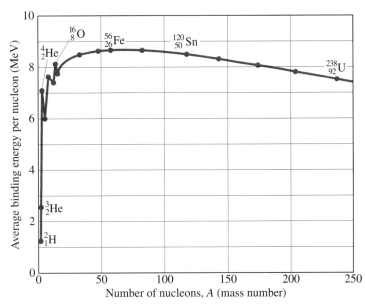

FIGURE 42–1 Average binding energy per nucleon as a function of mass number A for stable nuclei.

EXAMPLE 42–3 **Binding energy for iron.** Calculate the total binding energy and the average binding energy per nucleon for $^{56}_{26}$Fe, the most common stable isotope of iron.

SOLUTION $^{56}_{26}$Fe has 26 protons and 30 neutrons whose separate masses are

$$(26)(1.007825\,\text{u}) = \;\;26.2035\,\text{u (includes electrons)}$$
$$(30)(1.008665\,\text{u}) = \;\;\underline{30.2600\,\text{u}}$$
$$\text{Total} = \;\;56.4635\,\text{u}.$$
$$\text{Subtract mass of } ^{56}_{26}\text{Fe:} \;\; -55.9349\,\text{u (Appendix D)}$$
$$\Delta m = \;\;0.5286\,\text{u}.$$

The total binding energy is thus

$$(0.5286\,\text{u})(931.5\,\text{MeV/u}) = 492.4\,\text{MeV}$$

and the average binding energy per nucleon is

$$\frac{492.4\,\text{MeV}}{56\,\text{nucleons}} = 8.8\,\text{MeV}.$$

EXAMPLE 42–4 **Binding energy of last neutron.** What is the binding energy of the last neutron in $^{13}_6\text{C}$?

SOLUTION We compare the mass of $^{13}_6\text{C}$ to that of the atom with one less neutron, $^{12}_6\text{C}$, plus a free neutron (Appendix D):

$$
\begin{array}{rl}
\text{Mass } ^{12}_6\text{C} = & 12.000000 \text{ u} \\
\text{Mass } ^1_0\text{n} = & \underline{1.008665 \text{ u}} \\
\text{Total} = & 13.008665 \text{ u.} \\
\text{Subtract mass of } ^{13}_6\text{C:} & \underline{-13.003355 \text{ u}} \\
\Delta m = & 0.005310 \text{ u}
\end{array}
$$

which in energy is $(931.5 \text{ MeV/u})(0.005310 \text{ u}) = 4.95 \text{ MeV}$. That is, it would require 4.95 MeV input of energy to remove one neutron from $^{13}_6\text{C}$.

We can analyze nuclei not only from the point of view of energy, but also from the point of view of the forces that hold them together. We would not expect a collection of protons and neutrons to come together spontaneously, since protons are all positively charged and thus exert repulsive forces on each other. Indeed, the question arises as to how a nucleus stays together at all in view of the fact that the electric force between protons would tend to break it apart. Since stable nuclei *do* stay together, it is clear that another force must be acting. Because this new force is stronger than the electric force (which, in turn, is much stronger than gravity at the nuclear level) it is called the **strong nuclear force**. The strong nuclear force is an attractive force that acts between all nucleons—protons and neutrons alike. Thus protons attract each other via the nuclear force at the same time they repel each other via the electric force. Neutrons, since they are electrically neutral, only attract other neutrons or protons via the nuclear force.

Strong nuclear force

The nuclear force turns out to be far more complicated than the gravitational and electromagnetic forces. A precise mathematical description is not yet possible. Nonetheless, a great deal of work has been done to try to understand the nuclear force. One important aspect of the strong nuclear force is that it is a **short-range** force: it acts only over a very short distance. It is very strong between two nucleons if they are less than about 10^{-15} m apart, but it is essentially zero if they are separated by a distance greater than this. Compare this to electric and gravitational forces, which can act over great distances and are therefore called **long-range** forces. The strong nuclear force has some strange quirks. For example, if a nuclide contains too many or too few neutrons relative to the number of protons, the binding of the nucleons is reduced; nuclides that are too unbalanced in this regard are unstable. As shown in Fig. 42–2, stable nuclei tend to have the same number of protons as neutrons ($N = Z$) up to about $A \approx 30$ or 40. Beyond this, stable nuclei contain more neutrons than protons. This makes sense since, as Z increases, the electrical repulsion increases, so a greater number of neutrons—which exert only the attractive nuclear force—are required to maintain stability. For very large Z, no number of neutrons can overcome the greatly increased electric repulsion. Indeed, there are no completely stable nuclides above $Z = 82$.

Long- and short-range forces

FIGURE 42–2 Number of neutrons versus number of protons for stable nuclides, which are represented by dots. The straight line represents $N = Z$.

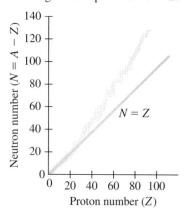

What we mean by a stable nucleus is one that stays together indefinitely. What then is an unstable nucleus? It is one that comes apart; and this results in radioactive decay. Before we discuss the important subject of radioactivity (the next Section), we note that there is a second type of nuclear force that is much weaker than the strong nuclear force. It is called the **weak nuclear force**, and we are aware of its existence only because it shows itself in certain types of radioactive decay. These two nuclear forces, the strong and the weak, together with the gravitational and electromagnetic forces, comprise the four known types of force in nature (more on this in Chapter 44).

Weak force

42–3 | Radioactivity

Discovery of radioactivity

Nuclear physics had its beginnings in 1896. In that year, Henri Becquerel (1852–1908) made an important discovery: in his studies of phosphorescence, he found that a certain mineral (which happened to contain uranium) would darken a photographic plate even when the plate was wrapped to exclude light. It was clear that the mineral emitted some new kind of radiation that, unlike X-rays, occurred without any external stimulus. This new phenomenon eventually came to be called **radioactivity**.

Soon after Becquerel's discovery, Marie Curie (1867–1934) and her husband, Pierre Curie (1859–1906), isolated two previously unknown elements that were very highly radioactive (Fig. 42–3). These were named polonium and radium. Other radioactive elements were soon discovered as well. The radioactivity was found in every case to be unaffected by the strongest physical and chemical treatments, including strong heating or cooling and the action of strong chemical reagents. It was clear that the source of radioactivity must be deep within the atom, that it must emanate from the nucleus. And it became apparent that radioactivity is the result of the *disintegration* or *decay* of an unstable nucleus. Certain isotopes are not stable under the action of the nuclear force, and they decay with the emission of some type of radiation or "rays."

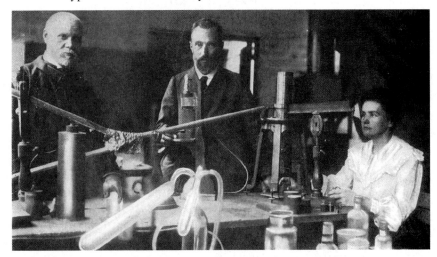

FIGURE 42–3 Marie and Pierre Curie in their laboratory (about 1906) where radium was discovered.

Many unstable isotopes occur in nature, and such radioactivity is called "natural radioactivity." Other unstable isotopes can be produced in the laboratory by nuclear reactions (Section 43–1); these are said to be produced "artificially" and to have "artificial radioactivity."

Rutherford and others began studying the nature of the rays emitted in radioactivity about 1898. They found that the rays could be classified into three distinct types according to their penetrating power. One type of radiation could barely penetrate a piece of paper. The second type could pass through as much as 3 mm of aluminum. The third was extremely penetrating: it could pass through several centimeters of lead and still be detected on the other side. They named these three types of radiation alpha (α), beta (β), and gamma (γ), respectively, after the first three letters of the Greek alphabet.

Each type of ray was found to have a different charge and hence is bent differently in a magnetic field, Fig. 42–4; α rays are positively charged, β rays are negatively charged, and γ rays are neutral. It was soon found that all three types of radiation consisted of familiar kinds of particles. Gamma rays are very high-energy photons whose energy is even higher than that of X-rays. Beta rays are electrons, identical to those that orbit the nucleus (but they are created within the nucleus itself). Alpha rays (or α particles) are simply the nuclei of helium atoms, ^4_2He; that is, an α ray consists of two protons and two neutrons bound together.

FIGURE 42–4 Alpha and beta rays are bent in opposite directions by a magnetic field, whereas gamma rays are not bent at all.

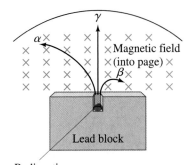

Radioactive sample (radium)

We now discuss each of these three types of radioactivity, or decay, in more detail.

42–4 | Alpha Decay

When a nucleus emits an α particle ($^4_2$He), it is clear that the remaining nucleus will be different from the original: for it has lost two protons and two neutrons. Radium 226 ($^{226}_{88}$Ra), for example, is an α emitter. It decays to a nucleus with $Z = 88 - 2 = 86$ and $A = 226 - 4 = 222$. The nucleus with $Z = 86$ is radon (Rn) —see Appendix D or the periodic table. Thus the radium decays to radon with the emission of an α particle. This is written

α decay

$$^{226}_{88}\text{Ra} \rightarrow {}^{222}_{86}\text{Rn} + {}^4_2\text{He}.$$

See Fig. 42–5.

FIGURE 42–5 Radioactive decay of radium to radon with emission of an alpha particle.

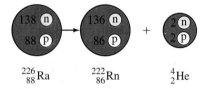

Daughter nucleus
Parent nucleus
Transmutation

It is clear that when α decay occurs, a new element is formed. The **daughter** nucleus ($^{222}_{86}$Rn in this case) is different from the **parent** nucleus ($^{226}_{88}$Ra in this case). This changing of one element into another is called **transmutation**.

Alpha decay can be written

$$^A_Z N \rightarrow {}^{A-4}_{Z-2} N' + {}^4_2\text{He} \qquad \text{[\alpha \text{ decay}]}$$

where N is the parent, N' the daughter, and Z and A are the atomic number and atomic mass number, respectively, of the parent.

Alpha decay occurs because the strong nuclear force is unable to hold very large nuclei together. Because the nuclear force is a short-range force, it acts only between neighboring nucleons. But the electric force can act all the way across a large nucleus. For very large nuclei, the large Z means the repulsive electric force becomes very large (Coulomb's law); and it acts between all protons. The strong nuclear force, since it acts only between neighboring nucleons, is overpowered and is unable to hold the nucleus together.

We can express the instability in terms of energy (or mass): the mass of the parent nucleus is greater than the mass of the daughter nucleus plus the mass of the α particle. The mass difference appears as kinetic energy, which is carried away by the α particle and the recoiling daughter nucleus. The total energy released is called the **disintegration energy**, Q, or the **Q-value** of the decay. From conservation of energy,

$$M_\text{P} c^2 = M_\text{D} c^2 + m_\alpha c^2 + Q,$$

so

Q-value

$$Q = M_\text{P} c^2 - (M_\text{D} + m_\alpha)c^2 \qquad \textbf{(42–3)}$$

where M_P, M_D, and m_α are the masses of the parent, daughter, and α particle, respectively. If the parent had *less* mass than the daughter plus the α particle (so $Q < 0$), the decay could not occur, for the conservation of energy law would be violated.

Why the strong nuclear force cannot hold a nucleus together

EXAMPLE 42-5 **Uranium decay energy release.** Calculate the disintegration energy when $^{232}_{92}U$ (mass = 232.037146 u) decays to $^{228}_{90}Th$ (228.028731 u) with the emission of an α particle. (As always, masses are for neutral atoms.)

SOLUTION Since the mass of the ^4_2He is 4.002603 u (Appendix D), the total mass in the final state is

$$228.028731 \text{ u} + 4.002603 \text{ u} = 232.031334 \text{ u}.$$

The mass lost when the $^{232}_{92}U$ decays is

$$232.037146 \text{ u} - 232.031334 \text{ u} = 0.005812 \text{ u}.$$

Since $1 \text{ u} = 931.5 \text{ MeV}$, the energy Q released is

$$Q = (0.005812 \text{ u})(931.5 \text{ MeV/u}) \approx 5.4 \text{ MeV},$$

and this energy appears as kinetic energy of the α particle and the daughter nucleus. (Using conservation of momentum, it can be shown that the α particle emitted by a $^{232}_{92}U$ nucleus at rest has a kinetic energy of about 5.3 MeV. Thus, the daughter nucleus—which recoils in the opposite direction from the emitted α particle—has about 0.1 MeV of kinetic energy. See Problem 68.)

If the mass of the daughter nucleus plus the mass of the α particle is less than the mass of the parent nucleus (so the parent is energetically allowed to decay), why are there any parent nuclei at all? That is, why haven't radioactive nuclei all decayed long ago, right after they were formed in the early stages of the universe? We can understand decay using a model of a nucleus inside of which there is an alpha particle (at least some of the time) bouncing around. The potential energy "seen" by the α particle would have a shape something like that shown in Fig. 42–6. The potential energy well (approximately square) between $r = 0$ and $r = R_0$ represents the short-range attractive nuclear force. Beyond the nuclear radius, R_0, the Coulomb repulsion dominates (since the nuclear force drops to zero) and we see the characteristic $1/r$ dependence of the Coulomb potential. The α particle, trapped within the nucleus, can be thought of as bouncing back and forth between the potential walls. Since the potential energy just beyond $r = R_0$ is greater than the energy of the α particle (dashed line), the α particle could not escape the nucleus if it were governed by classical physics. But according to quantum mechanics, there is a certain probability that the α particle can tunnel through the Coulomb barrier, from point A to point B in Fig. 42–6, as we discussed in Section 39–10. The height and width of the barrier affect the rate at which the nuclei decay (Section 42–8). Because of this barrier, the lifetimes of α-unstable nuclei can be quite long, from a fraction of a microsecond to over 10^{10} years. Note in Fig. 42–6 that the Q-value represents the total kinetic energy when the α particle is far from the nucleus.

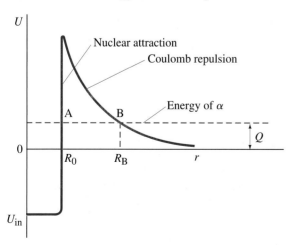

FIGURE 42–6 Potential energy for alpha particle and (daughter) nucleus, showing the Coulomb barrier through which the α particle must tunnel to escape. The Q-value of the reaction is also shown.

Why, you may wonder, do nuclei emit this combination of four nucleons called an α particle? Why not just four separate nucleons, or even one? The answer is that the α particle is very strongly bound, so that its mass is significantly less than that of four separate nucleons. As we saw in Example 42–2, two protons and two neutrons separately have a total mass of about 4.032980 u. The total mass of a $^{228}_{90}$Th nucleus plus four separate nucleons is 232.061711 u, which is greater than the mass of the parent nucleus. Such a decay could not occur because it would violate the conservation of energy. Similarly, it is almost always true that the emission of a single nucleon is energetically not possible. Or, to put it another way, for any nuclide for which it could be possible (mass of parent > mass of daughter + mass of nucleon), the decay happens so fast after formation of the parent that we don't see the parent in nature.

One widespread application of nuclear physics is present in nearly every home in the form of an ordinary **smoke detector**. The most common type of detector contains about 0.2 mg of the radioactive americium isotope, $^{241}_{95}$Am, in the form of AmO_2. The radiation ionizes the nitrogen and oxygen molecules in the air space between two oppositely charged plates. The resulting conductivity allows a small steady current. If smoke enters, the radiation is absorbed by the smoke particles rather than by the air molecules, thus reducing the current. The current drop is detected by the device's electronics and sets off the alarm. The radiation dose that escapes from an intact americium smoke detector is much less than the natural radioactive background, and so can be considered harmless. There is no question that smoke detectors save lives and reduce property damage.

42–5 Beta Decay

Transmutation of elements also occurs when a nucleus decays by β decay—that is, with the emission of an electron or $β^-$ particle. The nucleus $^{14}_{6}$C, for example, emits an electron when it decays:

$$^{14}_{6}\text{C} \rightarrow\, ^{14}_{7}\text{N} + \text{e}^- + \text{a neutrino},$$

where e^- is the symbol for the electron. (The symbol $^{0}_{-1}$e is sometimes used for the electron whose charge corresponds to $Z = -1$ and, since it is not a nucleon and has very small mass, has $A = 0$.) The particle known as the neutrino, with rest mass $m = 0$ and charge $q = 0$, was not initially detected and was only later hypothesized to exist, as we shall discuss later in this Section. No nucleons are lost when an electron is emitted, and the total number of nucleons, A, is the same in the daughter nucleus as in the parent. But because an electron has been emitted from the nucleus itself, the charge on the daughter nucleus is $+1e$ greater than that on the parent. The parent nucleus had $Z = +6$, so from charge conservation the nucleus remaining behind must have a charge of $+7e$. So the daughter nucleus has $Z = 7$, which is nitrogen.

It must be carefully noted that the electron emitted in β decay is *not* an orbital electron. Instead, the electron is created *within the nucleus itself*. What happens is that one of the neutrons changes to a proton and in the process (to conserve charge) throws off an electron. Indeed, free neutrons actually do decay in this fashion:

$$\text{n} \rightarrow \text{p} + \text{e}^- + \text{a neutrino}.$$

Because of their origin in the nucleus, the electrons emitted in β decay are often referred to as "β particles," rather than as electrons, to remind us of their origin. They are, nonetheless, indistinguishable from orbital electrons.

EXAMPLE 42–6 **Energy release in $^{14}_{6}$C decay.** How much energy is released when $^{14}_{6}$C decays to $^{14}_{7}$N by β emission? Use Appendix D.

SOLUTION The masses given in Appendix D are those of the neutral atom, and we have to keep track of the electrons involved. Assume the parent nucleus has six orbiting electrons so it is neutral, and its mass is 14.003242 u. The daughter, which in this decay is $^{14}_{7}$N, is not neutral since it has the same six electrons circling it but the nucleus has a charge of $+7e$. However, the mass of this daughter with its six electrons, plus the mass of the emitted electron (which makes a total of seven electrons), is just the mass of a neutral nitrogen atom. That is, the total mass in the final state is

$$\text{(mass of }^{14}_{7}\text{N nucleus + 6 electrons) + (mass of 1 electron)},$$

and this is equal to

$$\text{mass of neutral }^{14}_{7}\text{N (includes 7 electrons)},$$

which, from Appendix D is a mass of 14.003074 u. (Note that the neutrino doesn't contribute to either the mass or charge balance since it has $m = 0$ and $q = 0$.) Hence the mass after decay is 14.003074 u, whereas before decay, it was 14.003242 u. So the mass difference is 0.000168 u, which corresponds to 0.156 MeV or 156 keV.

According to this Example, we would expect the emitted electron to have a kinetic energy of 156 keV. (The daughter nucleus, because its mass is very much larger than that of the electron, recoils with very low velocity and hence gets very little of the kinetic energy.) Indeed, very careful measurements indicate that a few emitted β particles do have kinetic energy close to this calculated value. But the vast majority of emitted electrons have somewhat less energy. In fact, the energy of the emitted electron can be anywhere from zero up to the maximum value as calculated above. This range of electron kinetic energy was found for any β decay. It was as if the law of conservation of energy was being violated, and indeed Bohr actually considered this possibility. Careful experiments indicated that linear momentum and angular momentum also did not seem to be conserved. Physicists were troubled at the prospect of having to give up these laws, which had worked so well in all previous situations. In 1930, Wolfgang Pauli proposed an alternate solution: perhaps a new particle that was very difficult to detect was emitted during β decay in addition to the electron. This hypothesized particle could be carrying off the energy, momentum, and angular momentum required to maintain the conservation laws. This new particle was named the **neutrino**—meaning "little neutral one"—by the great Italian physicist Enrico Fermi (1901–1954; Fig. 42–7), who in 1934 worked out a detailed theory of β decay. (It was Fermi who, in this theory, postulated the existence of the fourth force in nature, which we call the weak nuclear force.) The electron neutrino has zero charge, spin of $\frac{1}{2}\hbar$, and seems to have zero rest mass, although we cannot yet rule out the possibility that it might have a very tiny rest mass. If its rest mass is zero, it is much like a photon in that it is neutral and travels at the speed of light. But the neutrino is far more difficult to detect. In 1956, complex experiments produced further evidence for the existence of the neutrino; but by then, most physicists had already accepted its existence.

The symbol for the neutrino is the Greek letter nu (ν). The correct way of writing the decay of $^{14}_{6}$C is then

$$^{14}_{6}\text{C} \rightarrow {}^{14}_{7}\text{N} + \text{e}^{-} + \bar{\nu}.$$

The bar ($^{-}$) over the neutrino symbol is to indicate that it is an "antineutrino." (Why this is called an antineutrino rather than simply a neutrino need not concern us now; it is discussed in Chapter 44.)

FIGURE 42–7 Enrico Fermi. Fermi contributed significantly to both theoretical and experimental physics, a feat almost unique in this century.

β^{-} decay

Many isotopes decay by electron emission. They are always isotopes that have too many neutrons compared to the number of protons. That is, they are isotopes that lie above the stable isotopes plotted in Fig. 42–2. But what about unstable isotopes that have too few neutrons compared to their number of protons—those that fall below the stable isotopes of Fig. 42–2? These, it turns out, decay by emitting a **positron** instead of an electron. A positron (sometimes called an e⁺ or β⁺ particle) has the same mass as the electron, but it has a positive charge of +1e. Because it is so like an electron, except for its charge, the positron is called the **antiparticle**† to the electron. An example of a β⁺ decay is that of $^{19}_{10}$Ne:

Positron (β^+) decay

$$^{19}_{10}\text{Ne} \rightarrow \, ^{19}_{9}\text{F} + \text{e}^+ + \nu,$$

where e⁺ (or $^{0}_{1}$e) stands for a positron. Note that the ν emitted here is a neutrino, whereas that emitted in β⁻ decay is called an antineutrino. Thus an antielectron (= positron) is emitted with a neutrino, whereas an antineutrino is emitted with an electron; this is discussed in Chapter 44.

We can write β⁻ and β⁺ decay, in general, as follows:

$$^{A}_{Z}N \rightarrow \, ^{A}_{Z+1}N' + \text{e}^- + \bar{\nu} \qquad\qquad [\beta^- \text{ decay}]$$

$$^{A}_{Z}N \rightarrow \, ^{A}_{Z-1}N' + \text{e}^+ + \nu, \qquad\qquad [\beta^+ \text{ decay}]$$

where N is the parent nucleus and N' is the daughter.

Besides β⁻ and β⁺ emission, there is a third related process. This is **electron capture** (abbreviated EC in Appendix D) and occurs when a nucleus absorbs one of its orbiting electrons. An example is $^{7}_{4}$Be, which as a result becomes $^{7}_{3}$Li. The process is written

Electron capture

$$^{7}_{4}\text{Be} + \text{e}^- \rightarrow \, ^{7}_{3}\text{Li} + \nu,$$

or, in general,

$$^{A}_{Z}N + \text{e}^- \rightarrow \, ^{A}_{Z-1}N' + \nu. \qquad\qquad [\text{electron capture}]$$

K-capture

Usually it is an electron in the innermost (K) shell that is captured, in which case it is called "K-capture." The electron disappears in the process and a proton in the nucleus becomes a neutron; a neutrino is emitted as a result. This process is inferred experimentally by detection of emitted X-rays (due to electrons jumping down to fill the empty state) of just the proper energy.

In β decay, it is the weak nuclear force that plays the crucial role. The neutrino is unique in that it interacts with matter only via the weak force, which is why it is so hard to detect.

42–6 Gamma Decay

Gamma rays are photons having very high energy. They have their origin in the decay of a nucleus, much like emission of photons by excited atoms. Like an atom, a nucleus itself can be in an excited state. When it jumps down to a lower energy state, or to the ground state, it emits a photon which we call a γ ray. The possible energy levels of a nucleus are much farther apart than those of an atom: on the order of keV or MeV, as compared to a few eV for electrons in an atom. Hence, the emitted photons have energies that can range from a few keV to several MeV. For a given decay, the γ ray always has the same energy. Since a γ ray carries no charge, there is no change in the element as a result of a γ decay.

---

†Discussed in Chapter 44. Briefly, an antiparticle has the same mass as its corresponding particle, but opposite charge.

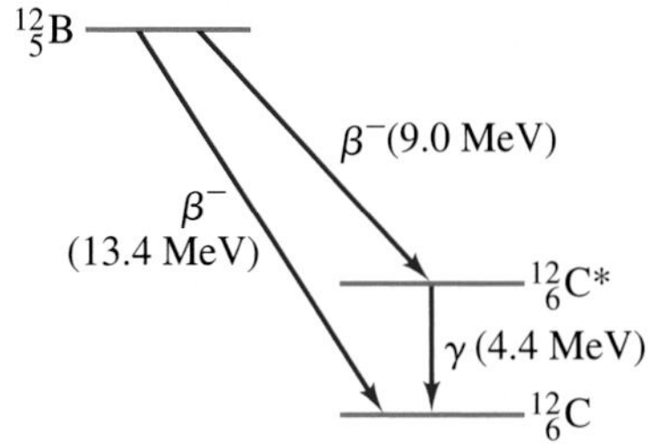

FIGURE 42–8 Energy-level diagram showing how $^{12}_{5}$B can decay to the ground state of $^{12}_{6}$C by β decay (total energy released = 13.4 MeV), or can instead β decay to an excited state of $^{12}_{6}$C (indicated by *), which subsequently decays to its ground state by emitting a 4.4-MeV γ ray.

How does a nucleus get into an excited state? It may occur because of a violent collision with another particle. More commonly, the nucleus remaining after a previous radioactive decay may be in an excited state. A typical example is shown in the energy-level diagram of Fig. 42–8. $^{12}_{5}$B can decay by β decay directly to the ground state of $^{12}_{6}$C; or it can go by β decay to an excited state of $^{12}_{6}$C, which then decays by emission of a 4.4 MeV γ ray to the ground state.

We can write γ decay as

$$^{A}_{Z}N^* \rightarrow \, ^{A}_{Z}N + \gamma, \qquad\qquad [\gamma \text{ decay}]$$

where the asterisk means "excited state" of that nucleus.

In some cases, a nucleus may remain in an excited state for some time before it emits a γ ray. The nucleus is then said to be in a **metastable state** and is called an **isomer**.

An excited nucleus can sometimes return to the ground state by another process known as **internal conversion** with no γ ray emitted. In this process, the excited nucleus interacts with one of the orbital electrons and ejects this electron from the atom with the same kinetic energy (minus the binding energy of the electron) that an emitted γ ray would have had.

What, you may wonder, is the difference between a γ ray and an X-ray? They both are electromagnetic radiation (photons) and, though γ rays usually have higher energy than X-rays, their range of energies overlap to some extent. The difference is not intrinsic. We use the term X-ray if the photon is produced by an electron–atom interaction, and γ ray if the photon is produced in a nuclear process.

Internal conversion

42–7 Conservation of Nucleon Number and Other Conservation Laws

In all three types of radioactive decay, the classical conservation laws hold. Energy, linear momentum, angular momentum, and electric charge are all conserved. These quantities are the same before the decay as after. But a new conservation law is also revealed, the **law of conservation of nucleon number**. According to this law, the total number of nucleons (A) remains constant in any process, although one type can change into the other type (protons into neutrons or vice versa). This law holds in all three types of decay. Table 42–2 gives a summary of α, β, and γ decay.

42–8 Half-Life and Rate of Decay

A macroscopic sample of any radioactive isotope consists of a vast number of radioactive nuclei. These nuclei do not all decay at one time. Rather, they decay one by one over a period of time. This is a random process: we can not predict exactly when a given nucleus will decay. But we can determine, on a probabilistic basis, approximately how many nuclei in a sample will decay over a given time period, by assuming that each nucleus has the same probability of decaying in each second that it exists.

The number of decays ΔN that occur in a very short time interval Δt is then proportional to Δt and to the total number N of radioactive nuclei present:

$$\Delta N = -\lambda N \, \Delta t. \qquad\qquad \textbf{(42–4)}$$

In this equation, λ is a constant of proportionality called the **decay constant**, which is different for different isotopes. The greater λ is, the greater the rate of decay and the more "radioactive" that isotope is said to be for a given number of nuclei. The number of decays that occur in the short time interval Δt is designated ΔN

TABLE 42–2 The Three Types of Radioactive Decay

α decay:
$$^{A}_{Z}N \rightarrow \, ^{A-4}_{Z-2}N' + \, ^{4}_{2}\text{He}$$

β decay:
$$^{A}_{Z}N \rightarrow \, ^{A}_{Z+1}N' + e^- + \bar{\nu}$$
$$^{A}_{Z}N \rightarrow \, ^{A}_{Z-1}N' + e^+ + \nu$$
$$^{A}_{Z}N + e^- \rightarrow \, ^{A}_{Z-1}N' + \nu \,\, [\text{EC}]^{\dagger}$$

γ decay:
$$^{A}_{Z}N^* \rightarrow \, ^{A}_{Z}N + \gamma$$

*Indicates the excited state of a nucleus.
†Electron capture.

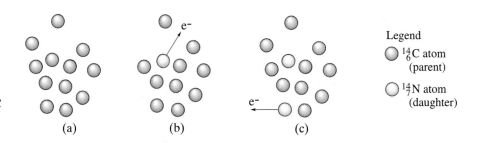

FIGURE 42–9 Radioactive nuclei decay one by one. Hence, the number of parent nuclei in a sample is continually decreasing. When a $^{14}_{6}C$ nucleus emits the electron, it becomes a $^{14}_{7}N$ nucleus.

Legend

$^{14}_{6}C$ atom (parent)

$^{14}_{7}N$ atom (daughter)

(a)   (b)   (c)

because each decay that occurs corresponds to a decrease by one in the number N of nuclei present. That is, radioactive decay is a "one-shot" process, Fig. 42–9. Once a particular parent nucleus decays into its daughter, it cannot do it again. The minus sign in Eq. 42–4 is needed to indicate that N is decreasing.

If we take the limit $\Delta t \to 0$ in Eq. 42–4, ΔN will be small compared to N, and we can write the equation in infinitesimal form as

$$dN = -\lambda N \, dt. \qquad (42\text{–}5)$$

We can determine N as a function of t by rearranging this equation to

$$\frac{dN}{N} = -\lambda \, dt$$

and then integrating from $t = 0$ to $t = t$:

$$\int_{N_0}^{N} \frac{dN}{N} = -\int_{0}^{t} \lambda \, dt,$$

where N_0 is the number of parent nuclei present at $t = 0$ and N is the number remaining at time t. The integration gives

$$\ln \frac{N}{N_0} = -\lambda t$$

Radioactive decay law

or

$$N = N_0 e^{-\lambda t}. \qquad (42\text{–}6)$$

Equation 42–6 is called the **radioactive decay law**. It tells us that the number of radioactive nuclei in a given sample decreases exponentially in time. This is shown in Fig. 42–10a, for the case of $^{14}_{6}C$ whose decay constant is $\lambda = 3.8 \times 10^{-12} \, \text{s}^{-1}$.

FIGURE 42–10 (a) The number N of parent nuclei in a given sample of $^{14}_{6}C$ decreases exponentially. (b) The number of decays per second also decreases exponentially. The half-life of $^{14}_{6}C$ is about 5730 yr, which means that the number of parent nuclei, N, and the rate of decay, dN/dt, decrease by half every 5730 yr.

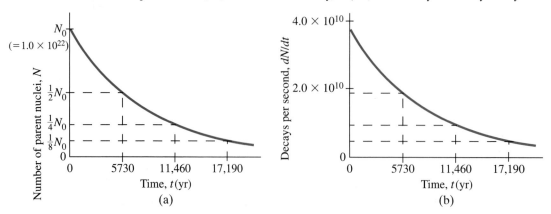

The rate of decay, or number of decays per second, in a pure sample is

$$\frac{dN}{dt}$$

which is called the **activity** of a given sample. From Eqs. 42–5 and 42–6,

$$\frac{dN}{dt} = -\lambda N = -\lambda N_0 e^{-\lambda t}. \qquad \textbf{(42–7a)}$$

At $t = 0$, the activity is

$$\left(\frac{dN}{dt}\right)_0 = -\lambda N_0. \qquad \textbf{(42–7b)}$$

Hence

$$\frac{dN}{dt} = \left(\frac{dN}{dt}\right)_0 e^{-\lambda t}, \qquad \textbf{(42–7c)} \qquad \textit{Activity}$$

so the activity also decreases exponentially in time at the same rate as for N (Fig. 42–10b).

The rate of decay of any isotope is often specified by giving its half-life rather than the decay constant λ. The **half-life** of an isotope is defined as the time it takes for half the original amount of isotope in a given sample to decay. For example, the half-life of $^{14}_{6}C$ is about 5730 yr. If at some time a piece of petrified wood contains, say, 1.00×10^{22} $^{14}_{6}C$ nuclei, then 5730 yr later it will contain only 0.50×10^{22} nuclei. After another 5730 yr it will contain 0.25×10^{22} nuclei, and so on. This is characteristic of the exponential function, and is shown in Fig. 42–10a. Since the rate of decay dN/dt is proportional to N, it too decreases by a factor of 2 every half-life, Fig. 42–10b.

The half-lives of known radioactive isotopes vary from as short as 10^{-22} s to about 10^{28} s (about 10^{21} yr). The half-lives of many isotopes[†] are given in Appendix D. It should be clear that the half-life (which we designate $T_{\frac{1}{2}}$) bears an inverse relationship to the decay constant. The longer the half-life of an isotope, the more slowly it decays, and hence λ is smaller. The precise relation is obtained from Eq. 42–6 by setting $N = N_0/2$ at $t = T_{\frac{1}{2}}$:

$$\frac{N_0}{2} = N_0 e^{-\lambda T_{\frac{1}{2}}} \qquad \text{or} \qquad e^{\lambda T_{\frac{1}{2}}} = 2.$$

We take natural logs of both sides ("ln" and "e" are inverse operations, meaning $\ln\left(e^x\right) = x$) and find

$$\ln\left(e^{\lambda T_{\frac{1}{2}}}\right) = \ln 2, \qquad \text{so} \qquad \lambda T_{\frac{1}{2}} = \ln 2$$

and

$$T_{\frac{1}{2}} = \frac{\ln 2}{\lambda} = \frac{0.693}{\lambda}. \qquad \textbf{(42–8)} \qquad \textit{Half-life}$$

We can then write Eq. 42–6 as

$$N = N_0 e^{-0.693 t / T_{\frac{1}{2}}}.$$

Now let us use these concepts and equations in worked-out Examples.

[†] You may find the *mean life* of an isotope quoted. The mean life τ is defined as $\tau = 1/\lambda$ (see Problem 76), so that Eq. 42–6 can be written $N = N_0 e^{-t/\tau}$ just as for RC and LR circuits (Chapters 26 and 30) where τ is called the time constant. Since

$$\tau = \frac{1}{\lambda} = \frac{T_{\frac{1}{2}}}{0.693}$$

the mean life and half-life differ significantly in numerical value, so confusing them can cause serious error.

EXAMPLE 42–7 **Sample activity.** The isotope $^{14}_6$C has a half-life of 5730 yr. If at some time a sample contains 1.00×10^{22} carbon-14 nuclei, what is the activity of the sample?

SOLUTION First we calculate the decay constant λ from Eq. 42–8, and obtain

$$\lambda = \frac{0.693}{T_{\frac{1}{2}}} = \frac{0.693}{(5730 \text{ yr})(3.156 \times 10^7 \text{ s/yr})} = 3.83 \times 10^{-12} \text{ s}^{-1},$$

since the number of seconds in a year is $(60)(60)(24)(365\frac{1}{4}) = 3.156 \times 10^7$ s. From Eq. 42–5, the magnitude of the activity or rate of decay is

$$\frac{dN}{dt} = \lambda N = (3.83 \times 10^{-12} \text{ s}^{-1})(1.00 \times 10^{22})$$

$$= 3.83 \times 10^{10} \text{ decays/s}.$$

(The unit "decays/s" is often written simply as s^{-1} since "decays" is not a unit but refers only to the number.) Note that the graph of Fig. 42–10b starts at this value, corresponding to the original value of $N = 1.0 \times 10^{22}$ nuclei in Fig. 42–10a.

EXAMPLE 42–8 **A sample of radioactive $^{13}_7$N.** A laboratory has 1.49 μg of pure $^{13}_7$N, which has a half-life of 10.0 min (600 s). (a) How many nuclei are present initially? (b) What is the activity initially? (c) What is the activity after 1.00 h? (d) After approximately how long will the activity drop to less than one per second?

SOLUTION (a) Since the atomic mass is 13.0, then 13.0 g will contain 6.02×10^{23} nuclei (Avogadro's number). Since we have only 1.49×10^{-6} g, the number of nuclei, N_0, that we have initially is given by the ratio

$$\frac{N_0}{1.49 \times 10^{-6} \text{ g}} = \frac{6.02 \times 10^{23}}{13.0 \text{ g}},$$

so $N_0 = 6.90 \times 10^{16}$ nuclei.
(b) From Eq. 42–8, $\lambda = (0.693)/(600 \text{ s}) = 1.16 \times 10^{-3} \text{ s}^{-1}$. Then, at $t = 0$ (Eq. 42–7b)

$$\left(\frac{dN}{dt}\right)_0 = \lambda N_0 = (1.16 \times 10^{-3} \text{ s}^{-1})(6.90 \times 10^{16}) = 8.00 \times 10^{13} \text{ s}^{-1}.$$

(c) After 1.00 h = 3600 s, the activity will be (Eq. 42–7c)

$$\frac{dN}{dt} = \left(\frac{dN}{dt}\right)_0 e^{-\lambda t}$$

$$= (8.00 \times 10^{13} \text{ s}^{-1})e^{-(1.16 \times 10^{-3} \text{ s}^{-1})(3600 \text{ s})} = 1.23 \times 10^{12} \text{ s}^{-1}.$$

This result can be obtained in another way: since 1.00 h represents six half-lives $(6 \times 10.0 \text{ min})$, the activity will decrease to $(\frac{1}{2})(\frac{1}{2})(\frac{1}{2})(\frac{1}{2})(\frac{1}{2})(\frac{1}{2}) = (\frac{1}{2})^6 = \frac{1}{64}$ of its original value, or $(8.00 \times 10^{13} \text{ s}^{-1})/64 = 1.25 \times 10^{12} \text{ s}^{-1}$. (The slight discrepancy between the two values arises because we kept only three significant figures.)
(d) We want to determine the time t when $dN/dt = 1.00 \text{ s}^{-1}$. From Eq. 42–7c, we have

$$e^{-\lambda t} = \frac{(dN/dt)}{(dN/dt)_0} = \frac{1.00 \text{ s}^{-1}}{8.00 \times 10^{13} \text{ s}^{-1}} = 1.25 \times 10^{-14}.$$

We take the natural log (ln) of both sides and divide by λ to find

$$t = -\frac{\ln(1.25 \times 10^{-14})}{\lambda} = \frac{32.0}{1.16 \times 10^{-3} \text{ s}^{-1}} = 2.76 \times 10^4 \text{ s} = 7.67 \text{ h}.$$

It is often the case that one radioactive isotope decays to another isotope that is also radioactive. Sometimes this daughter decays to yet a third isotope which also is radioactive. Such successive decays are said to form a **decay series**. An important example is illustrated in Fig. 42–11. As can be seen, $^{238}_{92}$U decays by α emission to $^{234}_{90}$Th, which in turn decays by β decay to $^{234}_{91}$Pa. The series continues as shown, with several possible branches near the bottom. For example, $^{218}_{84}$Po can decay either by α decay to $^{214}_{82}$Pb or by β decay to $^{218}_{85}$At. The series ends at the stable lead isotope $^{206}_{82}$Pb. Other radioactive series also exist.

Because of such decay series, certain radioactive elements are found in nature that otherwise would not be. For when the solar system acquired its present form about 5 billion years ago, it is believed that nearly all nuclides were formed (by the fusion process, Sections 43–4 and 45–2). Many isotopes with short half-lives decayed quickly and no longer exist in nature today. But long-lived isotopes, such as $^{238}_{92}$U with a half-life of 4.5×10^9 yr, still do exist in nature today. Indeed, about half of the original $^{238}_{92}$U still remains (assuming that the origin of the solar system was about 5×10^9 yr ago). We might expect, however, that radium ($^{226}_{88}$Ra), with a half-life of 1600 yr, would long since have disappeared from the Earth. Indeed, the original $^{226}_{88}$Ra nuclei must by now have all decayed. However, because $^{238}_{92}$U decays (in several steps) to $^{226}_{88}$Ra, the supply of $^{226}_{88}$Ra is continually replenished, which is why it is still found on Earth today. The same can be said for many other radioactive nuclides.

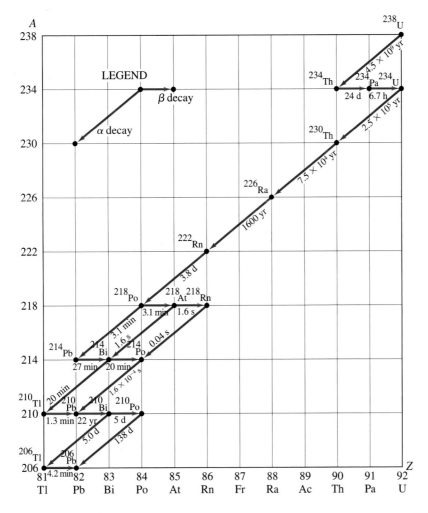

FIGURE 42–11 Decay series beginning with $^{238}_{92}$U. Nuclei in the series are specified by a dot representing A and Z values. Half-lives are given in seconds (s), minutes (min), hours (h), days (d), or years (yr). Note that a horizontal arrow represents β decay (A does not change), whereas a diagonal line represents α decay (A changes by 4, Z changes by 2).

CONCEPTUAL EXAMPLE 42-9 | **Decay chain.** The decay chain starting with ^{234}U in Fig. 42-11 has nuclides with half-lives of 250,000 yr, 75,000 yr, 1600 yr, and a little under 4 days, respectively. Each decay in the chain has an alpha particle of a characteristic energy, and so we can monitor the radioactive decay rate of each nuclide. Given a sample that was pure ^{234}U a million years ago, which alpha decay would you expect to have the highest activity rate in the sample?

RESPONSE The first instinct is to say that the process with the shortest half-life would show the highest activity. Surprisingly, however, the activity rates in this sample are all the same! The reason is that in each case the decay of the parent acts as a bottleneck to the decay of the daughter. Compared to the 1600-yr half-life of ^{226}Ra, for example, its daughter ^{222}Rn decays almost immediately, but it cannot decay until it is made. (This is like an automobile assembly line: If worker A takes 20 minutes to do a task and then worker B takes only 1 minute to do the next task, worker B still does only one car every 20 minutes).

42-10 | Radioactive Dating

Radioactive decay has many interesting applications. One is the technique of *radioactive dating* by which the age of ancient materials can be determined.

PHYSICS APPLIED

Carbon-14 dating

The age of any object made from once-living matter, such as wood, can be determined using the natural radioactivity of $^{14}_6$C. All living plants absorb carbon dioxide (CO_2) from the air and use it to synthesize organic molecules. The vast majority of these carbon atoms are $^{12}_6$C, but a small fraction, about 1.3×10^{-12}, is the radioactive isotope $^{14}_6$C. The ratio of $^{14}_6$C to $^{12}_6$C in the atmosphere has remained roughly constant over many thousands of years, in spite of the fact that $^{14}_6$C decays with a half-life of about 5730 yr. This is because neutrons in the cosmic radiation that impinges on the Earth from outer-space collide with atoms of the atmosphere. In particular, collisions with nitrogen nuclei produce the following nuclear transformation: $n + {}^{14}_7N \rightarrow {}^{14}_6C + p$. That is, a neutron strikes and is absorbed by a $^{14}_7$N nucleus, and a proton is knocked out in the process. The remaining nucleus is $^{14}_6$C. This continual production of $^{14}_6$C in the atmosphere roughly balances the loss of $^{14}_6$C by radioactive decay. As long as a plant or tree is alive, it continually uses the carbon from carbon dioxide in the air to build new tissue and to replace old. Animals eat plants, so they too are continually receiving a fresh supply of carbon for their tissues. Organisms cannot distinguish[†] $^{14}_6$C from $^{12}_6$C, and since the ratio of $^{14}_6$C to $^{12}_6$C in the atmosphere remains nearly constant, the ratio of the two isotopes within the living organism remains nearly constant as well. But when an organism dies, carbon dioxide is no longer absorbed and utilized. Because the $^{14}_6$C decays radioactively, the ratio of $^{14}_6$C to $^{12}_6$C in a dead organism decreases in time. Since the half-life of $^{14}_6$C is about 5730 yr, the $^{14}_6$C/$^{12}_6$C ratio decreases by half every 5730 yr. If, for example, the $^{14}_6$C/$^{12}_6$C ratio of an ancient wooden tool is half of what it is in living trees, then the object must have been made from a tree that was felled about 5700 years ago. Actually, corrections must be made for the fact that the $^{14}_6$C/$^{12}_6$C ratio in the atmosphere has not remained precisely constant over time. The determination of what this ratio has been over the centuries has required using techniques such as comparing the expected ratio to the actual ratio for objects whose age is known, such as very old trees whose annual rings can be counted.

[†]Organisms operate almost exclusively via chemical reactions—which involve only the outer orbital electrons of the atom; extra neutrons in the nucleus have essentially no effect.

EXAMPLE 42–10 **An ancient animal.** An animal bone fragment found in an archeological site has a carbon mass of 200 g. It registers an activity of 16 decays/s. What is the age of the bone?

➡ PHYSICS APPLIED

Archeological dating

SOLUTION When the animal was alive, the ratio of $^{14}_{6}C$ to $^{12}_{6}C$ in the 200-g piece of bone was 1.3×10^{-12}. The number of $^{14}_{6}C$ nuclei at that time was

$$N_0 = \left(\frac{6.02 \times 10^{23} \text{ atoms}}{12 \text{ g}} \right)(200 \text{ g})(1.3 \times 10^{-12}) = 1.3 \times 10^{13}.$$

From Eq. 42–7b, considering only the magnitude,

$$\left(\frac{dN}{dt} \right)_0 = \lambda N_0,$$

where $\lambda = 3.83 \times 10^{-12} \text{ s}^{-1}$ (Example 42–7). So the original activity was

$$\left(\frac{dN}{dt} \right)_0 = (3.83 \times 10^{-12} \text{ s}^{-1})(1.3 \times 10^{13}) = 50 \text{ s}^{-1}.$$

From Eq. 42–7c,

$$\frac{dN}{dt} = \left(\frac{dN}{dt} \right)_0 e^{-\lambda t},$$

and we rewrite this as

$$e^{\lambda t} = \left[\frac{(dN/dt)_0}{(dN/dt)} \right].$$

Now we take the natural log (ln) of both sides to get

$$t = \frac{1}{\lambda} \ln \left[\frac{(dN/dt)_0}{(dN/dt)} \right] = \frac{1}{3.83 \times 10^{-12} \text{ s}^{-1}} \ln \left[\frac{50 \text{ s}^{-1}}{16 \text{ s}^{-1}} \right]$$
$$= 2.98 \times 10^{11} \text{ s} = 9400 \text{ yr},$$

which is the time elapsed since the death of the animal.

Carbon dating is useful only for determining the age of objects less than about 60,000 years old. The amount of $^{14}_{6}C$ remaining in older objects is usually too small to measure accurately, although new techniques are allowing detection of even smaller amounts of $^{14}_{6}C$, pushing the time frame further back. On the other hand, radioactive isotopes with longer half-lives can be used in certain circumstances to obtain the age of older objects. For example, the decay of $^{238}_{92}U$, because of its long half-life of 4.5×10^9 years, is useful in determining the ages of rocks on a geologic time scale. When molten material solidified into rock as the temperature dropped, different compounds solidified according to the melting points, and thus different compounds separated to some extent. Uranium present in a material became fixed in position and the daughter nuclei that result from the decay of uranium were also fixed in that position. Thus, by measuring the amount of $^{238}_{92}U$ remaining in the material relative to the amount of daughter nuclei, the time when the rock solidified can be determined.

➡ PHYSICS APPLIED

Geological dating

Radioactive dating methods using $^{238}_{92}U$ and other isotopes have shown the age of the oldest Earth rocks to be about 4×10^9 yr. The age of rocks in which the oldest fossilized organisms are embedded indicates that life appeared at least 3 billion years ago. The earliest fossilized remains of mammals are found in rocks 200 million years old, and the first humanlike creatures seem to have appeared about 2 million years ago. Radioactive dating has been indispensable for the reconstruction of Earth's history.

➡ PHYSICS APPLIED

*Oldest Earth rocks
and
earliest life*

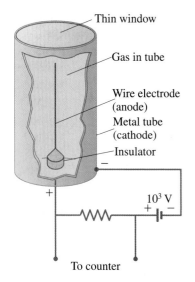

FIGURE 42–12 Diagram of a Geiger counter.

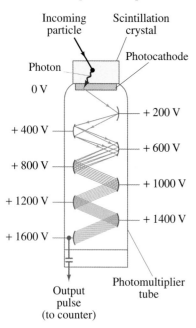

FIGURE 42–13 Scintillation counter with a photomultiplier tube.

Individual particles such as electrons, protons, α particles, neutrons, and γ rays are not detected directly by our senses. Consequently, a variety of instruments have been developed to detect them.

One of the most common is the **Geiger counter**. As shown in Fig. 42–12, it consists of a cylindrical metal tube filled with a certain type of gas. A long wire runs down the center and is kept at a high positive voltage $(\approx 10^3 \text{ V})$ with respect to the outer cylinder. The voltage is just slightly less than that required to ionize the gas atoms. When a charged particle enters through the thin "window" at one end of the tube, it ionizes a few atoms of the gas. The freed electrons are attracted toward the positive wire and as they are accelerated they strike and ionize additional atoms. An "avalanche" of electrons is quickly produced, and when it reaches the wire anode, it produces a voltage pulse. The pulse, after being amplified, can be sent to an electronic counter, which counts how many particles have been detected. Or the pulses can be sent to a loudspeaker and each detection of a particle is heard as a "click."

A **scintillation counter** makes use of a solid, liquid, or gas known as a **scintillator** or **phosphor**. The atoms of a scintillator are easily excited when struck by an incoming particle and emit visible light when they return to their ground states. Typical scintillators are crystals of NaI and certain plastics. One face of a solid scintillator is cemented to a photomultiplier tube, and the whole is wrapped with opaque material to keep it light tight or is placed within a light-tight container. The **photomultiplier (PM) tube** converts the energy of the scintillator-emitted photon(s) into an electric signal. A PM tube is a vacuum tube containing several electrodes (typically 8 to 14), called *dynodes*, which are maintained at successively higher voltages as shown in Fig. 42–13. At its top surface is a photoelectric surface, called the *photocathode*, whose work function (Section 38–2) is low enough that an electron is easily released when struck by a photon from the scintillator. Such an electron is accelerated toward the first dynode. When it strikes the first dynode, the electron has acquired sufficient kinetic energy so that it can eject two to five more electrons. These, in turn, are accelerated to the second dynode, and a multiplication process begins. The number of electrons striking the last dynode may be 10^6 or more. Thus the passage of a particle through the scintillator results in an electric signal at the output of the PM tube that can be sent to an electronic counter just as for a Geiger tube. Because a scintillator crystal is a solid and therefore much more dense than the gas of a Geiger counter, it is a more efficient detector—especially for γ rays, which interact less with matter than do β rays.

In tracer work (Section 43–8), **liquid scintillators** are often used. Radioactive samples taken at different times or from different parts of an organism are placed directly in small bottles containing the liquid scintillator. This is particularly convenient for detection of β rays from ${}_{1}^{3}\text{H}$ and ${}_{6}^{14}\text{C}$, which have very low energies and have difficulty passing through the outer covering of a crystal scintillator or Geiger tube. A PM tube is still used to produce the electric signal.

A **semiconductor detector** consists of a reverse-biased *p-n* junction diode (Sections 41–8 and 9). A particle passing through the junction can excite electrons into the conduction band, leaving holes in the valence band. The freed charges produce a short electrical pulse that can be counted just as for Geiger and scintillation counters.

In Chapter 44 we will discuss additional detectors, those that allow visualization of particle tracks and are used in connection with nuclear reactions and elementary particles.

Summary

Nuclear physics is the study of atomic nuclei. Nuclei contain **protons** and **neutrons**, which are collectively known as **nucleons**. The total number of nucleons, A, is the **atomic mass number**. The number of protons, Z, is the **atomic number**. The number of neutrons equals $A - Z$. **Isotopes** are nuclei with the same Z, but with different numbers of neutrons. For an element X, an isotope of given Z and A is represented by

$$^{A}_{Z}\text{X}.$$

The nuclear radius is proportional to $A^{\frac{1}{3}}$, indicating that all nuclei have about the same density. Nuclear masses are specified in **unified atomic mass units** (u), where the mass of $^{12}_{6}\text{C}$ (including its 6 electrons) is defined as exactly 12.000000 u, or in terms of their energy equivalent (because $E = mc^2$), where

$$1\,\text{u} = 931.5\,\text{MeV}/c^2 = 1.66 \times 10^{-27}\,\text{kg}.$$

The mass of a stable nucleus is less than the sum of the masses of its constituent nucleons. The difference in mass (times c^2) is the **total binding energy**. It represents the energy needed to break the nucleus into its constituent nucleons. The **binding energy per nucleon** averages about 8 MeV per nucleon, and is lowest for low mass and high mass nuclei.

Unstable nuclei undergo **radioactive decay**; they change into other nuclei with the emission of an α, β, or γ particle. An α particle is a $^{4}_{2}\text{He}$ nucleus; a β particle is an electron or positron; and a γ ray is a high-energy photon. In β decay, a **neutrino** is also emitted. The transformation of the parent into the daughter nucleus is called **transmutation** of the elements. Radioactive decay occurs spontaneously only when the rest mass of the products is less than the mass of the parent nucleus. The loss in mass appears as kinetic energy of the products.

Nuclei are held together by the **strong nuclear force**. The **weak nuclear force** makes itself apparent in β decay. These two forces, plus the gravitational and electromagnetic forces, are the four known types of force. Electric charge, linear and angular momentum, mass–energy, and **nucleon number** are **conserved** in all decays.

Radioactive decay is a statistical process. For a given type of radioactive nucleus, the number of nuclei that decay (ΔN) in a time Δt is proportional to the number N of parent nuclei present:

$$\Delta N = -\lambda N\,\Delta t.$$

The proportionality constant, λ, is called the **decay constant** and is characteristic of the given nucleus. The number N of nuclei remaining after a time t decreases exponentially

$$N \propto e^{-\lambda t}$$

as does the **activity**, dN/dt. The **half-life**, $T_{\frac{1}{2}}$, is the time required for half the nuclei of a radioactive sample to decay. It is related to the decay constant by $T_{\frac{1}{2}} = 0.693/\lambda$.

Questions

1. What do different isotopes of a given element have in common? How are they different?

2. What are the elements represented by the X in the following: (a) $^{232}_{92}\text{X}$; (b) $^{18}_{7}\text{X}$; (c) $^{1}_{1}\text{X}$; (d) $^{82}_{38}\text{X}$; (e) $^{247}_{97}\text{X}$?

3. How many protons and how many neutrons do each of the isotopes in Question 2 have?

4. Why are the atomic masses of many elements (see the periodic table) not close to whole numbers?

5. How do we know there is such a thing as the strong nuclear force?

6. What are the similarities and the differences between the strong nuclear force and the electric force?

7. What is the experimental evidence in favor of radioactivity being a nuclear process?

8. The isotope $^{64}_{29}\text{Cu}$ is unusual in that it can decay by γ, β^-, and β^+ emission. What is the resulting nuclide for each case?

9. A $^{238}_{92}\text{U}$ nucleus decays to a nucleus containing how many neutrons?

10. Describe, in as many ways as possible, the difference between α, β, and γ rays.

11. What element is formed by the radioactive decay of (a) $^{24}_{11}\text{Na}$ (β^-); (b) $^{22}_{11}\text{Na}$ (β^+); (c) $^{210}_{84}\text{Po}$ (α)?

12. What element is formed by the decay of (a) $^{32}_{15}\text{P}$ (β^-); (b) $^{35}_{16}\text{S}$ (β^-); (c) $^{211}_{83}\text{Bi}$ (α)?

13. Fill in the missing particle or nucleus:
    (a) $^{45}_{20}\text{Ca} \rightarrow ? + e^- + \bar{\nu}$
    (b) $^{58}_{29}\text{Cu} \rightarrow ? + \gamma$
    (c) $^{46}_{24}\text{Cr} \rightarrow ^{46}_{23}\text{V} + ?$
    (d) $^{234}_{94}\text{Pu} \rightarrow ? + \alpha$
    (e) $^{239}_{93}\text{Np} \rightarrow ^{239}_{94}\text{Pu} + ?$

14. Explain α decay by tunneling using the uncertainty principle.

15. Immediately after a $^{238}_{92}\text{U}$ nucleus decays to $^{234}_{90}\text{Th} + ^{4}_{2}\text{He}$, the daughter thorium nucleus still has 92 electrons circling it. Since thorium normally holds only 90 electrons, what do you suppose happens to the two extra ones?

16. When a nucleus undergoes either β^- or β^+ decay, what happens to the energy levels of the atomic electrons? What is likely to happen to these electrons following the decay?

17. The alpha particles from a given alpha-emitting nuclide are monoenergetic; that is, they all have the same kinetic energy. But the beta particles from a beta-emitting nuclide have a spectrum of energies. Explain the difference between these two cases.

18. Do isotopes that undergo electron capture generally lie above or below the line of stability in Fig. 42–2?

19. Can hydrogen or deuterium emit an α particle?

20. Why are many artificially produced radioactive isotopes rare in nature?

21. An isotope has a half-life of one month. After two months, will a given sample of this isotope have completely decayed? If not, how much remains?

22. Describe how the potential energy curve for an α particle in an α-emitting nucleus differs from that for a stable nucleus.

23. Explain the absence of β^+ emitters in the radioactive decay series of Fig. 42–11.

24. Can $^{14}_{6}C$ dating be used to measure the age of stone walls and tablets of ancient civilizations?

25. What assumptions are made in carbon-dating? What do you think could affect these assumptions?

Problems

Section 42–1

1. (I) What is the rest mass of an α particle in MeV/c^2?

2. (I) A pi meson has a mass of $139\,\text{MeV}/c^2$. What is this in atomic mass units?

3. (I) What is the approximate radius of an alpha particle $(^{4}_{2}\text{He})$?

4. (I) By what percentage is the radius of the isotope $^{14}_{6}C$ greater than that of its sister $^{12}_{6}C$?

5. (II) (a) Determine the density of nuclear matter in kg/m^3, and show that it is essentially the same for all nuclei. (b) What would be the radius of the Earth if it had its actual mass but had the density of nuclei? (c) What would be the radius of a $^{238}_{92}U$ nucleus if it had the density of the Earth?

6. (II) (a) What is the approximate radius of a $^{64}_{29}\text{Cu}$ nucleus? (b) Approximately what is the value of A for a nucleus whose radius is $3.9 \times 10^{-15}\,\text{m}$?

7. (II) How much energy must an α particle have to just "touch" the surface of a $^{238}_{92}U$ nucleus?

8. (II) If an alpha particle were released from rest near the surface of a $^{243}_{95}\text{Am}$ nucleus, what would its kinetic energy be when far away?

9. (II) What stable nucleus has approximately half the radius of a uranium nucleus? [Hint: Find A and use Appendix D to get Z.]

Section 42–2

10. (I) Estimate the total binding energy for $^{40}_{20}\text{Ca}$, using Fig. 42–1.

11. (I) Use Fig. 42–1 to estimate the total binding energy of (a) $^{238}_{92}U$, and (b) $^{84}_{36}\text{Kr}$.

12. (II) Use Appendix D to calculate the binding energy of $^{2}_{1}\text{H}$ (deuterium).

13. (II) Calculate the binding energy per nucleon for a $^{14}_{7}N$ nucleus.

14. (II) Determine the binding energy of the last neutron in a $^{40}_{19}\text{K}$ nucleus.

15. (II) Calculate the total binding energy, and the binding energy per nucleon, for (a) $^{6}_{3}\text{Li}$, (b) $^{208}_{82}\text{Pb}$. Use Appendix D.

16. (II) Calculate the binding energy of (a) the last proton, and (b) the last neutron, in a $^{12}_{6}C$ nucleus. Use Appendix D.

17. (II) Compare the binding energy of a neutron in $^{23}_{11}\text{Na}$ to that in $^{24}_{11}\text{Na}$.

18. (II) How much energy is required to remove (a) a proton, (b) a neutron, from $^{16}_{8}O$? Explain the difference in your answers.

19. (II) (a) Show that the nucleus $^{8}_{4}\text{Be}$ (mass = 8.005305 u) is unstable to decay into two α particles. (b) Is $^{12}_{6}C$ stable against decay into three α particles? Show why or why not.

Sections 42–3 to 42–7

20. (I) How much energy is released when tritium, $^{3}_{1}\text{H}$, decays by β^- emission?

21. (I) What is the maximum kinetic energy of an electron emitted in the β decay of a free neutron?

22. (I) Show that the decay $^{11}_{6}C \rightarrow {}^{10}_{5}B + p$ is not possible because energy would not be conserved.

23. (II) $^{22}_{11}\text{Na}$ is radioactive. Is it a β^- or β^+ emitter? Write down the decay reaction, and estimate the maximum kinetic energy of the emitted β.

24. (II) Give the result of a calculation that shows whether or not the following decays are possible:
(a) $^{236}_{92}U \rightarrow {}^{235}_{92}U + n$;   (b) $^{16}_{8}O \rightarrow {}^{15}_{8}O + n$;
(c) $^{23}_{11}\text{Na} \rightarrow {}^{22}_{11}\text{Na} + n$.

25. (II) A $^{232}_{92}U$ nucleus emits an α particle with kinetic energy = 5.32 MeV. What is the daughter nucleus and what is the approximate atomic mass (in u) of the daughter atom? Ignore recoil of the daughter nucleus.

26. (II) When $^{23}_{10}\text{Ne}$ (mass = 22.9945 u) decays to $^{23}_{11}\text{Na}$ (mass = 22.9898 u), what is the maximum kinetic energy of the emitted electron? What is its minimum energy? What is the energy of the neutrino in each case?

27. (II) The nuclide $^{32}_{15}\text{P}$ decays by emitting an electron whose maximum kinetic energy can be 1.71 MeV. (a) What is the daughter nucleus? (b) What is its atomic mass (in u)?

28. (II) The isotope $^{218}_{84}\text{Po}$ can decay by either α or β^- emission. What is the energy release in each case? The mass of $^{218}_{84}\text{Po}$ is 218.008965 u.

29. (II) How much energy is released in electron capture by beryllium: $^{7}_{4}\text{Be} + {}^{0}_{-1}e \rightarrow {}^{7}_{3}\text{Li} + \nu$?

30. (II) Decay series, such as that shown in Fig. 42–11, can be classified into four families, depending on whether the mass numbers have the form $4n$, $4n + 1$, $4n + 2$, or $4n + 3$, where n is an integer. Justify this statement and show that for a nuclide in any family, all its daughters will be in the same family.

31. (III) What is the energy of the α particle emitted in the decay $^{210}_{84}\text{Po} \rightarrow {}^{206}_{82}\text{Pb} + \alpha$? Take into account the recoil of the daughter nucleus.

32. (III) The α particle emitted when $^{238}_{92}U$ decays has 4.20 MeV of kinetic energy. Calculate the recoil kinetic energy of the daughter nucleus and the Q-value of the decay.

33. (III) (*a*) Show that when a nucleus decays by β^+ decay, the total energy released is equal to

$$(M_P - M_D - 2m_e)c^2,$$

where M_P and M_D are the masses of the parent and daughter atoms (neutral), and m_e is the mass of an electron or positron. (*b*) Determine the maximum kinetic energy of β^+ particles released when $^{11}_6C$ decays to $^{11}_5B$. What is the maximum energy the neutrino can have? What is its minimum energy?

34. (III) (*a*) Calculate the kinetic energy of the α particle emitted when $^{236}_{92}U$ decays. (*b*) Use Eq. 42–1 to estimate the radius of an α particle and $^{232}_{90}Th$ nucleus. Use this to estimate (*c*) the maximum height of the Coulomb barrier, and (*d*) its width AB in Fig. 42–6.

Sections 42–8 to 42–10

35. (I) A radioactive material produces 1280 decays per minute at one time, and 6 h later produces 320 decays per minute. What is its half-life?

36. (I) (*a*) What is the decay constant of $^{238}_{92}U$ whose half-life is 4.5×10^9 yr? (*b*) The decay constant of a given nucleus is $8.2 \times 10^{-5}\,s^{-1}$. What is its half-life?

37. (I) What is the activity of a sample of $^{14}_6C$ that contains 3.1×10^{20} nuclei?

38. (I) What fraction of a sample of $^{68}_{32}Ge$, whose half-life is about 9 months, will remain after 3.0 yr?

39. (I) How many nuclei of $^{238}_{92}U$ remain in a rock if the activity registers 875 decays per second?

40. (II) What fraction of a sample is left after (*a*) exactly 4 half-lives, (*b*) exactly $4\frac{1}{2}$ half-lives?

41. (II) In a series of decays, the nuclide $^{235}_{92}U$ becomes $^{207}_{82}Pb$. How many α and β^- particles are emitted in this series?

42. (II) The iodine isotope $^{131}_{53}I$ is used in hospitals for diagnosis of thyroid function. If $632\,\mu g$ are ingested by a patient, determine the activity (*a*) immediately, (*b*) 1.0 h later when the thyroid is being tested, and (*c*) 6 months later. Use Appendix D.

43. (II) $^{124}_{55}Cs$ has a half-life of 30.8 s. (*a*) If we have $9.8\,\mu g$ initially, how many Cs nuclei are present? (*b*) How many are present 2.0 min later? (*c*) What is the activity at this time? (*d*) After how much time will the activity drop to less than about 1 per second?

44. (II) Calculate the activity of a pure 7.7-μg sample of $^{32}_{15}P$ $\left(T_{\frac{1}{2}} = 1.23 \times 10^6\,s\right)$.

45. (II) The activity of a sample of $^{35}_{16}S$ $\left(T_{\frac{1}{2}} = 7.56 \times 10^6\,s\right)$ is 2.65×10^5 decays per second. What is the mass of the sample?

46. (II) A sample of $^{233}_{92}U$ $\left(T_{\frac{1}{2}} = 1.59 \times 10^5\,yr\right)$ contains 7.50×10^{19} nuclei. (*a*) What is the decay constant? (*b*) Approximately how many disintegrations will occur per minute?

47. (II) The activity of a sample drops by a factor of 10 in 8.6 minutes. What is its half-life?

48. (II) A 185-g sample of pure carbon contains 1.3 parts in 10^{12} (atoms) of $^{14}_6C$. How many disintegrations occur per second?

49. (II) A sample of $^{40}_{19}K$ is decaying at a rate of 8.70×10^2 decays/s. What is the mass of the sample?

50. (II) The rubidium isotope $^{87}_{37}Rb$, a β emitter with a half-life of 4.75×10^{10} yr, is used to determine the age of rocks and fossils. Rocks containing fossils of early animals contain a ratio of $^{87}_{38}Sr$ to $^{87}_{37}Rb$ of 0.0160. Assuming that there was no $^{87}_{38}Sr$ present when the rocks were formed, calculate the age of these fossils.

51. (II) Use Fig. 42–11 and calculate the relative decay rates for α decay of $^{218}_{84}Po$ and $^{214}_{84}Po$.

52. (II) ^7_4Be decays with a half-life of about 53 d. It is produced in the upper atmosphere, and filters down onto the Earth's surface. If a plant leaf is detected to have 250 decays/s of ^7_4Be, how long do we have to wait for the decay rate to drop to 10/s? Estimate the initial mass of ^7_4Be on the leaf.

53. (II) Which radioactive isotope of lead is being produced in a reaction where the measured activity of a sample drops to 1.050 percent of its original activity in 4.00 h?

54. (II) An ancient club is found that contains 190 g of carbon and has an activity of 5.0 decays per second. Determine its age assuming that in living trees the ratio of $^{14}C/^{12}C$ atoms is about 1.3×10^{-12}.

55. (II) At $t = 0$, a pure sample of radioactive nuclei contains N_0 nuclei whose decay constant is λ. Determine a formula for the number of daughter nuclei, N_D, as a function of time. Assume the daughter is stable and that $N_D = 0$ at $t = 0$.

56. (III) At $t = 0$, a pure sample of a radioactive nuclide (the parent) contains N_{P0} nuclei whose half-life is T_P. The daughter nuclide is also radioactive, with half-life T_D. (*a*) Determine the number of daughter nuclei, N_D, as a function of time, assuming that $N_D = 0$ at $t = 0$. (*b*) Plot N_D versus t for the cases $T_P = T_D$, $T_P = 3T_D$, $T_P = \frac{1}{3}T_D$. [*Hint*: Eq. 42–5 must be modified.]

☐ General Problems

57. (*a*) What is the fraction of the hydrogen atom's mass that is in the nucleus? (*b*) What is the fraction of the hydrogen atom's volume that is occupied by the nucleus? (*c*) What is the density of nuclear matter? Compare this with water.

58. Using the uncertainty principle and the radius of a nucleus, estimate the minimum possible kinetic energy of a nucleon in, say, iron. Ignore relativistic corrections. [*Hint*: A particle can have an energy at least as large as its uncertainty.]

59. How much recoil energy does a $^{40}_{19}K$ nucleus get when it emits a 1.46 MeV gamma ray?

60. An old wooden tool is found to contain only 9.0 percent of $^{14}_6C$ that a sample of fresh wood would. How old is the tool?

61. A neutron star consists of neutrons at approximately nuclear density. Estimate, for a 10-km-diameter neutron star, (*a*) its mass number, (*b*) its mass (kg), and (*c*) the acceleration of gravity at its surface.

62. The $_1^3$H isotope of hydrogen, which is called *tritium* (because it contains three nucleons), has a half-life of 12.33 yr. It can be used to measure the age of objects up to about 100 yr. It is produced in the upper atmosphere by cosmic rays and brought to Earth by rain. As an application, determine approximately the age of a bottle of wine whose $_1^3$H radiation is about $\frac{1}{10}$ that present in new wine.

63. Some elementary particle theories (Section 44–11) suggest that the proton may be unstable, with a half-life $\geq 10^{32}$ yr. How long would you expect to wait for one proton in your body to decay (consider that your body is all water)?

64. When water is placed near an intense neutron source, the neutrons can be slowed down by collisions with the water molecules and eventually captured by a hydrogen nucleus to form the stable isotope called deuterium, $_1^2$H, giving off a gamma ray. What is the energy of the gamma ray?

65. How long must you wait (in half-lives) for a radioactive sample to drop to 1.00 percent of its original activity?

66. If the potassium isotope $_{19}^{40}$K gives 60 decays/s in a liter of milk, estimate how much $_{19}^{40}$K and regular $_{19}^{39}$K are in a liter of milk. Use Appendix D.

67. Show, using the decays given in Section 42–5, that the neutrino must have spin $\frac{1}{2}$.

68. In α decay of, say, a $_{88}^{226}$Ra nucleus, show that the nucleus carries away a fraction $1/(1 + A_D/4)$ of the total energy available, where A_D is the mass number of the daughter nucleus. [*Hint*: Use conservation of momentum as well as conservation of energy.] Approximately what percentage of the energy available is thus carried off by the α particle in the case cited?

69. Strontium-90 is produced as a nuclear fission product of uranium in both reactors and atomic bombs. Look at its location in the periodic table to see what other elements it might be similar to chemically, and tell why you think it might be dangerous to ingest. It has too many neutrons, and it decays with a half-life of about 29 yr. How long will we have to wait for the amount of $_{38}^{90}$Sr on the Earth's surface to reach 1 percent of its current level, assuming no new material is scattered about? Write down the decay reaction, including the daughter nucleus. Is the daughter radioactive? If so, write down its decay. Finish the decay scheme until you reach a stable nucleus.

70. The nuclide $_{76}^{191}$Os decays with β^- energy of 0.14 MeV accompanied by γ rays of energy 0.042 MeV and 0.129 MeV. (*a*) What is the daughter nucleus? (*b*) Draw an energy-level diagram showing the ground states of the parent and daughter and excited states of the daughter. To which of the daughter states does β^- decay of $_{76}^{191}$Os occur?

71. Use the uncertainty principle to argue why electrons are unlikely to be found in the nucleus. Use relativity. [*Hint*: A particle can have an energy at least as large as its uncertainty.]

72. Estimate the total binding energy for copper and then estimate the energy, in joules, needed to break a 3.0-g copper penny into its constituent nucleons.

73. Instead of giving atomic masses for nuclides as in Appendix D, some tables give the *mass excess*, Δ, defined as $\Delta = M - A$, where A is the atomic number and M is the mass in u. Determine the mass excess, in u and in MeV/c^2, for: (*a*) $_2^4$He; (*b*) $_6^{12}$C; (*c*) $_{47}^{107}$Ag; (*d*) $_{92}^{235}$U. (*e*) From a glance at Appendix D, can you make a generalization about the sign of Δ as a function of Z or A?

74. (*a*) A 100-gram sample of natural carbon contains the usual fraction of $_6^{14}$C. How long will it take on average before there is only one $_6^{14}$C nucleus left? (*b*) How does the answer in (*a*) change if the sample is 200 grams? What does this tell you about the limits of carbon dating?

75. If the mass of the proton were just a little closer to the mass of the neutron, the following reaction would be possible even at low collision energies:
$$e^- + p \rightarrow n + \nu.$$
Why would this situation be catastrophic? By what percentage would the proton's mass have to be increased to make this reaction possible?

76. (*a*) Show that the *mean life*, or *average lifetime* of a radioactive nuclide, defined as
$$\tau = \frac{\int_0^\infty N(t)t\,dt}{\int_0^\infty N(t)\,dt}$$
is $\tau = 1/\lambda$. (*b*) What fraction of the original number of nuclei remains after one mean life?

77. Two of the naturally occurring radioactive decay sequences start with $_{90}^{232}$Th, and $_{92}^{235}$U. The first five decays of these two sequences are:
$$\alpha, \beta, \beta, \alpha, \alpha$$
and
$$\alpha, \beta, \alpha, \beta, \alpha.$$
With this information determine the resulting intermediate daughter nuclei in each case.

78. What is the ratio of the kinetic energies for an alpha particle and a beta particle if both make tracks with the same radius of curvature in a magnetic field, oriented perpendicular to the paths of the particles.

79. In a rough theoretical model of α decay, the α particle moves back and forth within the parent nucleus at speed v_{in}, making a collision with the Coulomb barrier (Fig. 42–6) every $2R_0/v_{in}$ seconds, where R_0 is the radius of the nucleus. (*a*) Show that the decay constant λ in this model can be written roughly (using Eq. 39–17) as $\lambda = (v_{in}/2R_0)e^{-2GL}$, where L is the "average" barrier thickness (because the barrier is not square), and $G = \sqrt{2m(U_0 - E)}/\hbar^2$. (*b*) Use this result, plus estimates of the barrier height and width (as in Problem 34) to estimate the halflife of $_{92}^{236}$U. For the barrier thickness (Fig. 42–6) try first $L = \frac{1}{2}(R_B - R_0)$, and then (*c*) set $L = \frac{1}{3}(R_B - R_0)$ as an average thickness since the barrier is not square. Choose the potential U_{in} inside the well to be zero. (*d*) What barrier width would provide, in this simple model, the measured half-life of 2×10^7 yr?

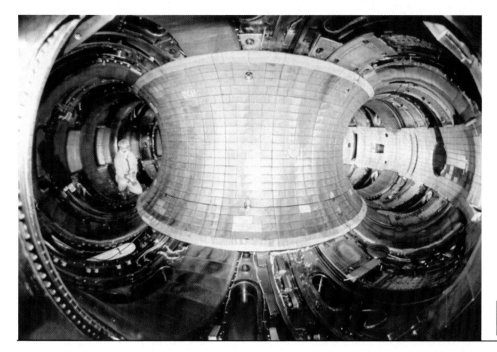

Interior of the Tokamak Fusion Test Reactor at Princeton, shown with a scientist inside (in yellow). A plasma of electrons and light nuclei can be heated to high temperatures that rival the Sun. Confining the plasma by magnetic fields has proved difficult, and intense research is needed if the fusion of light nuclei is to fulfill its great promise as a source of abundant and relatively clean power. In this chapter we also discuss fission of large nuclei which is used in present-day nuclear power plants. We also examine nuclear reactions, dosimetry, and uses of radiation including therapy and imaging.

CHAPTER 43

Nuclear Energy; Effects and Uses of Radiation

We continue our study of nuclear physics in this chapter. We begin with a discussion of nuclear reactions, after which we examine the important energy-releasing processes of fission and fusion. The remainder of the chapter deals with the effects of nuclear radiation when it passes through matter, including us, and how radiation is used medically for therapy and diagnosis, including valuable imaging techniques.

43–1 Nuclear Reactions and the Transmutation of Elements

When a nucleus undergoes α or β decay, the daughter nucleus is that of a different element from the parent. The transformation of one element into another, called *transmutation*, also occurs by means of nuclear reactions. A **nuclear reaction** is said to occur when a given nucleus is struck by another nucleus, or by a simpler particle such as a γ ray or neutron, so that an interaction takes place. Ernest Rutherford was the first to report seeing a nuclear reaction. In 1919 he observed that some of the α particles passing through nitrogen gas were absorbed and protons emitted. He concluded that nitrogen nuclei had been transformed into oxygen nuclei via the reaction

$$\,^4_2\text{He} + \,^{14}_7\text{N} \rightarrow \,^{17}_8\text{O} + \,^1_1\text{H},$$

where $\,^4_2\text{He}$ is an α particle, and $\,^1_1\text{H}$ is a proton.

Since then, a great many nuclear reactions have been observed. Indeed, many of the radioactive isotopes used in the laboratory are made by means of nuclear reactions. Nuclear reactions can be made to occur in the laboratory, but they also occur regularly in nature. In Chapter 42 we saw an example of this: $^{14}_{6}C$ is continually being made in the atmosphere via the reaction $n + {}^{14}_{7}N \rightarrow {}^{14}_{6}C + p$.

Nuclear reactions are sometimes written in a shortened form: for example, the reaction

$$n + {}^{14}_{7}N \rightarrow {}^{14}_{6}C + p$$

is written

$$^{14}_{7}N\,(n, p)\,^{14}_{6}C.$$

The symbols outside the parentheses on the left and right represent the initial and final nuclei, respectively. The symbols inside the parentheses represent the bombarding particle (first) and the emitted small particle (second).

In any nuclear reaction, both electric charge and nucleon number are conserved. These conservation laws are valuable in "balancing" nuclear reactions, as the following Example shows.

CONCEPTUAL EXAMPLE 43–1 | **Deuterium reaction.** A neutron is observed to strike an $^{16}_{8}O$ nucleus and a deuteron is given off. (A **deuteron** is the nucleus of *Deuterium* **deuterium**, the isotope of hydrogen containing one proton and one neutron, $^{2}_{1}H$; it is sometimes given the symbol d or D.) What is the nucleus that results?

RESPONSE We have the reaction $n + {}^{16}_{8}O \rightarrow ? + {}^{2}_{1}H$. The total number of nucleons initially is $1 + 16 = 17$, and the total charge is $0 + 8 = 8$. The same totals apply to the right side of the reaction. Hence the product nucleus must have $Z = 7$ and $A = 15$. From the periodic table, we find that it is nitrogen that has $Z = 7$, so the nucleus produced is $^{15}_{7}N$. The reaction can be written $^{16}_{8}O\,(n, d)\,^{15}_{7}N$, where d represents deuterium, $^{2}_{1}H$.

Energy (as well as momentum) is conserved in nuclear reactions, and we can use this to determine whether a given reaction can occur or not. For example, if the total mass of the products is less than the total mass of the initial particles, this decrease in mass (recall $E = mc^2$) appears as kinetic energy (K) of the outgoing particles. But if the total mass of the products is greater than the total mass of the initial reactants, the reaction requires an energy input. The reaction will then not occur unless the bombarding particle has sufficient kinetic energy. Consider a nuclear reaction of the general form

$$a + X \rightarrow Y + b,$$

where a is a projectile particle (or small nucleus) that strikes nucleus X, producing nucleus Y and particle b (typically, p, n, α, γ). We define the **reaction energy**, or **Q-value**, in terms of the masses involved, as

Q-value
$$Q = (M_a + M_X - M_b - M_Y)c^2. \tag{43–1}$$

Since energy is conserved, Q is equal to the change in kinetic energy (final minus initial):

$$Q = K_b + K_Y - K_a - K_X. \tag{43–2}$$

Normally, $K_X = 0$ since X is the target nucleus at rest (or nearly so) struck by an incoming particle a. For $Q > 0$, the reaction is said to be *exothermic* or *exoergic*. Energy is released in the reaction, so the total kinetic energy is greater after the reaction than before. If Q is negative ($Q < 0$), the reaction is said to be *endothermic* or *endoergic*. In this case the final total kinetic energy is less than the initial kinetic energy, and an energy input is required to make the reaction happen. The energy input comes from the kinetic energy of the initial colliding particles (a and X).

EXAMPLE 43–2 **A slow-neutron reaction.** The nuclear reaction

$$n + {}^{10}_{5}B \rightarrow {}^{7}_{3}Li + {}^{4}_{2}He$$

is observed to occur even when very slow-moving neutrons $(M_n = 1.0087\,u)$ strike a boron atom at rest. For a particular reaction in which $K_n \approx 0$, the helium $(M_{He} = 4.0026\,u)$ is observed to have a speed of $9.30 \times 10^6\,m/s$. Determine (*a*) the kinetic energy of the lithium $(M_{Li} = 7.0160\,u)$, and (*b*) the Q-value of the reaction.

SOLUTION (*a*) Since the neutron and boron are both essentially at rest, the total momentum before the reaction is zero, and afterward is also zero. Therefore,

$$M_{Li}v_{Li} = M_{He}v_{He}.$$

We solve this for v_{Li} and substitute it into the equation for kinetic energy. We can use classical kinetic energy with little error, rather than relativistic formulas, because $v_{He} = 9.30 \times 10^6\,m/s$ is not close to the speed of light c, and v_{Li} will be even less since $M_{Li} > M_{He}$. Thus we can write:

$$K_{Li} = \frac{1}{2}M_{Li}v_{Li}^2 = \frac{1}{2}M_{Li}\left(\frac{M_{He}v_{He}}{M_{Li}}\right)^2 = \frac{M_{He}^2 v_{He}^2}{2M_{Li}}.$$

We put in numbers, changing the mass in u to kg and recalling that $1.60 \times 10^{-13}\,J = 1\,MeV$:

$$K_{Li} = \frac{(4.0026\,u)^2(1.66 \times 10^{-27}\,kg/u)^2(9.30 \times 10^6\,m/s)^2}{2(7.0160\,u)(1.66 \times 10^{-27}\,kg/u)}$$

$$= 1.64 \times 10^{-13}\,J = 1.02\,MeV.$$

(*b*) We are given the data $K_a = K_X = 0$ in Eq. 43–2, so $Q = K_{Li} + K_{He}$, where

$$K_{He} = \tfrac{1}{2}M_{He}v_{He}^2 = \tfrac{1}{2}(4.0026\,u)(1.66 \times 10^{-27}\,kg/u)(9.30 \times 10^6\,m/s)^2$$

$$= 2.87 \times 10^{-13}\,J = 1.80\,MeV.$$

Hence, $Q = 1.02\,MeV + 1.80\,MeV = 2.82\,MeV.$

EXAMPLE 43–3 **Will the reaction "go"?** Can the reaction ${}^{13}_{6}C\,(p, n)\,{}^{13}_{7}N$ occur when ${}^{13}_{6}C$ is bombarded by 2.0-MeV protons?

SOLUTION We use Eq. 43–1, looking up the masses of the nuclei in Appendix D. The total masses before and after the reaction are:

Before	After
$M({}^{13}_{6}C) = 13.003355$	$M({}^{13}_{7}N) = 13.005739$
$M({}^{1}_{1}H) = 1.007825$	$M(n) = 1.008665$
14.011180	14.014404

(We must use the mass of the ${}^{1}_{1}H$ atom rather than that of the bare proton because the masses of ${}^{13}_{6}C$ and ${}^{13}_{7}N$ include the electrons, and we must include an equal number of electrons on each side of the equation since none are created or destroyed.) The products have an excess mass of $(14.014404 - 14.011180)u = 0.003224\,u \times 931.5\,MeV/u = 3.00\,MeV$. Thus $Q = -3.00\,MeV$, and the reaction is endoergic. This reaction requires energy, and the 2.0 MeV protons do not have enough to make it go.

The proton in this last Example would have to have somewhat more than 3.00 MeV of kinetic energy to make this reaction go; 3.00 MeV would be enough to conserve

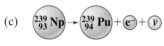

(a) Neutron captured by $^{238}_{92}U$.

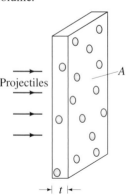

(b) $^{239}_{92}U$ decays by β decay to neptunium-239.

(c) $^{239}_{93}Np$ itself decays by β decay to produce plutonium-239.

FIGURE 43–1 Neptunium and plutonium are produced in this series of reactions, after bombardment of $^{238}_{92}U$ by neutrons.

FIGURE 43–2 Projectile particles fall on a target of area A and thickness t made up of n nuclei per unit volume.

energy, but a proton of this energy would produce the $^{13}_{7}N$ and n with no kinetic energy and hence no momentum. Since an incident 3.0-MeV proton has momentum, conservation of momentum would be violated. A more complicated calculation shows that to conserve both energy and momentum, the minimum proton energy, called the **threshold energy**, is 3.23 MeV in this case (see Problem 14).

The artificial transmutation of elements took a great leap forward in the 1930s when Enrico Fermi realized that neutrons would be the most effective projectiles for causing nuclear reactions and in particular for producing new elements. Because neutrons have no net electric charge, they are not repelled by positively charged nuclei as are protons or alpha particles. Hence the probability of a neutron reaching the nucleus and causing a reaction is much greater than for charged projectiles,[†] particularly at low energies. Between 1934 and 1936, Fermi and his co-workers in Rome produced many previously unknown isotopes by bombarding different elements with neutrons. Fermi realized that if the heaviest known element, uranium, were bombarded with neutrons, it might be possible to produce new elements whose atomic numbers were greater than that of uranium. After several years of hard work, it was suspected that two new elements had been produced, neptunium ($Z = 93$) and plutonium ($Z = 94$). The full confirmation that such "transuranic" elements could be produced came several years later at the University of California, Berkeley. The reactions are shown in Fig. 43–1.

It was soon shown that what Fermi actually had observed when he bombarded uranium was an even stranger process—one that was destined to play an extraordinary role in the world at large. We discuss it in Section 43–3.

43–2 | Cross Section

Some reactions have a higher probability of occurring than others. The reaction probability is specified by a quantity called the *cross section*. Although the size of a nucleus, like that of an atom, is not a clearly defined quantity since the edges are not distinct like those of a tennis ball or baseball, we can nonetheless define a **cross section** for nuclei undergoing collisions by using an analogy. Suppose that projectile particles strike a stationary target of total area A and thickness t, as shown in Fig. 43–2. Suppose the target is made up of identical objects (such as marbles or nuclei), each of which has a cross-sectional area σ. We assume that the objects are fairly far apart, so we don't have to worry about overlapping; this is often a reasonable assumption because nuclei have diameters on the order of 10^{-14} m but are at least 10^{-10} m (atomic size) apart even in solids. If there are n such objects per unit volume, the total area of all these tiny targets is

$$A' = nАt\sigma$$

since $nAt = (n)(\text{volume})$ is the total number of targets and σ is the area of each. If $A' \ll A$, most of the incident projectile particles will pass through the target without colliding. If R_0 is the rate at which the projectile particles strike the target (number/second), the rate at which collisions occur, R, is

$$R = R_0 \frac{A'}{A} = R_0 \frac{nАt\sigma}{A}$$

so

$$R = R_0 nt\sigma.$$

[†]That is, positively charged particles. Electrons rarely cause nuclear reactions because they do not interact via the strong nuclear force.

Thus, by measuring the collision rate, R, we can determine σ:

$$\sigma = \frac{R}{R_0\,nt}.$$ **(43–3)** *Cross section*

If nuclei were simple billiard balls, and R the number of particles that are deflected per second, σ would represent the cross-sectional area of each ball. But nuclei are complicated objects that cannot be considered to have distinct boundaries. Furthermore, collisions can be either elastic or inelastic, and reactions can occur in which the nature of the particles can change. By measuring R for each possible process, we can determine an **effective cross section**, σ, for each process. None of these cross sections is necessarily related to a geometric cross-sectional area. Rather, σ is an "effective" target area. It is a *measure of the probability of a collision or of a particular reaction occurring* per target nucleus, independent of the dimensions of the entire target. The concept of cross section is useful because σ depends only on the properties of the interacting particles, whereas R depends on the thickness and area of the physical (macroscopic) target, on the number of particles in the incident beam, and so on.

We can define the elastic cross section σ_{el}, using Eq. 43–3, where R for a given *Elastic scattering* experimental setup is the rate of elastic collisions (or **elastic scattering**), by which we mean collisions for which the final particles are the same as the initial particles $(a = b, X = Y)$ and $Q = 0$. Similarly, the inelastic cross section, σ_{inel}, is related to the rate of inelastic collisions, or **inelastic scattering**, which involves the same final *Inelastic scattering* and initial particles but $Q \neq 0$, usually because excited states are involved. Then for each possible reaction there is a particular cross section. For protons (p) incident on $^{13}_{6}C$, for example, we could have various reactions, such as $p + {}^{13}_{6}C \rightarrow {}^{13}_{7}N + n$ or $p + {}^{13}_{6}C \rightarrow {}^{10}_{5}B + {}^{4}_{2}He$, and so on. The sum of all the separate reaction cross sections is called the **total reaction cross section**, σ_R. The **total cross section**, σ_T, is

$$\sigma_T = \sigma_{el} + \sigma_{inel} + \sigma_R$$ *Total cross section*

and is a measure of all possible interactions or collisions. Said another way, σ_T is a measure of how many particles interact in some way and hence are eliminated from the incident beam. It is also possible to define "differential cross sections," which represent the probability of the deflected (or emitted) particles leaving at particular angles.

It is said that when one of the first nuclear cross sections was measured, a physicist, surprised that it was as large as it was $(\approx 10^{-28}\,\text{m}^2)$, remarked, "it's as big as a barn." Ever since then nuclear cross sections have been measured in "barns," *The barn* where 1 barn (bn) $= 10^{-28}\,\text{m}^2$.

The value of σ for a given reaction depends on, among other things, the incident kinetic energy. Typical nuclear cross sections are on the order of barns, but they can vary from millibarns to 1000 bn or more. Figure 43–3 shows the cross section for neutron capture in cadmium $(n + {}^{114}Cd \rightarrow {}^{115}Cd + \gamma)$ as a function of neutron kinetic energy. Neutron cross sections for most materials are greater at low energies, as in Fig. 43–3. To produce nuclear reactions at a high rate it is therefore desirable that the bombarding neutrons have low energy. Neutrons that have been slowed down and have reached equilibrium with matter at room temperature $(\frac{3}{2}kT \approx 0.04\,\text{eV}$ at $T = 300\,\text{K})$ are called **thermal neutrons**. *Thermal neutrons*

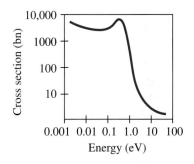

FIGURE 43–3 The neutron cross section for cadmium, which is extraordinarily large for $K \lesssim 1\,\text{eV}$. Note that both scales are logarithmic.

EXAMPLE 43–4 Using cross section. The reaction

$$p + {}^{56}_{26}Fe \rightarrow {}^{56}_{27}Co + n$$

has a cross section of 0.65 bn for a particular incident proton energy. Suppose the iron target has an area of 1.5 cm^2, and is $2.0 \, \mu\text{m}$ thick. The density of iron is $7.8 \times 10^3 \text{ kg/m}^3$. If the protons are incident at a rate of 2.0×10^{13} particles/s, calculate the rate at which neutrons are produced.

SOLUTION We use Eq. 43–3 in the form:

$$R = R_0 n t \sigma.$$

Here $R_0 = 2.0 \times 10^{13}$ particles/s, $t = 2.0 \times 10^{-6}$ m, and $\sigma = 0.65$ bn. Recalling from Chapter 17 that one mole (mass = 56 g for iron) contains 6.02×10^{23} molecules, then the number of iron atoms per unit volume is

$$n = \frac{(7.8 \times 10^3 \text{ kg/m}^3)(6.02 \times 10^{23} \text{ atoms/mole})}{(56 \times 10^{-3} \text{ kg/mole})} = 8.4 \times 10^{28} \text{ atoms/m}^3.$$

Then we find that the rate at which neutrons are produced is

$$R = (2.0 \times 10^{13} \text{ particles/s})(8.4 \times 10^{28} \text{ atoms/m}^3)(2.0 \times 10^{-6} \text{ m})(0.65 \times 10^{-28} \text{ m}^2)$$

$$= 2.2 \times 10^8 \text{ particles/s}.$$

FIGURE 43–4 Fission of a ${}^{235}_{92}U$ nucleus after capture of a neutron, according to the liquid-drop model.

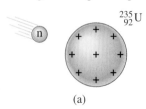

(a) ${}^{235}_{92}U$

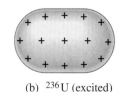

(b) ${}^{236}U$ (excited)

Liquid-drop model

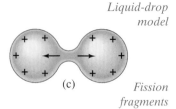

(c)

Fission fragments

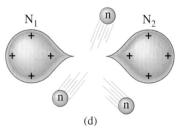

(d)

43–3 Nuclear Fission; Nuclear Reactors

In 1938, the German scientists Otto Hahn and Fritz Strassmann made an amazing discovery. Following up on Fermi's work, they found that uranium bombarded by neutrons sometimes produced smaller nuclei that were roughly half the size of the original uranium nucleus. Lise Meitner and Otto Frisch, two refugees from Nazi Germany working in Scandinavia, quickly realized what had happened: the uranium nucleus, after absorbing a neutron, actually had split into two roughly equal pieces. This was startling, for until then the known nuclear reactions involved knocking out only a tiny fragment (for example, n, p, or α) from a nucleus.

This new phenomenon was named **nuclear fission** because of its resemblance to biological fission (cell division). It occurs much more readily for ${}^{235}_{92}U$ than for the more common ${}^{238}_{92}U$. The process can be visualized by imaging the uranium nucleus to be like a liquid drop. According to this **liquid-drop model**, the neutron absorbed by the ${}^{235}_{92}U$ nucleus gives the nucleus extra internal energy (like heating a drop of water). This intermediate state, or **compound nucleus**, is ${}^{236}_{92}U$ (because of the absorbed neutron). The extra energy of this nucleus—it is in an excited state—appears as increased motion of the individual nucleons, which causes the nucleus to take on abnormal elongated shapes, Figure 43–4b. When the nucleus elongates (in this model) into the shape shown in Fig. 43–4c, the attraction of the two ends via the short-range nuclear force is greatly weakened by the increased separation distance, and the electric repulsive force becomes dominant. So the nucleus splits in two. The two resulting nuclei, N_1 and N_2, are called **fission fragments**, and in the process a number of neutrons (typically two or three) are also given off. The reaction can be written

$$n + {}^{235}_{92}U \rightarrow {}^{236}_{92}U \rightarrow N_1 + N_2 + \text{neutrons.} \quad \textbf{(43–4)}$$

The compound nucleus, ${}^{236}_{92}U$, exists for less than 10^{-12} s, so the process occurs very quickly. The two fission fragments more often split the original uranium mass as about 40 percent–60 percent rather than precisely half and half. A typical fission reaction is

$$n + {}^{235}_{92}U \rightarrow {}^{141}_{56}Ba + {}^{92}_{36}Kr + 3n, \qquad (43\text{--}5)$$

although many others also occur. Figure 43–5 shows the distribution of fission fragments according to mass. Note that only rarely $(\sim 1 \text{ in } 10^4)$ does a fission result in equal mass fragments (small arrow in Fig. 43–5).

A tremendous amount of energy is released in a fission reaction because the mass of ${}^{235}_{92}U$ is considerably greater than the total mass of the fission fragments plus neutrons. This can be seen from the binding-energy-per-nucleon curve of Fig. 42–1; the binding energy per nucleon for uranium is about 7.6 MeV/nucleon, but for fission fragments that have intermediate mass (in the center portion of the graph, $A \approx 100$), the average binding energy per nucleon is about 8.5 MeV/nucleon. Since the fission fragments are more tightly bound, they have less mass. The difference in mass, or energy, between the original uranium nucleus and the fission fragments is about $8.5 - 7.6 = 0.9$ MeV per nucleon. Since there are 236 nucleons involved in each fission, the total energy released per fission is

$$(0.9 \,\text{MeV/nucleon})(236 \,\text{nucleons}) \approx 200 \,\text{MeV}.$$

This is an enormous amount of energy for one single nuclear event. At a practical level, the energy from one fission is, of course, tiny. But if many such fissions could occur in a short time, an enormous amount of energy at the macroscopic level would be available. A number of physicists, including Fermi, recognized that the neutrons released in each fission (Eqs. 43–4 and 5) could be used to create a **chain reaction**. That is, one neutron initially causes one fission of a uranium nucleus; the two or three neutrons released can go on to cause additional fissions, so the process multiplies as shown schematically in Fig. 43–6. If a **self-sustaining chain reaction** was actually possible in practice, the enormous energy available in fission could be released on a larger scale. Fermi and his co-workers (at the University of

Chain reaction

FIGURE 43–5 Mass distribution of fission fragments from ${}^{235}_{92}U + n$. The small arrow indicates equal mass fragments $(\frac{1}{2} \times (236 - 2) = 117)$. Note that the vertical scale is logarithmic.

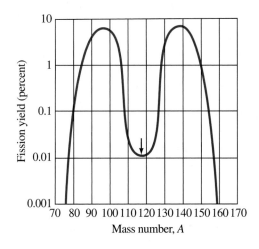

FIGURE 43–6 Chain reaction.

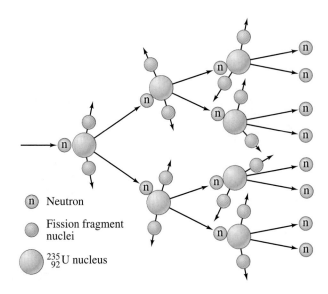

FIGURE 43–7 Color painting of the first nuclear reactor, built by Fermi under the grandstand of Stagg Field at the University of Chicago. (There are no photographs of the original reactor because of military secrecy.) Natural uranium was used with graphite as moderator. On December 2, 1942, Fermi slowly withdrew the cadmium control rods and the reactor went critical. This first self-sustaining chain reaction was announced to Washington, by telephone, by Arthur Compton who witnessed the event and reported: "The Italian navigator has just landed in the new world."

Chicago) showed it was possible by constructing the first **nuclear reactor** in 1942 (Fig. 43–7).

Several problems have to be overcome to make any nuclear reactor function. First, the probability that a $^{235}_{92}$U nucleus will absorb a neutron is large only for slow neutrons, but the neutrons emitted during a fission, and which are needed to sustain a chain reaction, are moving very fast. A substance known as a **moderator** must be used to slow down the neutrons. The most effective moderator will consist of atoms whose mass is as close as possible to that of the neutrons. (To see why this is true, recall from Chapter 9, especially Examples 9–7 and 9–8, that a billiard ball striking an equal mass ball at rest can itself be stopped in one collision; but a billiard ball striking a heavy object bounces off with nearly the same speed it had.) The best moderator would thus contain $^{1}_{1}$H atoms. Unfortunately, $^{1}_{1}$H tends to absorb neutrons. But the isotope of hydrogen called *deuterium*, $^{2}_{1}$H, does not absorb many neutrons and is thus an almost ideal moderator. Either $^{1}_{1}$H or $^{2}_{1}$H can be used in the form of water. In the latter case, it is **heavy water**, in which the hydrogen atoms have been replaced by deuterium. Another common moderator is *graphite*, which consists of $^{12}_{6}$C atoms.

A second problem is that the neutrons produced in one fission may be absorbed and produce other nuclear reactions with other nuclei in the reactor, rather than produce further fissions. In a "light-water" reactor, the $^{1}_{1}$H nuclei absorb neutrons, as does $^{238}_{92}$U to form $^{239}_{92}$U in the reaction n + $^{238}_{92}$U → $^{239}_{92}$U + γ. Naturally occurring uranium[†] contains 99.3 percent $^{238}_{92}$U and only 0.7 percent fissionable $^{235}_{92}$U. To increase the probability of fission of $^{235}_{92}$U nuclei, natural uranium can be **enriched** to increase the percentage of $^{235}_{92}$U using processes such as diffusion or centrifugation. Enrichment is not usually necessary for reactors using heavy water as moderator since heavy water doesn't absorb neutrons.

Moderator

FIGURE 43–8 If the amount of uranium exceeds the critical mass, as in (b), a sustained chain reaction is possible. If the mass is less than critical, as in (a), most neutrons escape before additional fissions occur, and the chain reaction is not sustained.

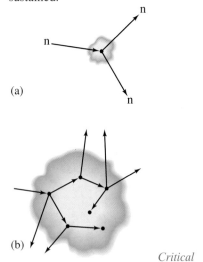

(a)

(b)

Critical mass

The third problem is that some neutrons will escape through the surface of the reactor core before they cause further fissions (Fig. 43–8). Thus the mass of fuel must be sufficiently large for a self-sustaining chain reaction to take place. The minimum mass of uranium needed is called the **critical mass**. The value of the critical mass depends on the moderator, the fuel ($^{239}_{94}$Pu may be used instead of $^{235}_{92}$U), and how much the fuel is enriched, if at all. Typical values are on the order of a few kilograms (that is, not grams nor thousands of kilograms).

To have a self-sustaining chain reaction, it is clear that on the average at least one neutron produced in each fission must go on to produce another fission. The

---

[†] $^{238}_{92}$U will fission, but only with fast neutrons ($^{238}_{92}$U is more stable than $^{235}_{92}$U). The probability of absorbing a fast neutron and producing a fission is too low to produce a self-sustaining chain reaction.

average number of neutrons per each fission that do go on to produce further fissions is called the **multiplication factor**, f. For a self-sustaining chain reaction, we must have $f \geq 1$. If $f < 1$, the reactor is "subcritical." If $f > 1$, it is "supercritical." Reactors are equipped with movable **control rods** (usually of cadmium or boron), whose function is to absorb neutrons and maintain the reactor at just barely "critical," $f = 1$. The release of neutrons and subsequent fissions caused by them occurs so quickly that manipulation of the control rods to maintain $f = 1$ would not be possible if it weren't for the small percentage (~ 1 percent) of so-called **delayed neutrons**. They come from the decay of neutron-rich fission fragments (or their daughters) having lifetimes on the order of seconds—sufficient to allow enough reaction time to operate the control rods and maintain $f = 1$.

Critical reaction

Delayed neutrons

Types of nuclear reactor

Nuclear reactors have been built for use in research and to produce electric power. Fission produces many neutrons and a "research reactor" is basically an intense source of neutrons. These neutrons can be used as projectiles in nuclear reactions to produce nuclides not found in nature, including isotopes used as tracers and for therapy. A "power reactor" is used to produce electric power. The energy released in the fission process appears as heat, which is used to boil water and produce steam to drive a turbine connected to an electric generator (Fig. 43–9). The **core** of a nuclear reactor consists of the fuel and a moderator (water in most U.S. commercial reactors). The fuel is usually uranium enriched so that it contains 2 to 4 percent $^{235}_{92}\text{U}$. Water at high pressure or other liquid (such as liquid sodium) is allowed to flow through the core. The thermal energy it absorbs is used to produce steam in the heat exchanger, so the fissionable fuel acts as the heat input for a heat engine (Chapter 20).

Many problems are associated with nuclear power plants. Besides the usual thermal pollution associated with any heat engine (page 525), there is the problem of disposal of the radioactive fission fragments produced in the reactor, plus radioactive nuclides produced by neutrons interacting with the structural parts of the reactor. Fission fragments, like their uranium or plutonium parents, have about 50 percent more neutrons than protons. Nuclei with atomic number in the typical range for fission fragments ($Z \approx 30$ to 60) are stable only if they have more nearly equal numbers of protons and neutrons (see Fig. 42–2). Hence the highly neutron-rich fission fragments are very unstable and decay radioactively. The accidental release of highly radioactive fission fragments into the atmosphere poses a serious threat to human health (Section 43–5), as does possible leakage of the radioactive

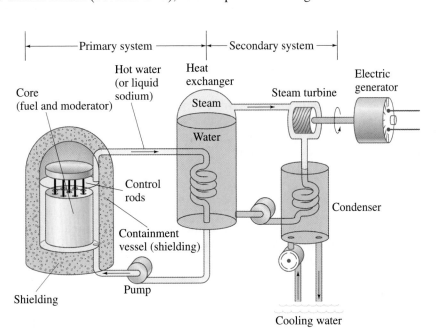

FIGURE 43–9 A nuclear reactor. The heat generated by the fission process in the fuel rods is carried off by hot water or liquid sodium and is used to boil water to steam in the heat exchanger. The steam drives a turbine to generate electricity and is then cooled in the condenser.

FIGURE 43–10 Devastation around Chernobyl in Russia, after the nuclear power plant disaster in 1986.

wastes when they are disposed of. The accidents at Three Mile Island (1979) and at Chernobyl (1986) have illustrated some of the dangers and have shown that nuclear plants must be constructed, maintained, and operated with great care and precision (Fig. 43–10). Finally, the lifetime of nuclear power plants is limited to 30-some years, due to buildup of radioactivity and the fact that the structural materials themselves are weakened by the intense conditions inside. Decommissioning of a power plant could take a number of forms, but the cost of any method of decommissioning a large plant will be very great.

So-called **breeder reactors** were proposed as a solution to the problem of limited supplies of fissionable uranium. A breeder reactor is one in which some of the neutrons produced in the fission of $^{235}_{92}$U are absorbed by $^{238}_{92}$U, and $^{239}_{94}$Pu is produced via the set of reactions shown in Fig. 43–1. $^{239}_{94}$Pu is fissionable with slow neutrons, so after separation it can be used as a fuel in a nuclear reactor. Thus a breeder reactor "breeds" new fuel† $\left(^{239}_{94}\text{Pu}\right)$ from otherwise useless $^{238}_{92}$U. Since natural uranium is 99.3 percent $^{238}_{92}$U, this means that the supply of fissionable fuel could be increased by more than a factor of 100. But breeder reactors not only have the same problems as other reactors, but in addition present other serious problems. Not only is plutonium considered by many to be a serious health hazard in itself (radioactive with a half-life of 24,000 years), but plutonium produced in a reactor can readily be used in a bomb. Thus the use of a breeder reactor, even more than a conventional uranium reactor, presents the danger of nuclear proliferation, and the possibility of theft of fuel by terrorists who could produce a bomb.

It is clear that nuclear power presents many risks. Other large-scale energy-conversion methods, such as conventional coal-burning steam plants, also present health and environmental hazards, some of which were discussed in Chapter 20, including air pollution, oil spills, and the release of CO_2 gas which may be trapping heat as in a greenhouse and raising the Earth's temperature. The solution to the world's needs for energy is not only technological, but economic and political as well. A major factor surely is to "conserve"—to not waste energy and use as little as possible.

EXAMPLE 43–5 **ESTIMATE** **Uranium fuel amount.** Estimate the minimum amount of $^{235}_{92}$U that needs to undergo fission in order to run a 1000-MW power reactor per year of continuous operation. Assume an efficiency (Chapter 20) of about 33 percent.

SOLUTION For 1000 MW output, the total power generation is 3000 MW, of which 2000 MW is dumped as "waste" heat. Thus the total energy release in 1 yr $\left(3 \times 10^7\,\text{s}\right)$ from fission is about

$$\left(3 \times 10^9\,\text{J/s}\right)\left(3 \times 10^7\,\text{s}\right) \approx 10^{17}\,\text{J}.$$

If each fission releases 200 MeV of energy, the number of fissions required is

$$\frac{\left(10^{17}\,\text{J}\right)}{\left(2 \times 10^8\,\text{eV/fission}\right)\left(1.6 \times 10^{-19}\,\text{J/eV}\right)} \approx 3 \times 10^{27}\,\text{fissions}.$$

The mass of a single uranium atom is about $\left(235\,\text{u}\right)\left(1.66 \times 10^{-27}\,\text{kg/u}\right) \approx 4 \times 10^{-25}\,\text{kg}$, so the total mass needed is $\left(4 \times 10^{-25}\,\text{kg}\right)\left(3 \times 10^{27}\,\text{fissions}\right) \approx 1000\,\text{kg}$, or about a ton. Since $^{235}_{92}$U is only a fraction of normal uranium, and even when enriched it is never more than 10 percent of the total, the yearly requirement for uranium is on the order of tens of tons. This is orders of magnitude less than coal, both in mass and volume (see Problem 18 in Chapter 20).

---

†A breeder reactor does *not* produce more fuel than it uses.

The first use of fission, however, was not to produce electric power. Instead, it was first used as a fission bomb (the "atomic bomb"). In early 1940, with Europe already at war, Hitler banned the sale of uranium from the Czech mines he had recently taken over. Research into the fission process suddenly was enshrouded in secrecy. Physicists in the United States were alarmed. A group of them approached Einstein—a man whose name was a household word—to send a letter to President Roosevelt about the possibilities of using nuclear fission for a bomb far more powerful than any previously known, and inform him that Germany might already have begun development of such a bomb. Roosevelt responded by authorizing the program known as the Manhattan Project, to see if a bomb could be built. Work began in earnest after Fermi's demonstration in 1942 that a sustained chain reaction was possible. A new secret laboratory was developed on an isolated mesa in New Mexico known as Los Alamos. Under the direction of J. Robert Oppenheimer (1904–1967; Fig. 43–11), it became the home of famous scientists from all over Europe and the United States.

To build a bomb that was subcritical during transport but that could be made supercritical (to produce a chain reaction) at just the right moment, two pieces of uranium were used, each less than the critical mass but together greater than the critical mass. The two masses would be kept separate until the moment of detonation arrived. Then a kind of gun would force the two pieces together very quickly, a chain reaction of explosive proportions would occur, and a tremendous amount of energy would be released very suddenly. The first fission bomb was tested in the New Mexico desert in July 1945. It was successful. In early August, a fission bomb using uranium was dropped on Hiroshima and a second, using plutonium, was dropped on Nagasaki (Fig. 43–12). World War II ended shortly thereafter.

Besides its great destructive power, a fission bomb produces many highly radioactive fission fragments, as does a nuclear reactor. When a fission bomb explodes, these radioactive isotopes are released into the atmosphere and are known as *radioactive fallout*.

Testing of nuclear bombs in the atmosphere after World War II was a cause of concern, for the movement of air masses spread the fallout all over the globe. Radioactive fallout eventually settles to the Earth, particularly in rainfall, and is absorbed by plants and grasses and enters the food chain. This is a far more serious problem than the same radioactivity on the exterior of our bodies, since α and β particles are largely absorbed by clothing and the outer (dead) layer of skin. But once inside our bodies via food, the isotopes are in direct contact with living cells. One particularly dangerous radioactive isotope is $^{90}_{38}$Sr, which is chemically much like calcium and becomes concentrated in bone, where it causes bone cancer and destruction of bone marrow. The 1963 treaty signed by over 100 nations that bans nuclear weapons testing in the atmosphere was motivated because of the hazards of fallout.

Atom bomb

FIGURE 43–11 J. Robert Oppenheimer, on the left, with General Leslie Groves, who was the administrative head of Los Alamos during the war. The photograph was taken at the Trinity site in the New Mexico desert, where the first atomic bomb was exploded.

FIGURE 43–12 Photo taken a month after the bomb was dropped on Nagasaki. The shacks were constructed afterwards from debris in the ruins.

43–4 Fusion

The mass of every stable nucleus is less than the sum of the masses of its constituent protons and neutrons. For example, the mass of the helium isotope $^4_2$He is less than the mass of two protons plus the mass of two neutrons, as we saw in Example 42–2. Thus, if two protons and two neutrons were to come together to form a helium nucleus there would be a loss of mass. This mass loss is manifested in the release of a large amount of energy.

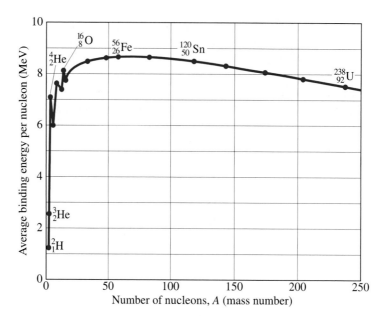

FIGURE 43–13 Average binding energy per nucleon as a function of mass number A for stable nuclei. Same as Fig. 42–1.

The process of building up nuclei by bringing together individual protons and neutrons, or building larger nuclei by combining small nuclei, is called **nuclear fusion**. A glance at Fig. 43–13 (same as Fig. 42–1) shows why small nuclei can combine to form larger ones with the release of energy: it is because the binding energy per nucleon is smaller for light nuclei than it is for those of increasing mass (up to about $A \approx 60$). It is believed that many of the elements in the universe were originally formed through the process of fusion (see Chapter 45), and that today, fusion is continually taking place within the stars, including our Sun, producing the prodigious amounts of radiant energy they emit.

EXAMPLE 43–6 **Fusion energy release.** One of the simplest fusion reactions involves the production of deuterium, ^2_1H, from a neutron and a proton: $^1_1H + n \rightarrow {}^2_1H + \gamma$. How much energy is released in this reaction?

SOLUTION From Appendix D, the initial rest mass is $1.007825\,u + 1.008665\,u = 2.016490\,u$, and after the reaction the mass is that of the ^2_1H, namely $2.014102\,u$. The mass difference is $0.002388\,u$, so the energy released is $(0.002388\,u)(931.5\,MeV/u) = 2.22\,MeV$, and it is carried off by the ^2_1H nucleus and the γ ray.

The energy output of our Sun is believed to be due principally to the following sequence of fusion reactions:

Fusion reactions
$$^1_1H + {}^1_1H \rightarrow {}^2_1H + e^+ + \nu \qquad (0.42\,MeV) \quad \textbf{(43–6a)}$$

in the Sun
$$^1_1H + {}^2_1H \rightarrow {}^3_2He + \gamma \qquad (5.49\,MeV) \quad \textbf{(43–6b)}$$

(proton–proton cycle)
$$^3_2He + {}^3_2He \rightarrow {}^4_2He + {}^1_1H + {}^1_1H \qquad (12.86\,MeV) \quad \textbf{(43–6c)}$$

where the energy released (Q-value) for each reaction is given in parentheses. The net effect of this sequence, which is called the **proton–proton cycle**, is for four protons to combine to form one ^4_2He nucleus plus two positrons, two neutrinos, and two gamma rays:

$$4\,{}^1_1H \rightarrow {}^4_2He + 2e^+ + 2\nu + 2\gamma. \qquad \textbf{(43–7)}$$

Note that it takes two of each of the first two reactions (Eqs. 43–6a and b) to produce the two ^3_2He for the third reaction, so the total energy release for the net reaction, Eq. 43–7, is $(2 \times 0.42\,MeV + 2 \times 5.49\,MeV + 12.86\,MeV) = 24.7\,MeV$.

However, each of the two e^+ (formed in reaction Eq. 43–6a) quickly annihilates with an electron to produce $2m_e c^2 = 1.02$ MeV, so the total energy released is $(24.7 \text{ MeV} + 2 \times 1.02 \text{ MeV}) = 26.7$ MeV. The first reaction, the formation of deuterium from two protons (Eq. 43–6a), has a very low probability, and the infrequency of that reaction serves to limit the rate at which the Sun produces energy.

In stars hotter than the Sun, it is more likely that the energy output comes principally from the **carbon cycle**, which comprises the following sequence of reactions:

$$^{12}_{6}\text{C} + {}^{1}_{1}\text{H} \rightarrow {}^{13}_{7}\text{N} + \gamma$$

$$^{13}_{7}\text{N} \rightarrow {}^{13}_{6}\text{C} + e^+ + \nu$$

$$^{13}_{6}\text{C} + {}^{1}_{1}\text{H} \rightarrow {}^{14}_{7}\text{N} + \gamma$$

$$^{14}_{7}\text{N} + {}^{1}_{1}\text{H} \rightarrow {}^{15}_{8}\text{O} + \gamma$$

$$^{15}_{8}\text{O} \rightarrow {}^{15}_{7}\text{N} + e^+ + \nu$$

$$^{15}_{7}\text{N} + {}^{1}_{1}\text{H} \rightarrow {}^{12}_{6}\text{C} + {}^{4}_{2}\text{He}.$$

Carbon

cycle

(some stars)

(See Problem 38.) The theory of the proton–proton cycle and of the carbon cycle as the source of energy for the Sun and stars was first worked out by Hans Bethe (1906–   ) in 1939.

CONCEPTUAL EXAMPLE 43–7 **Stellar fusion.** What is the heaviest element likely to be produced in fusion processes in stars?

RESPONSE Fusion is possible as long as the final product has more binding energy (less mass) than the reactants, for then there is net release of energy. Since the binding energy curve in Fig. 43–13 (or 42–1) peaks near $A \approx 56$ to 58, which corresponds to iron or nickel, it would not be energetically favorable to produce elements heavier than that. Nevertheless, in the center of massive stars or in supernova explosions, there is enough energy available to drive endothermic reactions that produce heavier elements, as well.

The possibility of utilizing the energy released in fusion to make a power reactor is very attractive. The fusion reactions most likely to succeed in a reactor involve the isotopes of hydrogen, $^{2}_{1}\text{H}$ (deuterium) and $^{3}_{1}\text{H}$ (tritium), and are as follows, with the energy released given in parentheses:

Fusion reactor

$$^{2}_{1}\text{H} + {}^{2}_{1}\text{H} \rightarrow {}^{3}_{1}\text{H} + {}^{1}_{1}\text{H} \qquad\qquad (4.03 \text{ MeV}) \quad \textbf{(43–8a)}$$

$$^{2}_{1}\text{H} + {}^{2}_{1}\text{H} \rightarrow {}^{3}_{2}\text{He} + n \qquad\qquad (3.27 \text{ MeV}) \quad \textbf{(43–8b)}$$

$$^{2}_{1}\text{H} + {}^{3}_{1}\text{H} \rightarrow {}^{4}_{2}\text{He} + n. \qquad\qquad (17.59 \text{ MeV}) \quad \textbf{(43–8c)}$$

Fusion reactions

for possible

reactor

Comparing these energy yields with that for the fission of $^{235}_{92}\text{U}$, we can see that the energy released in fusion reactions can be greater for a given mass of fuel than in fission. Furthermore, as fuel, a fusion reactor could use deuterium, which is very plentiful in the water of the oceans (the natural abundance of $^{2}_{1}\text{H}$ is 0.015 percent, or about 1 g of deuterium per 60 L of water). The simple proton–proton reaction of Eq. 43–6a, which could use a much more plentiful source of fuel, $^{1}_{1}\text{H}$, has such a small probability of occurring that it cannot be considered a possibility on Earth.

Although a successful fusion reactor has not yet been achieved, considerable progress has been made in overcoming the inherent difficulties. The problems are associated with the fact that all nuclei have a positive charge and thus repel each other. However, if they can be brought close enough together so that the short-range attractive nuclear force can come into play, the latter can pull the nuclei together and fusion will occur. Thus, in order for the nuclei to get close enough together, they must have rather large kinetic energy to overcome the electric repulsion. High kinetic energies are easily attainable with particle accelerators such as the cyclotron, but the number of particles involved is too small. To produce realistic amounts of energy, we must deal with matter in bulk, for which high kinetic energy means higher temperatures. Indeed, very high temperatures are required for fusion to occur, and fusion devices are often referred to as **thermonuclear devices**. The Sun and other stars are very hot, many millions of degrees, so the nuclei are moving fast enough for fusion to take place, and the energy released keeps the temperature high so that further fusion reactions can occur. The Sun and the stars represent huge self-sustaining thermonuclear reactors that stay together because of their great gravitational mass; but on Earth, containment of the plasma at the high temperatures and densities required has proven difficult.

It was realized after World War II that the temperature produced within a fission (or "atomic") bomb was close to 10^8 K. This suggested that a fission bomb could be used to ignite a fusion bomb (popularly known as a thermonuclear or hydrogen bomb) to release the vast energy of fusion. The uncontrollable release of fusion energy in an H-bomb was relatively easy to obtain. But to realize usable energy from fusion at a slow and controlled rate turned out to be difficult.

EXAMPLE 43–8 **ESTIMATE** **Temperature needed for d–t fusion.** Estimate the temperature required for deuterium–tritium fusion (d–t) to occur.

SOLUTION We assume the nuclei approach head-on, each with kinetic energy K, and that the nuclear force comes into play when the distance between their centers equals the sum of their radii as given by Eq. 42–1, namely $r_d \approx 1.5$ fm and $r_t \approx 1.7$ fm. The electrostatic potential energy (Chapter 23) of the two particles at this distance must equal the total kinetic energy of the two particles when far apart:

$$2K \approx \frac{1}{4\pi\epsilon_0} \frac{e^2}{(r_d + r_t)}$$

$$\approx \left(9.0 \times 10^9 \, \frac{\text{N} \cdot \text{m}^2}{\text{C}^2} \right) \frac{(1.6 \times 10^{-19} \, \text{C})^2}{(3.2 \times 10^{-15} \, \text{m})(1.6 \times 10^{-19} \, \text{J/eV})}$$

$$\approx 0.45 \, \text{MeV}.$$

Thus, $K \approx 0.22$ MeV, and if we ask that the average kinetic energy be this high, then from Eq. 18–4, $\frac{3}{2}kT = \overline{K}$, we have

$$T = \frac{2K}{3k} = \frac{2(0.22 \, \text{MeV})(1.6 \times 10^{-13} \, \text{J/MeV})}{3(1.38 \times 10^{-23} \, \text{J/K})} \approx 2 \times 10^9 \, \text{K}.$$

More careful calculations show that the temperature required for fusion is actually about an order of magnitude less than this rough estimate, partly because it is not necessary that the *average* kinetic energy be 0.22 MeV—a small percentage with this much energy (particles in the high-energy tail of the Maxwell distribution, Fig. 18–3) would be sufficient—and because of tunneling through the Coulomb barrier. Reasonable estimates for a usable fusion reactor are in the range $T \gtrsim 2$ to 4×10^8 K.

It is not only a high temperature that is required for a fusion reactor. There must also be a high density of nuclei to ensure a sufficiently high collision rate. A real difficulty with controlled fusion is to contain nuclei long enough and at a high enough density for sufficient reactions to occur that a usable amount of energy is obtained. At the temperatures needed for fusion, the atoms are ionized, and the resulting collection of nuclei and electrons is referred to as a **plasma**. Ordinary materials vaporize at a few thousand degrees at best, and hence cannot be used to contain a high-temperature plasma. Two major containment techniques are being investigated at present: *magnetic confinement* and *inertial confinement*.

Plasma

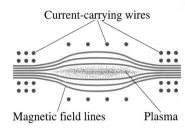

FIGURE 43–14 "Magnetic bottle" used to confine a plasma.

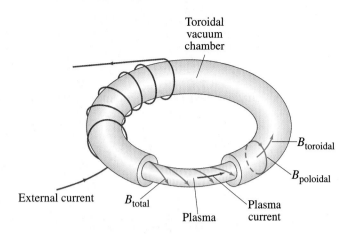

FIGURE 43–15 Tokamak configuration showing total **B** field due to external current plus current in the plasma.

In **magnetic confinement**, magnetic fields are used to try to contain the hot plasma. One possibility is the "magnetic bottle" shown in Fig. 43–14. The paths of the charged particles in the plasma are bent by the magnetic field, and where the lines are close together, the force on the particles is such that they are reflected back toward the center by this "magnetic mirror." Unfortunately, magnetic bottles develop "leaks" and the charged particles slip out before sufficient fusion takes place. A more complex device called the **tokamak**, first developed in the USSR, shows considerable promise. A tokamak (Fig. 43–15) is toroid-shaped and involves complicated magnetic fields: first, current-carrying conductors produce a magnetic field directed along the axis of the toroid ("toroidal" field); an additional field is produced by currents within the plasma itself ("poloidal" field). The combination of these fields produces a helical field as shown in Fig. 43–15, confining the plasma, at least briefly, so it doesn't touch the vacuum chamber's metal walls.

Magnetic confinement

In 1957, J. D. Lawson showed that the product of ion density n and confinement time τ must satisfy, at a minimum, approximately

$$n\tau \gtrsim 3 \times 10^{20} \, \text{s/m}^3.$$

Lawson criterion

This **Lawson criterion** must be reached to produce **ignition**, by which we mean a self-sustaining thermonuclear "burn" that continues after all external heating is turned off. To reach **break-even**, the point at which the energy output due to fusion is equal to the energy input to heat the plasma, requires an $n\tau$ about an order of magnitude less. The break-even point was very closely approached at the Tokamak Fusion Test Reactor (TFTR) at Princeton, and the very high temperature needed for ignition $(4 \times 10^8 \, \text{K})$ was exceeded—although not both of these at the same time. It is believed that a working fusion reactor might be a reality in the first decade of this century.

Ignition

Break-even

The second method for containing the fuel for fusion is **Inertial confinement**: a small pellet of deuterium and tritium is struck simultaneously from several directions by very intense laser beams. The intense influx of energy in such a laser fusion device heats and ionizes the pellet into a plasma. The outer layers evaporate, but the collisions they make with ions in the core of the pellet drive the latter inward. This implosion raises the density to about 10^3 times normal and further heats the core to temperatures at which fusion occurs. The confinement time is very short, on the order of 10^{-11} to 10^{-9} s, during which time the ions do not move appreciably because of their own inertia. Very soon thereafter fusion takes place and the pellet explodes.

Inertial confinement

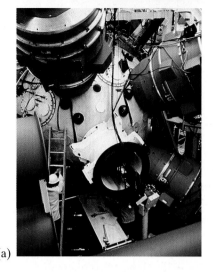

(a)   (b)

FIGURE 43–16 (a) Target chamber (5 m in diameter) of NOVA laser at Lawrence Livermore Laboratory, into which 10 laser beams converge on a target. (b) A 1-mm-diameter DT (deuterium–tritium) target, on its support, at the center of the target chamber.

The NOVA laser used for fusion research at the Lawrence Livermore Laboratory (Fig. 43–16) can deposit 10^5 J into the pellet over a 10^{-9} s pulse. This is a power input of 10^{14} W, more than the total electric-generator capacity of the United States! Of course, it is sustained only over an extremely short interval of time. A future reactor might implode 100 such pellets per second, thus requiring an input of 10^5 J $\times$ $100\,\text{s}^{-1} = 10^7$ W on the average. Inertial confinement depends on high particle density, n, over very short time intervals, τ. Recently, the Lawson criterion has been achieved, but at a temperature less than that needed for ignition.

Besides the problems of confinement, the building of a practical fusion reactor will require the development of materials used in its construction that can withstand high temperatures and high levels of radiation.

43–5 | Passage of Radiation Through Matter; Radiation Damage

When we speak of *radiation*, we include α, β, γ, and X-rays, as well as protons, neutrons, and other particles such as pions (see Chapter 44). Because charged particles can ionize the atoms or molecules of any material they pass through, they are referred to as **ionizing radiation**. And because radiation produces ionization, it can cause considerable damage to materials, particularly to biological tissue.

Charged particles, such as α and β rays and protons, cause ionization because of the electric force. That is, when they pass through a material, they can attract or repel electrons strongly enough to remove them from the atoms of the material. Since the α and β rays emitted by radioactive substances have energies on the order of a MeV (10^4 to 10^7 eV), whereas ionization of atoms and molecules requires on the order of 10 eV, it is clear that a single α or β particle can cause thousands of ionizations.

Neutral particles also give rise to ionization when they pass through materials. For example, X-ray and γ-ray photons can ionize atoms by knocking out electrons by means of the photoelectric and Compton effects (Chapter 38). Furthermore, if a γ ray has sufficient energy (greater than 1.02 MeV), it can undergo pair production: an electron and a positron are produced (Section 38–4). The charged particles produced in all three of these processes can themselves go on to produce further ionization. Neutrons, on the other hand, interact with matter mainly by collisions with nuclei, with which they interact strongly. Often the nucleus is broken apart by such a collision, altering the molecule of which it was a part. And the fragments produced can in turn cause ionization.

Radiation passing through matter can do considerable damage. Metals and other structural materials become brittle and their strength can be weakened if the radiation is very intense, as in nuclear reactor power plants and for space vehicles that must pass through areas of intense cosmic radiation.

The radiation damage produced in biological organisms is due primarily to ionization produced in cells. Several related processes can occur. Ions or radicals are produced that are highly reactive and take part in chemical reactions that interfere with the normal operation of the cell. All forms of radiation can ionize atoms by knocking out electrons. If these are bonding electrons, the molecule may break apart, or its structure may be altered so that it does not perform its normal function or performs a harmful function. If many cells die, the organism may not be able to recover. On the other hand, a cell may survive but be defective. It may go on dividing and produce many more defective cells, to the detriment of the whole organism. Thus radiation can cause cancer—the rapid production of defective cells.

Radiation damage to biological organisms is often separated into categories. *Somatic damage* refers to any part of the body except the reproductive organs, and affects that particular organism, causing cancer and, at high doses, radiation sickness (characterized by nausea, fatigue, loss of body hair, and other symptoms) or even death. *Genetic damage* refers to damage to reproductive cells, causing mutations, the majority of which are harmful, which are transmitted to future generations. The possible damage done by the medical use of X-rays and other radiation must be balanced against the medical benefits and prolongation of life as a result of their use.

Biological damage

43–6 | Measurement of Radiation—Dosimetry

Although the passage of ionizing radiation through the human body can cause considerable damage, radiation can also be used to treat certain diseases, particularly cancer, often by using very narrow beams directed at the cancerous tumor to destroy it (Section 43–7). It is therefore important to be able to quantify the amount, or **dose**, of radiation. This is the subject of **dosimetry**.

The strength of a source can be specified at a given time by stating the **source activity**, or how many disintegrations occur per second. The traditional unit is the **curie** (Ci), defined as

Source activity

$$1 \, \text{Ci} = 3.70 \times 10^{10} \text{ disintegrations per second.}$$

(This figure comes from the original definition as the activity of exactly 1 gram of radium.) Although the curie is still in common use, the proper SI unit for source activity is the **becquerel** (Bq), defined as

$$1 \, \text{Bq} = 1 \text{ disintegration/s.}$$

Commercial suppliers of **radionuclides** (radioactive nuclides) specify the activity at a given time. Since the activity decreases over time, particularly for short-lived isotopes, it is important to take this into account.

The source activity (dN/dt) is related to the half-life, $T_{\frac{1}{2}}$, by (see Section 42–8):

$$\left| \frac{dN}{dt} \right| = \lambda N = \frac{0.693}{T_{\frac{1}{2}}} N.$$

EXAMPLE 43–9 **Radioactivity taken up by cells.** In a certain experiment, $0.016 \, \mu\text{Ci}$ of $^{32}_{15}\text{P}$ is injected into a medium containing a culture of bacteria. After 1 h, the cells are washed and a detector that is 70 percent efficient (counts 70 percent of emitted β particles) records 720 counts per minute from all the cells. What percentage of the original $^{32}_{15}\text{P}$ was taken up by the cells?

SOLUTION The halflife of $^{32}_{15}\text{P}$ being about 14 days, the loss of activity over the 1 hour can be ignored. The total number of disintegrations per second originally was $(0.016 \times 10^{-6})(3.7 \times 10^{10}) = 590$. The counter could be expected to count 70 percent of this, or 410 per second. Since it counted $720/60 = 12$ per second, then $12/410 = 0.029$ or 2.9 percent was incorporated into the cells.

Another type of measurement is the exposure or **absorbed dose**—that is, the effect the radiation has on the absorbing material. The earliest unit of dosage was the **roentgen** (R), which was defined in terms of the amount of ionization produced by the radiation (1.6×10^{12} ion pairs per gram of dry air at standard conditions). Today, 1 R is defined as the amount of X- or γ radiation that deposits 0.878×10^{-2} J of energy per kilogram of air. The roentgen was largely superseded

by another unit of absorbed dose applicable to any type of radiation, the **rad**: *1 rad is that amount of radiation which deposits energy at a rate of 1.00×10^{-2} J/kg in any absorbing material.* (This is quite close to the roentgen for X- and γ rays.) The proper SI unit for absorbed dose is the **gray** (Gy):

$$1\,\text{Gy} = 1\,\text{J/kg} = 100\,\text{rad}, \tag{43–9}$$

and is now coming into use. The absorbed dose depends not only on the strength of a given radiation beam (number of particles per second) and the energy per particle, but also on the type of material that is absorbing the radiation. Since bone, for example, is denser than flesh and absorbs more of the radiation normally used, the same beam passing through a human body deposits a greater dose (in rads or grays) in bone than in flesh.

TABLE 43–1
Quality Factor (QF) of Different Kinds of Radiation

Type	QF
X and γ rays	≈ 1
β (electrons)	≈ 1
Fast protons	1
Slow neutrons	≈ 3
Fast neutrons	Up to 10
α particles and heavy ions	Up to 20

The gray and the rad are physical units of dose—the energy deposited per unit mass of material. They are, however, not the most meaningful units for measuring the biological damage produced by radiation. This is because equal doses of different types of radiation cause differing amounts of damage. For example, 1 rad of α radiation does 10 to 20 times the amount of damage as 1 rad of β or γ rays. This difference arises largely because α rays (and other heavy particles such as protons and neutrons) move much more slowly than β and γ rays of equal energy due to their greater mass. Hence, ionizing collisions occur closer together, so more

irreparable damage can be done. The **relative biological effectiveness** (RBE) or **quality factor** (QF) of a given type of radiation is defined as the number of rads of X or γ radiation that produces the same biological damage as 1 rad of the given radiation. Table 43–1 gives the QF for several types of radiation. The numbers are approximate since they depend somewhat on the energy of the particles and on the type of damage that is used as the criterion.

The **effective dose** can be given as the product of the dose in rads and the QF, and this unit is known as the **rem** (which stands for *rad equivalent man*):

$$\text{effective dose (in rem)} = \text{dose (in rad)} \times \text{QF.} \tag{43–10a}$$

This unit is being replaced by the SI unit for "effective dose," the **sievert** (Sv):

$$\text{effective dose (Sv)} = \text{dose (Gy)} \times \text{QF.} \tag{43–10b}$$

By this definition, 1 rem of any type of radiation does approximately the same amount of biological damage. For example, 50 rem of fast neutrons does the same damage as 50 rem of γ rays. But note that 50 rem of fast neutrons is only 5 rads, whereas 50 rem of γ rays is 50 rads.

We are constantly exposed to low-level radiation from natural sources: cosmic rays, natural radioactivity in rocks and soil, and naturally occurring radioactive isotopes in our food, such as $^{40}_{19}\text{K}$. Radon, $^{222}_{86}\text{Rn}$, is of considerable concern today. It is the product of radium decay and is an intermediate in the decay series from uranium (see Fig. 42–11). Most intermediates remain in the rocks where formed, but radon is a gas that can escape from rock (and from building material like concrete) to enter the atmosphere we breathe. Although radon is inert chemically (it is a noble gas), it is not inert physically—it decays by alpha emission, and its products, also radioactive, are *not* chemically inert and can attach to the interior of the lung.

The natural radioactive background averages about 0.36 rem (3.6 mSv) per year per person, although there are large variations. From medical X-rays, the average person receives about 40 mrem (0.4 mSv) per year. The U.S. government specifies the recommended upper limit of allowed radiation for an individual in the general populace at about 0.5 rem (5 mSv) per year, exclusive of natural sources. It is not known if low doses of radiation increase the chances of cancer or genetic defects, so the attitude today is to play safe and keep the radiation dose low.

The upper limit for people who work around radiation—in hospitals, power plants, research—has been set somewhat higher, on the order of 5 rem/yr whole-body dose (presumably because such people know what they are getting into). People who work around radiation generally carry some type of dosimeter, typically a **radiation film badge**, which is a piece of film wrapped in light-tight material. The passage of ionizing radiation through the film changes it so that the film is darkened upon development, and so indicates the received dose.

Large doses of radiation can cause a large number of unpleasant symptoms such as nausea, fatigue, and loss of body hair. Such effects are sometimes referred to as **radiation sickness**. Large doses can also be fatal, although the time span of the dose is important. A short dose of 1000 rem (10 Sv) is nearly always fatal. A 400-rem (4-Sv) dose in a short period of time is fatal in 50 percent of the cases. However, the body possesses remarkable repair processes, so that a 400-rem dose spread over several weeks is not usually fatal. It will, nonetheless, cause considerable damage to the body.

Radiation sickness

EXAMPLE 43–10 **Whole-body dose.** What whole-body dose is received by a 70-kg laboratory worker exposed to a 40 mCi $^{60}_{27}$Co source, assuming the person's body has cross-sectional area 1.5 m^2 and is, on average, 4.0 m from the source for 4.0 h per day? $^{60}_{27}$Co emits γ rays of energy 1.33 MeV and 1.17 MeV in quick succession. Approximately 50 percent of the γ rays interact in the body and deposit all their energy. (The rest pass through.)

SOLUTION The total γ-ray energy per decay is $(1.33 + 1.17)$MeV $= 2.50$ MeV, so the total energy emitted by the source is

$$(0.040 \, \text{Ci})(3.7 \times 10^{10} \, \text{decays/Ci·s})(2.50 \, \text{MeV}) = 3.7 \times 10^9 \, \text{MeV/s}.$$

The proportion of this intercepted by the body is its 1.5 m^2 area divided by the area of a sphere of radius 4.0 m (Fig. 43–17)

$$\frac{1.5 \, \text{m}^2}{4\pi r^2} = \frac{1.5 \, \text{m}^2}{4\pi (4.0 \, \text{m})^2} = 7.5 \times 10^{-3}.$$

So the rate energy is deposited in the body (remembering that only $\frac{1}{2}$ of the γ rays interact in the body) is

$$E = \left(\tfrac{1}{2}\right)(7.5 \times 10^{-3})(3.7 \times 10^9 \, \text{MeV/s})(1.6 \times 10^{-13} \, \text{J/MeV})$$
$$= 2.2 \times 10^{-6} \, \text{J/s}.$$

Since 1 Gy = 1 J/kg, the whole-body dose rate for this 70-kg person is $(2.2 \times 10^{-6} \, \text{J/s})/(70 \, \text{kg}) = 3.1 \times 10^{-8} \, \text{Gy/s}$. In the space of 4.0 h, this amounts to a dose of $(4.0 \, \text{h})(3600 \, \text{s/h})(3.1 \times 10^{-8} \, \text{Gy/s}) = 4.5 \times 10^{-4} \, \text{Gy}$. Since QF ≈ 1 for gammas, the effective dose (Eq. 43–10) is 450 μSv or (see Eq. 43–9):

$$(100 \, \text{rad/Gy})(4.5 \times 10^{-4} \, \text{Gy})(1) = 45 \, \text{mrem}.$$

This 45-mrem effective dose is nearly 10 percent of the normal allowed dose for a whole year (500 mrem/yr), or 1 percent of the yearly allowance for radiation workers. This worker should not receive such a dose every day and should seek ways to reduce it (shield the source, vary the work, work farther away, etc.).

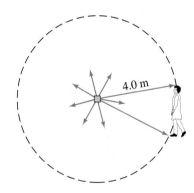

FIGURE 43–17 Radiation spreads out in all directions. A person 4.0 m away intercepts only a fraction: her cross-sectional area divided by the area of a sphere of radius 4.0 m.

The applications of radioactivity and radiation to human beings and other organisms is a vast field that has filled many books. In the medical field there are two basic aspects: (1) *radiation therapy*—the treatment of disease (mainly cancer)—which we discuss in this Section; and (2) the *diagnosis* of disease, which we discuss in the following Sections of this chapter.

Radiation can cause cancer. It can also be used to treat it. Rapidly growing cancer cells are especially susceptible to destruction by radiation. Nonetheless, large doses are needed to kill the cancer cells, and some of the surrounding normal cells are inevitably killed as well. It is for this reason that cancer patients receiving radiation therapy often suffer side effects characteristic of radiation sickness. To minimize the destruction of normal cells, a narrow beam of γ or X-rays is often used when the cancerous tumor is well localized. The beam is directed at the tumor, and the source (or body) is rotated so that the beam passes through various parts of the body to keep the dose at any one place as low as possible—except at the tumor and its immediate surroundings, where the beam passes at all times (Fig. 43–18). The radiation may be from a radioactive source such as $^{60}_{27}$Co, or it may be from an X-ray machine that produces photons in the range 200 keV to 5 MeV. Protons, neutrons, electrons, and pions, which are produced in particle accelerators (Section 44–2), are also being used in cancer therapy.

In some cases, a tiny radioactive source may be inserted directly inside a tumor, which will eventually kill the majority of the cells. A similar technique is used to treat cancer of the thyroid with the radioactive isotope $^{131}_{53}$I. The thyroid gland tends to concentrate any iodine present in the bloodstream; so when $^{131}_{53}$I is injected into the blood, it becomes concentrated in the thyroid, particularly in any area where abnormal growth is taking place. The intense radioactivity emitted can then destroy the defective cells.

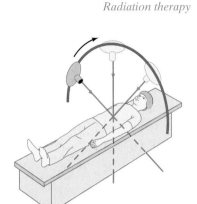

FIGURE 43–18 Radiation source rotates so that the beam always passes through the diseased tissue, but minimizing the dose in the rest of the body.

* | **43–8** | **Tracers**

Radioactive isotopes are commonly used in biological, chemical, and medical research as **tracers**. A given compound is artificially synthesized using a radioactive isotope such as $^{14}_{6}$C or $^{3}_{1}$H. Such "tagged" molecules can then be traced as they move through an organism or as they undergo chemical reaction. The presence of these tagged molecules (or parts of them, if they undergo chemical change) can be detected by a Geiger or scintillation counter (see Section 42–11).

For medical diagnosis, the radionuclide commonly used today is $^{99m}_{43}$Tc, a long-lived excited state of technetium-99 (the "m" in the symbol stands for "metastable" state). It is formed when $^{99}_{42}$Mo decays. The great usefulness of $^{99m}_{43}$Tc derives from its convenient half-life of 6 h (short, but not too short) and the fact that it can combine with a large variety of compounds. The compound to be labeled with the radionuclide is so chosen because it concentrates in the organ or region of the anatomy to be studied. Detectors outside the body then record, or image, the distribution of the radioactively labeled compound. The detection can be done by a single detector (Fig. 43–19) which is moved across the body, measuring the intensity of radioactivity at a large number of points. The image represents the relative intensity of radioactivity at each point. The relative radioactivity is a diagnostic tool. For example, high or low radioactivity may represent overactivity or underactivity of an organ or part of an organ, or in another case may represent a lesion or tumor. More complex *gamma cameras* make use of many detectors which simultaneously record the radioactivity at many points. The measured intensities can be displayed on a CRT (TV monitor or oscilloscope screen), and allow "dynamic" studies (that is, images that change in time) to be performed.

FIGURE 43–19 Collimated gamma-ray detector for scanning (moving) over patient. The collimator (to "collimate" means to "make straight") is necessary to select γ rays that come in a straight line from the patient. Without the collimator, γ rays from all parts of the body could strike the scintillator, producing a very poor image.

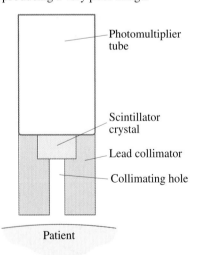

Photomultiplier tube

Scintillator crystal

Lead collimator

Collimating hole

Patient

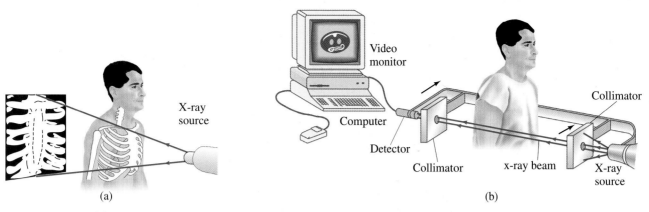

FIGURE 43–20 (a) Conventional X-ray imaging, which is essentially shadowing. (b) Tomographic imaging: the X-ray source and detector move together across the body, the transmitted intensity being measured at a large number of points. Then the source–detector assembly is rotated slightly (say, 1°) and another scan is made. This process is repeated for perhaps 180°. The computer reconstructs the image of the slice and it is presented on a TV monitor (cathode-ray tube).

*43–9 Imaging by Tomography: CAT Scans and Emission Tomography

Normal X-Ray Image

For a conventional medical (or dental) X-ray photograph, the X-rays emerging from the tube (Fig. 36–25, Section 36–10) pass through the body and are detected on photographic film or a fluorescent screen, Fig. 43–20a. The rays travel in very nearly straight lines through the body with minimal deviation since at X-ray wavelengths there is little diffraction or refraction. There is absorption (and scattering), however; and the difference in absorption by different structures in the body is what gives rise to the image produced by the transmitted rays. The less the absorption, the greater the transmission and the darker the film. The image is, in a sense, a "shadow" of what the rays have passed through.

Computerized Axial Tomography (CAT Scan)

In conventional X-ray images, the entire thickness of the body is projected onto the film; structures overlap and in many cases are difficult to distinguish. In the 1970s, a revolutionary new technique called **computerized tomography** (CT) using X-rays was developed, which produces an image of a *slice* through the body. (The word **tomography** comes from the Greek: *tomos* = slice, *graph* = picture.) Structures and lesions previously impossible to visualize can now be seen with remarkable clarity. The principle behind CT is shown in Fig. 43–20b: a thin collimated beam of X-rays passes through the body to a detector that measures the transmitted intensity. Measurements are made at a large number of points as the source and detector are moved past the body together. The apparatus is then rotated slightly about the body axis and again scanned; this is repeated at (perhaps) 1° intervals for 180°. The intensity of the transmitted beam for the many points of each scan, and for each angle, are sent to a computer that reconstructs the image of the slice. Note that the imaged slice is perpendicular to the long axis of the body. For this reason, CT is sometimes called **computerized axial tomography** (CAT), although the abbreviation CAT, as in CAT scan, can also be read as **computer-assisted tomography**.

    The use of a single detector as in Fig. 43–20b would require some time to make the many scans needed to form a complete image. Much faster scanners use

➡ **PHYSICS APPLIED**

CT images

CAT scans

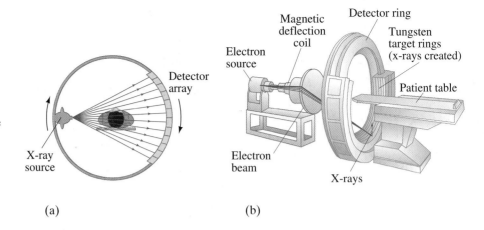

FIGURE 43–21 (a) Fan-beam scanner. Rays transmitted through the entire body are measured simultaneously at each angle. The source and detector rotate to take measurements at different angles. In another type of fan-beam scanner, there are detectors around the entire 360° of the circle which remain fixed as the source moves. (b) In another type, a beam of electrons from the source is directed by magnetic fields at tungsten targets surrounding the patient.

(a)   (b)

FIGURE 43–22 Two CT images, with different resolutions, each showing a cross section of a brain. Photo (a) is of low resolution; photo (b), of higher resolution, shows a brain tumor (dark area on the right).

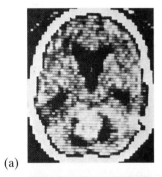

(a)

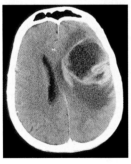

(b)

➡ **PHYSICS APPLIED**

Medical imaging

SPET
PET

a fan beam, Fig. 43–21a, in which beams passing through the entire cross section of the body are detected simultaneously by many detectors. The source and detectors are then rotated about the patient, and an image requires only a few seconds. Even faster, and therefore useful for heart scans, are fixed source machines wherein an electron beam is directed (by magnetic fields) to tungsten targets surrounding the patient, creating the X-rays. See Fig. 43–21b.

But how is the image formed? We can think of the slice to be imaged as being divided into many tiny picture elements (or **pixels**), which could be squares. For CT, the width of each pixel is chosen according to the width of the detectors and/or the width of the X-ray beams, and this determines the resolution of the image, which might be 1 mm. An X-ray detector measures the intensity of the transmitted beam. Subtracting this value from the intensity of the beam at the source, yields the total absorption (called a "projection") along that beam line. Complicated mathematical techniques, such as Fourier transforms, are used to analyze all the absorption projections for the huge number of beam scans measured, obtaining the absorption at each pixel and assigning each a "grayness value" according to how much radiation was absorbed. The image is made up of tiny spots (pixels) of varying shades of gray. Often the amount of absorption is color-coded. The colors in the resulting ("false-color") image have nothing to do, however, with the actual color of the object.

Figure 43–22 illustrates what actual CT images look like. It is generally agreed that CT scanning has revolutionized some areas of medicine by providing much less invasive, and/or more accurate, diagnosis.

Computerized tomography can also be applied to ultrasound imaging (Section 16–9) and to emissions from radioisotopes and nuclear magnetic resonance, which we discuss next.

Emission Tomography

It is possible to image the emissions of a radioactive tracer (see Section 43–8) in a single plane or slice through a body using computed tomography techniques. A basic gamma detector (Fig. 43–19) can be moved around the patient to measure the radioactive intensity from the tracer at many points and angles; the data are processed in much the same way as for X-ray CT scans. This technique is referred to as **single photon emission tomography** (SPET).[†]

Another important technique is **positron emission tomography** (PET), which makes use of positron emitters such as $^{11}_6C$, $^{13}_7N$, $^{15}_8O$, and $^{18}_9F$. These isotopes are incorporated into molecules that, when inhaled or injected, accumulate in the organ or region of the body to be studied. When such a nuclide undergoes β^+ decay, the emitted positron travels at most a few millimeters before it collides with a normal electron. In this collision, the positron and electron are annihilated, producing two γ rays $(e^+ + e^- \rightarrow 2\gamma)$, each having energy of 511 keV. The two γ rays fly off in oppo-

[†] Also known as SPECT, "single photon emission computed tomography."

site directions (180° ± 0.25°) since they must have almost exactly equal and opposite momenta to conserve momentum (the momenta of the e^+ and e^- are essentially zero compared to the momenta of the γ rays). Because the photons travel along the same line in opposite directions, their detection in coincidence by rings of detectors surrounding the patient (Fig. 43–23) readily establishes the line along which the emission took place. If the difference in time of arrival of the two photons could be determined accurately, the actual position of the emitting nuclide along that line could be calculated. Present-day electronics can measure times to at best ± 300 ps, so at the γ ray's speed $(c = 3 \times 10^8 \, \text{m/s})$, the actual position could be determined to an accuracy on the order of about $vt \approx (3 \times 10^8 \, \text{m/s})(300 \times 10^{-12} \, \text{s}) \approx 10 \, \text{cm}$, which is not very useful. Although there may be future potential for time-of-flight measurements to determine position, today computed tomography techniques are used instead, similar to those for X-ray CT, which can reconstruct PET images with a resolution on the order of 3–5 mm. One big advantage of PET is that no collimators are needed (as for detection of a single photon—see Fig. 43–19). Thus, fewer photons are "wasted" and lower doses can be administered to the patient with PET.

Both PET and SPET systems can give images related to biochemistry, metabolism, and function. This is to be compared to X-ray CT scans, whose images reflect shape and structure—that is, the anatomy of the imaged region.

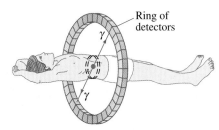

FIGURE 43–23 Positron emission tomography (PET) system showing a ring of detectors to detect the two annihilation γ rays $(e^+ + e^- \rightarrow 2\gamma)$ emitted at 180° to each other.

* 43–10 Nuclear Magnetic Resonance (NMR) and Magnetic Resonance Imaging (MRI)

Nuclear magnetic resonance (NMR) is a phenomenon which soon after its discovery in 1946 became a powerful research tool in a variety of fields from physics to chemistry and biochemistry. It is also an important medical imaging technique. We first briefly discuss the phenomenon, and then look at its applications.

Nuclear Magnetic Resonance (NMR)

We saw in Chapter 40 that the energy levels in atoms are split when they are placed in a magnetic field B (the Zeeman effect) according to the angular momentum or spin of the state. The splitting is proportional to B and to the magnetic moment, μ. Nuclei too have magnetic moments (Section 42–1), and we examine only the simplest, the hydrogen $({}_1^1 \text{H})$ nucleus which consists of a single proton. Its spin angular momentum (and its magnetic moment), like that of the electron, can take on only two values when placed in a magnetic field: spin up (parallel to the field) and spin down (antiparallel to the field) as suggested in Fig. 43–24. When a magnetic field is present, an energy state splits into two levels as shown in Fig. 43–25 with the spin-up state (parallel to field) having the lower energy. The spin-down state acquires an additional energy μB_T and the spin-up state $-\mu B_T$ (Eq. 27–12 and Section 40–7), where B_T is the total magnetic field at the nucleus. The difference in energy between the two states (Fig. 43–25) is thus

$$\Delta E = 2\mu_p B_T,$$

where μ_p is the magnetic moment of the proton.

In a standard nuclear magnetic resonance (NMR) setup, the sample to be examined is placed in a static magnetic field. A radiofrequency (RF) pulse of electromagnetic radiation (that is, photons) is applied to the sample. If the frequency, f, of this pulse corresponds precisely to the energy difference between the two energy levels (Fig. 43–25), so that

$$hf = \Delta E = 2\mu_p B_T \qquad \textbf{(43–11)}$$

then the photons of the RF beam will be absorbed, exciting many of the nuclei from the lower state to the upper state. This is a resonance phenomenon since there is significant absorption only if f is very near $f = 2\mu_p B_T/h$. Hence the name "nuclear magnetic resonance." For free ${}_1^1 \text{H}$ nuclei, the frequency is 42.58 MHz for a field $B_T = 1.0 \, \text{T}$ (Example 43–11). If the H atoms are bound in a

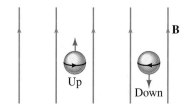

FIGURE 43–24 Schematic picture of a proton represented in a magnetic field **B** (pointing upward) with its two possible states of spin, up and down.

FIGURE 43–25 Energy E_0 in the absence of a magnetic field splits into two levels in the presence of a magnetic field.

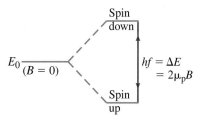

molecule, the total magnetic field B_T at the H nuclei will be the sum of the external applied field (B_{ext}) plus the local magnetic field (B_{loc}) due to electrons and nuclei of neighboring atoms. Since f is proportional to B_T, the value of f for a given external field will be slightly different for bound H atoms than for free atoms:

$$hf = 2\mu_p(B_{ext} + B_{loc}).$$

This change in frequency, which can be measured, is called the "chemical shift." A great deal has been learned about the structure of molecules and bonds using this NMR technique.

EXAMPLE 43–11 **NMR for free protons.** Calculate the resonant frequency for free protons in a 1.000-T magnetic field.

SOLUTION We saw in Section 42–1 that the magnetic moment of the proton is

$$\mu_p = 2.7928\mu_N = 2.7928\left(\frac{e\hbar}{2m_p}\right).$$

Then

$$f = \frac{\Delta E}{h} = \frac{2\mu_p B}{h} = (2.7928)\left(\frac{eB}{2\pi m_p}\right) = 2.7928\left[\frac{(1.6022 \times 10^{-19}\,\text{C})(1.000\,\text{T})}{2\pi(1.6726 \times 10^{-27}\,\text{kg})}\right]$$

$$= 42.58\,\text{MHz}.$$

Magnetic Resonance Imaging (MRI)

For producing medically useful NMR images—now commonly called MRI, or **magnetic resonance imaging**—the element most used is hydrogen since it is the commonest element in the human body and gives the strongest NMR signals. The experimental apparatus is shown in Fig. 43–26. The large coils set up the static magnetic field, and the RF coils produce the RF pulse of electromagnetic waves (photons) that cause the nuclei to jump from the lower state to the upper one (Fig. 43–25). These same coils (or another coil) can detect the absorption of energy or the emitted radiation (also of frequency $f = \Delta E/h$) when the nuclei jump back down to the lower state.

The formation of a two-dimensional or three-dimensional image can be done using techniques similar to those for computed tomography (Section 43–9). The simplest thing to measure for creating an image is the intensity of absorbed and/or reemitted radiation from many different points of the body, and this would be a measure of the density of H atoms at each point. But how do we determine from what part of the body a given photon comes? One technique is to give the static

FIGURE 43–26 Typical MRI imaging setup: (a) diagram; (b) photograph.

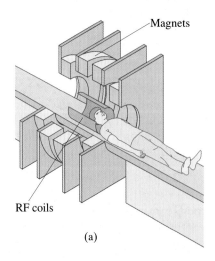

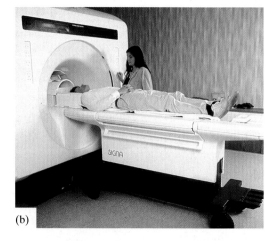

(a)   (b)

magnetic field a gradient; that is, instead of applying a uniform magnetic field, B_T, the field is made to vary with position across the width of the sample (or patient). Since the frequency absorbed by the H nuclei is proportional to B_T (Eq. 43–11), only one plane within the body will have the proper value of B_T to absorb photons of a particular frequency f. By varying f, absorption by different planes can be measured. Alternately, if the field gradient is applied *after* the RF pulse, the frequency of the emitted photons will be a measure of where they were emitted. See Fig. 43–27. If a magnetic field gradient in one direction is applied during excitation (absorption of photons) and photons of a single frequency are transmitted, only H nuclei in one thin slice will be excited. By applying a gradient in a different direction, perpendicular to the first, during reemission, the frequency f of the reemitted radiation will represent depth in that slice. Other ways of varying the magnetic field throughout the volume of the body can be used in order to correlate NMR frequency with position.

A reconstructed image based on the density of H atoms (that is, the intensity of absorbed or emitted radiation) is not very interesting. More useful are images based on the rate at which the nuclei decay back to the ground state, and such images can produce resolution of 1 mm or better. This NMR technique (sometimes called **spin-echo**) is producing images of great diagnostic value, both in the delineation of structure (anatomy) and in the study of metabolic processes. An NMR image is shown in Fig. 43–28.

NMR imaging is considered to be noninvasive. We can calculate the energy of the photons involved: as determined in Example 43–11, in a 1.0 T magnetic field, $f = 42.58$ MHz for $^1_1$H. This corresponds to an energy of $hf = (6.6 \times 10^{-34}\,\text{J}\cdot\text{s})(43 \times 10^6\,\text{Hz}) \approx 3 \times 10^{-26}$ J or about 10^{-7} eV. Since molecular bonds are on the order of 1 eV, it is clear that the RF photons can cause little cellular disruption. This should be compared to X- or γ rays whose energies are 10^4 to 10^6 eV and thus can cause significant damage. The static magnetic fields, though often large (~ 0.1 to 1 T), are believed to be harmless (except for people wearing heart pacemakers), although not a great deal of research has been done in this area.

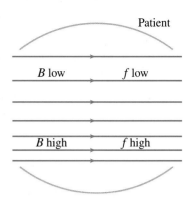

FIGURE 43–27 A static field that is stronger at the bottom than at the top. The frequency of absorbed or emitted radiation is proportional to B in NMR.

FIGURE 43–28 False-color NMR image (MRI) of a vertical section through the head showing structures in the normal brain.

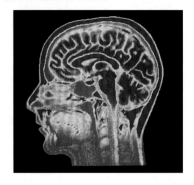

Summary

A **nuclear reaction** occurs when two nuclei collide and two or more other nuclei (or particles) are produced. In this process, as in radioactivity, **transmutation** (change) of elements occurs.

The **reaction energy** or **Q-value** of a reaction a + X → Y + b is

$$Q = (M_a + M_X - M_Y - M_b)c^2$$
$$= K_b + K_Y - K_a - K_X.$$

The effective **cross section** for a reaction is a measure of the reaction probability per target nucleus.

In **fission** a heavy nucleus such as uranium splits into two intermediate-sized nuclei after being struck by a neutron. $^{235}_{92}$U is fissionable by slow neutrons, whereas some fissionable nuclei require fast neutrons. Much energy is released in fission because the binding energy per nucleon is lower for heavy nuclei than it is for intermediate-sized nuclei, so the mass of a heavy nucleus is greater than the total mass of its fission products. The fission process releases neutrons, so that a **chain reaction** is possible. The **critical mass** is the minimum mass of fuel needed to sustain a chain reaction. In a **nuclear reactor** or nuclear bomb, a moderator is needed to slow down the released neutrons.

The **fusion** process, in which small nuclei combine to form larger ones, also releases energy. The energy from our Sun is believed to originate in the fusion reactions known as the **proton–proton cycle** in which four protons fuse to form a $^4_2$He nucleus producing over 25 MeV of energy. A working fusion reactor for power generation has not yet proved possible because of the difficulty in containing the fuel (e.g., deuterium) long enough at the high temperature required. Nonetheless, great progress has been made in confining the collection of charged ions known as a **plasma**. The two main methods are **magnetic confinement**, using a magnetic field in a device such as the toroidal-shaped **tokamak**, and **inertial confinement**, in which intense laser beams compress a fuel pellet of deuterium and tritium.

Radiation can cause damage to materials, including biological tissue. Quantifying amounts of radiation is the subject of **dosimetry**. The **curie** (Ci) and the **becquerel** (Bq) are units that measure the **source activity** or rate of decay of a sample: $1\,Ci = 3.70 \times 10^{10}$ disintegrations per second, whereas $1\,Bq = 1$ disintegration/s. The **absorbed dose**, often specified in **rads**, measures the amount of energy deposited per unit mass of absorbing material: 1 rad is the amount of radiation that deposits energy at the rate of $10^{-2}\,J/kg$ of material. The SI unit of absorbed dose is the **gray**: $1\,Gy = 1\,J/kg = 100\,rad$. The **effective dose** is often specified by the **rem** = rad × QF, where QF is the "quality factor" of a given type of radiation; 1 rem of any type of radiation does approximately the same amount of biological damage. The average dose received per person per year in the United States is about 0.36 rem. The SI unit for effective dose is the **sievert**: $1\,Sv = 10^2\,rem$.

Questions

1. Fill in the missing particles or nuclei: (a) $^{137}_{56}Ba(n, \gamma)$?; (b) $^{137}_{56}Ba(n, ?)^{137}_{55}Cs$; (c) $^{2}_{1}H(d, ?)^{4}_{2}He$; (d) $^{197}_{79}Au(\alpha, d)$? where d stands for deuterium.

2. The isotope $^{32}_{15}P$ is produced by an (n, p) reaction. What must be the target nucleus?

3. When $^{22}_{11}Na$ is bombarded by deuterons $(^{2}_{1}H)$, an α particle is emitted. What is the resulting nuclide?

4. Why are neutrons such good projectiles for producing nuclear reactions?

5. A proton strikes a $^{20}_{10}Ne$ nucleus, and an α particle is observed to emerge. What is the residual nucleus? Write down the reaction equation.

6. Are fission fragments β^+ or β^- emitters?

7. If $^{235}_{92}U$ released only 1.5 neutrons per fission on the average, would a chain reaction be possible? If so, what would be different?

8. $^{235}_{92}U$ releases an average of 2.5 neutrons per fission compared to 2.9 for $^{239}_{94}Pu$. Pure samples of which of these two nuclei do you think would have the smaller critical mass?

9. The energy from nuclear fission appears in the form of thermal energy—but the thermal energy of what?

10. Why can't uranium be enriched by chemical means?

11. How can a neutron, with practically no kinetic energy, excite a nucleus to the extent shown in Fig. 43–4?

12. Why would a porous block of uranium be more likely to explode if kept under water rather than in air?

13. A reactor that uses highly enriched uranium can use ordinary water (instead of heavy water) as a moderator and still have a self-sustaining chain reaction. Explain.

14. Why must the fission process release neutrons if it is to be useful?

15. Discuss the relative merits and disadvantages, including pollution and safety, of power generation by fossil fuels, nuclear fission, and nuclear fusion.

16. What is the reason for the "secondary system" in a nuclear reactor, Fig. 43–9? That is, why is the water heated by the fuel in a nuclear reactor not used directly to drive the turbines?

17. Discuss how the course of history might have been changed if, during World War II, scientists had refused to work on developing a nuclear bomb. Do you think it would have been possible to delay the building of a bomb indefinitely?

18. Research in molecular biology is moving toward the ability to perform genetic manipulations on human beings. The moral implications of future discoveries along these lines has led to a warning that this may be the molecular biologists' "Hiroshima." Discuss.

19. Does $E = mc^2$ apply in (a) fission, (b) fusion, (c) nuclear reactions?

20. Light energy emitted by the Sun and stars comes from the fusion process. What conditions in the interior of stars makes this possible?

21. How do stars, and our Sun, maintain confinement of the plasma for fusion?

22. What is the basic difference between fission and fusion?

23. People who work around metals that emit alpha particles are trained that there is little danger from proximity or even touching the material, but that they must take extreme precautions against ingesting it. Hence, there are strong rules against eating and drinking while working, and against machining the metal. Why?

24. Why is the recommended maximum radiation dose higher for women beyond the child-bearing age than for younger women?

25. A higher temperature is required for deuterium–deuterium ignition than for deuterium–tritium. Explain.

26. Radiation is sometimes used to sterilize medical supplies and even food. Explain how it works.

* 27. How might radioactive tracers be used to find a leak in a pipe?

Problems

Section 43–1

1. (I) Natural aluminum is all $^{27}_{13}Al$. If it absorbs a neutron, what does it become? Does it decay by β^+ or β^-? What will be the product nucleus?

2. (I) Determine whether the reaction $^{2}_{1}H(d, n)^{3}_{2}He$ requires a threshold energy. (d stands for deuterium, $^{2}_{1}H$.)

3. (I) Is the reaction $^{238}_{92}U(n, \gamma)^{239}_{92}U$ possible with slow neutrons? Explain.

4. (II) Does the reaction $^{7}_{3}\text{Li}(p, \alpha)^{4}_{2}\text{He}$ require energy or does it release energy? How much energy?

5. (II) Calculate the energy released (or energy input required) for the reaction $^{9}_{4}\text{Be}(\alpha, n)^{12}_{6}\text{C}$.

6. (II) (a) Can the reaction $^{24}_{12}\text{Mg}(n, d)^{23}_{11}\text{Na}$ occur if the bombarding particles have 10.00 MeV of kinetic energy? (b) If so, how much energy is released?

7. (II) (a) Can the reaction $^{7}_{3}\text{Li}(p, \alpha)^{4}_{2}\text{He}$ occur if the incident proton has kinetic energy = 2500 keV? (b) If so, what is the total kinetic energy of the products?

8. (II) In the reaction $^{14}_{7}\text{N}(\alpha, p)^{17}_{8}\text{O}$, the incident α particles have 7.68 MeV of kinetic energy. (a) Can this reaction occur? (b) If so, what is the total kinetic energy of the products? The mass of ^{17}O is 16.999131 u.

9. (II) Calculate the Q-value for the "capture" reaction $^{16}_{8}\text{O}(\alpha, \gamma)^{20}_{10}\text{Ne}$.

10. (II) Calculate the total kinetic energy of the products of the reaction $^{13}_{6}\text{C}(d, n)^{14}_{7}\text{N}$ if the incoming deuteron (d) has $K = 36.3$ MeV.

11. (II) An example of a "stripping" nuclear reaction is $^{6}_{3}\text{Li}(d, p)\text{X}$. (a) What is X, the resulting nucleus? (b) Why is it called a "stripping" reaction? (c) What is the Q-value of this reaction? Is the reaction endothermic or exothermic?

12. (II) An example of a "pick-up" nuclear reaction is $^{12}_{6}\text{C}(^{3}_{2}\text{He}, \alpha)\text{X}$. (a) Why is it called a "pickup" reaction? (b) What is the resulting nucleus? (c) What is the Q-value of this reaction? Is the reaction endothermic or exothermic?

13. (II) (a) Complete the following nuclear reaction, $?(p, \gamma)^{32}_{16}\text{S}$. (b) What is the Q-value?

14. (III) Use conservation of energy and momentum to show that a bombarding proton must have an energy of 3.23 MeV in order to make the reaction $^{13}_{6}\text{C}(p, n)^{13}_{7}\text{N}$ occur. (See Example 43–3.)

Section 43–2

15. (I) What is the effective cross section for the collision of two hard spheres of radius R_1 and R_2.

16. (I) The cross section for the reaction n $+$ $^{10}_{5}\text{B}$ $\rightarrow$ $^{7}_{3}\text{Li}$ $+$ $^{4}_{2}\text{He}$ is about 40 bn for an incident neutron of low energy (kinetic energy $\approx$ 0). The boron is contained in a gas with $n = 1.7 \times 10^{21}$ nuclei/m^3 and the target has thickness $t = 12.0$ cm. What fraction of incident neutrons will be scattered?

17. (II) When the target is thick, the rate at which projectile particles collide with nuclei in the rear of the target is less than in the front of the target, since some scattering (i.e., collisions) takes place in the front layers. Let R_0 be the rate at which incident particles strike the front of the target, and R_x be the rate at a distance x into the target ($R_x = R_0$ at $x = 0$). Then show that the rate at which particles are scattered (and therefore lost from the incident beam) in a thickness dx is $-dR_x = R_x n\sigma dx$, where the minus sign means that R_x is decreasing and n is the number of nuclei per unit volume. Then show that $R_x = R_0 e^{-n\sigma x}$, where σ is the total cross section. If the thickness of the target is t, what does $R_x = R_0 e^{-n\sigma t}$ represent?

18. (II) A 1.0-cm-thick lead target reduces a beam of gamma rays to 30 percent of its original intensity. What thickness of lead will allow only one γ in 10^6 to penetrate (see Problem 17)?

19. (II) Use Fig. 43–3 to estimate what thickness of Cd $(\rho = 8650 \text{ kg/m}^3)$ will cause a 1 percent reaction rate $(R/R_0 = 0.01)$ for (a) 0.1-eV neutrons (b) 10-eV neutrons.

Section 43–3

20. (I) Calculate the energy released in the fission reaction n $+$ $^{235}_{92}\text{U}$ $\rightarrow$ $^{88}_{38}\text{Sr}$ $+$ $^{136}_{54}\text{Xe}$ $+$ 12n. Use Appendix D, and assume the initial kinetic energy of the neutron is very small.

21. (I) What is the energy released in the fission reaction of Eq. 43–5? (The masses of $^{141}_{56}\text{Ba}$ and $^{92}_{36}\text{Kr}$ are 140.91440 u and 91.92630 u, respectively.)

22. (I) How many fissions take place per second in a 200-MW reactor? Assume 200 MeV is released per fission.

23. (II) Suppose that the average electric power consumption, day and night, in a typical house is 300 W. What initial mass of $^{235}_{92}\text{U}$ would have to undergo fission to supply the electrical needs of such a house for a year? (Assume 200 MeV is released per fission, as well as 100% efficiency.)

24. (II) What initial mass of $^{235}_{92}\text{U}$ is required to operate a 500-MW reactor for 1 yr? Assume 40 percent efficiency.

25. (II) If a 1.0-MeV neutron emitted in a fission reaction loses one-half of its kinetic energy in each collision with moderator nuclei, how many collisions must it make to reach thermal energy $(\frac{3}{2}kT = 0.040 \text{ eV})$?

26. (II) Suppose that the neutron multiplication factor is 1.0004. If the average time between successive fissions in a chain of reactions is 1.0 ms, by what factor will the reaction rate increase in 1.0 s?

27. (II) Estimate the ratio of the height of the Coulomb barrier for α decay to that for fission of $^{236}_{92}\text{U}$. (Both are described by a potential energy diagram of the shape shown in Fig. 42–6.)

28. (II) Assuming a fission of $^{236}_{92}\text{U}$ into two roughly equal fragments, estimate the electric potential energy just as the fragments separate from each other. Assume that the fragments are spherical (see Eq. 42–1) and compare your calculation to the nuclear fission energy released, about 200 MeV.

Section 43–4

29. (I) What is the average kinetic energy of protons at the center of a star where the temperature is 10^7 K?

30. (II) Show that the energy released in the fusion reaction $^{2}_{1}\text{H}$ $+$ $^{3}_{1}\text{H}$ $\rightarrow$ $^{4}_{2}\text{He}$ $+$ n is 17.59 MeV.

31. (II) Show that the energy released when two deuterium nuclei fuse to form $^{3}_{2}\text{He}$ with the release of a neutron is 3.27 MeV.

32. (II) Verify the Q-value stated for each of the reactions of Eqs. 43–6. [Hint: Be careful with electrons.]

33. (II) Calculate the energy release per gram of fuel for the reactions of Eqs. 43–8a, b, and c. Compare to the energy release per gram of uranium in fission.

34. (II) If a typical house requires 300 W of electric power on average, what minimum amount of deuterium fuel would have to be used in a year to supply these electrical needs? Assume the reaction of Eq. 43–8b.

35. (II) Show that the energies carried off by the $^4_2$He nucleus and the neutron for the reaction of Eq. 43–8c are about 3.5 MeV and 14 MeV, respectively. Are these fixed values, independent of the plasma temperature?

36. (II) Suppose a fusion reactor ran on "d–d" reactions, Eqs. 43–8a and b. Estimate how much water, for fuel, would be needed per hour to run a 1000-MW reactor, assuming 30 percent efficiency.

37. (II) How much energy (J) is contained in 1.00 kg of water if its natural deuterium is used in the fusion reaction of Eq. 43–8a? Compare to the energy obtained from the burning of 1.0 kg of gasoline, about 5×10^7 J.

38. (III) The energy output of massive stars is believed to be due to the *carbon cycle* (see text). (a) Show that no carbon is consumed in this cycle and that the net effect is the same as for the proton–proton cycle. (b) What is the total energy release? (c) Determine the energy output for each reaction and decay. (d) Why does the carbon cycle require a higher temperature $(\approx 2 \times 10^7\,\text{K})$ than the proton–proton cycle $(\approx 1.5 \times 10^7\,\text{K})$?

39. (III) (a) Compare the energy needed for the first reaction of the carbon cycle to that for a deuteron–tritium reaction (Example 43–8). (b) If a deuteron–tritium reaction requires $T \approx 3 \times 10^8\,\text{K}$, estimate the temperature needed for the first carbon-cycle reaction.

40. (III) The deuterium–tritium pellet in a laser fusion device contains equal numbers of $^2_1$H and $^3_1$H atoms raised to a density of $200 \times 10^3\,\text{kg/m}^3$ by the laser pulses. Estimate (a) the density of particles in this compressed state, and (b) the length of time τ the pellet must be confined to meet the Lawson criterion for ignition.

Section 43–6

41. (I) A dose of 4.0 Sv of γ rays in a short period would be lethal to about half the people subjected to it. How many grays is this?

42. (I) Fifty rads of α-particle radiation is equivalent to how many rads of X-rays in terms of biological damage?

43. (I) How many rads of slow neutrons will do as much biological damage as 50 rads of fast neutrons?

44. (I) How much energy is deposited in the body of a 70-kg adult exposed to a 50-rad dose?

45. (II) A 0.025-μCi sample of $^{32}_{15}$P is injected into an animal for tracer studies. If a Geiger counter intercepts 20 percent of the emitted β particles, what will be the counting rate, assumed 90 percent efficient.

46. (II) A 1.0-mCi source of $^{32}_{15}$P (in NaHPO$_4$), a β^- emitter, is implanted in an organ where it is to administer 36 Gy. The half-life of $^{32}_{15}$P is 14.3 days and 1 mCi delivers about 10 mGy/min. Approximately how long should the source remain implanted?

47. (II) About 35 eV is required to produce one ion pair in air. Show that this is consistent with the two definitions of the roentgen given in the text.

48. (II) $^{57}_{27}$Co emits 122-keV γ rays. If a 70-kg person swallowed 1.85 μCi of $^{57}_{27}$Co, what would be the dose rate (Gy/day) averaged over the whole body? Assume that 50 percent of the γ-ray energy is deposited in the body.

49. (II) What is the mass of a 1.00-μCi $^{14}_6$C source?

50. (II) Huge amounts of radioactive $^{131}_{53}$I were released in the accident at Chernobyl in 1986. Chemically, iodine goes to the human thyroid. (Doctors can use it for diagnosis and treatment of thyroid problems.) In a normal thyroid, $^{131}_{53}$I absorption can cause damage to the thyroid. (a) Write down the decay scheme for $^{131}_{53}$I. (b) Its half-life is 8.0 d; how long would it take for ingested $^{131}_{53}$I to become 10 percent of the initial value? (c) Absorbing 1 mCi of $^{131}_{53}$I can be harmful; what mass of iodine is this?

51. (II) Assume a liter of milk typically has an activity of 2000 pCi due to $^{40}_{19}$K. If a person drinks two glasses (0.5 L) per day, estimate the total dose (in Sv and in rem) received in a year. As a crude model, assume the milk stays in the stomach 12 hr and is then released. Assume also that very roughly 10 percent of the 1.5 MeV released per decay is absorbed by the body. Compare your result to the normal allowed dose per year. Make your estimate for (a) a 50-kg adult, and (b) a 5-kg baby.

52. (II) Radon gas, $^{222}_{86}$Rn, is considered a serious health hazard (see discussion in text). It decays by α-emission. (a) What is the daughter nucleus? (b) Is the daughter nucleus stable or radioactive? If the latter, how does it decay, and what is its lifetime? (c) Is the daughter nucleus also a noble gas, or is it chemically reacting? (d) Suppose 1.0 ng of $^{222}_{86}$Rn seeps into a basement. What will be its activity? If the basement is then sealed, what will be the activity 1 month later?

* Section 43–9

* 53. (II) (a) Suppose for a conventional X-ray image that the X-ray beam consisted of parallel rays. What would be the magnification of the image? (b) Suppose, instead, that the X-rays came from a point source (as in Fig. 43–20a) that is 15 cm in front of a human body 25 cm thick, and the film is pressed against the person's back. Determine and discuss the range of magnifications that result.

* Section 43–10

* 54. (I) Calculate the wavelength of photons needed to produce NMR transitions in free protons in a 1.000-T field. In what region of the spectrum does it lie?

General Problems

55. J. Chadwick discovered the neutron by bombarding $^9_4$Be with the popular projectile of the day, alpha particles. (*a*) If one of the reaction products was the then unknown neutron, what was the other product? (*b*) What is the *Q* of this reaction?

56. Fusion temperatures are often given in keV. Determine the conversion factor from kelvins to keV using, as is common in this field, $\bar{K} = kT$ without the factor $\frac{3}{2}$.

57. One means of enriching uranium is by diffusion of the gas UF$_6$. Calculate the ratio of the speeds of molecules of this gas containing $^{235}_{92}$U and $^{238}_{92}$U, on which this process depends.

58. (*a*) What mass of $^{235}_{92}$U was actually fissioned in the first atomic bomb, whose energy was the equivalent of about 20 kilotons of TNT (1 kiloton of TNT releases 5×10^{12} J)? (*b*) What was the actual mass transformed to energy?

59. In a certain town the average yearly background radiation consists of 21 mrad of X-rays and γ rays plus 3.0 mrad of particles having a QF of 10. How many rem will a person receive per year on the average?

60. Deuterium makes up 0.015 percent of natural hydrogen. Make a rough estimate of the total deuterium in the Earth's oceans and estimate the total energy released if all of it were used in fusion reactors.

61. A shielded γ-ray source yields a dose rate of 0.052 rad/h at a distance of 1.0 m for an average-sized person. If workers are allowed a maximum dose rate of 5.0 rem/yr, how close to the source may they operate, assuming a 40-h work week? Assume that the intensity of radiation falls off as the square of the distance. (It actually falls off more rapidly than $1/r^2$ because of absorption in the air, so the answer above will give a better-than-permissible value.)

62. Radon gas, $^{222}_{86}$Rn, is formed by α decay. (*a*) Write the decay equation. (*b*) Ignoring the kinetic energy of the daughter nucleus (it's so massive), estimate the kinetic energy of the α particle produced. (*c*) Estimate the momentum of the alpha and of the daughter nucleus. (*d*) Estimate the kinetic energy of the daughter, and show that your approximation in (*b*) was valid.

63. The reaction $^{18}_8$O(p, n)$^{18}_9$F requires an input of energy equal to 2.453 MeV. What is the mass of $^{18}_9$F?

64. Consider a system of nuclear power plants that produce 3400 MW. (*a*) What total mass of $^{235}_{92}$U fuel would be required to operate these plants for 1 yr, assuming that 200 MeV is released per fission? (*b*) Typically 6 percent of the $^{235}_{92}$U nuclei that fission produce $^{90}_{38}$Sr, a β^- emitter with a half-life of 29 yr. What is the total radioactivity of the $^{90}_{38}$Sr, in curies, produced in 1 yr? (Neglect the fact that some of it decays during the 1-yr period.)

65. In the net reaction, Eq. 43–7, for the proton–proton cycle in the Sun, the neutrinos escape from the Sun with energy of about 0.5 MeV. The remaining energy, 26.2 MeV, is available within the Sun. Use this value to calculate the "heat of combustion" per kilogram of hydrogen fuel and compare it to the heat of combustion of coal, about 3×10^7 J/kg.

66. Energy reaches the Earth from the Sun at a rate of about 1400 W/m^2. Calculate (*a*) the total energy output of the Sun, and (*b*) the number of protons consumed per second in the reaction of Eq. 43–7, assuming that this is the source of all the Sun's energy. (*c*) Assuming that the Sun's mass of 2.0×10^{30} kg was originally all protons and that all could be involved in nuclear reactions in the Sun's core, how long would you expect the Sun to "glow" at its present rate?

67. Estimate how many solar neutrinos pass through a 100 m^2 ceiling in a room at latitude 40°, in a year. [*Hint*: See Problems 65 and 66.]

68. Show, using the laws of conservation of energy and momentum, that for a nuclear reaction requiring energy, the minimum kinetic energy of the bombarding particle (the *threshold energy*) is equal to $[-Qm_{pr}/(m_{pr} - m_b)]$, where $-Q$ is the energy required (difference in total mass between products and reactants), m_b is the rest mass of the bombarding particle, and m_{pr} the total rest mass of the products. Assume the target nucleus is at rest before an interaction takes place, and that all particles are nonrelativistic.

69. The early scattering experiments performed around 1910 in Ernest Rutherford's laboratory in England produced the first evidence that an atom consists of a heavy nucleus surrounded by electrons. In one of these experiments, α particles struck a gold-foil target 4.0×10^{-5} cm thick in which there were 5.9×10^{28} gold atoms per cubic meter. Although most α particles either passed straight through the foil or were scattered at small angles, approximately 1.6×10^{-3} percent were scattered at angles greater than 90°—that is, in the backward direction. (*a*) Calculate the cross section, in barns, for backward scattering. (*b*) Rutherford concluded that such backward scattering could occur only if an atom consisted of a very tiny, massive, and positively charged nucleus with electrons orbiting some distance away. Assuming that backward scattering occurs for nearly direct collisions (i.e., $\sigma \approx$ area of nucleus), estimate a diameter of a gold nucleus.

70. Some stars, in a later stage of evolution, may begin to fuse two $^{12}_6$C nuclei into one $^{24}_{12}$Mg nucleus. (*a*) How much energy would be released in such a reaction? (*b*) What kinetic energy must two carbon nuclei each have when far apart, if they can then approach each other to within 6.0 fm, center-to-center? (*c*) Approximately what temperature would this require?

71. An average adult body contains about 0.10 μCi of $^{40}_{19}$K, which comes from food. (*a*) How many decays occur per second? (*b*) The potassium decays produce beta particles with energies of around 1.4 MeV. Calculate the dose per year in sieverts for a 50-kg adult. Is this a significant fraction of the 3.6 mSv/year background rate?

72. When the nuclear reactor accident occurred at Chernobyl in 1986, 2.0×10^7 Ci were released into the atmosphere. Assuming that this radiation was distributed uniformly over the surface of the Earth, what was the activity per square meter? (The actual activity was not uniform; even within Europe wet areas received more radioactivity from rainfall).

This computer-generated reconstruction of a proton–antiproton collision at Fermilab (Fig. 44–4) occurred at a combined energy of nearly 2 TeV. It is one of the events that provided evidence for the top quark, announced in 1995. The particle tracks are curved, due to a magnetic field, and the radius of curvature is a measure of each particle's momentum (Chapter 27). The top quark (t) has too brief a lifetime ($\approx 10^{-23}$ s) to be detected itself, so we look for its possible decay products. Analysis indicates the following interaction and subsequent decays:

The tracks in the photo include jets (groups of particles moving in roughly the same direction), and a muon (μ^-) whose track is the pink one enclosed by a yellow rectangle to make it stand out. After reading this chapter, try to give the name for each symbol above and comment on whether all conservation laws hold.

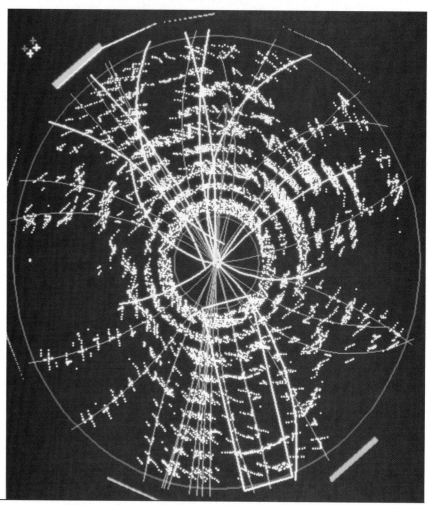

CHAPTER 44

Elementary Particles

In the final two chapters of this book we discuss two of the most exciting areas of contemporary physics: (1) elementary particles, and (2) cosmology and astrophysics. These are subjects at the forefront of knowledge—one treats the smallest objects in the universe, the other the largest (and oldest) aspects of the universe. The reader who wants an understanding of the great beauties of present-day science—and/or wants to be a good citizen—will want to read these chapters, even if there is not time to cover them in a physics course.

In this penultimate chapter we discuss *elementary particle* physics, which represents the human endeavor to understand the basic building blocks of all matter. In the years after World War II, it was found that if the incoming particle in a nuclear reaction has sufficient energy, new types of particles can be produced. The earliest experiments used **cosmic rays**—particles that impinge on the Earth from space. But in order to produce high-energy particles in the laboratory, various types of particle accelerators have been constructed. Most commonly they accelerate protons or electrons, although heavy ions can also be accelerated, depending on the design. These high-energy accelerators have been used to probe the nucleus more deeply, to produce and study new particles, and to give us information about the basic forces and constituents of nature. Because the projectile particles are at high-energy, this field is sometimes called **high-energy physics**.

44–1 | High-Energy Particles

Particles accelerated to high energy are projectiles which can probe the interior of nuclei and nucleons that they strike. An important factor is that faster-moving projectiles can reveal more detail. The wavelength of projectile particles is given by de Broglie's wavelength formula (Eq. 38–7),

$$\lambda = \frac{h}{p},$$

(44–1)

de Broglie
wavelength

showing that the greater the momentum p, of the bombarding particle, the shorter its wavelength. As discussed in Chapter 36 on diffraction, resolution of details in images is limited by the wavelength: the shorter the wavelength, the finer the detail that can be obtained. This is one reason why particle accelerators of higher and higher energy have been built in recent years.

EXAMPLE 44–1 **High resolution with electrons.** What is the wavelength, and hence the expected resolution, for a beam of 1.3-GeV electrons?

SOLUTION The 1.3 GeV = 1300 MeV refers to the kinetic energy, and is about 2500 times the rest energy (mass) of the electron ($mc^2 = 0.51$ MeV). We are clearly dealing with relativistic speeds here, and it is easily shown (see Eqs. 37–10 and 37–13) that the speed of the electron is nearly $c = 3.0 \times 10^8$ m/s. Therefore $K = (\gamma - 1)mc^2 \approx \gamma(mc^2) = 1300$ MeV, and

$$\lambda = \frac{h}{p} = \frac{h}{\gamma m v} \approx \frac{h}{\gamma m c} = \frac{hc}{\gamma m c^2} \approx \frac{hc}{K}$$

$$= \frac{(6.6 \times 10^{-34}\,\text{J·s})(3.0 \times 10^8\,\text{m/s})}{(1.3 \times 10^9\,\text{eV})(1.6 \times 10^{-19}\,\text{J/eV})} = 0.96 \times 10^{-15}\,\text{m},$$

or 0.96 fm. The maximum possible resolution of this beam of electrons is far greater than for a light beam in a light microscope ($\lambda \approx 500$ nm). Indeed, this resolution of about 1 fm is on the order of the size of nuclei (see Eq. 42–1).

Another major reason for building high-energy accelerators is that new particles of greater mass can be produced at higher energies, as we will discuss shortly. Now we look briefly at several types of particle accelerator.

44–2 | Particle Accelerators and Detectors

Van de Graaff Accelerator

The key component of a Van de Graaff accelerator (Fig. 44–1) is a Van de Graaff generator, invented in 1931. A large, hollow, spherical conductor is charged to a high potential by a nonconducting moving belt in the following way: A high voltage of typically 50,000 V is applied to a pointed conductor A, which "sprays" positive charge onto the moving belt. (Actually, electrons are pulled off the belt onto the electrode A.) The belt carries the positive charge into the interior of the sphere, where it is "wiped off" the belt at B and races to the outer surface of the spherical conductor. (Remember that charge collects on the outer surface of any conductor, since the charges repel each other and try to get as far from each other as possible.) Very high potential differences can be obtained in this way, often giving the accelerated particles as much as 30 MeV of kinetic energy when generators are used in tandem. Connected to the Van de Graaff generator is an evacuated tube which serves as the particle accelerator. A source of H or He ions (p or α) is located inside the tube, and the large positive voltage repels them so they are accelerated toward the grounded target at the far end of the tube (Fig. 44–1).

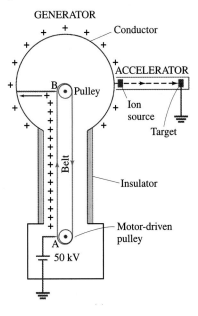

FIGURE 44–1 Van de Graaff generator and accelerator.

FIGURE 44–2 Ernest O. Lawrence, around 1930, holding the first cyclotron (we see the vacuum chamber enclosing it).

Cyclotron frequency

FIGURE 44–3 Diagram of a cyclotron. The magnetic field, applied by a large electromagnet, points into the page. A is the ion source. The field lines shown are for the electric field in the gap.

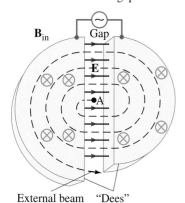

Cyclotron

The cyclotron, developed in 1930 by E. O. Lawrence (1901–1958; Fig. 44–2) at the University of California, Berkeley, uses a magnetic field to maintain charged ions—usually protons—in nearly circular paths (Chapter 27). The protons move within two D-shaped cavities, as shown in Fig. 44–3. Each time they pass into the gap between the "dees," a voltage accelerates them via the electric force, increasing their speed and increasing the radius of curvature of their path. After many revolutions, the protons acquire high kinetic energy and reach the outer edge of the cyclotron. They then either strike a target placed inside the cyclotron or leave the cyclotron with the help of a carefully placed "bending magnet" and are directed to an external target. Acceleration only occurs when the protons are in the gap *between* the dees, and the voltage must be alternating. When the protons are moving to the right across the gap in Fig. 44–3, the right dee must be electrically negative and the left one positive. A half-cycle later, the protons are moving to the left, so the left dee must be negative in order to accelerate them. The frequency, f, of the applied voltage must be equal to that of the circulating protons, and this is just the **cyclotron frequency** determined in Section 27–4, Eq. 27–6:

$$f = \frac{1}{T} = \frac{qB}{2\pi m},$$

(44–2)

where q and m are the charge and mass of the particles moving in the magnetic field B.

EXAMPLE 44–2 Cyclotron. A small cyclotron of maximum radius $R = 0.25\,\text{m}$ accelerates protons in a 1.7-T magnetic field. Calculate (*a*) what frequency is needed for the applied alternating voltage, and (*b*) the kinetic energy of protons when they leave the cyclotron.

SOLUTION (*a*) From Eq. 44–2,

$$f = \frac{qB}{2\pi m}$$

$$= \frac{(1.6 \times 10^{-19}\,\text{C})(1.7\,\text{T})}{(6.28)(1.67 \times 10^{-27}\,\text{kg})} = 2.6 \times 10^7\,\text{Hz} = 26\,\text{MHz},$$

which is in the radio wave region of the EM spectrum (Fig. 32–12).
(*b*) The protons leave the cyclotron at $r = R = 0.25\,\text{m}$. From $F = ma$, we have (Eq. 27–5) $qvB = mv^2/r$, so $v = qBr/m$, and the kinetic energy is

$$K = \frac{1}{2}mv^2 = \frac{1}{2}m\frac{q^2B^2R^2}{m^2} = \frac{q^2B^2R^2}{2m}$$

$$= \frac{(1.6 \times 10^{-19}\,\text{C})^2(1.7\,\text{T})^2(0.25\,\text{m})^2}{(2)(1.67 \times 10^{-27}\,\text{kg})} = 1.4 \times 10^{-12}\,\text{J} = 8.7\,\text{MeV}.$$

Note that the magnitude of the voltage applied to the dees does not affect the final energy. But the higher this voltage, the fewer revolutions are required to bring the protons to full energy.

An important aspect of the cyclotron is that the frequency of the applied voltage, as given by Eq. 44–2, does not depend on the radius r. That is, the frequency does not have to be changed as the ions start from the source and are accelerated to paths of larger and larger radii. But this is only true at nonrelativistic energies. At higher speeds, the momentum is (Eq. 37–7) $p = mv/\sqrt{1 - v^2/c^2}$ $= \gamma mv$, so we must replace m in Eq. 44–2 with γm. One way to build a machine for higher speeds is to use a magnetic field that increases with radius so as to keep $B/\gamma m$ constant as the particles move to larger radius and momentum, thus allowing the frequency to be kept constant. Another approach is to keep B uniform while decreasing the frequency in time as a packet of ions increase in speed and mass, and reach larger orbits. Such a machine is called a **synchrocyclotron**.

Synchrocyclotron

Synchrotron

Another way to deal with the relativistic increase in effective mass (γm) with speed is to increase the magnetic field B in time as the particles speed up. Such devices are called **synchrotrons**, and today they can be enormous. The *Tevatron* accelerator at Fermilab (the Fermi National Accelerator Laboratory) at Batavia, Illinois, has a radius of 1.0 km, and that at CERN (European Center for Nuclear Research) in Geneva, Switzerland, is 4.3 km in radius. The Tevatron uses superconducting magnets to accelerate protons to about 1000 GeV = 1 TeV (hence its name; 1 TeV = 10^{12} eV). These large synchrotrons use a narrow ring of magnets (see Fig. 44–4) with each magnet placed at the same radius from the center of the circle. The magnets are interrupted by gaps where high voltage accelerates the particles. Thus, once the particles are injected, they must move in a circle of constant radius. This is accomplished by giving them considerable energy initially in a much smaller accelerator, and then slowly increasing the magnetic field as they speed up in the large synchrotron.

Synchrotron

One problem of any accelerator is that accelerating electric charges radiate electromagnetic energy (see Chapter 32). Since ions or electrons are accelerated in an accelerator, we can expect considerable energy to be lost by radiation. The effect increases with speed and is especially important in circular machines where centripetal acceleration is present, particularly in synchrotrons, and hence is called **synchrotron radiation**. Synchrotron radiation can actually be useful, however. Intense beams of photons are sometimes needed, and they are usually obtained from an electron synchrotron.

FIGURE 44–4 (a) Aerial view of Fermilab at Batavia, Illinois; the accelerator is a circular ring 1.0 km in radius. (b) The interior of the tunnel of the main accelerator at Fermilab. The upper (rectangular-shaped) ring of magnets is for the older 500-GeV accelerator. Below it is the ring of superconducting magnets for the 1-TeV Tevatron.

(a)

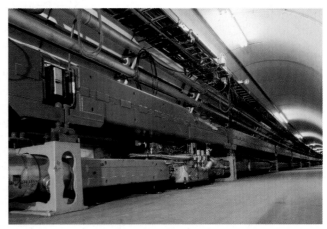

(b)

Linear Accelerators

Linac

A Van de Graaff accelerator is essentially a linear accelerator since the ions move in a linear path. But the name *linear accelerator* is usually reserved for a more complex arrangement in which particles are accelerated many, many times along a straight-line path. Figure 44–5a is a diagram of a simple "linac." The ions pass through a series of tubular conductors. The voltage applied to the tubes must be alternating so that when positive ions (say) reach a gap, the tube in front of them is negative and the one they just left is positive. This assures that they are accelerated at each gap. As the ions increase in speed, they cover more distance in the same amount of time. Consequently, the tubes must be longer the farther they are from the source. Linear accelerators are of particular importance for accelerating electrons. Because of their small mass, electrons reach high speeds very quickly; an electron linac such as the one shown in Fig. 44–5a would have tubes nearly equal in length, since the electrons would be traveling close to $c = 3.0 \times 10^8\,\text{m/s}$ for almost the entire distance. The amount of energy radiated away by electrons in a linear machine is much less than for a circular machine (see mention of synchrotron radiation above). The largest electron linear accelerator is that at Stanford (Stanford Linear Accelerator Center, or SLAC), Fig. 44–5b. It is about 3 km (2 miles) long and can accelerate electrons to 50 GeV. Many hospitals have 10-MeV electron linacs that produce photons to irradiate tumors.

FIGURE 44–5 (a) Diagram of a simple linear accelerator. (b) Photo of the Stanford Linear Accelerator (SLAC) in California.

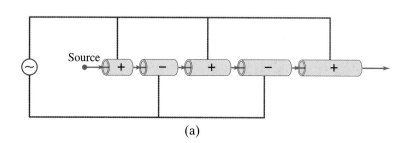

(a)

(b)

Colliding Beams

Colliders

Originally, high-energy physics experiments were carried out by allowing a beam of particles from an accelerator to strike a stationary target. A newer, very powerful technique allows us to obtain the maximum possible collision energy from a given accelerator: two beams of particles are accelerated to very high energy and are steered so that they collide head-on. One way to accomplish such **colliding beams** with a single accelerator is through the use of **storage rings**, in which the particles of one beam can continue to circulate while the second beam is being accelerated. The two beams are then steered so that they intersect and collide head-on. For example, in the experiments that provided strong evidence for the top quark (see chapter opening photo and Section 44–9), the Fermilab Tevatron accelerated protons and antiprotons each to 900 GeV, so that the combined energy of head-on collisions was 1.8 TeV. The largest collider accelerator today is the Large Electron-Positron (LEP) Collider at CERN, which has a circumference of 26.7 km (Fig. 44–6). It produces oppositely revolving beams of e^+ and e^- each of energy 95 GeV, for a total interaction energy of 190 GeV. The most powerful accelerator-collider will soon be the Large Hadron Collider (LHC) at CERN, scheduled to be completed about 2005. The two colliding beams will each carry 7 TeV protons for a total interaction energy of 14 TeV.

FIGURE 44–6 The large circle represents the position of the tunnel now containing the 8.5 km diameter LEP collider, about 100 m below the ground, at CERN (near Geneva) on the French-Swiss border. The LHC will use the same tunnel. The smaller circle shows the position of the Super Proton Synchrotron that will be used for accelerating protons in the LHC.

EXAMPLE 44–3 **Protons at relativistic speeds.** Determine the energy required to accelerate a proton in a high-energy accelerator (*a*) from rest to $v = 0.900c$, and (*b*) from $v = 0.900c$ to $v = 0.999c$. (*c*) What is the kinetic energy achieved by each proton?

SOLUTION We use the work-energy principle, which is still valid relativistically as we saw in Section 37–11: $W = \Delta K$. The kinetic energy of a proton is given by Eq. 37–10

$$K = (\gamma - 1)mc^2,$$

where the relativistic factor γ is

$$\gamma = \frac{1}{\sqrt{1 - v^2/c^2}}.$$

The work-energy theorem becomes

$$W = \Delta K = (\Delta\gamma)mc^2$$

since m (the rest mass) and c are constant.
(*a*) For $v = 0$, $\gamma = 1$; and for $v = 0.900c$

$$\gamma = \frac{1}{\sqrt{1 - (0.900)^2}} = 2.29.$$

Hence, with rest mass of the proton $mc^2 = 938 \text{ MeV}$, the work needed is

$$W = \Delta K = (\Delta\gamma)mc^2 = (2.29 - 1.00)(938 \text{ MeV}) = 1.21 \text{ GeV}.$$

(*b*) For $v = 0.999c$,

$$\gamma = \frac{1}{\sqrt{1 - (0.999)^2}} = 22.4.$$

So the work needed to accelerate a proton from $0.900c$ to $0.999c$ is

$$W = \Delta K = (\Delta\gamma)mc^2 = (22.4 - 2.29)(938 \text{ MeV}) = 18.9 \text{ GeV}.$$

(*c*) The kinetic energy reached by the proton in (*a*) is just equal to the work done on it, $K = 1.21 \text{ GeV}$. The final kinetic energy of the proton in (*b*), moving at $v = 0.999c$, is

$$K = (\gamma - 1)mc^2 = (21.4)(938 \text{ MeV}) = 20.1 \text{ GeV},$$

which makes sense since, starting from rest, we did work $W = 1.21 \text{ GeV} + 18.9 \text{ GeV} = 20.1 \text{ GeV}$ on it.

FIGURE 44–7 (a) In a cloud or bubble chamber, droplets or bubbles are formed around ions produced by the passage of a charged particle. (b) Bubble-chamber photo of particle tracks, here revealing evidence for the Ω^- particle. Note the curvature of the tracks in the magnetic field.

Path of particle

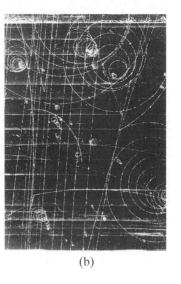

(a)   (b)

Particle Detectors

Detectors of particles such as scintillators, photomultipliers, and semiconductor detectors, which we discussed in Section 42–11, are also used in the study of nuclear reactions and in elementary particle physics. More sophisticated devices have been developed that allow the track of a charged particle to be *seen*. The simplest is the **photographic emulsion** (which, being small and simple and therefore portable, is now used mainly for cosmic-ray studies from balloons). A particle passing through a layer of photographic emulsion ionizes the atoms along its path. This results in a chemical change at these points, and when the emulsion is developed, the particle's path is revealed.

In a **cloud chamber**, a gas is cooled to a temperature slightly below its usual condensation point. (It is said to be "supercooled.") The gas molecules begin to condense on any ionized molecules present. Thus the ions produced when a charged particle passes through serve as centers on which tiny droplets form (Fig. 44–7a). Light scatters more from these droplets than from the gas background, so a photo of the cloud chamber at the right moment shows the track of the particle. An important instrument in the early days of nuclear physics, it is little used today.

The **bubble chamber**, invented in 1952 by D. A. Glaser (1926–), makes use of a superheated liquid. The liquid is kept close to its normal boiling point and the bubbles characteristic of boiling form around ions produced by the passage of a charged particle (Fig. 44–7b). A photograph of the interior of the chamber thus reveals the paths of particles that recently passed through. Because the bubble chamber uses a liquid—often liquid hydrogen—the density of atoms is much greater than in a cloud chamber. Hence it is a much more efficient device for observing the tracks of charged particles and their interactions with the nuclei of the liquid. Usually, a magnetic field is applied across the chamber and the momentum of the moving particles can be determined from the radius of curvature of their paths.

Much more used today is the **wire drift chamber**, which consists of a set of closely spaced fine wires immersed in a gas. Many wires are grounded, and others in between are kept at very high voltage. A charged particle passing through produces ions in the gas. The freed electrons drift toward the nearest high voltage wires, creating an "avalanche," and producing an electric pulse or signal at that wire. The positions of the particles can be determined both by the position of the wire and by the time it takes the pulses to reach detectors at the ends of the wires. The paths of the particles are reconstructed electronically with computers which can then "draw" a picture of the tracks, as shown in the photo at the start of the chapter: the white dots are the wires, and the colored lines are the particle tracks. The device that produced that photo is shown in Fig. 44–8. (Another example is shown in Fig. 44–12).

FIGURE 44–8 Photo of the large and complicated CDF detector at Fermilab which, utilizing wire drift chambers, detected the particles and produced the image of the tracks shown at the start of this chapter.

Beginnings of Elementary Particle Physics— Particle Exchange

By the mid-1930s, it was recognized that all atoms can be considered to be made up of neutrons, protons, and electrons. The basic constituents of the universe were no longer considered to be atoms but rather the proton, neutron, and electron. Besides these three *elementary particles*, several others were also known: the positron (a positive electron), the neutrino, and the γ particle (or photon), for a total of six elementary particles.

In the decades that followed, hundreds of other subnuclear particles were discovered. The properties and interactions of these particles, and which ones should be considered as fundamental or "elementary," became the substance of research in **elementary particle physics**.

Today, the standard model for elementary particles views even more fundamental entities, *quarks* and *leptons*, as the basic constituents of matter. In order to understand the standard model, we need to begin with the ideas leading up to its formulation.[†]

Elementary particle physics might be said to have begun in 1935 when the Japanese physicist Hideki Yukawa (1907–1981) predicted the existence of a new particle that would in some way mediate the strong nuclear force. To understand Yukawa's idea, we first consider the electromagnetic force. When we first discussed electricity, we saw that the electric force acts over a distance, without contact. To better perceive how a force can act over a distance, we saw that Faraday introduced the idea of a **field**. The force that one charged particle exerts on a second can be said to be due to the electric field set up by the first. Similarly, the magnetic field can be said to carry the magnetic force. Later (Chapter 32), we saw that electromagnetic (EM) fields can travel through space as waves. Finally, in Chapter 38, we saw that electromagnetic radiation (light) can be considered as either a wave or as a collection of particles called *photons*. Because of this wave–particle duality, it is possible to imagine that the electromagnetic force between charged particles is due to

(1) the EM field set up by one charged particle and felt by the other, or

(2) an exchange of photons or γ particles between them.

It is (2) that we want to concentrate on here, and a crude analogy for how an exchange of particles could give rise to a force is suggested in Fig. 44–9. In part (a), two children start throwing heavy pillows at each other; each catch results in the child being pushed backward by the impulse. This is the equivalent of a repulsive force. On the other hand, if the two children exchange pillows by grabbing them out of the other person's hand, they will be pulled toward each other, as when an attractive force acts.

For the electromagnetic force, it is photons that are exchanged between two charged particles that give rise to the force between them. A simple diagram describing this photon exchange is shown in Fig. 44–10. Such a diagram, called a **Feynman diagram** (after its inventor, the American physicist, Richard Feynman (1918–1988)), is based on the theory of **quantum electrodynamics** (QED).

Figure 44–10 represents the simplest case, in which a single photon is exchanged. One of the charged particles emits the photon and recoils somewhat as a result; and the second particle absorbs the photon. In any collision or *interaction*, energy and momentum are transferred from one charged particle to the other, carried by the photon. Because the photon is absorbed by the second particle very shortly after it is emitted by the first, it is not observable, and is referred to as a *virtual* photon, in contrast to one that is free and can be detected by instruments. The photon is said to *mediate*, or *carry*, the electromagnetic force.

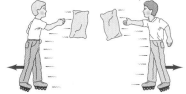

(a) Repulsive force (children throwing pillows)

(b) Attractive force (children grabbing pillows from each other's hands)

FIGURE 44–9 Forces equivalent to particle exchange. (a) Repulsive force (children throwing pillows at each other). (b) Attractive force (children grabbing pillows from each other's hands).

FIGURE 44–10 Feynman diagram showing a photon acting as the carrier of the electromagnetic force between two electrons. This is sort of an *x* vs. *t* graph, with *t* increasing upward. Starting at the bottom, two electrons approach each other (the distance between them decreases in time). As they get close, momentum and energy get transferred from one to the other, carried by a photon (or, perhaps, by more than one), and the two electrons bounce apart.

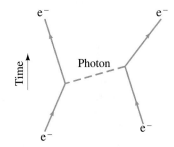

Particles that mediate or "carry" forces

[†] Just telling you how it is today would not be a scientific discussion but one of dogma—see footnote on p. 962.

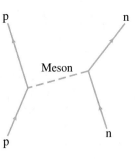

FIGURE 44–11 Meson exchange when a proton and neutron interact via the strong nuclear force.

Mass estimate of exchange particle

By analogy with photon exchange that mediates the electromagnetic force, Yukawa argued in this early theory that there ought to be a particle that mediates the strong nuclear force—the force that holds nucleons together in the nucleus.

Yukawa called this predicted particle a **meson** (meaning "of medium mass"), and Fig. 44–11 is a Feynman diagram of meson exchange. We can make a rough estimate of the mass of the meson as follows. Suppose the proton on the left in Fig. 44–11 is at rest. For it to emit a meson would require energy (to make the meson's mass) that would have to come from nowhere, which would violate conservation of energy. But the uncertainty principle allows nonconservation of energy by an amount ΔE if it occurs only for a time Δt given by $(\Delta E)(\Delta t) \approx h/2\pi$. We set ΔE equal to the energy needed to create the mass m of the meson: $\Delta E = mc^2$. Conservation of energy is violated only as long as the meson exists, which is the time Δt required for the meson to pass from one nucleon to the other. If we assume the meson travels at relativistic speed, close to the speed of light c, then Δt need be at most about $\Delta t = d/c$, where d is the maximum distance that can separate the interacting nucleons. Thus we can write

$$\Delta E \, \Delta t \approx \frac{h}{2\pi}$$

$$mc^2\left(\frac{d}{c}\right) \approx \frac{h}{2\pi}$$

or

Mass of exchange particle

$$mc^2 \approx \frac{hc}{2\pi d}. \qquad (44\text{–}3)$$

The range of the strong nuclear force (the maximum distance away it can be felt), is small—not much more than the size of a nucleon or small nucleus (see Eq. 42–1)—so let us take $d \approx 1.5 \times 10^{-15}$ m. Then from Eq. 44–3,

$$mc^2 \approx \frac{hc}{2\pi d} = \frac{(6.6 \times 10^{-34}\,\text{J}\cdot\text{s})(3.0 \times 10^{8}\,\text{m/s})}{(6.28)(1.5 \times 10^{-15}\,\text{m})}$$

$$\approx 2.1 \times 10^{-11}\,\text{J} = 130\,\text{MeV}.$$

The mass of the predicted meson is roughly $130\,\text{MeV}/c^2$ or about 250 times the electron mass of $0.51\,\text{MeV}/c^2$. [Note, incidentally, that since the electromagnetic force has infinite range ($d = \infty$), Eq. 44–3 tells us that the exchanged particle for the electromagnetic force, the photon, will have zero rest mass.]

Pion

The particle predicted by Yukawa was discovered in cosmic rays by C. F. Powell and G. Occhialini in 1947. It is called the "π" or pi meson, or simply the **pion**. It comes in three charge states: $+$, $-$, or 0. The π^+ and π^- have mass of $139.6\,\text{MeV}/c^2$ and the π^0 a mass of $135.0\,\text{MeV}/c^2$. All three interact strongly with matter. Reactions observed in the laboratory, using a particle accelerator, include

$$p + p \rightarrow p + p + \pi^0,$$
$$p + p \rightarrow p + n + \pi^+. \qquad (44\text{–}4)$$

The incident proton from the accelerator must have sufficient energy to produce the additional mass of the free pion.

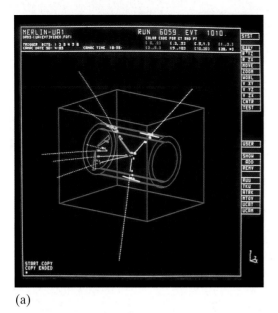
(a)

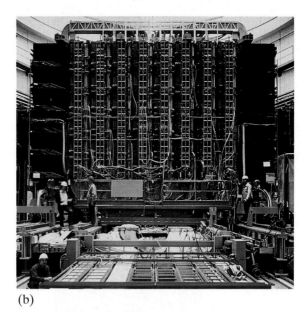
(b)

FIGURE 44–12 (a) Computer reconstruction of a Z-particle decay into an electron and a positron $\left(Z^0 \rightarrow e^+ + e^-\right)$ whose tracks are shown in white, which took place in the UA1 detector at CERN. (b) Photo of the UA1 detector at CERN as it was being built.

Yukawa's theory of pion exchange as carrier of the strong force is now out of date, and has been replaced by *quantum chromodynamics* in which the basic entities are *quarks*, and between them are exchanged *gluons* as the basic carriers of the strong force, as we shall discuss shortly. But the basic idea of the earlier theory, that forces can be understood as the exchange of particles, remains valid, as does Eq. 44–3 for estimating the mass of exchange particles.

There are four known types of force—or interaction—in nature. What about the other two: the weak nuclear force, and gravity? Theorists believe that these are also mediated by particles. The particles presumed to transmit the weak force are referred to as the W^+, W^-, and Z^0, and were only detected in 1983 (see Fig. 44–12). The quantum (or carrier) of the gravitational force is called the **graviton**, and if it exists it has not yet been observed. A comparison of the four forces is given in Table 44–1, where they are listed according to their (approximate) relative strengths. Notice that although gravity may be the most obvious force in daily life (because of the huge mass of the Earth), on a nuclear scale, it is much the weakest of the four forces and its effect at the nuclear or atomic level can nearly always be ignored.

Graviton

TABLE 44–1   The Four Forces in Nature

Type	Relative Strength (approx., for 2 protons in nucleus)	Field Particle
Strong nuclear	1	Gluons[†] (mesons)
Electromagnetic	10^{-2}	Photon
Weak nuclear	10^{-6}	$W^{\pm}$ and Z^0
Gravitational	10^{-38}	Graviton (?)

[†]Until the 1970s, thought to be mesons, but now gluons (see Section 44–10).

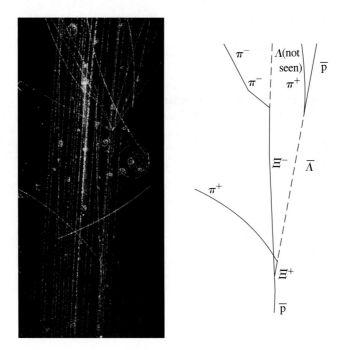

FIGURE 44–13 Liquid-hydrogen bubble-chamber photograph of an antiproton ($\bar{p}$) colliding with a proton, producing a Xi–anti-Xi pair ($\bar{p} + p \rightarrow \Xi^- + \Xi^+$) that subsequently decay into other particles. The drawing indicates the assignment of particles to each track, which is based on how or if that particle decays, and on mass values estimated from measurement of momentum (curvature of track in magnetic field) and energy (heaviness of track, for example). Neutral particle paths are shown by dashed lines since neutral particles produce no bubbles and hence no tracks.

44–4 | Particles and Antiparticles

The positron, as we have seen, is basically a positive electron. That is, many of its properties are the same as for the electron, such as mass, but it has the opposite charge. The positron is said to be the **antiparticle** to the electron. After the positron was discovered in 1932, it was predicted that other particles also ought to have antiparticles. In 1955 the antiparticle to the proton was found, the **antiproton** ($\bar{p}$), which carries a negative charge; see Fig. 44–13. (The bar over the p is used to indicate antiparticle.) Soon after, the antineutron ($\bar{n}$) was found. All other particles also have antiparticles. But a few particles, like the photon and the π^0, do not have distinct antiparticles—we say that they are their own antiparticles.

Antiparticles are produced in nuclear reactions when there is sufficient energy available, and they do not live very long in the presence of matter. For example, a positron is stable when by itself, but if it encounters an electron, the two annihilate each other. The energy of their vanished mass, plus any kinetic energy they possessed, is converted to the energy of γ rays or of other particles. Annihilation also occurs for all other particle–antiparticle pairs.

44–5 | Particle Interactions and Conservation Laws

One of the important uses of high-energy accelerators is to study the interactions of elementary particles with each other. As a means of ordering this subnuclear world, the conservation laws are indispensable. The laws of conservation of energy, of momentum, of angular momentum, and of electric charge are found to hold precisely in all particle interactions.

A study of particle interactions has revealed a number of new conservation laws which (just like the old ones) are ordering principles: they help to explain why some reactions occur and others do not. For example, the following reaction has never been found to occur:

$$p + n \nrightarrow p + p + \bar{p}$$

even though charge, energy, and so on, are conserved ($\bar{p}$ means an antiproton and $\nrightarrow$ means the reaction does not occur). To understand why such a reaction does not occur, physicists hypothesized a new conservation law, the conservation of **baryon number**. (Baryon number is a generalization of nucleon number, which we saw earlier is conserved in nuclear reactions and decays.) An important addition to this law is the proposal that whereas all nucleons have baryon number $B = +1$, all antinucleons (antiprotons, antineutrons) have $B = -1$. The reaction above does not conserve baryon number since on the left side we have $B = (+1) + (+1) = +2$, and on the right $B = (+1) + (+1) + (-1) = +1$. On the other hand, the following reaction does conserve B and *does* occur if the incoming proton has sufficient energy:

$$p + p \rightarrow p + p + \bar{p} + p,$$
$$B = +1 + 1 = +1 + 1 - 1 + 1.$$

As indicated, $B = +2$ on both sides of this equation. From these and other reactions, the conservation of baryon number has been established as a basic law of physics.

Also useful are the conservation laws for the three **lepton numbers**, associated with weak interactions including decays. In ordinary β decay, an electron or positron is emitted along with a neutrino or antineutrino. In a similar type of decay, a particle known as a **muon** can be emitted instead of an electron. A muon seems to be much like an electron, except its mass is 207 times as large ($106\,\text{MeV}/c^2$). The neutrino (ν_e) that accompanies an emitted electron is found to be different from the neutrino (ν_μ) that accompanies an emitted muon. Each of these neutrinos has an antiparticle: $\bar{\nu}_e$ and $\bar{\nu}_\mu$. In ordinary β decay we have, for example,

$$n \rightarrow p + e^- + \bar{\nu}_e$$

but never $n \rightarrow p + e^- + \bar{\nu}_\mu$ nor $n \rightarrow p + e^- + \bar{\nu}_e + \nu_e$. To explain why these do not occur, the concept of electron lepton number, L_e, was invented. If the electron (e^-) and the electron neutrino (ν_e) are assigned $L_e = +1$, and e^+ and $\bar{\nu}_e$ are assigned $L_e = -1$, whereas all other particles have $L_e = 0$, then all observed decays conserve L_e. For example, in $n \rightarrow p + e^- + \bar{\nu}_e$, $L_e = 0$ initially, and $L_e = 0 + (+1) + (-1) = 0$ after the decay. Decays that do not conserve L_e but would obey the other conservation laws are not observed to occur. Hence it is believed that L_e is conserved in all[†] interactions.

In a decay involving muons, such as

$$\pi^+ \rightarrow \mu^+ + \nu_\mu,$$

a second quantum number, muon lepton number (L_μ), is conserved. The μ^- and ν_μ are assigned $L_\mu = +1$, and μ^+ and $\bar{\nu}_\mu$ have $L_\mu = -1$, whereas other particles have $L_\mu = 0$. It is believed that L_μ is also conserved in all interactions or decays. Similar assignments can be made for a third lepton number, L_τ, associated with the more recently discovered τ lepton and its neutrino, ν_τ.

Keep in mind that antiparticles have not only opposite electric charge from their particles, but also opposite B, L_e, L_μ, and L_τ.

CONCEPTUAL EXAMPLE 44–4 **Lepton number in muon decay.** Which of the following decay schemes is possible for muon decay: (a) $\mu^- \rightarrow e^- + \bar{\nu}_e$; (b) $\mu^- \rightarrow e^- + \bar{\nu}_e + \nu_\mu$; (c) $\mu^- \rightarrow e^- + \nu_e$? All of these particles have $L_\tau = 0$.

RESPONSE A μ^- has $L_\mu = +1$ and $L_e = 0$. This is the initial state, and the final state (after decay) must also have $L_\mu = +1$, $L_e = 0$. In (a), the final state has $L_\mu = 0 + 0 = 0$, and $L_e = +1 - 1 = 0$; L_μ would not be conserved and indeed this decay is not observed to occur. The final state of (b) has $L_\mu = 0 + 0 + 1 = +1$ and $L_e = +1 - 1 + 0 = 0$, so both L_μ and L_e are conserved. This is in fact the most common decay mode of the μ^-. Finally, (c) does not occur because L_e ($= +2$ in the final state) is not conserved, nor is L_μ.

[†] Recent evidence hints that under certain circumstances, lepton number may not always be conserved.

Baryon number

Lepton numbers

Antiparticles have opposite Q, B, L

EXAMPLE 44–5 **Energy and momentum are conserved.** In addition to the "number" conservation laws which help explain the decay schemes of particles, we can also apply the laws of conservation of energy and momentum. The decay of a Σ^+ particle with a rest mass of $1189\,\text{MeV}/c^2$ (see Table 44–2 in the next Section) commonly yields a proton (rest mass of $938\,\text{MeV}/c^2$) and a neutral pion, π^0 (rest mass of $135\,\text{MeV}/c^2$). What are the kinetic energies of the decay products, assuming the Σ^+ parent particle was at rest?

SOLUTION The energy released through the decay can be calculated from the change in the rest energies, as we did for nuclear processes (Eq. 42–3 or 43–1):

$$Q = \left[m_{\Sigma^+} - (m_{\text{p}} + m_{\pi^0})\right]c^2 = \left[1189 - (938 + 135)\right]\text{MeV} = 116\,\text{MeV}.$$

This energy Q becomes the kinetic energy of the resulting decay particles, p and π^0:

$$Q = K_{\text{p}} + K_{\pi^0}$$

with each particle's kinetic energy related to its momentum by (Eqs. 37–11 and 13):

$$K_{\text{p}} = E_{\text{p}} - m_{\text{p}}c^2 = \sqrt{(p_{\text{p}}c)^2 + (m_{\text{p}}c^2)^2} - m_{\text{p}}c^2,$$

and similarly for the pion. From momentum conservation, the proton and pion have the same magnitude of momentum since the original particle was at rest: $p_{\text{p}} = p_{\pi} = p$. Then

$$Q = 116\,\text{MeV} = \left[\sqrt{(pc)^2 + (938\,\text{MeV})^2} - 938\,\text{MeV}\right]$$
$$+ \left[\sqrt{(pc)^2 + (135\,\text{MeV})^2} - 135\,\text{MeV}\right].$$

We solved this for pc, which gives $pc = 189\,\text{MeV}$. Substituting into the expression for the kinetic energy, first for the proton, then for the pion we obtain $K_{\text{p}} = 19\,\text{MeV}$ and $K_{\pi^0} = 97\,\text{MeV}$.

44–6 Particle Classification

In the decades following the discovery of the π meson in the late 1940s, a great many other subnuclear particles were discovered, now numbering in the hundreds. It is useful to arrange the particles in categories according to their properties. One way of doing this is according to their interactions, since not all particles interact by means of all four of the forces known in nature (though all interact via gravity). Table 44–2 lists some of the more common particles classified in this way along with many of their properties. The particles listed are those that are stable, and many that are unstable. At the top of the table are the **gauge bosons** (so-named[†] after the theory that describes them, "gauge theory"), which include the *photon*, and the W and Z particles, that mediate the electromagnetic and weak interactions, respectively.

Gauge bosons

Leptons

Next in Table 44–2 are the **leptons**, which are particles that do not interact via the strong force but do interact via the weak nuclear force (as well as the much weaker gravitational force); those that carry electric charge also interact via the electromagnetic force. The leptons include the electron, the muon, and the tau (or τ, discovered in 1976 and more than 3000 times heavier than the electron), and three types of neutrino: the electron neutrino (ν_e), the muon neutrino (ν_μ), and the tau neutrino (ν_τ). They each have antiparticles, as indicated in Table 44–2.

Hadrons

The third category of particle in Table 44–2 is the **hadron**. Hadrons are those particles that interact via the strong nuclear force. Hence they are said to be **strongly interacting particles**. They also interact via the other forces, but the strong

[†]Bosons are particles that are not governed by the Pauli exclusion principle (see Sections 40–4, 41–6, and 44–10).

TABLE 44–2 Particles (stable under strong decay)[†]

Category	Forces involved	Particle name	Symbol	Anti-particle	Spin	Rest Mass (MeV/c^2)	B	L_e	L_μ	L_τ	S	Lifetime (s)	Principal Decay Modes
Gauge bosons	em	Photon	γ	Self	1	0	0	0	0	0	0	Stable	
	w, em	W	W^+	W^-	1	80.41×10^3	0	0	0	0	0	3×10^{-25}	$e\nu_e, \mu\nu_\mu, \tau\nu_\tau$, hadrons
	w	Z	Z^0	Self	1	91.19×10^3	0	0	0	0	0	3×10^{-25}	$e^+e^-, \mu^+\mu^-, \tau^+\tau^-$, hadrons
Leptons	w, em[‡]	Electron	e^-	e^+	$\frac{1}{2}$	0.511	0	+1	0	0	0	Stable	
		Neutrino (e)	ν_e	$\bar{\nu}_e$	$\frac{1}{2}$	$0(<4 \times 10^{-6})$[‡]	0	+1	0	0	0	Stable	
		Muon	μ^-	μ^+	$\frac{1}{2}$	105.7	0	0	+1	0	0	2.20×10^{-6}	$e^-\bar{\nu}_e\nu_\mu$
		Neutrino (μ)	ν_μ	$\bar{\nu}_\mu$	$\frac{1}{2}$	$0(<0.17)$[‡]	0	0	+1	0	0	Stable	
		Tau	τ^-	τ^+	$\frac{1}{2}$	1777	0	0	0	+1	0	2.91×10^{-13}	$\mu^-\bar{\nu}_\mu\nu_\tau, e^-\bar{\nu}_e\nu_\tau$, hadrons $+\nu_\tau$
		Neutrino (τ)	ν_τ	$\bar{\nu}_\tau$	$\frac{1}{2}$	$0(<18)$[‡]	0	0	0	+1	0	Stable	
Hadrons (selected)													
Mesons	s, em, w	Pion	π^+	π^-	0	139.6	0	0	0	0	0	2.60×10^{-8}	$\mu^+\nu_\mu$
			π^0	Self	0	135.0	0	0	0	0	0	0.84×10^{-16}	2γ
		Kaon	K^+	K^-	0	493.7	0	0	0	0	+1	1.24×10^{-8}	$\mu^+\nu_\mu, \pi^+\pi^0$
			K_S^0	$\overline{K}_S^0$	0	497.7	0	0	0	0	+1	0.89×10^{-10}	$\pi^+\pi^-, 2\pi^0$
			K_L^0	$\overline{K}_L^0$	0	497.7	0	0	0	0	+1	5.17×10^{-8}	$\pi^\pm e^\mp \overset{(-)}{\nu}_e, \pi^\pm \mu^\mp \overset{(-)}{\nu}_\mu, 3\pi$
		Eta	η^0	Self	0	547.3	0	0	0	0	0	5×10^{-19}	$2\gamma, 3\pi^0, \pi^+\pi^-\pi^0$
		and others											
Baryons	s, em, w	Proton	p	$\bar{p}$	$\frac{1}{2}$	938.3	+1	0	0	0	0	Stable	
		Neutron	n	$\bar{n}$	$\frac{1}{2}$	939.6	+1	0	0	0	0	887	$pe^-\bar{\nu}_e$
		Lambda	Λ^0	$\overline{\Lambda}^0$	$\frac{1}{2}$	1115.7	+1	0	0	0	−1	2.63×10^{-10}	$p\pi^-, n\pi^0$
		Sigma	Σ^+	$\overline{\Sigma}^-$	$\frac{1}{2}$	1189.4	+1	0	0	0	−1	0.80×10^{-10}	$p\pi^0, n\pi^+$
			Σ^0	$\overline{\Sigma}^0$	$\frac{1}{2}$	1192.6	+1	0	0	0	−1	7.4×10^{-20}	$\Lambda^0\gamma$
			Σ^-	$\overline{\Sigma}^+$	$\frac{1}{2}$	1197.4	+1	0	0	0	−1	1.48×10^{-10}	$n\pi^-$
		Xi	Ξ^0	$\overline{\Xi}^0$	$\frac{1}{2}$	1314.9	+1	0	0	0	−2	2.90×10^{-10}	$\Lambda^0\pi^0$
			Ξ^-	$\overline{\Xi}^+$	$\frac{1}{2}$	1321.3	+1	0	0	0	−2	1.64×10^{-10}	$\Lambda^0\pi^-$
		Omega	Ω^-	Ω^+	$\frac{1}{2}$	1672.5	+1	0	0	0	−3	0.82×10^{-10}	$\Xi^0\pi^-, \Lambda^0 K^-, \Xi^-\pi^0$
		and others											

[†] See also Table 44–4 for particles with charm and bottomness.
[‡] Neutrinos partake only in the weak interaction. Experimental upper limits on neutrino masses are given in parentheses. Recent evidence (2000) suggests that at least one of the neutrinos may have mass, although probably quite small ($<1\,\text{eV}$).

force predominates at short distances. The hadrons include nucleons, pions, and a large number of other particles. They are divided into two subgroups: **baryons**, which are those particles that have baryon number +1 (or −1 in the case of their antiparticles); and **mesons**, which have baryon number = 0.

Baryons

Mesons

Notice that the baryons $\Lambda, \Sigma, \Xi,$ and Ω all decay to lighter-mass baryons, and eventually to a proton or neutron. All these processes conserve baryon number. Since there is no lighter particle than the proton with $B = +1$, if baryon number is strictly conserved, the proton itself cannot decay and is stable. (But see Section 44–11.)

44–7 | Particle Stability and Resonances

Many of the particles listed in Table 44–2 are unstable. The lifetime of an unstable particle depends on which force is most active in causing the decay. When we say the strong nuclear force is stronger than the electromagnetic, we mean that two particles will interact more strongly and more quickly if this force is acting. When a stronger force influences a decay, that decay occurs more quickly. Decays caused by the weak force typically have lifetimes of 10^{-13} s or longer. Particles that decay via the electromagnetic force have much shorter lifetimes, typically about 10^{-16} to 10^{-19} s. (Exceptions to this scheme are the W and Z particles, which decay via the

Lifetime depends on which force is acting

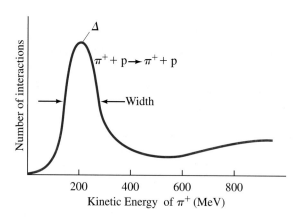

FIGURE 44–14 Number of π^+ particles scattered by a proton target as a function of the incident π^+ kinetic energy. The resonance shape represents the formation of a short-lived particle, the Δ, which has a charge in this case of $+2e\,(\Delta^{++})$.

weak interaction but have very short lifetimes due to their special nature as exchange particles.) The unstable particles listed in Table 44–2 decay either via the weak or the electromagnetic interaction. Decays that involve a γ (photon) are electromagnetic (such as $\pi^0 \rightarrow 2\gamma$). The other decays shown take place via the weak interaction, often accompanied by a neutrino which interacts only via the weak interaction: examples are $\pi^+ \rightarrow \mu^+ \nu_\mu$ and $\Sigma^- \rightarrow n\pi^-$.

A great many particles have been found that decay via the strong interaction, and these are not listed in Table 44–2. Such particles decay into other strongly interacting particles (say, n, p, π, but not involving γ, e, ν, and so on) and their lifetimes are very short, typically about 10^{-23} s. In fact their lifetimes are so short that they do not travel far enough to be detected before decaying. Their decay products can be detected, however, and it is from them that the existence of such short-lived particles is inferred. To see how this is done, we consider the first such particle discovered (by Fermi). Fermi used a beam of π^+ directed through a hydrogen target (protons) with varying amounts of energy. A graph of the number of interactions (π^+ scattered) versus the pion's kinetic energy is shown in Fig. 44–14. The large peak around 200 MeV was much higher than expected and certainly much higher than the number of interactions at neighboring energies. This led Fermi to conclude that the π^+ and proton combined momentarily to form a short-lived particle before coming apart again, or at least that they resonated together for a short time. Indeed, the large peak in Fig. 44–14 resembles a resonance curve (see Figs. 14–24, 14–27, and 31–9), and this new "particle"—now called the Δ—is referred to as a **resonance**. Hundreds of other resonances have been found in a similar way. Many resonances are regarded as excited states of other particles such as of the nucleon (proton or neutron).

The width of a resonance—in Fig. 44–14 the width of the Δ peak is on the order of 100 MeV—is an interesting application of the uncertainty principle. If a particle lives only 10^{-23} s, then its mass (i.e., its rest energy) will be uncertain by an amount $\Delta E \approx h/2\pi\Delta t \approx (6.6 \times 10^{-34}\,\text{J}\cdot\text{s})/(6)(10^{-23}\,\text{s}) \approx 10^{-11}\text{J} \approx 100\,\text{MeV}$, which is what is observed. Actually, the lifetimes of $\approx 10^{-23}$ s for such resonances are inferred by the reverse process: from the measured width being ≈ 100 MeV.

Very short-lived particles are inferred from their decay products

Resonance

44–8 | Strange Particles

In the early 1950s, certain of the newly found particles, namely, the K, Λ, and Σ, were found to behave rather strangely in two ways. First, they were always produced in pairs. For example, the reaction

$$\pi^- + p \rightarrow K^0 + \Lambda^0$$

occurred with high probability, but the reaction $\pi^- + p \rightarrow K^0 + n$ was never observed to occur. This seemed strange because the unobserved reaction would not have violated any known conservation law, and plenty of energy was available. The second feature of these **strange particles** (as they came to be called) was that, although they were clearly produced via the strong interaction (that is, at a high rate), they did not decay at a rate characteristic of the strong interaction even though they decayed into strongly interacting particles (for example, $K \rightarrow 2\pi$, $\Sigma^+ \rightarrow p + \pi^0$). Instead of lifetimes of 10^{-23} s as expected for strongly interacting particles, strange particles have lifetimes of 10^{-10} to 10^{-8} s, which are characteristic of the weak interaction.

To explain these observations, a new quantum number, **strangeness**, and a new conservation law, conservation of strangeness, were introduced. By assigning the strangeness numbers (S) indicated in Table 44–2, the production of strange particles in pairs was readily explained. Antiparticles were assigned opposite strangeness from their particles: one of each pair was assigned $S = +1$ and the other $S = -1$ (see Table 44–2). For example, in the reaction $\pi^- + p \rightarrow K^0 + \Lambda^0$, the initial state has strangeness $S = 0 + 0 = 0$, and the final state has $S = +1 - 1 = 0$, so strangeness is conserved. But for $\pi^- + p \rightarrow K^0 + n$, the initial state has $S = 0$ and the final state has $S = +1 + 0 = +1$, so strangeness would not be conserved; and this reaction is not observed.

Strangeness and its conservation

To explain the decay of strange particles, it is assumed that strangeness is conserved in the strong interaction but is *not* conserved in the weak interaction. Thus, although strange particles were forbidden by strangeness conservation to decay to nonstrange particles of lower mass via the strong interaction, they could undergo such decay by means of the weak interaction. This would occur much more slowly, of course, which accounts for their longer lifetimes of 10^{-10} to 10^{-8} s.

Strangeness conserved in strong interactions but not in weak

The conservation of strangeness was the first example of a "partially conserved" quantity. In this case, the quantity strangeness is conserved by strong interactions but not by weak.

CONCEPTUAL EXAMPLE 44–6 **Guess the missing particle.** Using the conservation laws for particle interactions, determine the other particle as a result of the reaction

$$\pi^- + p \rightarrow K^0 + ?,$$

in addition to $K^0 + \Lambda^0$.

SOLUTION We write equations for the conserved numbers in this reaction, with B, L_e, S, and Q as unknowns whose determination will reveal what the possible particle might be:

Baryon number:	$0 + 1 = 0 + B$
Lepton number:	$0 + 0 = 0 + L_e$
Charge:	$-1 + 1 = 0 + Q$
Strangeness:	$0 + 0 = 1 + S.$

The unknown product particle would have to have these characteristics:

$$B = +1 \qquad L_e = 0 \qquad Q = 0 \qquad S = -1.$$

In addition to Λ^0, a neutral sigma particle, Σ^0, is also consistent with these numbers.

44–9 | Quarks

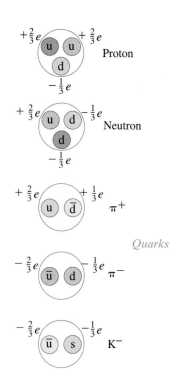

Quarks

FIGURE 44–15 Quark compositions for several particles.

We saw in our discussion of Table 44–2 that all particles, except the gauge bosons, fall into either of two categories: leptons and hadrons. The principal difference between these two groups is that the hadrons interact via the strong interaction, whereas the leptons do not. Another important difference that physicists had to deal with in the 1960s was that there were only four known leptons (e^-, μ^-, ν_e, ν_μ; the τ and ν_τ were not yet discovered), but there were well over a hundred hadrons.

The leptons are considered to be truly elementary particles since they do not seem to break down into smaller entities, do not show any internal structure, and have no measurable size. (Attempts to determine the size of leptons have put an upper limit of about 10^{-18} m.)

The hadrons, on the other hand, are more complex. Experiments indicate they do have an internal structure. And the fact that there are so many of them suggests that they can't all be elementary. To deal with this problem, M. Gell-Mann and G. Zweig in 1963 independently proposed that none of the hadrons so far observed, not even the proton and neutron, was elementary. Instead, they proposed that the hadrons are made up of combinations of three, more fundamental, pointlike entities called **quarks**.[†] Today, the quark theory is well-accepted, and quarks are considered the truly elementary particles, like leptons. The three quarks were labeled u, d, s, and given the names *up*, *down*, and *strange*. They were assumed to have fractional charge ($\frac{1}{3}$ or $\frac{2}{3}$ the charge on the electron—that is, less than the previously thought smallest charge). Other properties of quarks and antiquarks are indicated in Table 44–3. All hadrons known at the time could be constructed in theory from these three types of quark. Mesons would consist of a quark–antiquark pair. For example, a π^+ meson is considered a $u\bar{d}$ pair (note that

[†] Gell-Mann chose the word from a phrase in James Joyce's *Finnegans Wake*.

TABLE 44–3 Properties of Quarks and Antiquarks

Quarks

Name	Symbol	Spin	Charge	Baryon Number	Strangeness	Charm	Bottomness	Topness
Up	u	$\frac{1}{2}$	$+\frac{2}{3}e$	$\frac{1}{3}$	0	0	0	0
Down	d	$\frac{1}{2}$	$-\frac{1}{3}e$	$\frac{1}{3}$	0	0	0	0
Strange	s	$\frac{1}{2}$	$-\frac{1}{3}e$	$\frac{1}{3}$	−1	0	0	0
Charmed	c	$\frac{1}{2}$	$+\frac{2}{3}e$	$\frac{1}{3}$	0	+1	0	0
Bottom	b	$\frac{1}{2}$	$-\frac{1}{3}e$	$\frac{1}{3}$	0	0	−1	0
Top	t	$\frac{1}{2}$	$+\frac{2}{3}e$	$\frac{1}{3}$	0	0	0	+1

Antiquarks

Name	Symbol	Spin	Charge	Baryon Number	Strangeness	Charm	Bottomness	Topness
Up	$\bar{u}$	$\frac{1}{2}$	$-\frac{2}{3}e$	$-\frac{1}{3}$	0	0	0	0
Down	$\bar{d}$	$\frac{1}{2}$	$+\frac{1}{3}e$	$-\frac{1}{3}$	0	0	0	0
Strange	$\bar{s}$	$\frac{1}{2}$	$+\frac{1}{3}e$	$-\frac{1}{3}$	+1	0	0	0
Charmed	$\bar{c}$	$\frac{1}{2}$	$-\frac{2}{3}e$	$-\frac{1}{3}$	0	−1	0	0
Bottom	$\bar{b}$	$\frac{1}{2}$	$+\frac{1}{3}e$	$-\frac{1}{3}$	0	0	+1	0
Top	$\bar{t}$	$\frac{1}{2}$	$-\frac{2}{3}e$	$-\frac{1}{3}$	0	0	0	−1

TABLE 44–4 Partial List of Hadrons Associated with Charm and Bottomness ($L_e = L_\mu = L_\tau = 0$)

Category	Particle	Anti-particle	Spin	Rest Mass (MeV/c^2)	Baryon Number	Strange-ness	Charm	Bottom-ness	Lifetime (s)	Principal Decay Modes
Mesons	D^+	D^-	0	1869.4	0	0	+1	0	10.6×10^{-13}	K + others, e + others
	D^0	$\overline{D}^0$	0	1864.6	0	0	+1	0	4.2×10^{-13}	K + others, μ or e + others
	D_S^+	D_S^-	0	1969	0	+1	+1	0	4.7×10^{-13}	K + others
	J/ψ (3097)	Self	1	3096.9	0	0	0	0	0.8×10^{-20}	Hadrons, e^+e^-, $\mu^+\mu^-$
	Υ (9460)	Self	1	9460.4	0	0	0	0	1.3×10^{-20}	Hadrons, $\mu^+\mu^-$, e^+e^-, $\tau^+\tau^-$
	B^-	B^+	0	5279	0	0	0	−1	1.5×10^{-12}	D^0 + others
	B^0	$\overline{B}^0$	0	5279	0	0	0	−1	1.5×10^{-12}	D^0 + others
Baryons	Λ_c^+	Λ_c^-	$\frac{1}{2}$	2285	+1	0	+1	0	2.0×10^{-13}	Hadrons (e.g., Λ + others)
	Σ_c^{++}	Σ_c^{--}	$\frac{1}{2}$	2453	+1	0	+1	0	?	$\Lambda_c^+ \pi^+$
	Σ_c^+	Σ_c^-	$\frac{1}{2}$	2454	+1	0	+1	0	?	$\Lambda_c^+ \pi^0$
	Σ_c^0	$\overline{\Sigma}_c^0$	$\frac{1}{2}$	2452	+1	0	+1	0	?	$\Lambda_c^+ \pi^-$
	Λ_b^0	$\overline{\Lambda}_b^0$	$\frac{1}{2}$	5640	+1	0	0	−1	1.1×10^{-12}	$J/\psi\Lambda^0$, $pD^0\pi^-$, $\Lambda_c^+\pi^+\pi^-\pi^-$

for the $u\bar{d}$ pair, $Q = \frac{2}{3}e + \frac{1}{3}e = +1e$, $B = \frac{1}{3} - \frac{1}{3} = 0$, $S = 0 + 0 = 0$, as they must for a π^+); and a $K^+ = u\bar{s}$, with $Q = +1$, $B = 0$, $S = +1$. Baryons, on the other hand, would consist of three quarks. For example, a neutron is n = ddu, whereas an antiproton is $\bar{p} = \bar{u}\bar{u}\bar{d}$. See Fig. 44–15.

Soon after the quark theory was proposed, physicists began looking for these fractionally charged particles, but direct detection has not been successful. Indeed, current models suggest that quarks may be so tightly bound together that they may not ever exist singly in the free state.

In 1964 several physicists proposed that there ought to be a fourth quark. Their argument was based on the expectation that there exists a deep symmetry in nature, including a connection between quarks and leptons. If there are four leptons (as was thought in the 1960s), then symmetry in nature would suggest there should also be four quarks. The fourth quark was said to be *charmed*. Its charge would be $+\frac{2}{3}e$ and it would have another property to distinguish it from the other three quarks. This new property, or quantum number, was called **charm** (see Table 44–3). Charm was assumed to be like strangeness: it would be conserved in strong and electromagnetic interactions, but would not be conserved by the weak. The new charmed quark would have charm $C = +1$ and its antiquark $C = -1$. The first charmed particle, the J/ψ meson, was discovered in 1974.

Charm

Charmed quark

Also in the 1970s strong evidence appeared for the tau (τ) lepton, with a mass of 1777 MeV/c^2. This lepton, like the electron and muon, presumably has a neutrino associated with it. Thus, the family of leptons is at present believed to have six members. This would upset the balance between leptons and quarks, the presumed basic building blocks of matter, unless two additional quarks also exist. Indeed, theoretical physicists postulated the existence of a fifth and sixth quark, named **top** and **bottom**. The names apply also to the new properties (quantum numbers) that distinguish the new quarks from the old quarks (see Table 44–3), and which (like strangeness) are conserved in strong, but not weak, interactions. New mesons involving b quarks were soon detected (Table 44–4). Convincing evidence for the top quark came in 1995 (see photo at start of this chapter) after years of searching. Its extremely high mass of about 175 GeV/c^2 contributed to its elusiveness because of the huge energy required to produce it.

Two more leptons

t and b quarks

TABLE 44–5   The Elementary Particles† as Seen Today

	First generation	Second generation	Third generation
Quarks	u, d	s, c	b, t
Leptons	e, ν_e	μ, ν_μ	τ, ν_τ
Gauge bosons	γ(photon)	$W^\pm$, Z^0	gluons

†Note that the quarks and leptons are arranged into three generations each, and the gauge particles are arranged in groups for the forces they mediate.

The elementary particles

Today, the truly elementary particles are considered to be the six quarks, the six leptons, and the gauge bosons that carry the fundamental forces. See Table 44–5, where the quarks and leptons are arranged in three groups (generations).

CONCEPTUAL EXAMPLE 44–7 | Quark combinations. Find the baryon number, charge, and strangeness for the following quark combinations and identify the hadron particle that is made up of these quark combinations: (*a*) udd, (*b*) u$\bar{u}$, (*c*) uss, (*d*) sdd, and (*e*) b$\bar{u}$.

RESPONSE We use Table 44–3 to get the properties of the quarks, then Table 44–2 or 44–4 to find the particle that has these properties. (*a*) udd has

$$Q = +\tfrac{2}{3}e - \tfrac{1}{3}e - \tfrac{1}{3}e = 0,$$
$$B = \tfrac{1}{3} + \tfrac{1}{3} + \tfrac{1}{3} = 1,$$
$$S = 0 + 0 + 0 = 0,$$

as well as $C = 0$, bottomness = 0, topness = 0. The only baryon ($B = +1$) that has $Q = 0$, $S = 0$, etc., is the neutron (Table 44–2).
(*b*) u$\bar{u}$ has $Q = \tfrac{2}{3}e - \tfrac{2}{3}e = 0$, $B = 0$, and all other quantum numbers = 0. Sounds like a π^0.
(*c*) uss has $Q = 0$, $B = +1$, $S = -2$, others = 0. This is a Ξ^0.
(*d*) sdd has $Q = -1$, $B = +1$, $S = -1$, so must be a Σ^-.
(*e*) b$\bar{u}$ has $Q = -1$, $B = 0$, $S = 0$, $C = 0$, bottomness = -1, topness = 0. This must be a B$^-$ meson (Table 44–4).

44–10 The "Standard Model": Quantum Chromodynamics (QCD) and the Electroweak Theory

Not long after the quark theory was proposed, it was suggested that quarks have another property (or quality) called **color**. The distinction between the six quarks (u, d, s, c, b, t) was referred to as **flavor**. According to theory, each of the flavors of quark can have three colors, usually designated red, green, and blue. (These are the three primary colors which, when added together in equal amounts, as on a TV screen, produce white.) Note that the names "color" and "flavor" have nothing to do with our senses, but are purely whimsical—as are other names, such as charm, in this new field. (We did, however, "color" the quarks in Fig. 44–15.) The antiquarks are colored antired, antigreen, and antiblue. Baryons are made up of three quarks, one of each color. Mesons consist of a quark–antiquark pair of a particular color and its anticolor. Thus both baryons and mesons are colorless.

Originally, the idea of quark color was proposed to preserve the Pauli exclusion principle (Section 40–4). Not all particles obey the exclusion principle. Those that do, such as electrons, protons, and neutrons, are called **fermions**. Those that don't are called **bosons**. These two categories are distinguished also in their spin (Section 40–2): bosons have integer spin (0, 1, 2, etc.) whereas fermions have half-integer spin, usually $\tfrac{1}{2}$, as for electrons and nucleons, but can also be $\tfrac{3}{2}$, $\tfrac{5}{2}$, etc. Matter is made up mainly of fermions, but the carriers of the forces (γ, W, Z, and gluons, as we'll see) are all bosons. Quarks are fermions (they have spin $\tfrac{1}{2}$) and

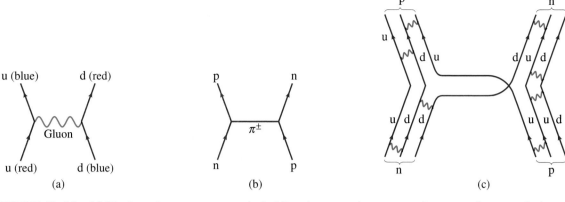

FIGURE 44–16 (a) The force between two quarks holding them together as part of a proton, for example, is carried by a gluon, which in this case involves a change in color. (b) Strong interaction n + p → n + p with the exchange of a charged π meson (+ or −, depending on whether it is considered moving to the left or to the right). (c) Quark representation of the same interaction n + p → n + p. The wavy lines between quarks represent gluon exchanges holding the hadrons together.

therefore should obey the exclusion principle. Yet for three particular baryons (uuu, ddd, and sss), all three quarks would have the same quantum numbers, and at least two of them have their spin in the same direction (since there are only two choices, spin up $[m_s = +\frac{1}{2}]$ or spin down $[m_s = -\frac{1}{2}]$). This would seem to violate the exclusion principle; but if quarks have an additional quantum number (color), which could be different for each quark, it would serve to distinguish them and the exclusion principle would hold. Although quark color, and the resulting threefold increase in the number of quarks, was thus originally an *ad hoc* idea, it also served to bring the theory into better agreement with experiment, such as predicting the correct lifetime of the π^0 meson. The idea of color soon became, in addition, a central feature of the theory as determining the force binding quarks together in a hadron. Each quark is assumed to carry a *color charge*, analogous to electric charge, and the strong force between quarks is often referred to as the **color force**. This new theory of the strong force is called **quantum chromodynamics** (*chroma* = *QCD* color in Greek), or **QCD**, to indicate that the force acts between color charges (and not between, say, electric charges). The strong force between two hadrons[†] is considered to be a force between the quarks that make them up, as suggested in Fig. 44–16. The particles that transmit the color force (analogous to photons for the EM force) are called **gluons** (a play on "glue"). They are included in *Gluons* Table 44–5. There are eight gluons, according to the theory, all massless, and six of them have color charge.[‡] Thus gluons have replaced mesons (Table 44–1) as the particles carrying the strong (color) force.

The color force has the interesting property that its strength increases with increasing distance (as does the force exerted by a coiled spring, Chapter 14); as two quarks approach each other very closely (equivalently, have high energy), the force between them becomes very small. This aspect is referred to as **asymptotic freedom**.

The weak force, as we have seen, is thought to be mediated by the W^+, W^-, and Z^0 particles. It acts between the "weak charges" that each particle has. Each elementary particle can thus have electric charge, weak charge, color charge, and gravitational mass, although one or more of these could be zero. For example, all leptons have color charge of zero, so they do not interact via the strong force.

[†] The strong force between hadrons appears feeble, however, in comparison to the force directly between quarks within hadrons.

[‡] Compare to the EM interaction, where the photon has no electric charge. Because gluons have color charge, they could attract each other and form composite particles (photons cannot). Such "glueballs" are being searched for.

To summarize, the standard model says that the truly elementary particles (Table 44–5) are the leptons, the quarks, and the gauge bosons (photon, W and Z, and the gluons). Some theories suggest there may be other bosons as well. The photon and the leptons are observed in experiments, and finally so too were the W^+, W^-, and Z^0. But so far only combinations of quarks (baryons and mesons) have been observed, and it seems likely that free quarks and gluons are unobservable.

Electroweak theory

One important aspect of new theoretical work is the attempt to find a unified basis for the different forces in nature. This was a long-held hope of Einstein, which he was never able to fulfill. A so-called **gauge theory** that unifies the weak and electromagnetic interactions was put forward in the 1960s by S. Weinberg, S. Glashow, and A. Salam. In this **electroweak theory**, the weak and electromagnetic forces are seen as two different manifestations of a single, more fundamental, *electroweak* interaction. The electroweak theory has had many successes, including the prediction of the $W^\pm$ particles as carriers of the weak force, with masses of $81 \pm 2 \, \text{GeV}/c^2$ in excellent agreement with the measured values of $80.41 \pm 0.10 \, \text{GeV}/c^2$ (and similar accuracy for the Z^0). The electroweak theory plus QCD for the strong interaction are often referred to today as the **Standard Model**.

Standard Model

Theoreticians have wondered why the W and Z have large masses rather than being massless like the photon. Electroweak theory suggests an explanation by means of a new **Higgs field** and its particle, the **Higgs boson**, which interact with the W and Z to "slow them down." In being forced to go slower than the speed of light, they must acquire mass.

44–11 | Grand Unified Theories

With the success of the unified electroweak theory, attempts have recently been made to incorporate it and QCD for the strong (color) force into a so-called **grand unified theory** (GUT). One type of such a grand unified theory of the electromagnetic, weak, and strong forces has been worked out in which there is only one class of particle—leptons and quarks belong to the same family and are able to change freely from one type to the other—and the three forces are different aspects of a single underlying force. The unity is predicted to occur, however, only on a scale of less than about 10^{-30} m. If two elementary particles (leptons or quarks) approach each other to within this **unification scale**, the apparently fundamental distinction between them would not exist at this level, and a quark could readily change to a lepton, or vice versa. Baryon and lepton numbers would not be conserved. The weak, electromagnetic, and strong (color) force would blend to a force of a single strength.

GUT

Unification of forces

How could a lepton become a quark, or vice versa? The theory predicts the existence of particles, called X bosons, that can be exchanged between a quark and a lepton allowing one to change into the other, somewhat like the charged pion exchanged between the p and n in Fig. 44–16b allows the proton on the right to become a neutron. The mass of the proposed X boson, consistent with the uncertainty principle as applied earlier in this chapter (see Eq. 44–3 with $d \approx 10^{-30}$ m), would be about $10^{14} \, \text{GeV}/c^2$, or 10^{14} times the proton mass. With such an incredibly large mass, there is little hope of seeing them in the laboratory. It is also this huge mass that would keep baryon and lepton numbers conserved in observed reactions, since the likelihood of producing such a massive particle, even as a virtual exchange particle, is extremely small at even the highest laboratory energies.

Symmetry breaking

What happens between the unification distance of 10^{-30} m and more normal (larger) distances is referred to as **symmetry breaking**. As an analogy, consider an atom in a crystal. Deep within the atom, there is much symmetry—in the innermost regions the electron cloud is spherically symmetric (Chapter 40). Farther out, this symmetry breaks down—the electron clouds are distributed preferentially along the lines (bonds) joining the atoms in the crystal. In a similar way, at 10^{-30} m the force between elementary particles is theorized to be a single force—it is symmetrical and does not single out one type of "charge" over another. But at larger distances, that symmetry is broken and we see three distinct forces. (In the "standard model" of electroweak interactions, Section 44–10, the symmetry breaking between the electromagnetic and the weak interactions occurs at about 10^{-18} m.)

CONCEPTUAL EXAMPLE 44–8 **Symmetry.** The table in Fig. 44–17 has four identical place settings. Four people sit down to eat. Describe the symmetry of this table and what happens to it when someone starts the meal.

RESPONSE The table has several kinds of symmetry. It is symmetric to rotations of 90°: that is, the table will look the same if everyone moved one chair to the left or to the right. It is also north–south symmetric and east–west symmetric, so that swaps across the table don't affect the way the table looks. It also doesn't matter whether any person picks up the fork to the left of the plate or the fork to the right. But once that first person picks up either fork, the choice is set for all the rest at the table as well. The symmetry has been *broken*. The underlying symmetry is still there—the water glasses could still be chosen either way—but some choice must get made and at that moment the symmetry of the diners is broken.

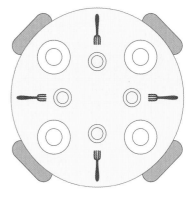

FIGURE 44–17 Symmetry around a table. Example 44–8.

Since unification occurs at such tiny distances and huge energies, the theory is difficult to test experimentally. But it is not completely impossible. One possibly testable prediction is the basis for the idea that the proton might decay (via, for example, $p \rightarrow \pi^0 + e^+$) and violate conservation of baryon number. This could happen if two quarks approached to within 10^{-31} m of each other. But it is very unlikely at normal temperature and energy, so the decay of a proton can only be an unlikely process. In the simplest form of GUT, the theoretical estimate of the proton lifetime for the decay mode $p \rightarrow \pi^0 + e^+$ is $\approx 10^{31}$ yr, and this has just come within the realm of testability. Proton decays have still not been seen and experiments put the lower limit on the proton lifetime for the above mode to be about 2.5×10^{32} yr, an order of magnitude greater than this prediction. This may seem a disappointment, but on the other hand, it presents a challenge. Indeed more complex GUTs are not affected by this result.

Proton decay?

EXAMPLE 44–9 ESTIMATE **Proton decay.** An experiment uses 3300 tons of water waiting to see a proton decay of the type $p \rightarrow \pi^0 + e^+$. If the experiment is run for four years without detecting a decay, estimate the lower limit on the proton half-life.

SOLUTION As with radioactive decay, the number of decays is proportional to the number of parent species (N), the time interval (Δt), and the decay constant (λ) which is related to the half-life $T_{\frac{1}{2}}$ by (see Eqs. 42–4 and 42–8):

$$\Delta N = -\lambda N \Delta t = -\frac{\ln 2}{T_{\frac{1}{2}}} N \Delta t.$$

Thus, dealing only with magnitudes,

$$T_{\frac{1}{2}} = \frac{N}{\Delta N} \Delta t \ln 2$$

which tells us that, for $\Delta N < 1$ over the four-year trial:

$$T_{\frac{1}{2}} > N(4\,\text{yr})(0.69)$$

where N is the number of protons in 3300 tons of water. To determine N, we note that each molecule of H_2O contains ($2 + 8 =$) 10 protons. So one mole of water (18 g) contains $10 \times 6 \times 10^{23}$ protons in 18 g of water, or about 3×10^{26} protons per kilogram. One ton is 10^3 kg, so the chamber contains $(3.3 \times 10^6\,\text{kg}) \cdot (3 \times 10^{26}\,\text{protons/kg}) \approx 1 \times 10^{33}$ protons. Then our very rough estimate for a lower limit on the proton half-life is $T_{\frac{1}{2}} > (10^{33})(4\,\text{yr})(0.7) \approx 3 \times 10^{33}$ yr.

An interesting prediction of unified theories relates to cosmology (see Chapter 45). It is thought that during the first 10^{-35} s after the theorized big bang that created the universe, the temperature was so extremely high that particles had energies corresponding to the unification scale. Baryon number would not have been conserved then, thus allowing an imbalance that might account for the observed predominance of matter ($B > 0$) over antimatter ($B < 0$) in the universe.

This last example is interesting, for it illustrates a deep connection between investigations at either end of the size scale: theories about the tiniest objects (elementary particles) have a strong bearing on the understanding of the universe as a whole. We will look at this more in the next chapter.

Even more ambitious than grand unified theories are attempts to also incorporate gravity, and thus unify all four forces in nature into a single theory. (Such theories are sometimes referred to misleadingly as **theories of everything**.) The only consistent theory so far that attempts to unify all four forces is called **string theory**, in which the elementary particles (Table 44–5) are imagined not as points but as one-dimensional strings about 10^{-35} m long.

A related idea is **supersymmetry**, which applied to strings is known as **superstring theory**. Supersymmetry predicts that interactions exist that would change fermions into bosons and vice versa, and that all known fermions have supersymmetric boson partners. Thus, for every quark there would be a *squark*, and for every lepton there would be a *slepton*. Likewise, for every known boson (photons and gluons, for example), there would be a supersymmetric fermion (*photinos* and *gluinos*). But why hasn't this "missing half" of the universe ever been detected? The best guess is that supersymmetric particles might be heavier than their conventional counterparts, perhaps too heavy to have been produced in today's accelerators. Until a supersymmetric particle is found, however, supersymmetry is just an elegant guess.

The world of elementary particles is opening new vistas. What happens in the near future is bound to be exciting.

Summary

Particle accelerators are used to accelerate charged particles, such as electrons and protons, to very high energy. High-energy particles have short wavelength and so can be used to probe the structure of matter at very small distances in great detail. High kinetic energy also allows the creation of new particles through collision (via $E = mc^2$).

Van De Graaff and linear accelerators use high voltage to accelerate particles along a line. Cyclotrons and synchrotrons use a magnetic field to keep the particles in a circular path and accelerate them at intervals by high voltage. **Colliding beams** allow higher interaction energy.

Particle **detectors** that can image particle tracks include photographic emulsions, cloud chambers, bubble chambers, and today wire drift chambers.

An **antiparticle** has the same mass as a particle but opposite charge. Certain other properties may also be opposite: for example, the antiproton has **baryon number** (nucleon number) opposite to that for the proton.

In all nuclear and particle reactions, the following conservation laws hold: momentum, angular momentum, mass-energy, electric charge, baryon number, and the three **lepton numbers**.

Certain particles have a property, called **strangeness**, which is conserved by the strong force but not by the weak force. The more recently noted properties, **charm**, **bottomness**, and **topness**, also are believed to be conserved by the strong force but not by the weak.

Just as the electromagnetic force can be said to be due to an exchange of photons, the strong nuclear force was first thought to be carried by *mesons* that have rest mass, or, according to more recent theory, by massless **gluons**. The W and Z particles carry the weak force. These fundamental force carriers (photon, W and Z, gluons) are called **gauge bosons**.

Other particles can be classified as either *leptons* or *hadrons*. **Leptons** participate in the weak and electromagnetic interactions. **Hadrons**, which today are considered to be made up of **quarks**, participate in the strong interaction as well. The hadrons can be classified as **mesons**, with baryon number zero, and **baryons**, with nonzero baryon number.

All particles, except for the photon, electron, neutrinos, and proton, decay with measurable half-lives varying from 10^{-25} s to 10^3 s. The half-life depends on which force is predominant in the decay. Weak decays usually have half-lives greater than about 10^{-13} s. Electromagnetic decays have half-lives on the order of 10^{-16} to 10^{-19} s. The shortest lived particles, called **resonances**, decay via the strong interaction and live typically for only about 10^{-23} s.

Today's standard model of elementary particles considers **quarks** as the basic building blocks of the hadrons. The six quarks are called **up**, **down**, **strange**, **charmed**, **bottom**, and **top** quarks. It is expected that there are the same number of quarks as leptons (six of each), and that quarks and leptons are the truly elementary particles along with the gauge bosons (γ, W, Z, gluons). Quarks are said to have **color**, and, according to **quantum chromodynamics** (QCD), the strong color force acts between their color charges and is transmitted by **gluons**. **Electroweak theory** views the weak and electromagnetic forces as two aspects of a single underlying interaction. QCD plus the electroweak theory are referred to as the **Standard Model**.

Grand unified theories of forces suggest that at very short distances $\left(10^{-30}\,\text{m}\right)$ and very high energy, the weak, electromagnetic, and strong forces appear as a single force, and the fundamental difference between quarks and leptons disappears.

Questions

1. Give a reaction between two nucleons, similar to Eq. 44–4, that could produce a π^-.
2. If a proton is moving at very high speed, so that its kinetic energy is much greater than its rest energy $(m_0 c^2)$, can it then decay via $\text{p} \to \text{n} + \pi^+$?
3. What would an "antiatom," made up of the antiparticles to the constituents of normal atoms, consist of? What might happen if *antimatter*, made of such antiatoms, came in contact with our normal world of matter?
4. What particle in a decay signals the electromagnetic interaction?
5. Does the presence of a neutrino among the decay products of a particle necessarily mean that the decay occurs via the weak interaction? Do all decays via the weak interaction produce a neutrino?
6. Why is it that a neutron decays via the weak interaction even though the neutron and one of its decay products (proton) are strongly interacting?
7. Which of the four interactions (strong, electromagnetic, weak, gravitational) does an electron take part in? A neutrino? A proton?

8. Check that charge and baryon number are conserved in each of the decays in Table 44–2.
9. Which of the particle decays in Table 44–2 occur via the electromagnetic interaction?
10. Which of the particle decays in Table 44–2 occur by the weak interaction?
11. By what interaction, and why, does $\Sigma_{\pm}^{\pm}$ decay to Λ^0? What about Σ^0 decaying to Λ^0?
12. The Δ baryon has spin $\frac{3}{2}$, baryon number 1, and charge $Q = +2$, $+1$, 0, or -1. Why is there no charge state $Q = -2$?
13. Which of the particle decays in Table 44–4 occur via the electromagnetic interaction?
14. Which of the particle decays in Table 44–4 occur by the weak interaction?
15. Quarks have spin $\frac{1}{2}$. How do you account for the fact that baryons have spin $\frac{1}{2}$ or $\frac{3}{2}$, and mesons have spin 0 or 1?
16. Suppose there were a kind of "neutrinolet" that was massless, had no color charge or electrical charge, and did not feel the weak force. Could you say that this particle even exists?

Problems

Sections 44–1 and 44–2

1. (I) What is the total energy of a proton whose kinetic energy is 6.35 GeV?
2. (I) Calculate the wavelength of 35-GeV electrons.
3. (I) What strength of magnetic field is used in a cyclotron in which protons make 2.8×10^7 revolutions per second?
4. (I) What is the time for one complete revolution for a very high-energy proton in the 1.0-km-radius Fermilab accelerator?
5. (I) If α particles are accelerated by the cyclotron of Example 44–2, what must be the frequency of the voltage applied to the dees?
6. (II) (a) If the cyclotron of Example 44–2 accelerated α particles, what maximum energy could they attain? What would their speed be? (b) Repeat for deuterons $\left(^2_1 \text{H}\right)$. (c) In each case, what frequency of voltage is required?

7. (II) Which is better for picking out details of the nucleus: 30-MeV alpha particles or 30-MeV protons? Compare each of their wavelengths with the size of a nucleon in a nucleus.
8. (II) The voltage across the dees of a cyclotron is 55 kV. How many revolutions do protons make to reach a kinetic energy of 25 MeV?
9. (II) What is the wavelength, and maximum resolving power attainable, using 900-GeV protons at Fermilab?
10. (II) A cyclotron with a radius of 1.0 m is to accelerate deuterons $\left(^2_1 \text{H}\right)$ to an energy of 10 MeV. (a) What is the required magnetic field? (b) What frequency is needed for the voltage between the dees? (c) If the potential difference between the dees averages 22 kV, how many revolutions will the particles make before exiting? (d) How much time does it take for one deuteron to go from start to exit, and (e) how far does it travel during this time?

11. (II) Protons are injected into the 1.0-km-radius Fermilab Tevatron with an energy of 8.0 GeV. If they are accelerated by 2.5 MV each revolution, how far do they travel and approximately how long does it take for them to reach 900 GeV?

12. (II) The Fermilab Tevatron takes about 20 seconds to bring the energies of the stored protons from 150 GeV to 900 GeV. The acceleration is done once per turn. Estimate the energy given to the protons on each turn. (You can assume that the speed of the protons is essentially c the whole time.)

13. (II) Show that the energy of a particle (charge e) in a synchrotron, in the relativistic limit ($v \approx c$), is given by E(in eV) $= Brc$, where B is magnetic field strength and r the radius of the orbit (SI units).

14. (II) What magnetic field intensity is needed at the 1.0-km-radius Fermilab synchrotron for 900-GeV protons?

Sections 44–3 to 44–6

15. (I) How much energy is released in the decay
$$\pi^+ \rightarrow \mu^+ + \nu_\mu ?$$

16. (I) About how much energy is released when a Λ^0 decays to n + π^0? (See Table 44–2.)

17. (I) How much energy is required to produce a neutron–antineutron pair?

18. (I) Estimate the range of the strong force if the mediating particle were the kaon instead of the pion.

19. (II) Two protons are heading toward each other with equal speeds. What minimum kinetic energy must each have if a π^0 meson is to be created in the process? (See Table 44–2.)

20. (II) Which of the following decays are possible? For those that are forbidden, explain which laws are violated.
(a) $\Xi^0 \rightarrow \Sigma^+ + \pi^-$ (b) $\Omega^- \rightarrow \Sigma^0 + \pi^- + \nu$
(c) $\Sigma^0 \rightarrow \Lambda^0 + \gamma + \gamma$.

21. (II) Estimate the range of the weak force using Eq. 44–3, given the masses of the W and Z particles as about 80 to 90 GeV/c^2.

22. (II) What are the wavelengths of the two photons produced when a proton and antiproton at rest annihilate?

23. (II) (a) Show, by conserving momentum and energy, that it is impossible for an isolated electron to radiate only a single photon. (b) With this result in mind, how can you defend the photon exchange diagram in Fig. 44–10?

24. (II) What would be the wavelengths of the two photons produced when an electron and a positron, each with 420 keV of kinetic energy, annihilate?

25. (II) In the rare decay $\pi^+ \rightarrow e^+ + \nu_e$, what is the kinetic energy of the positron? Assume the π^+ decays from rest.

26. (II) What minimum kinetic energy must a neutron and proton each have if they are traveling at the same speed toward each other, collide, and produce a K^+K^- pair in addition to themselves? (See Table 44–2.)

27. (II) Calculate the kinetic energy of each of the two products in the decay $\Xi^- \rightarrow \Lambda^0 + \pi^-$. Assume the Ξ^- decays from rest.

28. (III) Could a π^+ meson be produced if a 100-MeV proton struck a proton at rest? What minimum kinetic energy must the incoming proton have?

29. (III) Calculate the maximum kinetic energy of the electron in the decay $\mu^- \rightarrow e^- + \bar{\nu}_e + \nu_\mu$. [Hint: In what direction do the two neutrinos move relative to the electron in order to give the latter the maximum kinetic energy? Both energy and momentum are conserved; use relativistic formulas.]

Sections 44–7 to 44–11

30. (I) Use Fig. 44–14 to estimate the energy width and then the lifetime of the Δ resonance using the uncertainty principle.

31. (I) The measured width of the J/ψ meson is 88 keV. Estimate its lifetime.

32. (I) The measured width of the ψ(3685) meson is 277 keV. Estimate its lifetime.

33. (I) What is the energy width (or uncertainty) of (a) η^0, and (b) Σ^0?

34. (I) The B$^-$ meson is presumed to be a b$\bar{u}$ quark combination. (a) Show that this is consistent for all quantum numbers. (b) What are the quark combinations for B$^+$, B^0, $\bar{B}^0$?

35. (II) What are the quark combinations that can form (a) a neutron, (b) an antineutron, (c) a Λ^0, (d) a $\bar{\Sigma}^0$?

36. (II) What particles do the following quark combinations produce? (a) uud, (b) $\bar{u}\bar{u}\bar{s}$, (c) $\bar{u}$s, (d) d$\bar{u}$, (e) $\bar{c}$s.

37. (II) What is the quark combination needed to produce a D^0 meson ($Q = B = S = 0$, $C = +1$)?

38. (II) The D$_S^+$ meson has $S = C = +1$, $B = 0$. What quark combination would produce it?

39. (II) Draw possible Feynman diagrams using quarks (as in Fig. 44–16c) for the reactions (a) $\pi^- + p \rightarrow \pi^0 + n$, (b) $\bar{p} + p \rightarrow 2\pi^0$.

40. (II) Draw a possible quark Feynman diagram (see Fig. 44–16c) for reaction K$^-$ + p $\rightarrow$ K$^-$ + p.

General Problems

41. What is the total energy of a proton whose kinetic energy is 25 GeV? What is its wavelength?

42. Assume there are 5×10^{13} protons at 900 GeV stored in the 1.0-km-radius ring of the Tevatron. (a) How much current (amperes) is carried by this beam? (b) How fast would a 1500 kg car have to move to carry the same kinetic energy as this beam?

43. The 4.25-km-radius LEP tunnel will be reused to house the magnets for the Large Hadron Collider (LHC). If the design calls for proton beams of energy 7.0 TeV, what magnetic field will be required?

44. (a) How much energy is released when an electron and a positron annihilate each other? (b) How much energy is released when a proton and an antiproton annihilate each other?

45. Which of the following reactions are possible, and by what interaction could they occur? For those forbidden, explain why.
   (a) $\pi^- + p \rightarrow K^+ + \Sigma^-$
   (b) $\pi^+ + p \rightarrow K^+ + \Sigma^+$
   (c) $\pi^- + p \rightarrow \Lambda^0 + K^0 + \pi^0$
   (d) $\pi^+ + p \rightarrow \Sigma^0 + \pi^0$
   (e) $\pi^- + p \rightarrow p + e^- + \bar{\nu}_e$
   (f) $\pi^- + p \rightarrow K^0 + p + \pi^0$
   (g) $K^- + p \rightarrow \Lambda^0 + \pi^0$
   (h) $K^+ + n \rightarrow \Sigma^+ + \pi^0 + \gamma$
   (i) $K^+ \rightarrow \pi^0 + \pi^0 + \pi^+$
   (j) $\pi^+ \rightarrow e^+ + \nu_e$

46. For the decay $\Lambda^0 \rightarrow p + \pi^-$, calculate (a) the Q-value (energy released), and (b) the kinetic energy of the p and π^-, assuming the Λ^0 decays from rest. (Use relativistic formulas.)

47. Symmetry breaking occurs in the electroweak theory at about 10^{-18} m. Show that this corresponds to an energy that is on the order of the mass of the $W^\pm$.

48. The mass of a π^0 can be measured by observing the reaction $\pi^- + p \rightarrow \pi^0 + n$ at very low incident π^- kinetic energy (assume it is zero). The neutron is observed to be emitted with a kinetic energy of 0.60 MeV. Use conservation of energy and momentum to determine the π^0 mass.

49. Calculate the Q-value for each of the reactions, Eq. 44–4, for producing a pion.

50. Calculate the Q-value for the reaction $\pi^- + p \rightarrow \Lambda^0 + K^0$, when negative pions strike stationary protons. Estimate the minimum pion kinetic energy needed to produce this reaction.

51. How many fundamental fermions are there in a water molecule?

52. Determine the maximum kinetic energy of (a) the positron, and (b) the π^-, in the decay $K^0 \rightarrow \pi^- + e^+ + \nu_e$.

53. (a) Show that the so-called unification distance of 10^{-30} m in grand unified theory is equivalent to an energy of about 10^{14} GeV. Use either the uncertainty principle or de Broglie's wavelength formula, and explain how they apply. (b) Calculate what temperature this corresponds to.

54. For the reaction $p + p \rightarrow 3p + \bar{p}$, where one of the initial protons is at rest, use relativistic formulas to show that the threshold energy is $6m_p c^2$, equal to three times the Q-value of the reaction, where m_p is the proton mass.

55. The lifetimes listed in Table 44–2 are in terms of *proper time*, measured in a reference frame where the particle is at rest. If a tau-lepton is created with a kinetic energy of 450 MeV, how long would its track be as measured in the lab, on average, ignoring any collisions?

56. A particle at rest, with a rest-energy of $m_p c^2$, decays into two fragments with rest energies of $m_{D1} c^2$ and $m_{D2} c^2$. Show that the kinetic energy of fragment D1 is

$$K_{D1} = \frac{1}{2m_p c^2}\left[\left(m_p c^2 - m_{D1} c^2\right)^2 - \left(m_{D2} c^2\right)^2\right]$$

57. Use the quark model to describe the reaction

$$\bar{p} + n \rightarrow \pi^- + \pi^0$$

58. What fraction of the speed of light c is the speed of a 7.0 TeV proton?

Hubble Space Telescope photo of columns of gas and dust inside M16, the Eagle Nebula. The Eagle Nebula is thought to be a site of recent star formation. To discuss the nature of the universe as we understand it today, we first discuss stars and galaxies and how they form and evolve, including the role of nucleosynthesis. We briefly discuss Einstein's general theory of relativity, which deals with gravity and curvature of space. Finally we look at the evidence for the expansion of the universe, and the standard moded of the univese evolving from an initial Big Bang.

CHAPTER 45

Astrophysics and Cosmology

In the previous chapter, we studied the tiniest objects in the universe—the elementary particles. Now we leap to the largest—stars and galaxies. These two extreme realms, elementary particles and the cosmos, are among the most intriguing and exciting subjects in science. And, surprising though it may seem, these distant realms are related in a fundamental way, as already hinted in Chapter 44.

Use of the techniques and ideas of physics to study the heavens is often referred to as **astrophysics**. At the base of our present theoretical understanding of the universe is Einstein's *general theory of relativity* and its theory of gravitation—for in the large-scale structure of the universe, gravity is the dominant force. General relativity serves also as the foundation for modern **cosmology**, which is the study of the universe as a whole. Cosmology deals especially with the search for a theoretical framework to understand the observed universe, its origin, and its future. The questions posed by cosmology are complex and difficult; the possible answers are often unimaginable. They are questions like "Has the universe always existed, or did it have a beginning in time?" Either alternative is difficult to imagine: time going back indefinitely into the past, or an actual moment when the universe began (but, then, what was there before?). And what about the size of the universe? Is it infinite in size? It is hard to imagine infinity. Or is it finite in size? This is also hard to imagine, for if the universe is finite, it does not make sense to ask what is beyond it, because the universe is all there is.

Our survey of astrophysics and cosmology will be necessarily brief and qualitative, but we will nonetheless touch on the major ideas. We begin with a look at what can be seen beyond the Earth.

45-1 | Stars and Galaxies

According to the ancients, the stars, except for the few that seemed to move (the planets), were fixed on a sphere beyond the last planet. The universe was neatly self-contained, and we on Earth were at or near its center. But in the centuries following Galileo's first telescopic observations of the heavens in 1610, our view of the universe has changed dramatically. We no longer place ourselves at the center, and we view the universe as vastly larger. The distances involved are so great that we specify them in terms of the time it takes light to travel the given distance: for example, 1 light-second $= (3.0 \times 10^8 \text{ m/s})(1.0 \text{ s}) = 3.0 \times 10^8 \text{ m} = 3.0 \times 10^5 \text{ km}$; 1 light-minute $= 18 \times 10^6$ km; and 1 **light-year** (ly) is

$$\begin{aligned} 1 \text{ ly} &= (2.998 \times 10^8 \text{ m/s})(3.156 \times 10^7 \text{ s/yr}) \\ &= 9.46 \times 10^{15} \text{ m} \approx 10^{13} \text{ km}. \end{aligned}$$

Light-year

For specifying distances to the Sun and Moon, we usually use meters or kilometers, but we could specify them in terms of light. The Earth–Moon distance is 384,000 km, which is 1.28 light-seconds. The Earth–Sun distance is 1.50×10^{11} m, or 150,000,000 km; this is equal to 8.3 light-minutes. The most distant planet in the solar system, Pluto, is about 6×10^9 km from the Sun, or 6×10^{-4} ly. The nearest star to us, other than the Sun, is Proxima Centauri, about 4.3 ly away. (Note that the nearest star is 10,000 times farther from us than the farthest planet.)

On a clear moonless night, thousands of stars of varying degrees of brightness can be seen, as well as the elongated cloudy stripe known as the Milky Way (Fig. 45–1). It was Galileo who first observed, about 1610, that the Milky Way is comprised of countless individual stars. A century and a half later (about 1750), Thomas Wright suggested that the Milky Way was a flat disc of stars extending to great distances in a plane, which we call the **Galaxy** (Greek for "milky way").

FIGURE 45-1 A section of the Milky Way. The thin line is the trail of an artificial Earth satellite.

FIGURE 45–2 Our Galaxy, as it would appear from the outside: (a) "end view," in the plane of the disc; (b) "top view," looking down on the disc. (If only we could see it like this—from the outside!) (c) Infrared photograph of the inner reaches of the Milky Way, showing the central bulge of our Galaxy. This very wide angle photo extends over 180° of sky, and to be viewed properly it should be wrapped in a semicircle with your eyes at the center. The white dots are nearby stars.

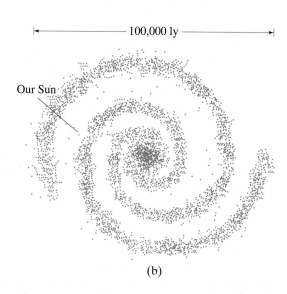

— 100,000 ly —

Our Sun

(b)

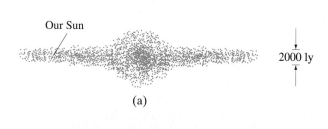

Our Sun

2000 ly

(a)

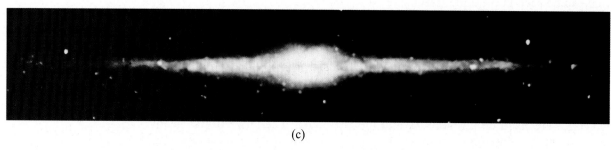

(c)

Our Galaxy has a diameter of almost 100,000 light-years and a thickness of very roughly 2000 light-years. It has a bulging central nucleus and spiral arms (Fig. 45–2). Our Sun, which seems to be just another star, is located more than halfway from the center to the edge, about 28,000 ly from the center. Our Galaxy contains about 10^{11} stars. The Sun orbits the galactic center approximately once every 200 million years, so its speed is about 250 km/s relative to the center of the Galaxy. The total mass of all the stars in our Galaxy is about 3×10^{41} kg.

EXAMPLE 45–1 ESTIMATE Our Galaxy's mass. Estimate the total mass of our Galaxy using the orbital data of the Sun (including our solar system) about the center of the Galaxy. Assume that most of the mass of the Galaxy can be approximated as a uniform sphere of mass (the central bulge, Fig. 45–2a).

SOLUTION Our Sun and solar system orbit the center of the Galaxy, according to the best measurements as mentioned above, with a speed of about $v = 250$ km/s at a distance from the Galaxy center of about $r = 28,000$ ly. We use Newton's second law, $F = ma$, with a being the centripetal acceleration, $a = v^2/r$, and F being the universal law of gravitation (Chapter 6)

$$F = ma$$
$$G \frac{Mm}{r^2} = m \frac{v^2}{r}$$

where M is the mass of the Galaxy and m is the mass of our Sun and solar system. Solving this, we find

$$M = \frac{rv^2}{G} \approx \frac{(28,000 \text{ ly})(10^{16} \text{ m/ly})(2.5 \times 10^5 \text{ m/s})^2}{6.67 \times 10^{-11} \text{ N} \cdot \text{m}^2/\text{kg}^2} \approx 3 \times 10^{41} \text{ kg}.$$

In terms of *numbers* of stars, if they are like our Sun ($m = 2.0 \times 10^{30}$ kg), there would be about $(3 \times 10^{41} \text{ kg})/(2 \times 10^{30} \text{ kg}) \approx 10^{11}$ or about 100 billion stars.

FIGURE 45–3 This globular star cluster is located in the constellation Hercules.

We can see by telescope, in addition to stars both within and outside the Milky Way, many faint cloudy patches in the sky which were all referred to once as "nebulae" (Latin for "clouds"). A few of these, such as those in the constellations Andromeda and Orion, can actually be discerned with the naked eye on a clear night. Some are **star clusters** (Fig. 45–3), groups of stars that are so numerous they appear to be a cloud. Others are glowing clouds of gas or dust (Fig. 45–4), and it is for these that we now mainly reserve the word **nebula**. Most fascinating are those that belong to a third category: they often have fairly regular elliptical shapes and seem to be a great distance beyond our Galaxy. Immanuel Kant (about 1755) seems to have been the first to suggest that these latter might be circular discs, but appear elliptical because we see them at an angle, and are faint because they are so distant. At first it was not universally accepted that these objects were **extragalactic**—that is, outside our Galaxy. The very large telescopes constructed in the twentieth century revealed that individual stars could be resolved within these extragalactic objects and that many contained spiral arms. Edwin Hubble (1889–1953), who did much of this observational work in the 1920s using the 2.5-m (100-inch) telescope[†] on Mt. Wilson near Los Angeles, California, was also able to demonstrate that these objects were indeed extragalactic because of their great distances. The distance to the Andromeda nebula (a galaxy), for example, is over 2 million light-years, a distance 20 times greater than the diameter of our Galaxy. Thus it was determined that these nebulae are **galaxies** similar to ours. Today, the largest telescopes can see about 10^{11} galaxies. See Fig. 45–5. (Note that it is usual to capitalize the word galaxy only when it refers to our own.)

Galaxies tend to be grouped in **galaxy clusters**, with anywhere from a few to many thousands of galaxies in each cluster. Furthermore, clusters themselves seem to be organized into even larger aggregates: clusters of clusters of galaxies, or **superclusters**. The galaxies nearest us are about 2 million light-years away. The farthest detectable galaxies are thousands of times farther away, on the order of 10^{10} ly. (See Table 45–1.)

[†]2.5 m (= 100 inches) refers to the diameter of the curved objective mirror. The bigger the mirror, the more light it collects and the less diffraction there is, so more and fainter stars can be seen. See Chapters 34 and 36.

FIGURE 45–4 This gaseous nebula, found in the constellation Carina, is about 9000 light-years from us.

TABLE 45–1
Heavenly Distances

Object	Approx. Distance from Earth (ly)
Moon	4×10^{-8}
Sun	1.6×10^{-5}
Farthest planet in our solar system (Pluto)	6×10^{-4}
Nearest star (Proxima Centauri)	4.3
Center of our Galaxy	3×10^4
Nearest galaxy	2×10^6
Farthest galaxies	10^{10}

FIGURE 45–5 Photographs of galaxies. (a) Spiral galaxy in the constellation Hydra. (b) Two galaxies: the larger and more dramatic one is known as the Whirlpool galaxy. (c) The same galaxies as in (b), but this is a false-color infrared image which shows the arms of the spiral as being more regular than in the visible light photo (b); the different colors correspond to different light intensities.

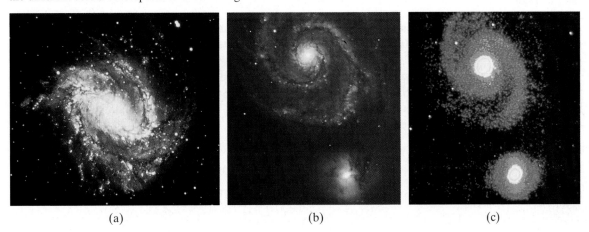

(a)　　　　　　(b)　　　　　　(c)

CONCEPTUAL EXAMPLE 45–2 | **Looking back in time.** Astronomers often think of their telescopes as time machines, looking back toward the origin of the universe. How far back do they look?

RESPONSE The distance in light-years measures exactly how long in years the light has been traveling to reach us, so Table 45–1 tells us also how far back in time we are looking. For example, if we saw Proxima Centauri explode into a supernova today, then the event would have really occurred 4.3 years ago. The most distant objects, galaxies 10^{10} ly away, emitted the light we see now 10^{10} years ago, so what we see was how they were then, close to the beginning of the universe.

Besides the usual stars, clusters of stars, galaxies, and clusters and superclusters of galaxies, the universe contains a number of other interesting objects. Among these are stars known as *red giants*, *white dwarfs*, *neutron stars*, *black holes* (at least theoretically), and exploding stars called *novae* and *supernovae*. In addition there are *quasars* ("quasistellar radio sources"), which, if we judge their distance correctly, are galaxies thousands of times brighter than ordinary galaxies. Furthermore, there is radiation that reaches the Earth but does not emanate from the bright pointlike objects we call stars: it is a background radiation that seems to arrive uniformly from all directions in the universe. We discuss all these phenomena in this chapter.

How astronomical distances are measured

We have talked about the vast distances of objects in the universe. But how do we measure these distances? One basic technique employs simple geometry to measure the **parallax** of a star. By parallax we mean the apparent motion of a star, against the background of more distant stars, due to the Earth's motion about the Sun. As shown in Fig. 45–6, the sighting angle of a star relative to the plane of Earth's orbit (angle θ) is measured at different times of the year. Since we know the distance d from Earth to Sun, we can reconstruct the right triangles shown in the figure and can determine[†] the distance D to the star.

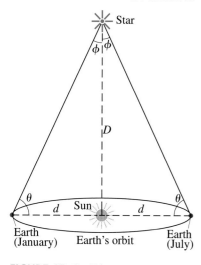

FIGURE 45–6 Distance to a star determined by parallax. (Not to scale.)

EXAMPLE 45–3 **ESTIMATE** **Distance to a star using parallax.** Estimate the distance D to a star if the angle θ in Fig. 45–6 is measured to be $89.99994°$.

SOLUTION The angle ϕ in Figure 45–6 is $0.00006°$, or about 1.0×10^{-6} radians. Since $\tan \phi = d/D$, then the distance D to the star is

$$D = \frac{d}{\tan \phi} = \frac{d}{\phi} = \frac{1.5 \times 10^8 \text{ km}}{1.0 \times 10^{-6} \text{ rad}} = 1.5 \times 10^{14} \text{ km},$$

or about 15 ly. (We can use $\tan \phi \approx \phi$ since ϕ is very small.)

The distances to stars are often specified in terms of parallax angle given in seconds of arc: 1 second ($1''$) is $\frac{1}{60}$ of a minute ($'$) of arc, which is $\frac{1}{60}$ of a degree, so $1'' = \frac{1}{3600}$ of a degree. The distance is then specified in parsecs (meaning *par*allax angle in *sec*onds of arc), where the **parsec** (pc) is defined as $1/\phi$ and ϕ is given in seconds. In the Example we just did, $\phi = (6 \times 10^{-5})°(3600) = 0.22''$ of arc, so we would say the star is at a distance of $1/0.22'' = 4.5$ pc. It is easy to show that the parsec is given by

Parsec

$$1 \text{ pc} = 3.26 \text{ ly}$$
$$= (3.26 \text{ ly})(9.46 \times 10^{15} \text{ m}) = 3.08 \times 10^{16} \text{ m}.$$

Parallax can be used to determine the distance to stars as far away as

[†]This is essentially the way the heights of mountains are determined, by "triangulation." See Example 1–7.

1144 CHAPTER 45 Astrophysics and Cosmology

100 light-years ($\approx$30 parsecs). Beyond that distance, parallax angles are too small to measure. For greater distances, more subtle techniques must be employed. We can, for example, compare the apparent brightnesses of galaxies, and—using the inverse square law (intensity drops off as the square of the distance)—estimate their relative distances. This method cannot be very precise since we cannot expect all galaxies to have the same intrinsic brightness. A better estimate compares the brightest stars in galaxies or the brightest galaxies in galaxy clusters. It is reasonable to assume, for example, that the brightest stars in all galaxies are similar and have about the same intrinsic luminosity. Consequently, their *apparent* brightness would be a measure of how far away they are.

Another important technique for estimating distance is via the "redshift" in the line spectra of elements, which is related to the expansion of the universe, as will be discussed in Section 45–4.

As we look farther and farther away, the measurement techniques are less and less reliable, so there is more and more uncertainty in the measurements of large distances.

45–2 | Stellar Evolution: The Birth and Death of Stars

The stars appear unchanging. Night after night the heavens reveal no significant variations. Indeed, on a human time scale, the vast majority of stars (except novae and supernovae) change very little. Although stars *seem* fixed in relation to each other, many move sufficiently for the motion to be detected. Indeed, the speeds of stars relative to neighboring stars can be hundreds of km/s, but at their great distance from us, this motion is detectable only by careful measurement. Furthermore, there is a great range of brightness among stars. The differences in brightness are due both to differences in the amount of light stars emit and to their distances from us.

A useful parameter for a star or galaxy is its **absolute luminosity**, L, by which we mean the total power radiated in watts. Also important is the **apparent brightness**, l, defined as the power crossing unit area perpendicular to the path of the light at the Earth. Given that energy is conserved, and ignoring any absorption in space, the total emitted power L at a distance d from the star will be spread over a sphere of surface area $4\pi d^2$. If we let d be the distance from the star to the Earth, then L must be equal to $4\pi d^2$ times l (power per unit area at Earth). That is,

Luminosity and brightness of stars

$$l = \frac{L}{4\pi d^2}. \qquad\qquad \textbf{(45–1)}$$

EXAMPLE 45–4 **Apparent brightness.** Suppose a particular star has absolute luminosity equal to that of the Sun but is 10 pc away from Earth. By what factor will it appear dimmer than the Sun?

SOLUTION The luminosity L is the same for both stars, so the apparent brightness depends only on their relative distances. We use the inverse square law as stated above in Eq. 45–1: the star appears dimmer by a factor

$$\frac{l_1}{l_2} = \frac{d_2^2}{d_1^2} = \frac{\left(1.5 \times 10^8\,\text{km}\right)^2}{\left(10\,\text{pc}\right)^2\left(3.26\,\text{ly/pc}\right)^2\left(10^{13}\,\text{km/ly}\right)^2} \approx 2 \times 10^{-13},$$

where d_1 and d_2 are the distances from Earth to the star and to the Sun, respectively.

Careful study of nearby stars has shown that for most stars, the absolute luminosity depends on the mass:[†] *the more massive the star, the greater its luminosity.* Another important parameter of a star is its surface temperature, which can be determined from the spectrum of electromagnetic frequencies it emits, just as for a blackbody (Section 38–1). As we saw in Chapter 38, the spectrum of hotter and hotter bodies shifts from predominantly lower frequencies (and longer wavelengths, such as red) to higher frequencies (and shorter wavelengths such as blue). Quantitatively, the relation is given by Wien's law (Eq. 38–1): the peak wavelength λ_P in the spectrum of light emitted by a blackbody (and stars are fairly good approximations to blackbodies) is inversely proportional to its kelvin temperature T; that is, $\lambda_P T = 2.90 \times 10^{-3}\,\text{m·K}$. The surface temperatures of stars typically range from about 3500 K (reddish) to perhaps 50,000 K (bluish).

An important astronomical discovery, made around 1900, was that for most stars, the color is related to the absolute luminosity and therefore to the mass. A useful way to present this relationship is by the so-called Hertzsprung–Russell (H–R) diagram. On the H–R diagram, the horizontal axis shows the temperature T whereas the vertical axis is the luminosity L, and each star is represented by a point on the diagram, Fig. 45–7. Most stars fall along the diagonal band termed the **main sequence**. Starting at the lower right we find the coolest stars, reddish in color; they are the least luminous and therefore of low mass. Farther up toward the left we find hotter and more luminous stars that are yellowish-white, like our Sun. Still farther up we find still more massive and more luminous stars, bluish in color. Stars that fall on this diagonal band are called *main-sequence stars*. There are also stars that fall outside the main sequence. Above and to the right we find extremely large stars, with high luminosities but with low (reddish) color temperature: these are called *red giants*. At the lower left, there are a few stars of low luminosity but with high temperature: these are the *white dwarfs*.

[†]The mass of a star can be determined by observing its gravitational effects. Most stars are part of a cluster, the simplest being a binary star in which two stars orbit around each other, allowing their masses to be determined using Newtonian mechanics.

FIGURE 45–7
Hertzsprung–Russell (H–R) diagram.

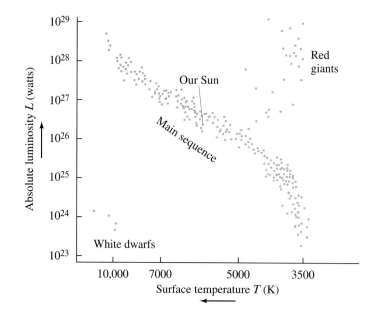

Before considering the different types of stars, let us first look at a couple of Examples that show how physical principles can be used to gain information about stars.

EXAMPLE 45–5 **Determining star temperatures and size.** Suppose that the distances to two nearby stars can be reasonably estimated and this data, together with their measured apparent brightnesses, suggests that the two stars have about the same absolute luminosity, L. The spectrum of one of the stars peaks at about 700 nm (so it is reddish). The spectrum of the other peaks at about 350 nm (bluish). Use Wien's law (Eq. 38–1) and the Stefan-Boltzmann equation (Section 19–10) to determine (a) the surface temperature of each star, and (b) how much larger one star is than the other.

SOLUTION (a) Wien's law states that $\lambda_P T = 2.90 \times 10^{-3}\,\text{m}\cdot\text{K}$. So the temperature of the reddish star is

$$T_r = \frac{2.90 \times 10^{-3}\,\text{m}\cdot\text{K}}{700 \times 10^{-9}\,\text{m}} = 4140\,\text{K}.$$

The temperature of the bluish star will be double this since its peak wavelength is half (350 nm vs. 700 nm); just to check:

$$T_b = \frac{2.90 \times 10^{-3}\,\text{m}\cdot\text{K}}{350 \times 10^{-9}\,\text{m}} = 8280\,\text{K}.$$

(b) The Stefan-Boltzmann equation, which we discussed in Chapter 19 (see Eq. 19–15), states that the power radiated *per unit area* of surface from a body is proportional to the fourth power of the kelvin temperature, T^4. Now the temperature of the bluish star is double that of the reddish star, so the bluish one must radiate $(2^4) = 16$ times as much energy per unit area. But we are given that they have the same luminosity (the same total power output), so the surface area of the blue star must be $\frac{1}{16}$ that of the red one. Since the surface area of a sphere is $4\pi r^2$, we conclude that the radius of the reddish star is $\sqrt{16} = 4$ times larger than the radius of the bluish star (or $4^3 = 64$ times the volume).

EXAMPLE 45–6 **ESTIMATE** **Distance to a star using H–R and color.** Suppose that detailed study of a certain star suggests that it most likely fits on the main sequence of an H–R diagram. Its measured apparent brightness is $l = 1.0 \times 10^{-12}\,\text{W/m}^2$, and the peak wavelength of its spectrum is $\lambda_P \approx 600\,\text{nm}$. Estimate its distance from us.

SOLUTION The star's temperature, from Wien's law (Eq. 38–1), is

$$T \approx \frac{2.90 \times 10^{-3}\,\text{m}\cdot\text{K}}{600 \times 10^{-9}\,\text{m}} \approx 4800\,\text{K}.$$

A star on the main sequence of an H–R diagram at this temperature has absolute luminosity of about $L \approx 1 \times 10^{26}\,\text{W}$, read off of Fig. 45–7. Then, from Eq. 45–1,

$$d = \sqrt{\frac{L}{4\pi l}} \approx \sqrt{\frac{1 \times 10^{26}\,\text{W}}{4(3.14)(1.0 \times 10^{-12}\,\text{W/m}^2)}} \approx 3 \times 10^{18}\,\text{m}.$$

Its distance from us in light-years and parsecs is

$$d = \frac{3 \times 10^{18}\,\text{m}}{10^{16}\,\text{m/ly}} \approx 300\,\text{ly} \approx \frac{300\,\text{ly}}{3.26\,\text{ly/pc}} \approx 90\,\text{pc}.$$

Why are there different types of stars, such as red giants and white dwarfs, as well as main-sequence stars? Were they all born this way, in the beginning? Or might each different type represent a different age in the life cycle of a star? Astronomers and astrophysicists today believe the latter is most likely the case. Note, however, that we cannot actually follow any but the tiniest part of the life cycle of any given star since they live for ages vastly greater than ours, on the order of millions or billions of years. Nonetheless, let us follow the process of **stellar evolution** from the birth to the death of a star, as astrophysicists have theoretically reconstructed it today.

Birth of a star

Contraction due to gravity

Stars are born, it is believed, when gaseous clouds (mostly hydrogen) contract due to the pull of gravity. A huge gas cloud might fragment into numerous contracting masses, each mass centered in an area where the density was only slightly greater than that at nearby points. Once such "globules" formed, gravity would cause each to contract in toward its center of mass. As the particles of such a *protostar* accelerate inward, their kinetic energy increases. When the kinetic energy is sufficiently high, the Coulomb repulsion that keeps the hydrogen nuclei apart can be overcome and nuclear fusion can take place. In a star like our Sun, the "burning" of hydrogen[†] occurs via the proton–proton cycle (Section 43–4, Eqs. 43–6), in which four protons fuse to form a $^4_2$He nucleus with the release of γ rays and neutrinos. These reactions require a temperature of about 10^7 K, corresponding to an average kinetic energy (kT) of about 1 keV. In more massive stars, the carbon cycle produces the same net effect: Four $^1_1$H produce a $^4_2$He—see Section 43–4. The fusion reactions take place primarily in the core of a star, where T is sufficiently high. (The surface temperature is, of course, much lower—on the order of a few thousand kelvins.) The tremendous release of energy in these fusion reactions produces a pressure sufficient to halt the gravitational contraction, and our protostar, now really a young *star*, stabilizes on the main sequence. Exactly where the star falls along the main sequence depends on its mass. The more massive the star, the farther up (and to the left) it falls on the diagram. To reach the main sequence requires perhaps 30 million years, if it is a star like our Sun, and (assuming our theory is right) it will remain there[‡] about 10 billion years (10^{10} yr). Although most stars are billions of years old, there is evidence that stars are actually being born at this moment.

Fusion begins when T (and $\bar{K}$) is large enough

Reaching the main sequence

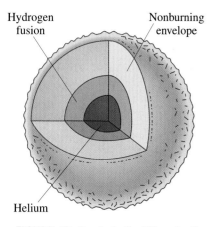

Hydrogen fusion

Nonburning envelope

Helium

FIGURE 45–8 A shell of "burning" hydrogen (fusing to become helium) surrounds the core where the newly formed helium gravitates.

Red giants

As hydrogen "burns"—that is, fuses to form helium—the helium that is formed is denser and tends to accumulate in the central core where it was formed. As the core of helium grows, hydrogen continues to "burn" in a shell around it, Fig. 45–8. When much of the hydrogen within the core has been consumed, the production of energy decreases and is no longer sufficient to prevent the huge gravitational forces from once again causing the core to contract and heat up. The hydrogen in the shell around the core then "burns" even more fiercely because of this rise in temperature, causing the outer envelope of the star to expand and to cool. The surface temperature, thus reduced, produces a spectrum of light that peaks at longer wavelength. By this time the star has left the main sequence. It has become redder, and as it has grown in size, it has become more luminous. So it will have moved to the right and upward on the H–R diagram, as shown in Fig. 45–9. As it moves upward, it enters the **red giant** stage. Thus, theory explains the origin of red giants as a natural step in a star's evolution. Our Sun, for example, has been on the main sequence for about $4\frac{1}{2}$ billion years. It will probably remain there another 4 or 5 billion years. (We can take comfort in that!) When our Sun leaves the main sequence, it is expected to grow in size (as a red giant) until it occupies all the volume out to approximately the present orbit of Earth.

[†] The word "burn" is put in quotation marks because these high-temperature fusion reactions occur via a *nuclear* process, and must not be confused with ordinary burning (of, say, paper, wood, or coal) in air, which is a chemical reaction at the *atomic* level (and at a much lower temperature).

[‡] More massive stars, since they are hotter and the Coulomb repulsion is more easily overcome, "burn" much more quickly, and so use up their fuel faster, resulting in shorter lives. A star 10 times more massive than our Sun, for example, will remain on the main sequence only for about 10^7 years. Stars less massive than our Sun live much longer than our Sun's 10^{10} yr.

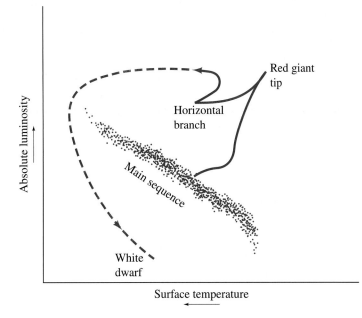

FIGURE 45–9 Evolution of a star like our Sun represented on an H–R diagram.

As a star's envelope expands, the core is shrinking and heating up. When the temperature reaches about 10^8 K, even helium nuclei, in spite of their greater charge and hence greater electrical repulsion, can then undergo fusion. The reactions are

$$\begin{aligned} {}^{4}_{2}\text{He} + {}^{4}_{2}\text{He} &\longrightarrow {}^{8}_{4}\text{Be} + \gamma &\quad (Q = -95\,\text{keV}) \\ {}^{4}_{2}\text{He} + {}^{8}_{4}\text{Be} &\longrightarrow {}^{12}_{6}\text{C} + \gamma. &\quad (Q = 7.4\,\text{MeV}) \end{aligned} \qquad (45\text{–}2)$$

Nucleosynthesis

The first reaction is (slightly) endothermic, and the ${}^{8}_{4}\text{Be}$ formed is unstable and can decay quickly $(10^{-16}\,\text{s})$ back to two alpha particles. So the second reaction must occur very quickly after the first, and this is aided by the particularly large cross section for the second reaction. The net effect is

$$3\,{}^{4}_{2}\text{He} \longrightarrow {}^{12}_{6}\text{C}. \qquad (Q = 7.3\,\text{MeV}).$$

The burning of helium causes a rapid and major change in the star, and the star moves rapidly to the "horizontal branch" on the H–R diagram (see Fig. 45–9). Further reactions are possible, creating elements of higher and higher Z, up to $Z = 10$ or 12. If the mass of the star is greater than about 0.7 solar masses, then still higher Z elements can be formed. The star can get even hotter and in the range $T = 2.5\text{–}5 \times 10^9$ K, nuclei as heavy as ${}^{56}_{26}\text{Fe}$ and ${}^{56}_{28}\text{Ni}$ can be made. But here the process of **nucleosynthesis**, the formation of heavy nuclei from lighter ones by fusion, ends. As we saw in Fig. 42–1, the average binding energy per nucleon begins to decrease for A greater than about 60. Further fusions would *require* energy, rather than release it. (Elements heavier than Ni are probably formed mainly by neutron capture, in supernova explosions, as we shall discuss shortly.)

Production of heavy elements

What happens next depends on the mass of the star. If the star has a residual mass less than about 1.4 solar masses[†] (known as the *Chandrasekhar limit*), no further fusion energy can be obtained and the star collapses under the action of gravity. As it shrinks, the star cools and typically follows the route shown in Fig. 45–9, according to theory, descending from the upper left downward, becoming a **white dwarf**. A white dwarf with a mass equal to that of the Sun would be about the size of the Earth. What determines its size is that it collapses to the point at which the electron clouds of the atoms start to overlap: at this point it collapses no further because, as the Pauli exclusion principle claims, no two electrons can be in the same quantum state. A white dwarf continues to lose internal energy, decreasing in temperature and becoming dimmer and dimmer until its light goes out. It has then become a *black dwarf*, a dark cold chunk of ash.

White dwarfs

[†] This cutoff refers to the mass of the star remaining *after* the red giant stage.

FIGURE 45-10 These glowing, expanding gas rings, a light-year in diameter, were observed by the Hubble Space Telescope a decade after the explosion of supernova SN1987a.

Neutron stars

Supernovae

FIGURE 45-11 SN1987a, the supernova that appeared in February 1987, is the very bright star on the right. The bright nebulous object is the Tarantula Nebula.

More massive stars (residual mass greater than 1.4 solar masses) are thought to follow a quite different scenario. A star with this great a mass can contract under gravity and heat up even further, reaching extremely high temperatures. The kinetic energy of the nuclei is then so high that fusion of elements heavier than iron is possible even though the reactions require energy input ($Q < 0$). Furthermore the high-energy collisions can also cause the breaking apart of iron and nickel nuclei into He nuclei, and eventually into protons and neutrons:

$$^{56}_{26}\text{Fe} \rightarrow 13\,^4_2\text{He} + 4\text{n}$$
$$^4_2\text{He} \rightarrow 2\text{p} + 2\text{n}.$$

These are energy-requiring (endoergic) reactions, but at such extremely high temperature and pressure there is plenty of energy available, enough even to force electrons and protons together to form neutrons in inverse β decay:

$$\text{e}^- + \text{p} \rightarrow \text{n} + \nu.$$

As the core contracts under the huge gravitational forces, the tremendous mass becomes essentially an enormous nucleus made up almost exclusively of neutrons. The size of the star is no longer limited by the exclusion principle applied to electrons, but rather applied to neutrons, and the star begins to contract rapidly toward forming an enormously dense **neutron star**. The contraction of the core would mean a great reduction in gravitational potential energy. Somehow this energy would have to be released. Indeed, it was suggested in the 1930s that the final core collapse to a neutron star may be accompanied by a catastrophic explosion whose tremendous energy could form virtually all elements of the periodic table and blow away the entire outer envelope of the star (Fig. 45–10), spreading its contents into interstellar space. Such explosions are believed to produce one type of observed **supernova** (there may be other mechanisms as well). Indeed, the presence of heavy elements on Earth and in our solar system suggests that our solar system may have formed from the debris from at least one supernova.

In a supernova explosion, a star's brightness is observed to suddenly increase billions of times in a period of just a few days and then fade away over the next few months. In February 1987 such a supernova occurred in an appendage to our Galaxy known as the Large Magellanic Cloud, only 170,000 ly away, and was visible to the naked eye in the southern hemisphere (Fig. 45–11). Supernovae are seen relatively often in distant galaxies, but this recent one, called SN1987a, is the first to have occurred close enough to be visible to the naked eye since the famous one of 1604 observed by Kepler and Galileo (and thus SN1987a is the first since the invention of the telescope). A supernova visible to the naked eye was observed by Chinese astronomers in A.D. 1054, and its remains are still visible in the sky (in the Crab Nebula), in the midst of which there is now a **pulsar**. Pulsars are astronomical objects that emit sharp pulses of radiation at regular intervals on the order of a second. They are now believed to be neutron stars which, because of conservation of angular momentum, increase greatly in rotational speed as their moment of inertia decreases during their contraction. The intense magnetic field of such a rapidly rotating star (perhaps as high as 10^6 to 10^{10} T) could trap and accelerate charged particles, which then give off radiation in a beam that rotates with the star. The discovery of pulsars in 1967 has lent support to the theory that neutron stars are the result of supernova explosions.

The core of a neutron star contracts to the point at which all neutrons are as close together as they are in a nucleus. That is, the density of a neutron star is on the order of 10^{14} times that of normal solids and liquids on Earth. A neutron star that has a mass 1.5 times that of our Sun would have a diameter of only about 10 km.

If the mass of a neutron star is less than about two or three solar masses, its subsequent evolution is thought to be similar to that of a white dwarf. If the mass is greater than this, the gravitational force may become so strong that the star contracts to an even smaller diameter and an even greater density, overcoming, in a sense, even the neutron exclusion principle. Gravity would then be so strong that even light emitted from it could not escape—it would be pulled back in by the force of gravity. In other words, the escape velocity from such a star is greater than c, the speed of light. Since no radiation could escape from such a star, we could not see it—it would be black. A body may pass by it and be deflected by its gravitational field, but if it came too close, it would be swallowed up, never to escape. This is a **black hole**. Black holes are predicted by theory to exist. Experimentally, evidence for their existence is strong but not everyone agrees it is sufficient to fully confirm the existence of black holes. One possibility is there may be a giant black hole at the center of our Galaxy—its mass is estimated at several million times that of the Sun—and it seems possible that many or all galaxies have black holes at their centers.

Black holes

45–3 General Relativity: Gravity and the Curvature of Space

We have seen that the force of gravity plays a dominant role in the processes that occur in stars and, in general, in the evolution of the universe as a whole. The reasons gravity, and not one of the other of the four forces in nature, plays the dominant role in the universe are (1) it is long-range and (2) it is always attractive. The strong and weak nuclear forces act over very short distances only, on the order of the size of a nucleus; hence they do not act over astronomical distances (although of course they act between nuclei and nucleons in stars to produce nuclear reactions). The electromagnetic force, like gravity, acts over great distances; but it can be either attractive or repulsive. And since the universe does not seem to contain large areas of net electric charge, a large net force does not occur. But gravity acts as an attractive force between *all* masses, and there are large accumulations in the universe of only the one "sign" of mass (not + and − as with electric charge).

However, the force of gravity as Newton described it in his law of universal gravitation shows discrepancies on a cosmological scale. Einstein, in his general theory of relativity, developed a theory of gravity that resolves these problems and forms the basis of cosmological dynamics.

In the *special theory of relativity* (Chapter 37), Einstein concluded that there is no way for an observer to determine whether a given frame of reference is at rest or is moving at constant velocity in a straight line. Thus the laws of physics must be the same in different inertial reference frames. But what about the more general case of motion where reference frames can be *accelerating*?

It is in the **general theory of relativity** that Einstein tackled the problem of accelerating reference frames and developed a theory of gravity. The mathematics of general relativity is very complex, so our discussion will be mainly qualitative.

We begin with Einstein's famous **principle of equivalence** (briefly discussed in Chapter 6), which states that

> **No experiment can be performed that could distinguish between a uniform gravitational field and an equivalent uniform acceleration.**

If observers sensed that they were accelerating (as in a vehicle speeding around a sharp curve), they could not prove by any experiment that in fact they weren't simply experiencing the pull of a gravitational field. Conversely, we might think we are being pulled by gravity when in fact we are undergoing an "inertial" acceleration having nothing to do with gravity.

As a thought experiment, consider a person in a freely falling elevator near the Earth's surface. If our observer held out a book and let go of it, what would happen? Gravity would pull it downward toward the Earth, but at the same rate $(g = 9.8 \text{ m/s}^2)$ at which the person and elevator were falling. So the book would hover right next to the person's hand (Fig. 45–12). The effect is exactly the same as if this reference frame were at rest and *no* forces were acting. On the other hand, suppose the elevator were far out in space where there is no gravitational field. If the person released the book, it would float, just as it does in Fig. 45–12.

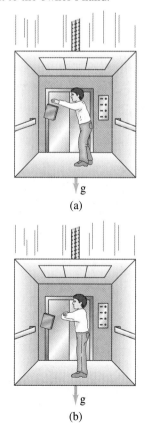

FIGURE 45–12 A freely falling elevator. The released book hovers next to the owner's hand.

(a)

(b)

If, instead, the elevator (out in space) were accelerating upward at an acceleration of $9.8 \, \text{m/s}^2$, the book as seen by our observer would fall to the floor with an acceleration of $9.8 \, \text{m/s}^2$, just as if it were falling because of gravity. According to the principle of equivalence, the observer could do no experiment to determine whether the book fell because the elevator was accelerating upward at $a = 9.8 \, \text{m/s}^2$ in the absence of gravity, or because a gravitational field with $g = 9.8 \, \text{m/s}^2$ was acting downward and he was at rest (say on the Earth). The two descriptions are equivalent.

Mass: inertial vs. gravitational

The principle of equivalence is related to the concept that there are two types of mass. Newton's second law, $F = ma$, uses **inertial mass**. We might say that inertial mass represents "resistance" to any type of force whatever. The second type of mass is **gravitational mass**. When one body attracts another by the gravitational force (Newton's law of universal gravitation, $F = Gm_1m_2/r^2$, Chapter 6), the strength of the force is proportional to the product of the gravitational masses of the two bodies. This is much like the electric force between two bodies which is proportional to the product of their electric charges. The electric charge of a body is not related to its inertial mass; so why should we expect that a body's gravitational mass (call it gravitational charge if you like) be related to its inertial mass? We have assumed they were the same. Why? Because no experiment—not even high-precision experiments—has been able to discern any measurable difference between inertial and gravitational mass. This is another way to state the equivalence principle: *gravitational mass is equivalent to inertial mass.*

Light bending

The principle of equivalence can be used to show that light ought to be deflected due to the gravitational force of a massive body. Consider another thought experiment, in which an elevator is in free space where no gravity acts. If a light beam enters a hole in the side of the elevator, the beam travels straight across the elevator and makes a spot on the opposite side if the elevator is at rest (Fig. 45–13a). If the elevator is accelerating upward as in Fig. 45–13b, the light beam still travels straight across in a reference frame at rest. In the upwardly accelerating elevator, however, the beam is observed to curve downward. Why? Because during the time the light travels from one side of the elevator to the other, the elevator is moving upward at ever-increasing speed. Now according to the equivalence principle, an upwardly accelerating reference frame is equivalent to a downward gravitational field. Hence, we can picture the curved light path in Fig. 45–13b as being the effect of a gravitational field. Thus we expect gravity to exert a force on a beam of light and to bend it out of a straight-line path.

That light is affected by gravity is an important prediction of Einstein's general theory of relativity. And it can be tested. The amount a light beam would be

FIGURE 45–13 (a) Light beam goes straight across an elevator not accelerating. (b) The light beam bends (exaggerated) in an elevator accelerating in an upward direction.

(a)

(b)

deflected from a straight-line path must be small even when passing a massive body. (For example, light near the Earth's surface after traveling 1 km is predicted to drop only about 10^{-10} m, which is equal to the diameter of a small atom and not detectable.) The most massive body near us is the Sun, and it was calculated that light from a distant star would be deflected by 1.75″ of arc (tiny but detectable) as it passed near the Sun (Fig. 45–14). However, such a measurement could be made only during a total eclipse of the Sun, so that the Sun's tremendous brightness would not overwhelm the starlight passing near its edge. An opportune eclipse occurred in 1919 and scientists journeyed to the South Atlantic to observe it. Their photos of stars around the Sun revealed shifts in accordance with Einstein's prediction. (See Fig. 45–15.)

A light beam must travel by the shortest, most direct, path between two points. If it didn't, some other object could travel between the two points in a shorter time and thus have a greater speed than the speed of light—a clear contradiction of the special theory of relativity. If a light beam can follow a curved path (as discussed above), then this curved path must be the shortest distance between the two points—which suggests that *space itself is curved* and that it is the gravitational field that causes the curvature. Indeed, the curvature of space—or rather, of four-dimensional space-time—is a basic aspect of general relativity.

What is meant by *curved space*? To understand, let us recall that our normal method of viewing the world is via Euclidean plane geometry. In Euclidean geometry, there are many axioms and theorems we take for granted, such as that the sum of the angles of any triangle is 180°. Other geometries, non-Euclidean which involve curved space, have also been imagined by mathematicians. Now it is hard enough to imagine three-dimensional curved space, much less curved four-dimensional space-time. So let us explain the idea of curved space by using two-dimensional surfaces.

Consider, for example, the two-dimensional surface of a sphere. It is clearly curved, Fig. 45–16, at least to us who view it from the outside—from our three-dimensional world. But how would hypothetical two-dimensional creatures determine whether their two-dimensional space were flat (a plane) or curved? One way would be to measure the sum of the angles of a triangle. If the surface is a plane, the sum of the angles is 180°, as we learn in plane geometry. But if the space is curved, and a sufficiently large triangle is constructed, the sum of the angles will *not* be 180°. To construct a triangle on a curved surface, say a sphere, we must use the equivalent of a straight line: that is, the shortest distance between two points, which is called a **geodesic**. On a sphere, a geodesic is an arc of a great circle (an arc contained in a plane passing through the center of the sphere) such as the Earth's equator and the Earth's longitude lines. Consider, for example, the large triangle of Fig. 45–16 whose sides are two "longitude" lines passing from the "north pole" to the equator, a part of which forms the third side. The two longitude lines make 90° angles with the equator (look at a world globe to see this more clearly). If they make, say, a 90° angle with each other at the north pole, the sum of these angles is $90° + 90° + 90° = 270°$. This is clearly *not* a Euclidean space. Note, however, that if the triangle is small in comparison to the radius of the sphere, the angles will add up to nearly 180°, and the triangle (and space) will seem flat.

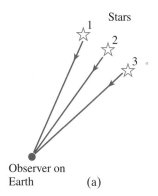

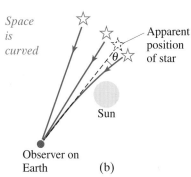

FIGURE 45–14 (a) Three stars in the sky. (b) If the light from one of these stars passes very near the Sun, whose gravity bends the rays, the star will appear higher than it actually is.

Geodesic

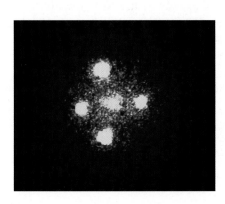

FIGURE 45–15 Photograph, taken by the Hubble Space Telescope, of the so-called "Einstein cross" (or Huchra's cross after its discoverer, John Huchra). It is thought to represent "gravitational lensing": The central spot is a relatively nearby galaxy, whereas the four other spots are thought to be images of a single quasar *behind* the galaxy, and the galaxy bends the light coming from the quasar to produce the four images. See also Fig. 45–14. (If the shape of the nearby galaxy were perfectly symmetric, we would expect the "image" of the distant quasar to be a circular ring or halo, instead of the four spots seen here.)

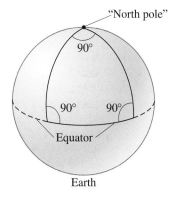

FIGURE 45–16 On a two-dimensional curved surface, the sum of the angles of a triangle may not be 180°.

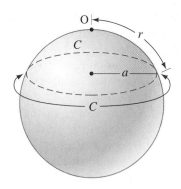

FIGURE 45–17 On a spherical surface, a circle of circumference C is drawn about point O as the center. The radius of the circle (not the sphere) is the distance r along the surface. (Note that in our three-dimensional view, we can tell that $2\pi a = C$; since $r > a$, then $2\pi r > C$.)

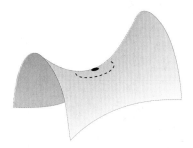

FIGURE 45–18 Example of a two-dimensional surface with negative curvature.

The universe: open or closed?

FIGURE 45–19 Rubber-sheet analogy for space-time curved by matter.

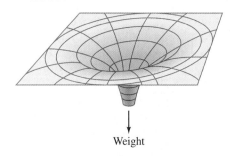

Weight

Another way to test the curvature of space is to measure the radius r and circumference C of a large circle. On a plane surface, $C = 2\pi r$. But on a two-dimensional spherical surface, C is *less* than $2\pi r$, as can be seen in Fig. 45–17. The proportionality constant between C and r is *less* than 2π. Such a surface is said to have *positive curvature*. On the saddlelike surface of Fig. 45–18, the circumference of a circle is greater than $2\pi r$, and the sum of the angles of a triangle is less than 180°. Such a surface is said to have a *negative curvature*.

Now, what about our universe? On a large scale (not just near a large mass), what is the overall curvature of the universe? Does it have positive curvature, negative curvature, or is it flat (zero curvature)? In the nineteenth century, Carl Friedrich Gauss (1777–1855) tried to determine whether our natural three-dimensional space deviated from Euclidean space by measuring the angles of a triangle formed by three mountain peaks using light rays as sides of the triangle. He was unable to detect any deviation from 180°, nor have experiments today detected any deviation.

Nonetheless, the question of the curvature of space is an important one in cosmology. If the universe has a positive curvature, then the universe would be *finite*, or *closed*. This does not mean that in such a universe the stars and galaxies would extend out to a certain boundary, beyond which there is empty space. Rather, galaxies would be spread throughout the space, and the space would fold back and "close on itself." There is no boundary or edge in such a universe. If a particle were to move in a straight line in a particular direction, it would eventually return to the starting point—albeit eons of time later. To ask "What is beyond such a closed universe?" is futile, since the space of the universe is all there is. If the curvature of space is zero or negative, on the other hand, the universe would be *open*. It would just go on and on and never fold back on itself. An open universe would be *infinite*. Whether the universe is open or closed depends, in part, on how much total mass there is in the universe, as we will discuss in Section 45–7. If the mass is great enough, it would bend space into a positively curved, closed, and finite space.

According to Einstein's theory, space-time is curved, especially near massive bodies. To visualize this, we might think of space as being like a thin rubber sheet; if a heavy weight is hung from it, it curves as shown in Fig. 45–19. The weight corresponds to a huge mass that causes space (space itself!) to curve. Thus, in Einstein's theory[†] we do not speak of the "force" of gravity acting on bodies. Instead we say that bodies and light rays move along geodesics (the equivalent of straight lines in plane geometry) in curved space-time. Thus, a body at rest or moving slowly near the great mass of Fig. 45–19 would follow a geodesic toward that body.

[†] Alexander Pope (1688–1744) wrote an epitaph for Newton:
"Nature, and Nature's laws lay hid in night:
God said, *Let Newton be!* and all was light."
Sir John Squire (1884–1958), perhaps uncomfortable with Einstein's profound thoughts, added:
"It did not last: the Devil howling '*Ho!*
Let Einstein be!' restored the status quo."

The extreme curvature of space-time shown in Fig. 45–19 could be produced by a **black hole**. A black hole, as we saw in the previous Section, is so dense that even light cannot escape from it. To become a black hole, a body of mass M must undergo **gravitational collapse**, contracting by gravitational self-attraction to within a radius called the **Schwarzschild radius**:

Black holes

$$R = \frac{2GM}{c^2}$$

where G is the gravitational constant and c the speed of light. If a body collapses to within this radius, it is predicted by general relativity to rapidly ($\approx 10^{-5}$ s) collapse to a point at $r = 0$, forming an infinitely dense **singularity**. This prediction is uncertain, however, since in this realm we need to combine quantum mechanics with gravity, a unification of theories not yet fully achieved.

The Schwarzschild radius also represents the **event horizon** of a black hole. By event horizon we mean the surface beyond which no signals can ever reach us, and thus inform us of events that happen. As a star collapses toward a black hole, the light it emits is pulled harder and harder by gravity, but we can still see it. Once the matter passes within the event horizon the emitted light cannot escape, but is pulled back in by gravity.[†]

Event horizon

All we can know about a black hole is its mass, its angular momentum (there could be rotating black holes), and its electric charge. No other information, no details of its structure or the kind of matter it was formed of, can be known.

How might we observe black holes? We cannot see them because no light can escape from them. They would be black objects against a black sky. But they do exert a gravitational force on nearby bodies. The suspected black hole at the center of our Galaxy was discovered by examining the motion of matter in its vicinity. Another technique is to examine stars which appear to be rotating as if they were members of a *binary system* (two stars rotating about their common center of mass), although the companion is invisible. If the unseen star is a black hole, it might be expected to pull off gaseous material from its visible companion (Fig. 45–20). As this matter approached the black hole, it would be highly accelerated and should emit X-rays of a characteristic type. One of the candidates for a black hole is in the binary-star system Cygnus X-1.

[†]According to quantum mechanics, matter and radiation could escape from a black hole by tunneling, but the rate at which this could happen would be expected to be extremely low.

FIGURE 45–20 Artist's conception of how a black hole, which is one star of a binary pair, might pull matter from the other (more normal) star.

45–4 | The Expanding Universe

We discussed in Section 45–2 how individual stars evolve from their birth to their death as white dwarfs, neutron stars, and black holes. But what about the universe as a whole: is it static, or does it evolve? The evolution of stars suggests that the universe as a whole evolves. One of the most important scientific ideas of this century proposes that distant galaxies are racing away from us, and that the farther they are from us, the faster they are moving away. How astronomers arrived at this astonishing idea and what it means for the past history of the universe as well as its future, will occupy us for the remainder of the book.

Hubble and the redshift

The idea that the universe is expanding was first put forth by Edwin Hubble in 1929. It was based on observations of the Doppler shift of light emitted by stars. In Chapter 16 we discussed how the frequency and wavelength of sound are altered if the source is moving toward or away from an observer. If the source moves toward us, the frequency is higher and the wavelength is shorter. If the source moves away from us, the frequency is lower and the wavelength is longer. The **Doppler effect** occurs also for light, but the shifted wavelength or frequency is given by a formula slightly different from that for sound because for light (according to special relativity), we can make no distinction between motion of the source and motion of the observer. (Recall that sound travels in a medium, such as air, but light does not; according to relativity, there is no ether.) According to special relativity, the formula is (see Section 37–12)

$$\lambda = \lambda_0 \sqrt{\frac{1 + v/c}{1 - v/c}}, \qquad \left[\begin{matrix} \text{source and observer moving} \\ \text{away from each other} \end{matrix}\right] \quad \textbf{(45–3)}$$

where λ_0 is the emitted wavelength as seen in a reference frame at rest with respect to the source, and λ is the wavelength measured in a frame moving with velocity v away from the source along the line of sight. Note that this relation depends only on the relative velocity v. (For relative motion toward each other, $v < 0$ in this formula.) Thus, when a source emits light of a particular wavelength and the source is moving away from us, the wavelength appears longer to us: the color of the light (if it is visible) is shifted toward the red end of the visible spectrum, an effect known as a **redshift**. If the source moves toward us, the color shifts toward the blue (shorter wavelength) end of the spectrum. The amount of shift depends on the velocity of the source (Eq. 45–3). For speeds not too close to the speed of light, it is easy to show (Problem 30) that the fractional change in wavelength is proportional to the speed of the source to or away from us (as was the case for sound):

Redshift

$$\frac{\Delta\lambda}{\lambda_0} = \frac{\lambda - \lambda_0}{\lambda_0} \approx \frac{v}{c}. \qquad\qquad [v \ll c] \quad \textbf{(45–4)}$$

In the spectra of stars and galaxies, lines are observed that correspond to lines in the known spectra of particular atoms (see Section 38–9). What Hubble found was that the lines seen in the spectra of galaxies were generally *redshifted*, and that the amount of shift seemed to be approximately proportional to the distance of the galaxy from us. That is, the velocity, v, of a galaxy moving away from us is proportional to its distance, d, from us:

Hubble's law

$$v = Hd. \qquad\qquad\qquad\qquad\qquad\qquad \textbf{(45–5)}$$

This is **Hubble's law**, one of the most important astronomical ideas of this century. The constant H is called the **Hubble parameter**. Hubble's law does not work well for nearby galaxies—in fact some are actually moving toward us ("blueshifted"); but this is believed to merely represent random motion of the galaxies. For more distant galaxies, the velocity of recession (Hubble's law) is much greater than that of random motion, and so is dominant.

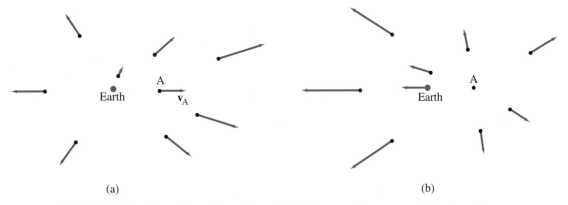

FIGURE 45–21 Expansion of the universe looks the same from any point in the universe.

The value of H is not known very precisely. It is now estimated to be about

$$H \approx 70 \, \text{km/s/Mpc}$$

Hubble's constant

(that is, 70 km/s per megaparsec of distance) with an uncertainty of about 20 percent. If we use light-years for distance, then $H \approx 20$ km/s per million light-years of distance.

What does it mean that distant galaxies are all moving away from us, and with ever greater speed the farther they are from us? It is as if there had been a great explosion at some distant time in the past. And at first sight we seem to be in the middle of it all. But we aren't, necessarily. The expansion appears the same from any other point in the universe. To understand why, see Fig. 45–21. In Fig. 45–21a we have the view from Earth (or from our Galaxy). The velocities of surrounding galaxies are indicated by arrows, pointing away from us, and greater for galaxies more distant from us. Now, what if we were on the galaxy labeled A in Fig. 45–21a? From Earth, galaxy A appears to be moving to the right at a velocity, call it $\mathbf{v}_A$, represented by the arrow pointing to the right. If we were *on* galaxy A, Earth would appear to be moving to the left at velocity $-\mathbf{v}_A$. To determine the velocities of other galaxies relative to A, we vectorially add the velocity vector, $-\mathbf{v}_A$, to all the velocity arrows shown in Fig. 45–21a. This yields Fig. 45–21b, where we see clearly that the universe is expanding away from galaxy A as well; and the velocities of galaxies receding from A are proportional to their distance from A.

Thus the expansion of the universe can be stated as follows: All galaxies are racing away from *each other* at an average rate of about 70 km/s per megaparsec of distance between them. The ramifications of this idea are profound, and we discuss them in a moment.

There is, however, a class of objects called **quasars**, or "quasistellar radio sources," that do not seem to conform to Hubble's law. Quasars are as bright as nearby stars but display very large redshifts. If quasars are normal participants in the general expansion of the universe according to Hubble's law, their large redshifts suggest they are very distant galaxies. If they are so far away, they must be incredibly bright—sometimes thousands of times brighter than normal galaxies. On the other hand, it has been suggested that some quasars, at least, are not of abnormal brightness because they are nearer than their redshifts suggest. In the first case we would have an unresolved brightness problem, in the second case an unresolved redshift problem. An interesting piece of evidence is that the population density of quasars seems to increase with distance from us. If they are at great distances from us—as their redshifts suggest—this would merely mean they were more common in the early universe than they are now. But if instead they are much closer to us—as their brightness suggests—it would seem that *we* are in a special place in the universe, the place where quasars are least populous. Most

Quasars

astronomers are unwilling to accept the latter possibility, for it would violate a basic assumption of uniformity, the so-called *cosmological principle*, which we discuss in the next paragraph. A recent suggestion that could resolve the quasar question is that these mysterious objects may be cores of galaxies that emit enormous amounts of energy when they draw in nearby material, for example when galaxies "collide." If the center of a galaxy is a black hole, a large amount of matter accelerated toward it would give off a prodigious amount of energy. It seems possible, then, that quasars are powered by black holes.

Cosmological principle

A basic assumption in cosmology has been that on a large scale, the universe looks the same to observers at different places at the same time. In other words, the universe is both isotropic (looks the same in all directions) and homogeneous (would look the same if we were located elsewhere, say in another galaxy). This assumption is called the **cosmological principle**. On a local scale, say in our solar system or within our Galaxy, it clearly does not apply (the sky looks different in different directions). But it has long been thought to be valid if we look on a large enough scale, so that the average population density of galaxies and clusters of galaxies ought to be the same in different areas of the sky. The expansion of the universe (Fig. 45–21) is consistent with the cosmological principle; and the uniformity of the cosmic microwave background radiation (discussed in the next Section) supports it. But matter (i.e., galaxies), even on the largest scale, seems not to be homogeneous but tends to clump, as already discussed. Although astronomers are reluctant to give up the cosmological principle, there is now some doubt about its validity. One possible resolution might be that over 90 percent of the universe may be nonluminous dark matter, which might be uniformly distributed. In any case, the cosmological principle has aided us in treating the universe as a single evolving entity and not as a random collection of material bodies. Another way of stating the cosmological principle is this: There is nothing special about the Earth on a cosmological scale; our large-scale observations are no different from those that might be made elsewhere in the universe.

The expansion of the universe, as described by Hubble's law, strongly suggests that galaxies must have been closer together in the past than they are now. Further, Hubble's law is consistent with all galaxies having been quite close together at some time in the past. This is, in fact, the basis of the *Big Bang* theory of the origin of the universe (discussed in the next Section) which pictures the beginning of the universe as a great explosion. Let us see what can then be said about the age of the universe.

Age of the universe

We can estimate the age of the universe using the Hubble parameter. With $H \approx 20 \, \text{km/s}$ per million light-years, then the time required for the galaxies to arrive at their present separations would be approximately (starting with $v = d/t$ and using Eq. 45–5):

$$t = \frac{d}{v} = \frac{d}{Hd} = \frac{1}{H} \approx \frac{(10^6 \, \text{ly})(10^{13} \, \text{km/ly})}{(20 \, \text{km/s})(3 \times 10^7 \, \text{s/y})} \approx 15 \times 10^9 \, \text{yr},$$

or roughly 15 billion years. The age of the universe calculated in this way is called the *characteristic expansion time* or "Hubble age." It is not very precise, in part because we don't know the value of H precisely.

There are two other independent checks on the age of the universe. The first is determination of the age of the Earth (and solar system) from radioactivity, primarily using uranium, which places the age of the solar system at about $4\frac{1}{2}$ billion years. Second, by using the theory of stellar evolution, the ages of stars have been estimated to be about 10 to 15 billion years. These independent and unrelated determinations are consistent with a Big Bang occurring, according to our best estimates today, 10 to 15 billion years ago. The lower value determined from radioactivity is consistent since we would expect the origin of the Earth (and the solidification of rocks) to have occurred somewhat after the origin of the universe as a whole.

Before discussing the Big Bang theory in more detail, we briefly discuss one of the alternatives to the Big Bang—the **steady-state model**—which assumes that the universe is infinitely old and on the average looks the same now as it always has.[†] According to the steady-state model, no large-scale changes have taken place in the universe as a whole, particularly no Big Bang. To maintain this view in the face of the recession of galaxies away from each other, mass–energy conservation must be violated. That is, matter is assumed to be created continuously, keeping the density of the universe constant. The rate of mass creation required is very small—about one nucleon per cubic meter every 10^9 years.

Steady-state model

The steady-state model provided the Big Bang model with healthy competition a half century ago. However, the discovery of the cosmic microwave background radiation (next Section), as well as the observed expansion of the universe, has made the Big Bang model the almost universally accepted model.

45–5 | The Big Bang and the Cosmic Microwave Background

The expansion of the universe seems to suggest, as we have seen, that the matter of the universe was once much closer together than it is now. This is the basis for the idea that the universe began about 15 billion years ago with a huge explosion known affectionately as the **Big Bang**.

The Big Bang

If there was a Big Bang, it must have occurred simultaneously at all points in the universe. If the universe is *finite*, the explosion would have taken place in a tiny volume approaching a point. However, this point of extremely dense matter is not to be thought of as a concentrated mass in the midst of a much larger space around it. Rather, the initial dense point *was* the universe—the entire universe. There wouldn't have been anything else. If, on the other hand, the universe is *infinite*, then the explosion would have occurred at *all* points in the universe since an infinite universe, even if smaller at an earlier time, would still have been infinite. In either case, when we say that the universe some time after the Big Bang was smaller than it is now, we mean that the average separation between galaxies was less. Thus, it is the *size of the universe itself* that has increased since the Big Bang.

What is the evidence supporting the Big Bang? First, the age of the universe as calculated from the Hubble expansion, from stellar evolution, and from radioactivity all point to a consistent time of origin for the universe, as we saw in the last Section. Another, and crucial, piece of evidence was the discovery in the 1960s of the **cosmic microwave background** radiation (or CMB), which came about as follows.

In 1964, Arno Penzias and Robert Wilson were experiencing difficulty with what they assumed to be background noise, or "static," in their radio telescope (a large antenna device for detecting radio waves from the heavens, Fig. 45–22).

FIGURE 45–22 Robert Wilson (left) and Arno Penzias, and behind them their "horn antenna."

[†] That the universe should be uniform in time as well as in space is essentially an extension of the cosmological principle and is sometimes referred to as the *perfect cosmological principle.*

The 2.7 K cosmic microwave background radiation

Eventually, they became convinced that it was real and that it was coming from outside our Galaxy. They made precise measurements at a wavelength $\lambda = 7.35\,\text{cm}$, which is in the microwave region of the electromagnetic spectrum. (See Fig. 32–12; this radiation is called "microwave" because the wavelength, though much greater than that of visible light, is somewhat smaller than wavelengths for ordinary radio waves, which are typically meters or hundred of meters.) The intensity of this radiation was found initially not to vary by day or night or time of year, nor to depend on direction. It came from all directions in the universe with equal intensity (within less than one part per thousand). It could only be concluded that this radiation came from beyond our Galaxy, from the universe as a whole. The remarkable uniformity of the cosmic microwave background radiation was in accordance with the cosmological principle. But theorists also felt that there needed to be some small inhomogeneities in the CMB that would have provided "seeds" around which galaxy formation could have started. Such small inhomogeneities, on the order of parts per million, were finally detected in 1992.

The intensity of this radiation as measured at $\lambda = 7.35\,\text{cm}$ corresponds to blackbody radiation (see Section 38–1) at a temperature of about 3 K, now measured more precisely to be 2.73 ± 0.01 K. When radiation at other wavelengths was measured, the intensities were found to fall on a blackbody curve as shown in Fig. 45–23, corresponding to a temperature of 2.73 K.

Importance of CMB: the Big Bang

The discovery of the CMB at a temperature of 2.7 K ranks as one of the two most significant cosmological discoveries of the twentieth century. (The other was Hubble's expanding universe.) It is highly significant because it provides strong evidence in support of the Big Bang, and it gives us some idea of conditions in the very early universe. In fact, in the late 1940s, George Gamow and his collaborators calculated that a Big Bang origin of the universe should have generated just such a microwave background radiation.

To understand why, let us look at what a Big Bang might have been like. There must have been a tremendous release of energy. The temperature must have been extremely high, so high that there could not have been any atoms in the very early stages of the universe. Instead, the universe must have consisted solely of radiation (photons) and elementary particles. The universe would have been opaque—the photons in a sense "trapped," since as soon as they were emitted they would have been scattered or absorbed, primarily by electrons. Indeed, the microwave background radiation is strong evidence that matter and radiation were once in equilibrium at a very high temperature. As the universe expanded, the energy would have spread out over an increasingly larger volume and the temperature would have dropped. Only when the temperature had reached about 3000 K some 300,000 years later, could nuclei and electrons have stayed together as atoms.

FIGURE 45–23 Spectrum of cosmic microwave background radiation, showing blackbody curve and experimental measurements including that of Penzias and Wilson. (Thanks to G. F. Smoot and D. Scott, 3/00. The vertical bars represent the experimental uncertainty in a measurement.)

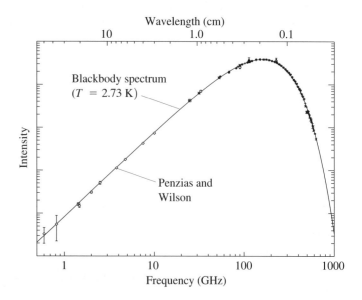

With the disappearance of free electrons, as they combined with nuclei to form atoms, the radiation would have been freed—"decoupled" from matter, we could say—to spread throughout the universe. As the universe expanded, so too the wavelengths of the radiation lengthened (you might think of standing waves, Section 15–9), thus redshifting to longer wavelengths that correspond to lower temperature (recall Wien's law, $\lambda_P T$ = constant, Section 38–1), until they would have reached the 2.7 K background radiation we observe today.

Although the total energy associated with the cosmic microwave background is much larger than that from other radiation sources (such as the light emitted by stars), it is small compared to the energy, and mass (remember $E = mc^2$), associated with matter. In fact today, radiation is believed to make up less than $\frac{1}{1000}$ of the energy of the universe: today the universe is *matter-dominated*. But it was not always so. The cosmic microwave background radiation strongly suggests that early in its history the universe was *radiation-dominated*.[†] But, as we shall see in the next Section, that period lasted less than $\frac{1}{10,000}$ of the history of the universe (thus far).

45–6 | The Standard Cosmological Model: the Early History of the Universe

It is now generally agreed that the evolution of the universe must have been determined in the first few moments of the Big Bang. In the last decade or two, a convincing theory of the origin and evolution of the universe has developed, now known as the **standard model**. Much of this theory is based on recent theoretical and experimental advances in elementary particle physics. Indeed, cosmology and elementary particle physics have cross-fertilized to a surprising extent.

Let us go back now to the earliest of times—as close as possible to the Big Bang—and follow a standard model scenario of events as the universe expanded and cooled after the Big Bang. Initially, we will be talking of extremely small time intervals, as well as extremely high temperatures, far beyond anything in the universe today. Figure 45–24 is a graphical representation of the events, and it may be helpful to consult it as we go along.

We begin at a time only a minuscule fraction of a second after the Big Bang, 10^{-43} s. Although this is an unimaginably short time, predictions as early as this can be made based on present theory, albeit somewhat speculatively. Earlier than this instant, we can say nothing since we do not yet have a theory of quantum effects on gravity which would be needed for the incredibly high densities and temperatures then.

The standard-model "scenario" of the history of the universe after the Big Bang

[†]If there had not been such intense radiation in the first few minutes of the universe, nuclear reactions might have produced a much larger percentage of heavy nuclei than we see. Instead, nearly 75 percent of visible matter is hydrogen, presumably because the intense radiation immediately blasted apart any heavy nuclei that formed into their constituent protons and neutrons.

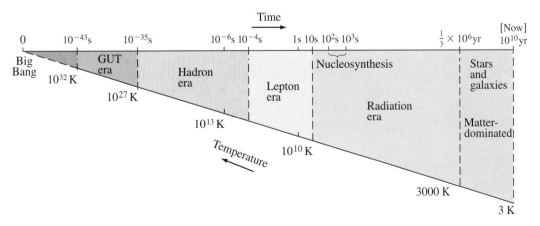

FIGURE 45–24 Graphical representation of the development of the universe after the Big Bang, according to the standard model.

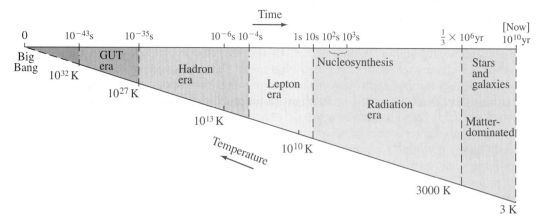

FIGURE 45-24 (repeated)
Graphical representation of
the development of the
universe after the Big Bang,
according to the standard
model.

All four forces unified

It is imagined, however, that prior to 10^{-43} s, the four forces in nature were uni-fied—there was only one force. The temperature would have been about 10^{32} K, corresponding to particles moving about every which way with an average kinetic energy K of 10^{19} GeV (see Eq. 18–4):

$$K \approx kT \approx \frac{(1.4 \times 10^{-23} \text{ J/K})(10^{32} \text{ K})}{1.6 \times 10^{-19} \text{ J/eV}} \approx 10^{28} \text{ eV} = 10^{19} \text{ GeV}.$$

*Symmetry broken
(gravity condensed out)*

(Note that the factor $\frac{3}{2}$ in Eq. 18–4 is usually ignored in such calculations.) At $t = 10^{-43}$ s, a kind of "phase transition" is believed to have occurred during which the gravitational force, in effect, "condensed out" as a separate force. This, and subsequent phase transitions, are somewhat analogous to the phase transitions water undergoes as it cools from a gas, condenses into a liquid, and with further cooling freezes into ice. The symmetry of the four forces was broken, but the strong, weak, and electromagnetic forces were still unified. At this point, the uni-verse entered the so-called **grand unified era** (after grand unified theory—see Chapter 44). There was no distinction between quarks and leptons; baryon and lepton numbers were not conserved. Very shortly thereafter, as the universe expanded considerably and the temperature had dropped to about 10^{27} K, there was another phase transition during which the strong force condensed out. This probably occurred about 10^{-35} s after the Big Bang. Now the universe was filled with a soup of leptons and quarks. The leptons included electrons, muons, taus, neutrinos, and all their antiparticles. The quarks were initially free (something we have not seen in our present universe), but soon they began to "condense" into more normal particles: nucleons and the other hadrons and their antiparticles.

*Quark confinement
(hadron era)*

With this **confinement of quarks**, the universe entered the **hadron era**.

We can think of this "soup" as a grand mixture of particles and antiparticles, as well as photons—all in roughly equal numbers—colliding with one another fre-quently and exchanging energy.

By the time the universe was only about a microsecond (10^{-6} s) old, it had cooled to about 10^{13} K, corresponding to an average kinetic energy of 1 GeV, and the vast majority of hadrons disappeared. To see why, let us focus on the most familiar hadrons: nucleons and their antiparticles. When the average kinetic energy of particles was somewhat higher than 1 GeV, protons, neutrons, and their antipar-ticles were continually being created out of the energies of collisions involving photons and other particles, such as

*Most hadrons
disappear*

$$\text{photons} \rightarrow p + \bar{p},$$
$$\rightarrow n + \bar{n}.$$

But just as quickly, particles and antiparticles would annihilate: for example

$$p + \bar{p} \rightarrow \text{photons or leptons}.$$

So the processes of creation and annihilation of nucleons were in equilibrium. The numbers of nucleons and antinucleons were high—roughly as many as there were electrons, positrons, or photons. But as the universe expanded and cooled, and the average kinetic energy of particles dropped below about 1 GeV, which is the minimum energy needed to create nucleons and antinucleons (about 940 MeV each) in a typical collision, the process of nucleon creation could not continue. The process of annihilation could continue, however, with antinucleons annihilating nucleons, until there were almost no nucleons left. But not quite zero. To explain our present world, which consists mainly of matter (nucleons and electrons) with very little antimatter in sight, we must suppose that earlier in the universe, perhaps around 10^{-35} s after the Big Bang, a slight excess of quarks over antiquarks was formed.[†] This would have resulted in a slight excess of nucleons over antinucleons. And it is these "leftover" nucleons that we are made of today. The excess of nucleons over antinucleons was about one part in 10^9. Earlier, during the hadron era, there should have been about as many nucleons as photons. After it ended, the "leftover" nucleons thus numbered only about one nucleon per 10^9 photons, and this ratio has persisted to this day. Protons, neutrons, and all other heavier particles were thus tremendously reduced in number by about 10^{-6} s after the Big Bang. The lightest hadrons, the pions, disappeared as the nucleons had; because they are the lightest mass hadrons (140 MeV), they were the last hadrons to go, around 10^{-4} s after the Big Bang. Lighter particles, including electrons, positrons, neutrinos, photons—in roughly equal numbers—dominated, and the universe entered the **lepton era**.

Why is there matter now?

Lepton era

By the time the first full second had passed (clearly the most eventful second in history!), the universe had cooled to about 10 billion degrees, 10^{10} K. The average kinetic energy was about 1 MeV. This was still sufficient energy to create electrons and positrons and balance their annihilation reactions, since their masses correspond to about 0.5 MeV. So there were about as many e^+ and e^- as there were photons. But within a few more seconds, the temperature had dropped sufficiently so that e^+ and e^- could no longer be formed. Annihilation ($e^+ + e^- \rightarrow$ photons) continued. And, like nucleons before them, electrons and positrons all but disappeared from the universe—except for a slight excess of electrons over positrons (later to join with nuclei to form atoms). Thus, about $t = 10$ s after the Big Bang, the universe entered the **radiation era**. Its major constituents were now photons and neutrinos. But the neutrinos, partaking only in the weak force, rarely interacted. So the universe, until then experiencing an energy balance between matter and radiation, became **radiation-dominated**: much more energy was contained in radiation than in matter, a situation that would last hundreds of thousands of years (Fig. 45–24).

Radiation-dominated universe

Meanwhile, during the next few minutes, crucial events were taking place. Beginning about 2 or 3 minutes after the Big Bang, nuclear fusion began to occur. The temperature had dropped to about 10^9 K, corresponding to an average kinetic energy $\overline{K} \approx 100$ keV, where nucleons could strike each other and be able to fuse (Section 43–4), but now cool enough that newly formed nuclei would not be immediately broken apart by subsequent collisions. Deuterium, helium, and very tiny amounts of lithium nuclei were probably made. But the universe was cooling too quickly, and larger nuclei were not made. After only a few minutes, probably not even a quarter of an hour after the Big Bang, the temperature dropped far enough that nucleosynthesis stopped, not to start again for millions of years (in stars).

Making He nuclei

[†]An alternative possibility is that there was perfect symmetry between quarks and antiquarks, matter and antimatter, but that the universe somehow separated into domains, some containing only matter, others only antimatter. If this were true, we would expect antiparticles from such distant domains to reach us, at least occasionally, in cosmic rays; but none has ever been detected.

Thus, after the first hour or so of the universe, matter consisted mainly of bare nuclei of hydrogen (about 75 percent) and helium (about 25 percent)[†] and electrons. But radiation (photons) continued to dominate.

Our story is almost complete. The next important event is presumed to have occurred about 300,000 years later. The universe had expanded to about $\frac{1}{1000}$ of its present size, and the temperature had cooled to about 3000 K. The average kinetic energy of nuclei, electrons, and photons was less than an electron volt. Since ionization energies of atoms are on the order of eV, then as the temperature dropped below this point, electrons could orbit the bare nuclei and remain there (without being ejected by collisions), thus forming atoms. With the birth of atoms, the photons—which had been continually scattering from the free electrons—now became much freer to spread nearly unhindered throughout the universe. We say that the photons become decoupled from matter. The total energy contained in radiation had been decreasing (lengthening in wavelength as the universe expanded), until at this point it was about equal to the total energy contained in matter. As the universe continued to expand, the radiation cooled further (to 2.7 K today, forming the cosmic microwave background radiation we detect from everywhere in the universe), and lost energy. But the mass of material particles did not decrease, so beginning at about this time, the energy of the universe became increasingly concentrated in matter rather than in radiation: the universe became **matter-dominated**, as it remains today.[‡]

Birth of stable atoms

Matter-dominated universe

After the birth of atoms, stars and galaxies could begin to form—probably by self-gravitation around mass concentrations (inhomogeneities). This began perhaps a million years after the Big Bang. The universe continued to evolve (see Section 45–2) until today, some 15 billion years later.

This scenario is by no means "proven" in any sense. Nor does it answer all questions. But it does provide a tentative picture, for the first time, of how the universe may have begun and evolved. It does have problems, however, some of which have been resolved by a modification first proposed in the early 1980s known as the **inflationary scenario**. According to the inflationary scenario, at the earliest stages of cosmological evolution, around 10^{-35} s after the Big Bang, the universe underwent a very rapid exponential expansion associated with the phase transition (symmetry breaking) that separated the strong force from the electroweak (as discussed earlier in this Section). After the brief inflationary period, the scenario settles back to the standard expansion already discussed. The appeal of the inflationary scenario is that it provides natural explanations for a number of problems not resolved by the standard model, such as why the universe is as close to being as flat as it seems to be, and why the cosmic microwave background radiation is so uniform.

Inflation

There are, however, questions we have not yet treated such as what is the future of the universe, which we will look at next, in the final Section of this book.

[†]This standard model prediction of a 25 percent primordial production of helium is fully in accord with what we observe today—the universe *does* contain about 25 percent He—and it is strong evidence in support of the standard Big Bang model. Furthermore, the theory says that 25 percent He abundance is fully consistent with there being three neutrino types, which is the number we observe so far. And it sets an upper limit of four to the maximum number of possible neutrino types. Actually, the fourth could be another type of low-mass particle, a *photino* or a *gravitino*, for example. Here we have a situation where cosmology actually makes a specific prediction about fundamental physics.

[‡]Although today matter contains more of the energy of the universe than does radiation, there are many more photons (perhaps 10^9 times more) than atoms, nuclei, and electrons. But each photon (at $T \approx 2.7$ K) has very little energy.

45–7 | The Future of the Universe?

According to the standard Big Bang model, the universe is evolving and changing. Individual stars are evolving and dying as white dwarfs, neutron stars, black holes. At the same time, the universe as a whole is expanding. One important question is whether the universe will continue to expand forever. This question is connected to the curvature of space-time (Section 45–3) and to whether the universe is open (and infinite) or closed (and finite). There are three possibilities as shown in Fig. 45–25. If the curvature of the universe is *negative*, the expansion of the universe will never stop, although the rate of expansion might be expected to decrease due to the gravitational attraction of its parts. Such a universe would be *open* and infinite. If the universe is *flat* (no curvature), it would still be open and infinite but its expansion would slowly approach a zero rate. Finally, if the universe has *positive* curvature, it would be *closed* and finite. The effect of gravity in this case would be strong enough so that the expansion would eventually stop and the universe would then begin to contract. All matter eventually would collapse back onto itself in a **big crunch**. If, in this last case, the maximum expansion of the universe corresponded to, say, an intergalactic separation twice what it is now, the maximum expansion would occur about 30 or 40 billion years from now. Then, as the universe began to contract, the big crunch would occur about 100 billion years after the Big Bang.

An open or closed universe?

A big crunch?

Whether we live in an open and continually expanding universe, or a closed one that eventually will contract, is connected also to the average mass density in the universe. If the average mass density is above a critical value known as the **critical density**, estimated to be about

$$\rho_c \approx 10^{-26}\,\text{kg/m}^3$$

Critical density of the universe

(i.e., a few nucleons/m^3 on average throughout the universe), then gravity will prevent expansion from continuing forever, and will eventually pull the universe back into a big crunch. To say it another way, if $\rho > \rho_c$, there would be sufficient mass that gravity would give space-time a positive curvature. If instead the actual density is equal to the critical density, $\rho = \rho_c$, the universe will be flat and open. If the actual density is less than the critical density, $\rho < \rho_c$, the universe will have negative curvature and be open, expanding forever.

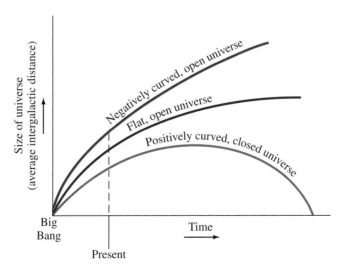

FIGURE 45–25 Three future possibilities for the universe.

EXAMPLE 45–7 ESTIMATE **Critical density of the universe.** Use energy conservation and the concept of escape velocity (Section 8–7) to estimate the critical density of the universe.

SOLUTION At the critical density, ρ_c, any given galaxy of mass m will just be able to "escape" away from our Galaxy. As we saw in Section 8–7, escape can just occur if the total energy E of the galaxy satisfies

$$E = K + U = \tfrac{1}{2}mv^2 - G\frac{mM}{R} = 0.$$

Here R is the distance of that galaxy m from our Galaxy, and we assume on the enormous scale of the universe that each galaxy is essentially a point. The total mass M that pulls inward on m is the total mass of all galaxies within a sphere of radius R (see Appendix C). If we assume the density of galaxies is roughly constant, then $M = \tfrac{4}{3}\pi\rho_c R^3$. Substituting this M into the above equation, and setting $v = HR$ (Hubble's law, Eq. 45–5), we obtain

$$\frac{GM}{R} = \tfrac{1}{2}v^2$$

or

$$\frac{G(\tfrac{4}{3}\pi\rho_c R^3)}{R} = \tfrac{1}{2}(HR)^2.$$

We solve for ρ_c:

$$\rho_c = \frac{3H^2}{8\pi G} \approx \frac{3[(70 \times 10^3 \, \text{m/s/Mpc})(1 \, \text{Mpc}/3.08 \times 10^{22} \, \text{m})]^2}{8(3.14)(6.67 \times 10^{-11} \, \text{N} \cdot \text{m}^2/\text{kg}^2)}$$

$$\approx 10^{-26} \, \text{kg/m}^3.$$

Great efforts have gone into measuring the actual density of the universe. Estimates of the amount of visible matter in the universe put the actual density at between one and two orders of magnitude less than the critical density, thereby suggesting an open universe. However, there is evidence for a significant amount of nonluminous matter in the universe, often referred to as the **missing mass or dark matter**, enough to bring the density to almost exactly ρ_c. For example, observations of the rotation of galaxies suggest that they rotate as if they had considerably more mass than we can see. And observations of the motion of galaxies within clusters also suggest that they have considerably more mass than can be seen. Furthermore, the highly regarded inflationary theory offers a strong argument in favor of ρ being closely equal to ρ_c, and thus that space on a grand scale is Euclidean. If there is nonluminous matter in the universe, what might it be? One suggestion is that the missing mass consists of previously unknown weakly interacting massive particles (WIMPS), possibly supersymmetric particles known as "neutralinos," or perhaps small primordial black holes made in the early stages of the universe. Another possibility for the missing mass (or part of it) are MACHOS (massive compact halo objects), which would be chunks of matter in the form of large planets (like Jupiter) or stars that are too small to sustain fusion; they might glow faintly due to energy released from gravitational contraction (thus they are sometimes referred to as *brown dwarfs*), but too faintly to be seen. Evidence for MACHOS was found in 1993, inferred from their (assumed) gravitational effect on light passing by them from a distant star.

Another proposal for the missing mass is that neutrinos, once believed to be massless, may actually have nonzero rest mass. Since the universe probably contains roughly as many neutrinos of each type as it does photons (that is, about 10^9 times the number of nucleons, although this neutrino background has yet to be detected), neutrino masses of only a few eV could help to bring the actual density

Missing mass or dark matter

of the universe up to the critical density. The supernova of 1987 offered an opportunity to estimate the electron neutrino mass. If neutrinos have nonzero rest mass, then their velocities are less than the speed of light ($v < c$). High-energy neutrinos emitted from the supernova should in this case have arrived at Earth earlier than lower-energy (and therefore slower) neutrinos emitted at the same instant. Since they traveled a distance of about 170,000 ly from SN1987a, the time difference ought to be measurable. To get an idea of the size of the effect, let us consider the following Example.

EXAMPLE 45–8 **ESTIMATE** **Neutrino mass estimate from a supernova.**
Suppose two neutrinos from SN1987a were detected on Earth (via the reaction $\bar{\nu}_e + p \rightarrow n + e^+$) 10 seconds apart, with measured kinetic energies of about 20 MeV and 10 MeV. (a) Estimate the rest mass of the neutrino, m_ν, using this data, assuming both neutrinos were emitted at the same time. (b) Theoretical models of supernova explosions suggest that the neutrinos are emitted in a burst that lasts from a second or two up to perhaps 10 s. If we assume the neutrinos are not emitted simultaneously but rather at any time over a 10-s interval, what then can we say about the neutrino mass based on the data given above?

SOLUTION (a) We expect the neutrino mass to be less than 100 eV (from laboratory measurements). Since our neutrinos have kinetic energy of 20 MeV and 10 MeV, we can make the approximation $m_\nu c^2 \ll E$, where E (the total energy) is essentially equal to the kinetic energy. From Eq. 37–11, we have

$$E = \frac{m_\nu c^2}{\sqrt{1 - v^2/c^2}}.$$

We solve this for v, the velocity of a neutrino with energy E:

$$v = c\left(1 - \frac{m_\nu^2 c^4}{E^2}\right)^{\frac{1}{2}} = c\left(1 - \frac{m_\nu^2 c^4}{2E^2} + \cdots\right),$$

where we have used the binomial expansion $(1 + x)^{\frac{1}{2}} = 1 + \frac{1}{2}x + \cdots$, and we ignore higher-order terms since $m_\nu^2 c^4 \ll E^2$. The time t for a neutrino to travel a distance $d (= 170,000 \text{ ly})$ is

$$t = \frac{d}{v} = \frac{d}{c\left(1 - \frac{m_\nu^2 c^4}{2E^2}\right)} \approx \frac{d}{c}\left(1 + \frac{m_\nu^2 c^4}{2E^2}\right),$$

where again we used the binomial expansion $\left[(1 + x)^{-1} = 1 - x + \cdots\right]$. The difference in arrival times for our two neutrinos of energies $E_1 = 20$ MeV and $E_2 = 10$ MeV is

$$t_2 - t_1 = \frac{d}{c}\frac{m_\nu^2 c^4}{2}\left(\frac{1}{E_2^2} - \frac{1}{E_1^2}\right).$$

We solve this for $m_\nu c^2$ and set $t_2 - t_1 = 10$ s:

$$m_\nu c^2 = \left[\frac{2c(t_2 - t_1)}{d}\frac{E_1^2 E_2^2}{E_1^2 - E_2^2}\right]^{\frac{1}{2}}$$

$$= \left[\frac{2(3 \times 10^8 \text{ m/s})(10 \text{ s})}{(1.7 \times 10^5 \text{ ly})(10^{16} \text{ m/ly})}\frac{(400 \text{ MeV}^2)(100 \text{ MeV}^2)}{(400 \text{ MeV}^2 - 100 \text{ MeV}^2)}\right]^{\frac{1}{2}}$$

$$= 22 \times 10^{-6} \text{ MeV} = 22 \text{ eV}.$$

We thus estimate the mass of the neutrino to be $22 \text{ eV}/c^2$, but there would of course be experimental uncertainties.
(b) If the two neutrinos of energy $E_1 = 20$ MeV and $E_2 = 10$ MeV were emitted at unknown times over a 10-s interval, then the 10-s difference in their arrival times could be due to a 10-s difference in their emission time. In this case our data are consistent with zero rest mass and it puts an approximate *upper limit* on the neutrino mass of $22 \text{ eV}/c^2$.

In the actual experiments, the most sensitive detector consisted of several thousand tons of water in an underground chamber. It detected 11 events in 12 seconds, probably via the reaction $\bar{\nu}_e + p \rightarrow n + e^+$. There was not a clear correlation between energy and time of arrival. Nonetheless, a careful analysis of this experiment and others, has set a rough upper limit on the electron anti-neutrino mass of about 4 eV. The upper limits for the masses of the other neutrinos are much higher (see Table 44-2). Very recent experiments that investigate mass differences among neutrinos suggest that at least one of the neutrinos may have nonzero mass.

The scenarios we have been describing are based on some well-founded assumptions. As of this writing (year 2000) one of those basic assumptions—that gravity should be slowing down the expansion of the universe—has been called into question, and there are suggestions that there might be a previously undetected repulsive component of gravity.

Latest experiments

Finally, very recent experiments on the early CMB are consistent with the universe being flat, and fully in accord with the inflation theory.

The questions raised by cosmology can seem absurd at times, they are so removed from everyday "reality." We can always say, "the Sun is shining, it's going to burn on for an unimaginably long time, all is well." Nonetheless, the questions of cosmology are deep ones that fascinate the human intellect. One aspect that is especially intriguing is this: calculations on the formation and evolution of the universe have been performed that deliberately varied the values—just slightly—of certain fundamental physical constants. The result? A universe in which life as we know it could not exist. [For example, if the difference in mass between proton and neutron were zero, or small (less than the mass of the electron, $0.511 \text{ MeV}/c^2$), there would be no atoms: electrons would be captured by protons never to be freed again.] Such results have given rise to the so-called **Anthropic principle**, which says that if the universe were even a little different than it is, we could not be here. It seems as if the universe is exquisitely tuned, almost as if to accommodate us.

☐ Summary

The night sky contains myriads of stars including those in the Milky Way, which is a "side view" of our **Galaxy** looking along the plane of the disc. Our Galaxy includes about 10^{11} stars. Beyond our Galaxy are billions of other galaxies.

Astronomical distances are measured in **light-years** $(1 \text{ ly} \approx 10^{13} \text{ km})$. The nearest star is about 4 ly away and the nearest other galaxy is 2 million ly away. Our galactic disc has a diameter of about 100,000 ly. Distances are often specified in **parsecs**, where 1 parsec = 3.26 ly.

Stars are believed to begin life as collapsing masses of hydrogen gas (protostars). As they contract, they heat up (potential energy is transformed to kinetic energy). When the temperature reaches 10 million degrees, nuclear fusion begins and forms heavier elements (**nucleosynthesis**), mainly helium at first. The energy released during these reactions balances the gravitational force, and the young star stabilizes as a **main-sequence** star. The tremendous luminosity of stars comes from the energy released during these thermonuclear reactions. After billions of years, as helium is collected in the core and hydrogen is used up, the core contracts and heats further. The envelope expands and cools, and the star becomes a **red giant** (larger diameter, redder color). The next stage of stellar evolution depends on the mass of the star. Stars of residual mass less than about 1.4 solar masses cool further and become **white dwarfs**, eventually fading and going out altogether. Heavier stars contract further due to their greater gravity: the density approaches nuclear density, the huge pressure forces electrons to combine with protons to form neutrons, and the star becomes essentially a huge nucleus of neutrons. This is a **neutron star**, and the energy released from its final core contraction is believed to produce **supernovae** explosions. If the star's residual mass is greater than two or three solar masses, it may contract even further and form a **black hole**, which is so dense that no matter or light can escape from it.

In the **general theory of relativity**, the **equivalence principle** states that an observer cannot distinguish acceleration from a gravitational field. Said another way, gravitational and inertial mass are the same. The theory predicts gravitational bending of light rays to a degree consistent with experiment. Gravity is treated as a curvature in space and time, the curvature being greater near massive bodies. The universe as a whole may be curved. If there is sufficient mass, the curvature of the universe is positive, and the universe is **closed** and **finite**; otherwise, it is **open** and **infinite**.

Distant galaxies display a **redshift** in their spectral lines, interpreted as a Doppler shift. The universe seems to be **expanding**, its galaxies racing away from each other at speeds (v) proportional to the distance (d) between them:

$$v = Hd,$$

which is known as **Hubble's law** (H is the **Hubble parameter**). This expansion of the universe suggests an explosive origin, the **Big Bang**, which probably occurred about 15 billion years ago.

Quasars are objects with a large redshift (suggesting great distance) and high luminosity (suggesting closeness or, more likely, extraordinary energy output).

The **cosmological principle** assumes that the universe, on a large scale, is homogeneous and isotropic.

Important evidence for the Big Bang model of the universe was the discovery of the **cosmic microwave background** radiation (CMB), which conforms to a blackbody radiation curve at a temperature of 2.7 K.

The **standard model** of the Big Bang provides a possible scenario as to how the universe developed as it expanded and cooled after the Big Bang. Starting at 10^{-43} seconds after the Big Bang, according to this model, there was a series of **phase transitions** during which previously unified forces of nature "condensed out" one by one. The **inflationary scenario** assumes that during one of these phase transitions, the universe underwent a brief but rapid exponential expansion. Until about 10^{-35} s, there was no distinction between quarks and leptons. Shortly thereafter, quarks were **confined** into hadrons (the **hadron era**). About 10^{-4} s after the Big Bang, the majority of hadrons disappeared, having combined with anti-hadrons, producing photons, leptons and energy, leaving mainly photons and leptons to freely move, thus introducing the **lepton era**. By the time the universe was about 10 s old, the electrons too had mostly disappeared, having combined with their antiparticles, and the universe became **radiation-dominated**. A couple of minutes later, nucleosynthesis began, but lasted only a few minutes. It then took several hundred thousand years before the universe was cool enough for electrons to combine with nuclei to form atoms. Also about this time, the background radiation had expanded and cooled so much that its total energy equaled the energy in matter. As the radiation cooled further, losing energy, the universe became **matter-dominated** (not in numbers, but in energy). Then stars and galaxies formed, producing a universe not much different than it is today—some 15 billion years later.

If the universe is open, it will continue to expand indefinitely. If it is closed, gravity is sufficiently strong to halt expansion and the universe will eventually begin to collapse back on itself, ending in a *big crunch*. Whether the universe is open or closed depends on whether its average mass density is above or below a critical density. The evidence today suggests that the universe is flat.

Questions

1. The Milky Way was once thought to be "cloudy," but no longer is considered so. Explain.

2. If you were measuring star parallaxes from the Moon instead of Earth, what corrections would you have to make? What changes would occur if you were measuring parallaxes from Mars?

3. Certain stars called *cepheid variables* change in luminosity with a period of typically several days. The period has been found to have a definite relationship with the absolute luminosity of the star. How could these stars be used to measure the distance to galaxies?

4. A star is in equilibrium when it radiates at its surface all the energy generated at its core. What happens when it begins to generate more energy than it radiates? Less energy? Explain.

5. Describe a red giant star. List some of its properties.

6. Select a point on the H–R diagram. Mark several directions away from this point. Now describe the changes that would take place in a star moving in each of these directions.

7. Does the H–R diagram reveal anything about the core of a star?

8. Why do some stars end up as white dwarfs, and others as neutron stars or black holes?

9. Can we tell, by looking at the population on the H–R diagram, that hotter suns have shorter lives? Explain.

10. What is a geodesic? What is its role in general relativity?

11. If it were discovered that the redshift of spectral lines of galaxies was due to something other than expansion, how might our view of the universe change? Would there be conflicting evidence? Discuss.

12. All galaxies appear to be moving away from us. Are we therefore at the center of the universe? Explain.

13. If you were located in a galaxy near the boundary of our observable universe, would galaxies in the direction of the Milky Way appear to be approaching you or receding from you? Explain.

14. What is the difference between the Hubble age of the universe and the actual age? Which is greater?

15. Compare an explosion on Earth to the Big Bang. Consider such questions as: Would the debris spread at a higher speed for more distant particles, as in the Big Bang? Would the debris come to rest? What type of universe would this correspond to, open or closed?

16. When the primordial nucleus exploded, thus creating the universe, into what did it expand? Discuss.

17. If nothing, not even light, escapes from a black hole, then how can we tell if one is there?

18. Explain what the 2.7 K cosmic microwave background radiation is. Where does it come from? Why is its temperature now so low?

19. The birth of atoms—that is, the combination of electrons with nuclei about which they orbit—occurred when the universe had cooled to about 3000 K and is generally called *recombination*. Why is this term misleading?

20. Why were atoms unable to exist until hundreds of thousands of years after the Big Bang?

21. Explain why today the universe is said to be matter-dominated, yet there are probably 10^9 times as many photons as there are massive particles.

22. If the universe is open, what will eventually happen to the cosmic microwave background radiation? If it is closed, what will happen to it?

23. Under what circumstances would the universe eventually collapse in on itself?

Problems

Sections 45–1 and 45–2

1. (I) Using the definitions of the parsec and the light-year, show that $1\,\text{pc} = 3.26\,\text{ly}$.

2. (I) A star exhibits a parallax of 0.38 seconds of arc. How far away is it?

3. (I) The parallax angle of a star is $0.00019°$. How far away is the star?

4. (I) A star is 36 parsecs away. What is its parallax angle? State (a) in seconds of arc, and (b) in degrees.

5. (I) What is the parallax angle for a star that is 55 light-years away? How many parsecs is this?

6. (I) A star is 35 parsecs away. How long does it take for its light to reach us?

7. (I) If one star is twice as far away from us as a second star, will the parallax angle of the first star be greater or less than that of the second star? By what factor?

8. (II) We saw earlier (Chapter 19) that the rate energy reaches the Earth from the Sun (the "solar constant") is about $1.3 \times 10^3\,\text{W/m}^2$. What is (a) the apparent brightness l of the Sun, and (b) the absolute luminosity L of the Sun?

9. (II) What is the apparent brightness of the Sun as seen on Jupiter? (Jupiter is 5.2 times farther from the Sun than the Earth.)

10. (II) Estimate the angular width that our Galaxy would subtend if observed from the nearest galaxy to us (Table 45–1). Compare to the angular width of the Moon from Earth.

11. (II) When our Sun becomes a red giant, what will be its average density if it expands out to the orbit of Earth ($1.5 \times 10^{11}\,\text{m}$ from the Sun)?

12. (II) When our Sun becomes a white dwarf, it is expected to be about the size of the Moon. What angular width will it subtend from Earth?

13. (II) Calculate the density of a white dwarf whose mass is equal to the Sun's and whose radius is equal to the Earth's. How many times larger than Earth's density is this?

14. (II) A neutron star whose mass is 1.5 solar masses has a radius of about 11 km. Calculate its average density and compare to that for a white dwarf (Problem 13) and to that of nuclear matter.

15. (II) Calculate the Q-values for the He burning reactions of Eq. 45–2. (The mass of ^8_4Be is 8.005305 u.)

16. (II) In the later stages of stellar evolution, a star (if massive enough) will begin fusing carbon nuclei to form, for example, magnesium:

$$^{12}_{6}\text{C} + {}^{12}_{6}\text{C} \rightarrow {}^{24}_{12}\text{Mg} + \gamma.$$

(a) How much energy is released in this reaction (see Appendix D). (b) How much kinetic energy must each carbon nucleus have (assume equal) in a head-on collision if they are just to touch (use Eq. 42–1) so that the strong force can come into play? (c) What temperature does this kinetic energy correspond to?

17. (II) Repeat Problem 16 for the reaction

$$^{16}_{8}\text{O} + {}^{16}_{8}\text{O} \rightarrow {}^{28}_{14}\text{Si} + {}^{4}_{2}\text{He}.$$

18. (II) Suppose two stars of the same apparent brightness l are also believed to be the same size. The spectrum of one star peaks at 800 nm whereas that of the other peaks at 400 nm. Use Wien's law (Section 38–1) and the Stefan-Boltzmann equation (Eq. 19–15) to estimate their relative distances from us. [*Hint*: See Examples 45–5 and 45–6.]

19. (III) Stars located in a certain cluster are assumed to be about the same distance from us. Two such stars have spectra that peak at $\lambda_1 = 500\,\text{nm}$ and $\lambda_2 = 700\,\text{nm}$, and the ratio of their apparent brightness is $l_1/l_2 = 0.091$. Estimate their relative sizes (give ratio of their diameters). [*Hint*: Use the Stefan-Boltzmann equation, Eq. 19–15.]

Section 45–3

20. (I) Describe a triangle, drawn on the surface of a sphere, for which the sum of the angles is (a) 360°, and (b) 180°.

21. (I) Show that the Schwarzschild radius for a star with mass equal to that (a) of our Sun is 2.95 km, and (b) of Earth is 8.9 mm.

22. (II) What is the Schwarzschild radius for a typical galaxy?

23. (II) What is the maximum sum-of-the-angles for a triangle on a sphere?

24. (II) Calculate the escape velocity, using Newtonian mechanics, from a body that has collapsed to its Schwarzschild radius.

Section 45–4

25. (I) If a galaxy is traveling away from us at 1.0 percent of the speed of light, roughly how far away is it?

26. (I) The redshift of a galaxy indicates a velocity of 3500 km/s. How far away is it?

27. (I) Estimate the speed of a galaxy (relative to us) that is near the "edge" of the universe, say 12 billion light-years away.

28. (II) Estimate the observed wavelength for the 656-nm line in the Balmer series of hydrogen in the spectrum of a galaxy whose distance from us is (a) 1.0×10^6 ly, (b) 1.0×10^8 ly, (c) 1.0×10^{10} ly.

29. (II) Estimate the speed of a galaxy, and its distance from us, if the wavelength for the hydrogen line at 434 nm is measured on Earth as being 610 nm.

30. (II) Starting from Eq. 45–3, show that the Doppler shift in wavelength is $\Delta\lambda/\lambda_0 \approx v/c$ (Eq. 45–4) for $v \ll c$. [Hint: Use the binomial expansion.]

Sections 45–5 to 45–7

31. (I) Calculate the wavelength at the peak of the blackbody radiation distribution at 2.7 K using the Wien law.

32. (II) The critical density for closure of the universe is $\rho_c \approx 10^{-26}$ kg/m³. State ρ_c in terms of the average number of nucleons per cubic meter.

33. (II) The size of the universe (the average distance between galaxies) at any one moment is believed to have been inversely proportional to the absolute temperature. Estimate the size of the universe, compared to today, at (a) $t = 10^6$ yr, (b) $t = 1$ s, (c) $t = 10^{-6}$ s, and (d) $t = 10^{-35}$ s.

34. (II) At approximately what time had the universe cooled below the threshold temperature for producing (a) kaons $(M \approx 500\ \text{MeV}/c^2)$, (b) Υ $(M \approx 9500\ \text{MeV}/c^2)$, and (c) muons $(M \approx 100\ \text{MeV}/c^2)$?

General Problems

35. Suppose that three main-sequence stars could undergo the three changes represented by the three arrows, A, B, and C, in the H–R diagram of Fig. 45–26. For each case, describe the changes in temperature, luminosity, and size.

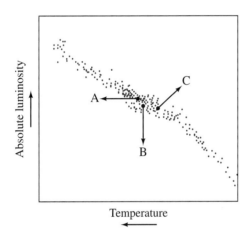

FIGURE 45–26 Problem 35.

36. Assume that the nearest stars to us have an intrinsic luminosity about the same as the Sun's. Their apparent brightness, however, is about 10^{11} times fainter than the Sun. From this, estimate the distance to the nearest stars. (Newton did this calculation, although he made a numerical error of a factor of 100.)

37. Use conservation of angular momentum to estimate the angular velocity of a neutron star which has collapsed to a diameter of 10 km, from a star whose radius was equal to that of our Sun $(7 \times 10^8\ \text{m})$, of mass 1.5 times that of the Sun, and which rotated (like our Sun) about once a month.

38. By what factor does the rotational kinetic energy change when the star in Problem 37 collapses to a neutron star?

39. A certain pulsar, believed to be a neutron star of mass 1.5 times that of the Sun, with diameter 10 km, is observed to have a rotation speed of 1.0 rev/s. If it loses rotational kinetic energy at the rate of 1 part in 10^9 per day, which is all transformed into radiation, what is the power output of the star?

40. Estimate the rate of which hydrogen atoms would have to be created, according to the steady-state model, to maintain the present density of the universe of about 10^{-27} kg/m³, assuming the universe is expanding with Hubble constant $H = 70$ km/s/Mpc.

41. Estimate what neutrino rest mass (in eV) would provide the critical density to close the universe. Assume the neutrino density is, like photons, about 10^9 times that of nucleons, and that nucleons make up only (a) 2 percent of the mass needed, or (b) 5 percent of the mass needed.

42. Two stars, whose spectra peak at 600 nm and 400 nm, respectively, both lie on the main sequence. Use Wien's law, the Stefan-Boltzmann equation, and the H–R diagram (Fig. 45–7) to estimate the ratio of their diameters. [Hint: See Examples 45–5 and 45–6.]

43. The farthest we can measure with parallax is about 30 parsecs. What is our minimum angular resolution (in degrees), based on this information?

44. Through some coincidence, the Balmer lines from singly ionized helium in a distant star happen to overlap with the Balmer lines from hydrogen (Fig. 38–20) in the Sun. How fast is the star receding from us?

45. What is the temperature that corresponds to the 1.8 TeV collisions at the Fermilab collider? To what era in cosmological history does this correspond?

46. Astronomers have recently measured the rotation of gas around what might be a supermassive black hole of about 2 billion solar masses at the center of a galaxy. If the radius from the galactic center to the gas clouds is 60 light-years, what Doppler shift $\Delta\lambda/\lambda_0$ do you estimate they saw?

47. Astronomers use an *apparent magnitude* (m) scale to describe apparent brightness. It uses a logarithmic scale, where a higher number corresponds to a less bright star. (For example, the Sun has magnitude -27, whereas most stars have positive magnitudes.) On this scale, a change in apparent magnitude by $+5$ corresponds to a decrease in apparent brightness by a factor of 100. If Venus has an apparent magnitude of -4.4, whereas Sirius has an apparent magnitude of -1.4, which is brighter? What is the ratio of the apparent brightness of these two objects?

48. How large would the Sun be if its density equaled the critical density of the universe, $\rho_c \approx 10^{-26}\,\text{kg/m}^3$? Express your answer in light-years and compare with the Earth–Sun distance and the size of our Galaxy.

49. Calculate the Schwarzschild radius using classical (Newtonian) gravitational theory, by calculating the minimum radius R for a sphere of mass M such that a photon (rest mass = 0) can escape from the surface. (General relativity gives $R = 2GM/c^2$.)

50. Estimate the radius of a white dwarf whose mass is equal to that of the Sun by the following method, assuming there are N nucleons and $\frac{1}{2}N$ electrons (why $\frac{1}{2}$?): (a) Use Fermi-Dirac statistics (Section 41–6) to show that the total energy of all the electrons is

$$E_e = \frac{3}{5}\left(\frac{1}{2}N\right)\frac{h^2}{8m_e}\left(\frac{3}{\pi}\frac{N}{2V}\right)^{\frac{2}{3}}.$$

[Hint: See Eqs. 41–12 and 41–13; we assume electrons fill energy levels from 0 up to the Fermi energy.] (b) The nucleons contribute to the total energy mainly via the gravitational force (note that the Fermi energy for nucleons is negligible compared to that for electrons—why?). Use a gravitational form of Gauss's law to show that the total gravitational potential energy of a uniform sphere of radius R is

$$-\frac{3}{5}\frac{GM^2}{R},$$

by considering the potential energy of a spherical shell of radius r due only to the mass inside it (why?) and integrate from $r = 0$ to $r = R$. (See also Appendix C.) (c) Write the total energy as a sum of these two terms, and set $dE/dR = 0$ to find the equilibrium radius, and evaluate it for a mass equal to the Sun's $(2.0 \times 10^{30}\,\text{kg})$.

51. Determine the radius of a neutron star using the same argument as in Problem 50 but for N neutrons only. Show that the radius of a neutron star, of 1.5 solar masses, is about 11 km.

52. (a) Use special relativity and Newton's law of gravitation to show that a photon of mass $m = E/c^2$ just grazing the Sun will be deflected by an angle $\Delta\theta$ given by

$$\Delta\theta = \frac{2GM}{c^2 R}$$

where G is the gravitational constant, R and M are the radius and mass of the Sun, and c is the speed of light. (b) Put in values and show $\Delta\theta = 0.87''$. (General relativity predicts an angle twice as large, $1.74''$.)

APPENDIX

Mathematical Formulas

A–1 Quadratic Formula

If

$$ax^2 + bx + c = 0$$

then

$$x = \frac{-b \pm \sqrt{b^2 - 4ac}}{2a}$$

A–2 Binomial Expansion

$$(1 \pm x)^n = 1 \pm nx + \frac{n(n-1)}{2!} x^2 \pm \frac{n(n-1)(n-2)}{3!} x^3 + \cdots$$

$$(x + y)^n = x^n \left(1 + \frac{y}{x}\right)^n = x^n \left(1 + n\frac{y}{x} + \frac{n(n-1)}{2!} \frac{y^2}{x^2} + \cdots\right)$$

A–3 Other Expansions

$$e^x = 1 + x + \frac{x^2}{2!} + \frac{x^3}{3!} + \cdots$$

$$\ln(1 + x) = x - \frac{x^2}{2} + \frac{x^3}{3} - \frac{x^4}{4} + \cdots$$

$$\sin \theta = \theta - \frac{\theta^3}{3!} + \frac{\theta^5}{5!} - \cdots$$

$$\cos \theta = 1 - \frac{\theta^2}{2!} + \frac{\theta^4}{4!} - \cdots$$

$$\tan \theta = \theta + \frac{\theta^3}{3} + \frac{2}{15} \theta^5 + \cdots \qquad |\theta| < \frac{\pi}{2}$$

In general:

$$f(x) = f(0) + \left(\frac{df}{dx}\right)_0 x + \left(\frac{d^2f}{dx^2}\right)_0 \frac{x^2}{2!} + \cdots$$

A–4 Areas and Volumes

Object	Surface area	Volume
Circle, radius r	πr^2	—
Sphere, radius r	$4\pi r^2$	$\frac{4}{3}\pi r^3$
Right circular cylinder, radius r, height h	$2\pi r^2 + 2\pi rh$	$\pi r^2 h$
Right circular cone, radius r, height h	$\pi r^2 + \pi r \sqrt{r^2 + h^2}$	$\frac{1}{3}\pi r^2 h$

A–5 Plane Geometry

1.

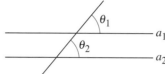

FIGURE A–1

If line a_1 is parallel to line a_2, then $\theta_1 = \theta_2$.

2.

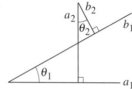

FIGURE A–2

If $a_1 \perp a_2$ and $b_1 \perp b_2$, then $\theta_1 = \theta_2$.

3. The sum of the angles in any plane triangle is 180°.

4. *Pythagorean theorem:*

FIGURE A–3

In any right triangle (one angle = 90°) of sides a, b, and c:

$$a^2 + b^2 = c^2$$

where c is the length of the hypotenuse (opposite the 90° angle).

A–6 Trigonometric Functions and Identities

(See Fig. A–4.)

$$\sin \theta = \frac{o}{h} \qquad\qquad \csc \theta = \frac{1}{\sin \theta} = \frac{h}{o}$$

$$\cos \theta = \frac{a}{h} \qquad\qquad \sec \theta = \frac{1}{\cos \theta} = \frac{h}{a}$$

$$\tan \theta = \frac{o}{a} = \frac{\sin \theta}{\cos \theta} \qquad\qquad \cot \theta = \frac{1}{\tan \theta} = \frac{a}{o}$$

$$a^2 + o^2 = h^2 \qquad\qquad \text{[Pythagorean theorem]}.$$

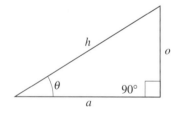

FIGURE A–4

Figure A–5 shows the signs (+ or −) that cosine, sine, and tangent take on for angles θ in the four quadrants (0° to 360°). Note that angles are measured counterclockwise from the x axis as shown; negative angles are measured from *below* the x axis, clockwise: for example, −30° = +330°, and so on.

FIGURE A–5

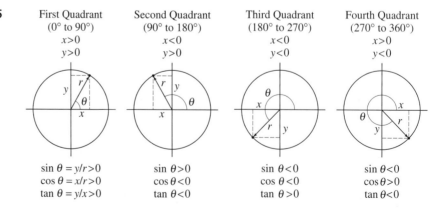

First Quadrant (0° to 90°)	Second Quadrant (90° to 180°)	Third Quadrant (180° to 270°)	Fourth Quadrant (270° to 360°)
$x > 0$	$x < 0$	$x < 0$	$x > 0$
$y > 0$	$y > 0$	$y < 0$	$y < 0$
$\sin \theta = y/r > 0$	$\sin \theta > 0$	$\sin \theta < 0$	$\sin \theta < 0$
$\cos \theta = x/r > 0$	$\cos \theta < 0$	$\cos \theta < 0$	$\cos \theta > 0$
$\tan \theta = y/x > 0$	$\tan \theta < 0$	$\tan \theta > 0$	$\tan \theta < 0$

The following are some useful identities among the trigonometric functions:

$$\sin^2\theta + \cos^2\theta = 1, \quad \sec^2\theta - \tan^2\theta = 1, \quad \csc^2\theta - \cot^2\theta = 1$$

$$\sin 2\theta = 2\sin\theta\cos\theta$$

$$\cos 2\theta = \cos^2\theta - \sin^2\theta = 2\cos^2\theta - 1 = 1 - 2\sin^2\theta$$

$$\tan 2\theta = \frac{2\tan\theta}{1 - \tan^2\theta}$$

$$\sin(A \pm B) = \sin A \cos B \pm \cos A \sin B$$

$$\cos(A \pm B) = \cos A \cos B \mp \sin A \sin B$$

$$\tan(A \pm B) = \frac{\tan A \pm \tan B}{1 \mp \tan A \tan B}$$

$$\sin(180° - \theta) = \sin\theta$$

$$\cos(180° - \theta) = -\cos\theta$$

$$\sin(90° - \theta) = \cos\theta$$

$$\cos(90° - \theta) = \sin\theta$$

$$\cos(-\theta) = \cos\theta$$

$$\sin(-\theta) = -\sin\theta$$

$$\tan(-\theta) = -\tan\theta$$

$$\sin\tfrac{1}{2}\theta = \sqrt{\frac{1 - \cos\theta}{2}}, \quad \cos\tfrac{1}{2}\theta = \sqrt{\frac{1 + \cos\theta}{2}}, \quad \tan\tfrac{1}{2}\theta = \sqrt{\frac{1 - \cos\theta}{1 + \cos\theta}}$$

$$\sin A \pm \sin B = 2\sin\left(\frac{A \pm B}{2}\right)\cos\left(\frac{A \mp B}{2}\right).$$

For any triangle (see Fig. A–6):

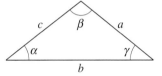

$$\frac{\sin\alpha}{a} = \frac{\sin\beta}{b} = \frac{\sin\gamma}{c} \qquad \text{[Law of sines]}$$

$$c^2 = a^2 + b^2 - 2ab\cos\gamma. \qquad \text{[Law of cosines]}$$

FIGURE A–6

A–7 Logarithms

The following identities apply to common logs (base 10), natural logs (base e) which are often abbreviated ln, or logs to any other base.

$$\log(ab) = \log a + \log b$$

$$\log\left(\frac{a}{b}\right) = \log a - \log b$$

$$\log a^n = n\log a.$$

A–8 Vectors

Vector addition is covered in Sections 3–2 to 3–5.
Vector multiplication is covered in Sections 3–3, 7–2 and 11–1.

Derivatives and Integrals

B–1 Derivatives: General Rules

(See also Section 2–3.)

$$\frac{dx}{dx} = 1$$

$$\frac{d}{dx}[af(x)] = a\frac{df}{dx} \qquad (a = \text{constant})$$

$$\frac{d}{dx}[f(x) + g(x)] = \frac{df}{dx} + \frac{dg}{dx}$$

$$\frac{d}{dx}[f(x)g(x)] = \frac{df}{dx}g + f\frac{dg}{dx}$$

$$\frac{d}{dx}[f(y)] = \frac{df}{dy}\frac{dy}{dx} \qquad \text{[chain rule]}$$

$$\frac{dx}{dy} = \frac{1}{\left(\dfrac{dy}{dx}\right)} \qquad \text{if } \frac{dy}{dx} \neq 0.$$

B–2 Derivatives: Particular Functions

$$\frac{da}{dx} = 0 \qquad (a = \text{constant})$$

$$\frac{d}{dx}x^n = nx^{n-1}$$

$$\frac{d}{dx}\sin ax = a\cos ax$$

$$\frac{d}{dx}\cos ax = -a\sin ax$$

$$\frac{d}{dx}\tan ax = a\sec^2 ax$$

$$\frac{d}{dx}\ln ax = \frac{1}{x}$$

$$\frac{d}{dx}e^{ax} = ae^{ax}$$

B–3 Indefinite Integrals: General Rules

(See also Section 7–3.)

$$\int dx = x$$

$$\int a f(x)\, dx = a \int f(x)\, dx \qquad (a = \text{constant})$$

$$\int \left[f(x) + g(x) \right] dx = \int f(x)\, dx + \int g(x)\, dx$$

$$\int u\, dv = uv - \int v\, du \qquad \text{[integration by parts]}$$

B–4 Indefinite Integrals: Particular Functions

(An arbitrary constant can be added to the right side of each equation.)

$$\int a\, dx = ax \qquad (a = \text{constant})$$

$$\int x^m\, dx = \frac{1}{m+1} x^{m+1} \qquad (m \neq -1)$$

$$\int \sin ax\, dx = -\frac{1}{a} \cos ax$$

$$\int \cos ax\, dx = \frac{1}{a} \sin ax$$

$$\int \tan ax\, dx = \frac{1}{a} \ln|\sec ax|$$

$$\int \frac{1}{x}\, dx = \ln x$$

$$\int e^{ax}\, dx = \frac{1}{a} e^{ax}$$

$$\int \frac{dx}{x^2 + a^2} = \frac{1}{a} \tan^{-1} \frac{x}{a}$$

$$\int \frac{dx}{x^2 - a^2} = \frac{1}{2a} \ln\left(\frac{x - a}{x + a} \right) \qquad (x^2 > a^2)$$

$$= -\frac{1}{2a} \ln\left(\frac{a + x}{a - x} \right) \qquad (x^2 < a^2)$$

$$\int \frac{dx}{\sqrt{x^2 \pm a^2}} = \ln(x + \sqrt{x^2 \pm a^2})$$

$$\int \frac{dx}{(x^2 \pm a^2)^{\frac{3}{2}}} = \frac{\pm x}{a^2 \sqrt{x^2 \pm a^2}}$$

$$\int \frac{x\, dx}{(x^2 \pm a^2)^{\frac{3}{2}}} = \frac{-1}{\sqrt{x^2 \pm a^2}}$$

$$\int \sin^2 ax\, dx = \frac{x}{2} - \frac{\sin 2ax}{4a}$$

$$\int x e^{-ax}\, dx = -\frac{e^{-ax}}{a^2} (ax + 1)$$

$$\int x^2 e^{-ax}\, dx = -\frac{e^{-ax}}{a^3} (a^2 x^2 + 2ax + 2)$$

B–5 A few Definite Integrals

$$\int_0^\infty x^n e^{-ax}\, dx = \frac{n!}{a^{n+1}}$$

$$\int_0^\infty e^{-ax^2}\, dx = \sqrt{\frac{\pi}{4a}}$$

$$\int_0^\infty x e^{-ax^2}\, dx = \frac{1}{2a}$$

$$\int_0^\infty x^2 e^{-ax^2}\, dx = \sqrt{\frac{\pi}{16a^3}}$$

$$\int_0^\infty x^3 e^{-ax^2}\, dx = \frac{1}{2a^2}$$

$$\int_0^\infty x^{2n} e^{-ax^2}\, dx = \frac{1 \cdot 3 \cdot 5 \cdots (2n - 1)}{2^{n+1} a^n} \sqrt{\frac{\pi}{a}}$$

APPENDIX C

Gravitational Force due to a Spherical Mass Distribution

In Chapter 6 (Section 6–1), we stated that the gravitational force exerted by or on a uniform sphere acts as if all the mass of the sphere were concentrated at its center, if the other mass is outside the sphere. In other words, the gravitational force that a uniform sphere exerts on a particle outside it is

$$F = G\frac{mM}{r^2}, \qquad\qquad [m \text{ outside sphere of mass } M]$$

where m is the mass of the particle, M the mass of the sphere, and r the distance of m from the center of the sphere. Now we will derive this result. We will use the concepts of infinitesimally small quantities and integration.

First we consider a very thin, uniform spherical shell (like a thin-walled basketball) of mass M whose thickness t is small compared to its radius R (Fig. C–1). The force on a particle of mass m at a distance r from the center of the shell can be calculated as the vector sum of the forces due to all the particles of the shell. We imagine the shell divided up into thin (infinitesimal) circular strips so that all points on a strip are equidistant from our particle m. One of these circular strips, labeled AB, is shown in Fig. C–1. It is $Rd\theta$ wide, t thick, and has a radius $R\sin\theta$. The force on our particle m due to a tiny piece of the strip at point A is represented by the vector $\mathbf{F}_A$ shown. The force due to a tiny piece of the strip at point B, which is diametrically opposite A, is the force $\mathbf{F}_B$. We take the two pieces at A and B to be of equal mass, so $F_A = F_B$. The horizontal components of $\mathbf{F}_A$ and $\mathbf{F}_B$ are each equal to

$$F_A \cos\phi$$

and point toward the center of the shell. The vertical components of $\mathbf{F}_A$ and $\mathbf{F}_B$ are of equal magnitude and point in opposite directions, and so cancel. Since for every point on the strip there is a corresponding point diametrically opposite (as with A and B), we see that the net force due to the entire strip points toward the center of the shell. Its magnitude will be

$$dF = G\frac{m\,dM}{l^2}\cos\phi,$$

where dM is the mass of the entire circular strip and l is the distance from all

FIGURE C–1 Calculating the gravitational force on a particle of mass m due to a uniform spherical shell of radius R and mass M.

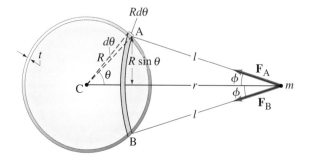

points on the strip to m, as shown. We write dM in terms of the density ρ; by density we mean the mass per unit volume (Section 13–1). Hence, $dM = \rho \, dV$, where dV is the volume of the strip and equals $(2\pi R \sin\theta)(t)(R \, d\theta)$. Then the force dF due to the circular strip shown is

$$dF = G\frac{m\rho 2\pi R^2 t \sin\theta \, d\theta}{l^2} \cos\phi. \qquad (C\text{–}1)$$

To get the total force F that the entire shell exerts on the particle m, we must integrate over all the circular strips: that is, from $\theta = 0°$ to $\theta = 180°$. But our expression for dF contains l and ϕ, which are functions of θ. From Fig. C–1 we can see that

$$l \cos\phi = r - R\cos\theta.$$

Furthermore, we can write the law of cosines for triangle CmA:

$$\cos\theta = \frac{r^2 + R^2 - l^2}{2rR}. \qquad (C\text{–}2)$$

With these two expressions we can reduce our three variables (l, θ, ϕ) to only one, which we take to be l. We do two things with Eq. C–2: (1) We put it into the equation for $l \cos\phi$ above:

$$\cos\phi = \frac{1}{l}(r - R\cos\theta) = \frac{r^2 + l^2 - R^2}{2rl};$$

and (2) we take the differential of both sides of Eq. C–2 (because $\sin\theta \, d\theta$ appears in the expression for dF, Eq. C–1):

$$-\sin\theta \, d\theta = -\frac{2l \, dl}{2rR} \qquad \text{or} \qquad \sin\theta \, d\theta = \frac{l \, dl}{rR},$$

since r and R are considered constants when summing over the strips. Now we insert these into Eq. C–1 for dF and find

$$dF = Gm\rho\pi t \frac{R}{r^2}\left(1 + \frac{r^2 - R^2}{l^2}\right) dl.$$

Now we integrate to get the net force on our thin shell of radius R. To integrate over all the strips ($\theta = 0°$ to $180°$), we must go from $l = r - R$ to $l = r + R$ (see Fig. C–1). Thus,

$$F = Gm\rho\pi t \frac{R}{r^2}\left[l - \frac{r^2 - R^2}{l}\right]_{l=r-R}^{l=r+R}$$

$$= Gm\rho\pi t \frac{R}{r^2}(4R).$$

The volume V of the spherical shell is its area $(4\pi R^2)$ times the thickness t. Hence the mass $M = \rho V = \rho 4\pi R^2 t$, and finally

$$F = G\frac{mM}{r^2}. \qquad \left[\begin{array}{c}\text{particle of mass } m \text{ outside a}\\ \text{thin uniform spherical shell of mass } M\end{array}\right]$$

This result gives us the force a thin shell exerts on a particle of mass m a distance r from the center of the shell, and *outside* the shell. We see that the force is the same as that between m and a particle of mass M at the center of the shell. In other words, for purposes of calculating the gravitational force exerted on or by a uniform spherical shell, we can consider all its mass concentrated at its center.

What we have derived for a shell holds also for a solid sphere, since a solid sphere can be considered as made up of many concentric shells, from $R = 0$ to $R = R_0$, where R_0 is the radius of the solid sphere. Why? Because if each shell has mass dM, we write for each shell, $dF = Gm\,dM/r^2$, where r is the distance from the center C to mass m and is the same for all shells. Then the total force equals the sum or integral over dM, which gives the total mass M. Thus the result

$$F = G\frac{mM}{r^2} \qquad \begin{bmatrix} \text{particle of mass } m \text{ outside} \\ \text{solid sphere of mass } M \end{bmatrix} \quad \text{(C–3)}$$

is valid for a solid sphere of mass M even if the density varies with distance from the center. (It is not valid if the density varies within each shell—that is, depends not only on R.) Thus the gravitational force exerted on or by spherical objects, including nearly spherical objects like the Earth, Sun, and Moon, can be considered to act as if the objects were point particles.

This result, Eq. C–3, is true only if the mass m is outside the sphere. Let us next consider a point mass m that is located inside the spherical shell of Fig. C–1. Here, r would be less than R, and the integration over l would be from $l = R - r$ to $l = R + r$, so

$$\left[l - \frac{r^2 - R^2}{l} \right]_{R-r}^{R+r} = 0.$$

Thus the force on any mass inside the shell would be zero. This result has particular importance for the electrostatic force, which is also an inverse square law. For the gravitational situation, we see that at points within a solid sphere, say 1000 km below the earth's surface, only the mass up to that radius contributes to the net force. The outer shells beyond the point in question contribute no net gravitational effect.

The results we have obtained here can also be reached using the gravitational analog of Gauss's law for electrostatics (Chapter 22).

Selected Isotopes

(1) Atomic Number Z	(2) Element	(3) Symbol	(4) Mass Number A	(5) Atomic Mass†	(6) % Abundance (or Radioactive Decay Mode)	(7) Half-life (if radioactive)
0	(Neutron)	n	1	1.008665	β^-	10.4 min
1	Hydrogen	H	1	1.007825	99.985%	
	Deuterium	D	2	2.014102	0.015%	
	Tritium	T	3	3.016049	β^-	12.33 yr
2	Helium	He	3	3.016029	0.000137%	
			4	4.002603	99.999863%	
3	Lithium	Li	6	6.015122	7.5%	
			7	7.016004	92.5%	
4	Beryllium	Be	7	7.016929	EC, γ	53.12 days
			9	9.012182	100%	
5	Boron	B	10	10.012937	19.9%	
			11	11.009305	80.1%	
6	Carbon	C	11	11.011434	β^+, EC	20.39 min
			12	12.000000	98.90%	
			13	13.003355	1.10%	
			14	14.003242	β^-	5730 yr
7	Nitrogen	N	13	13.005739	β^+	9.965 min
			14	14.003074	99.63%	
			15	15.000108	0.37%	
8	Oxygen	O	15	15.003065	β^+, EC	122.24 s
			16	15.994915	99.76%	
			18	17.999160	0.20%	
9	Fluorine	F	19	18.998403	100%	
10	Neon	Ne	20	19.992440	90.48%	
			22	21.991386	9.25%	
11	Sodium	Na	22	21.994437	β^+, EC, γ	2.6019 yr
			23	22.989770	100%	
			24	23.990963	β^-, γ	14.9590 h
12	Magnesium	Mg	24	23.985042	78.99%	
13	Aluminum	Al	27	26.981538	100%	

†The masses given in column (5) are those for the neutral atom, including the Z electrons.

(1) Atomic Number Z	(2) Element	(3) Symbol	(4) Mass Number A	(5) Atomic Mass†	(6) % Abundance (or Radioactive Decay Mode)	(7) Half-life (if radioactive)
14	Silicon	Si	28	27.976927	92.23%	
			31	30.975363	β^-, γ	157.3 min
15	Phosphorus	P	31	30.973762	100%	
			32	31.973907	β^-	14.262 days
16	Sulfur	S	32	31.972071	95.02%	
			35	34.969032	β^-	87.32 days
17	Chlorine	Cl	35	34.968853	75.77%	
			37	36.965903	24.23%	
18	Argon	Ar	40	39.962383	99.600%	
19	Potassium	K	39	38.963707	93.2581%	
			40	39.963999	0.0117%	
					$\beta^-, EC, \gamma, \beta^+$	1.28×10^9 yr
20	Calcium	Ca	40	39.962591	96.941%	
21	Scandium	Sc	45	44.955910	100%	
22	Titanium	Ti	48	47.947947	73.8%	
23	Vanadium	V	51	50.943964	99.750%	
24	Chromium	Cr	52	51.940512	83.79%	
25	Manganese	Mn	55	54.938049	100%	
26	Iron	Fe	56	55.934942	91.72%	
27	Cobalt	Co	59	58.933200	100%	
			60	59.933822	β^-, γ	5.2714 yr
28	Nickel	Ni	58	57.935348	68.077%	
			60	59.930791	26.233%	
29	Copper	Cu	63	62.929601	69.17%	
			65	64.927794	30.83%	
30	Zinc	Zn	64	63.929147	48.6%	
			66	65.926037	27.9%	
31	Gallium	Ga	69	68.925581	60.108%	
32	Germanium	Ge	72	71.922076	27.66%	
			74	73.921178	35.94%	
33	Arsenic	As	75	74.921596	100%	
34	Selenium	Se	80	79.916522	49.61%	
35	Bromine	Br	79	78.918338	50.69%	
36	Krypton	Kr	84	83.911507	57.0%	
37	Rubidium	Rb	85	84.911789	72.17%	
38	Strontium	Sr	86	85.909262	9.86%	
			88	87.905614	82.58%	
			90	89.907737	β^-	28.79 yr
39	Yttrium	Y	89	88.905848	100%	
40	Zirconium	Zr	90	89.904704	51.45%	
41	Niobium	Nb	93	92.906377	100%	
42	Molybdenum	Mo	98	97.905408	24.13%	

† The masses given in column (5) are those for the neutral atom, including the Z electrons.

(1) Atomic Number Z	(2) Element	(3) Symbol	(4) Mass Number A	(5) Atomic Mass†	(6) % Abundance (or Radioactive Decay Mode)	(7) Half-life (if radioactive)
43	Technetium	Tc	98	97.907216	β^-, γ	4.2×10^6 yr
44	Ruthenium	Ru	102	101.904349	31.6%	
45	Rhodium	Rh	103	102.905504	100%	
46	Palladium	Pd	106	105.903483	27.33%	
47	Silver	Ag	107	106.905093	51.839%	
			109	108.904756	48.161%	
48	Cadmium	Cd	114	113.903358	28.73%	
49	Indium	In	115	114.903878	95.7%; β^-, γ	4.41×10^{14} yr
50	Tin	Sn	120	119.902197	32.59%	
51	Antimony	Sb	121	120.903818	57.36%	
52	Tellurium	Te	130	129.906223	33.80%	7.9×10^{20} yr
53	Iodine	I	127	126.904468	100%	
			131	130.906124	β^-, γ	8.0207 days
54	Xenon	Xe	132	131.904154	26.9%	
			136	135.907220	8.9%	
55	Cesium	Cs	133	132.905446	100%	
56	Barium	Ba	137	136.905821	11.23%	
			138	137.905241	71.70%	
57	Lanthanum	La	139	138.906348	99.9098%	
58	Cerium	Ce	140	139.905434	88.48%	
59	Praseodymium	Pr	141	140.907647	100%	
60	Neodymium	Nd	142	141.907718	27.13%	
61	Promethium	Pm	145	144.912744	EC, γ, α	17.7 yr
62	Samarium	Sm	152	151.919728	26.7%	
63	Europium	Eu	153	152.921226	52.2%	
64	Gadolinium	Gd	158	157.924101	24.84%	
65	Terbium	Tb	159	158.925343	100%	
66	Dysprosium	Dy	164	163.929171	28.2%	
67	Holmium	Ho	165	164.930319	100%	
68	Erbium	Er	166	165.930290	33.6%	
69	Thulium	Tm	169	168.934211	100%	
70	Ytterbium	Yb	174	173.938858	31.8%	
71	Lutecium	Lu	175	174.940767	97.4%	
72	Hafnium	Hf	180	179.946549	35.100%	
73	Tantalum	Ta	181	180.947996	99.988%	
74	Tungsten (wolfram)	W	184	183.950933	30.67%	
75	Rhenium	Re	187	186.955751	62.60%; β^-	4.35×10^{10} yr
76	Osmium	Os	191	190.960927	β^-, γ	15.4 days
			192	191.961479	41.0%	
77	Iridium	Ir	191	190.960591	37.3%	
			193	192.962923	62.7%	
78	Platinum	Pt	195	194.964774	33.8%	

† The masses given in column (5) are those for the neutral atom, including the Z electrons.

(1) Atomic Number Z	(2) Element	(3) Symbol	(4) Mass Number A	(5) Atomic Mass[†]	(6) % Abundance (or Radioactive Decay Mode)	(7) Half-life (if radioactive)
79	Gold	Au	197	196.966551	100%	
80	Mercury	Hg	199	198.968262	16.87%	
			202	201.970625	29.86%	
81	Thallium	Tl	205	204.974412	70.476%	
82	Lead	Pb	206	205.974449	24.1%	
			207	206.975880	22.1%	
			208	207.976635	52.4%	
			210	209.984173	β^-, γ, α	22.3 yr
			211	210.988731	β^-, γ	36.1 min
			212	211.991887	β^-, γ	10.64 h
			214	213.999798	β^-, γ	26.8 min
83	Bismuth	Bi	209	208.980383	100%	
			211	210.987258	α, γ, β^-	2.14 min
84	Polonium	Po	210	209.982857	α, γ	138.376 days
			214	213.995185	α, γ	164.3 μs
85	Astatine	At	218	218.008681	α, β^-	1.5 s
86	Radon	Rn	222	222.017570	α, γ	3.8235 days
87	Francium	Fr	223	223.019731	β^-, γ, α	21.8 min
88	Radium	Ra	226	226.025402	α, γ	1600 yr
89	Actinium	Ac	227	227.027746	β^-, γ, α	21.773 yr
90	Thorium	Th	228	228.028731	α, γ	1.9116 yr
			232	232.038050	100%; α, γ	1.405×10^{10} yr
91	Protactinium	Pa	231	231.035878	α, γ	3.276×10^4 yr
92	Uranium	U	232	232.037146	α, γ	68.9 yr
			233	233.039628	α, γ	1.592×10^5 yr
			235	235.043923	0.720%, α, γ	7.038×10^8 yr
			236	236.045561	α, γ	2.342×10^7 yr
			238	238.050782	99.2745%; α, γ	4.468×10^9 yr
			239	239.054287	β^-, γ	23.45 min
93	Neptunium	Np	237	237.048166	α, γ	2.144×10^6 yr
			239	239.052931	β^-, γ	2.3565 d
94	Plutonium	Pu	239	239.052157	α, γ	24,110 yr
			244	244.064197	α	8.08×10^7 yr
95	Americium	Am	243	243.061373	α, γ	7370 yr
96	Curium	Cm	247	247.070346	α, γ	1.56×10^7 yr
97	Berkelium	Bk	247	247.070298	α, γ	1380 yr
98	Californium	Cf	251	251.079580	α, γ	898 yr
99	Einsteinium	Es	252	252.082972	α, EC, γ	472 d
100	Fermium	Fm	257	257.095099	α, γ	101 d
101	Mendelevium	Md	258	258.098425	α, γ	51.5 d
102	Nobelium	No	259	259.10102	α, EC	58 min
103	Lawrencium	Lr	262	262.10969	α, EC, fission	216 min

[†]The masses given in column (5) are those for the neutral atom, including the Z electrons.

(1) Atomic Number Z	(2) Element	(3) Symbol	(4) Mass Number A	(5) Atomic Mass†	(6) % Abundance (or Radioactive Decay Mode)	(7) Half-life (if radioactive)
104	Rutherfordium	Rf	261	261.10875	α	65 s
105	Dubnium	Db	262	262.11415	α, fission, EC	34 s
106	Seaborgium	Sg	266	266.12193	α, fission	21 s
107	Bohrium	Bh	264	264.12473	α	0.44 s
108	Hassium	Hs	269	269.13411	α	9 s
109	Meitnerium	Mt	268	268.13882	α	0.07 s
110			271	271.14608	α	0.06 s
111			272	272.15348	α	1.5 ms
112			277	277	α	0.24 ms
114			289	289	α	20 s
116			289	289	α	0.6 ms
118			293	293	α	0.1 ms

† The masses given in column (5) are those for the neutral atom, including the Z electrons.

Answers to Odd-Numbered Problems

CHAPTER 1

1. (a) 1×10^{10} yr; (b) 3×10^{17} s.

3. (a) 1.156×10^3; (b) 2.18×10^1;
(c) 6.8×10^{-3}; (d) 2.7635×10^1;
(e) 2.19×10^{-1}; (f) 2.2×10^1.

5. 7.7%.

7. (a) 4%; (b) 0.4%; (c) 0.07%.

9. 1.0×10^5 s.

11. 9%.

13. (a) 0.286 6 m; (b) 0.000 085 V;
(c) 0.000 760 kg;
(d) 0.000 000 000 060 0 s;
(e) 0.000 000 000 000 022 5 m;
(f) 2,500,000,000 volts.

15. 1.8 m.

17. (a) 0.111 yd²; (b) 10.76 ft².

19. (a) 3.9×10^{-9} in;
(b) 1.0×10^8 atoms.

21. (a) 0.621 mi/h;
(b) 1 m/s = 3.28 ft/s; (c) 0.278 m/s.

23. (a) 9.46×10^{15} m;
(b) 6.31×10^4 AU; (c) 7.20 AU/h.

25. (a) 10^3; (b) 10^4; (c) 10^{-2}; (d) 10^9.

27. ≈20%.

29. 1×10^5 cm³.

31. (a) ≈600 dentists.

33. ≈3×10^8 kg/yr.

35. 51 km.

37. $A = \left[L/T^4 \right] = $ m/s⁴,
$B = \left[L/T^2 \right] = $ m/s².

39. (a) 0.10 nm; (b) 1.0×10^5 fm;
(c) 1.0×10^{10} Å; (d) 9.5×10^{25} Å.

41. (a) 3.16×10^7 s; (b) 3.16×10^{16} ns;
(c) 3.17×10^{-8} yr.

43. (a) 1,000 drivers.

45. 1×10^{11} gal/yr.

47. 9 cm.

49. 4×10^5 t.

51. ≈4 yr.

53. 1.9×10^2 m.

55. (a) 3%, 3%; (b) 0.7%, 0.2%.

CHAPTER 2

1. 5.0 h.

3. 61 m.

5. 0.78 cm/s (toward $+x$).

7. ≈300 m/s.

9. (a) 10.1 m/s;
(b) +3.4 m/s, away from trainer.

11. (a) 0.28 m/s; (b) 1.2 m/s;
(c) 0.28 m/s; (d) 1.6 m/s;
(e) −1.0 m/s.

13. (a) 13.4 m/s;
(b) +4.5 m/s, away from master.

15. 24 s.

17. 55 km/h, 0.

19. 6.73 m/s.

21. 5.2 s

23. -7.0 m/s², 0.72.

25. (a) 4.7 m/s²; (b) 2.2 m/s²;
(c) 0.3 m/s²; (d) 1.6 m/s².

27. $v = (6.0$ m/s$) + (17$ m/s²$)t$,
$a = 17$ m/s².

29. 1.5 m/s², 99 m.

31. 1.7×10^2 m.

33. 4.41 m/s², $t = 2.61$ s.

35. 55.0 m.

37. (a) 2.3×10^2 m; (b) 31 s;
(c) 15 m, 13 m.

39. (a) 103 m; (b) 64 m.

41. 31 m/s.

43. (b) 3.45 s.

45. 32 m/s (110 km/h).

47. 2.83 s.

49. (a) 8.81 s; (b) 86.3 m/s.

51. 1.44 s.

53. 15 m/s.

55. 5.44 s.

59. 0.035 s.

61. 1.8 m above the top of the window.

63. 52 m.

65. 19.8 m/s, 20.0 m.

67. (a) $v = (g/k)(1 - e^{-kt})$;
(b) $v_{\text{term}} = g/k$.

69. $6h_{\text{Earth}}$.

71. 1.3 m.

73. (b) $H_{50} = 9.8$ m; (c) $H_{100} = 39$ m.

75. (a) 1.3 m; (b) 6.1 m/s; (c) 1.2 s.

77. (a) 3.88 s; (b) 73.9 m; (c) 38.0 m/s,
48.4 m/s.

79. (a) 52 min; (b) 31 min.

81. (a) $v_0 = 26$ m/s; (b) 35 m; (c) 1.2 s;
(d) 4.1 s.

83. (a) 4.80 s; (b) 37.0 m/s; (c) 75.2 m.

85. She should decide to stop!

87. $\Delta v_{0\text{down}} = 0.8$ m/s, $\Delta v_{0\text{up}} = 0.9$ m/s.

89. 29.0 m.

CHAPTER 3

1. 263 km, 13° S of W.

3. $\mathbf{V}_{\text{wrong}} = \mathbf{V}_2 - \mathbf{V}_1$.

5. 13.6 m, 18° N of E,

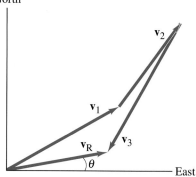

7. (a)

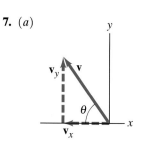

(b) $V_x = -11.7$, $V_y = 8.16$;
(c) 14.3, 34.9° above $- x$-axis.

9. (a) $V_{\text{N}} = 476$ km/h,
$V_{\text{W}} = 421$ km/h;
(b) $d_{\text{N}} = 1.43 \times 10^3$ km,
$d_{\text{W}} = 1.26 \times 10^3$ km.

11. (a) 4.2, 45° below $+x$-axis;
(b) 5.1, 79° below $+x$-axis.

13. (a) 53.7, 1.40° above $-x$-axis;
(b) 53.7, 1.40° below $+x$-axis.

15. (a) 94.5, 11.8° below $-x$-axis;
(b) 150, 35.3° below $+x$-axis.

17. (a) $A_x = \pm 82.9$; (b) 166.6, 12.1°
above $-x$-axis.

19. $(7.60$ m/s$)\mathbf{i} - (4.00$ m/s$)\mathbf{k}$; 8.59 m/s.

21. (a) Unknown; (b) 4.11 m/s², 33.2°
north of east; (c) unknown.

23. (a) $\mathbf{v} = (4.0 \text{ m/s}^2)t\mathbf{i} + (3.0 \text{ m/s}^2)t\mathbf{j}$;
(b) $(5.0 \text{ m/s}^2)t$;
(c) $\mathbf{r} = (2.0 \text{ m/s}^2)t^2\mathbf{i} + (1.5 \text{ m/s}^2)t^2\mathbf{j}$;
(d) $\mathbf{v} = (8.0 \text{ m/s})\mathbf{i} + (6.0 \text{ m/s})\mathbf{j}$,
$|\mathbf{v}| = 10.0 \text{ m/s}$,
$\mathbf{r} = (8.0 \text{ m})\mathbf{i} + (6.0 \text{ m})\mathbf{j}$.

25. (a) $-(18.0 \text{ m/s})\sin(3.0 \text{ s}^{-1})t\mathbf{i}$
$+ (18.0 \text{ m/s})\cos(3.0 \text{ s}^{-1})t\mathbf{j}$;
(b) $-(54.0 \text{ m/s}^2)\cos(3.0 \text{ s}^{-1})t\mathbf{i}$
$- (54.0 \text{ m/s}^2)\sin(3.0 \text{ s}^{-1})t\mathbf{j}$;
(c) circle; (d) $a = (9.0 \text{ s}^{-2})r$, $180°$.

27. 44 m, 6.3 m.

29. $38°$ and $52°$.

31. 1.95 s.

33. 22 m.

35.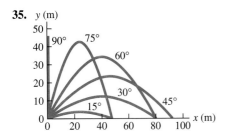

37. 5.71 s.

39. (a) 65.7 m; (b) 7.32 s; (c) 267 m;
(d) 42.2 m/s, $30.1°$ above the horizontal.

43. Unsuccessful, 34.7 m.

45. (a) $\mathbf{v}_0 = 3.42$ m/s, $47.5°$ above the horizontal; (b) 5.32 m above the water; (c) $\mathbf{v}_f = 10.5$ m/s, $77°$ below the horizontal.

47. $\theta = \tan^{-1}(gt/v_0)$ below the horizontal.

49. $\theta = \frac{1}{2}\tan^{-1}(-\cot\phi)$.

51. $7.29g$ up.

53. 5.9×10^{-3} m/s² toward the Sun.

55. $0.94g$.

59. 2.7 m/s, $22°$ from the river bank.

61. 23.1 s.

63. 1.41 m/s.

65. (a) 1.82 m/s; (b) 3.22 m/s.

67. (a) 60 m; (b) 75 s.

69. 58 km/h, $31°$, 58 km/h opposite to $\mathbf{v}_{12}$.

71. 0.0889 m/s².

73. $D_x = 60$ m, $D_y = -35$ m, $D_z = -12$ m; 70 m; $\theta_h = 30°$ from the x-axis toward the $-y$-axis, $\theta_v = 9.8°$ below the horizontal.

75. 7.0 m/s.

77. $\pm 28.5°$, ± 25.2.

79. 170 km/h, $41.5°$ N of E.

81. 1.6 m/s².

83. 2.7 s, 1.9 m/s.

85. (a) $Dv/(v^2 - u^2)$;
(b) $D/(v^2 - u^2)^{1/2}$.

87. $54.6°$ below the horizontal.

89. (a) 464 m/s; (b) 355 m/s.

91. Row at an angle of $23°$ upstream and run 243 m in a total time of 20.7 min.

93. 1.8×10^3 rev/day.

CHAPTER 4

3. 6.9×10^2 N.

5. (a) 5.7×10^2 N; (b) 99 N;
(c) 2.1×10^2 N; (d) 0.

7. 107 N.

9. -9.3×10^5 N, 25% of the weight of the train.

11. $m > 1.9$ kg.

13. 2.1×10^2 N.

15. -1.40 m/s² (down).

17. a (downward) ≥ 1.2 m/s².

19. $a_{\max} = 0.557$ m/s².

21. (a) 2.2 m/s²; (b) 18 m/s; (c) 93 s.

23. 3.0×10^3 N downward.

25. (a) 1.4×10^2 N; (b) 14.5 m/s.

27. Southwesterly direction.

29.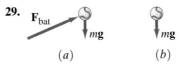

31. (a) 1.13 m/s², $52.2°$ below $-x$-axis;
(b) 0.814 m/s², $42.3°$ above $+x$-axis.

33. (a) $m_1 g - F_T = m_1 a$, $F_T - m_2 g = m_2 a$.

35. (a) lower bucket $= 34$ N, upper bucket $= 68$N; (b) lower bucket $= 40$ N, upper bucket $= 80$ N.

37. 1.4×10^3 N.

39. $F_B = 6890$ N, $\mathbf{F}_A + \mathbf{F}_B = 8860$ N.

41. (a) 2.2 m up the plane; (b) 2.2 s.

43. $\frac{5}{2}(F_0/m)t_0^2$.

47. (a)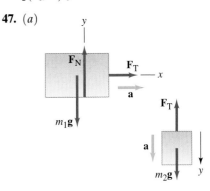

(b) $a = m_2 g/(m_1 + m_2)$,
$F_T = m_1 m_2 g/(m_1 + m_2)$.

49. $a = [m_2 + m_C(\ell_2/\ell)]g/$
$(m_1 + m_2 + m_C)$.

51. 1.74 m/s², $F_{T1} = 22.6$ N,
$F_{T2} = 20.9$ N.

53. $(m + M)g\tan\theta$.

55. $F_{T1} = [4m_1 m_2 m_3/$
$(m_1 m_3 + m_2 m_3 + 4m_1 m_2)]g$,
$F_{T3} = [8m_1 m_2 m_3/$
$(m_1 m_3 + m_2 m_3 + 4m_1 m_2)]g$,
$a_1 = [(m_1 m_3 - 3m_2 m_3 + 4m_1 m_2)/$
$(m_1 m_3 + m_2 m_3 + 4m_1 m_2)]g$,
$a_2 = [(-3m_1 m_3 + m_2 m_3 + 4m_1 m_2)/$
$(m_1 m_3 + m_2 m_3 + 4m_1 m_2)]g$,
$a_3 = [(m_1 m_3 + m_2 m_3 - 4m_1 m_2)/$
$(m_1 m_3 + m_2 m_3 + 4m_1 m_2)]g$.

57. $v = \{[2m_2 \ell_2 + m_C(\ell_2^2/\ell)]g/$
$(m_1 + m_2 + m_C)\}^{1/2}$.

59. 2.0×10^{-2} N.

61. 4.3 N.

63. 1.5×10^4 N.

65. 1.2 s, no change.

67. (a) 2.45 m/s² (up the incline);
(b) 0.50 kg; (c) 7.35 N, 4.9 N.

69. 1.3×10^2 N.

71. $8.8°$.

73. 82 m/s (300 km/h).

75. (a) $F = \frac{1}{2}Mg$;
(b) $F_{T1} = F_{T2} = \frac{1}{2}Mg$, $F_{T3} = \frac{3}{2}Mg$,
$F_{T4} = Mg$.

77. -8.3×10^2 N.

79. (a) 0.606 m/s²; (b) 150 kN.

CHAPTER 5

1. 35 N, no force.

3. (a)

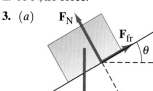

(b)

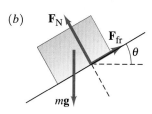

(c)

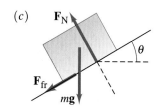

5. 0.20.

7. 69 N, $\mu_k = 0.54$.

9. 8.0 kg.

11. 1.3 m.

13. 1.3×10^3 N.

15. (a) 0.58; (b) 5.7 m/s; (c) 15 m/s.

17. (a) 1.4 m/s^2; (b) 5.4×10^2 N; (c) 1.41 m/s^2, 2.1×10^2 N.

19. (a) 86 cm up the plane; (b) 1.5 s.

21. (a) 2.8 m/s^2; (b) 2.1 N.

23. $a = \{\sin\theta - [(\mu_1 m_1 + \mu_2 m_2)/ (m_1 + m_2)]\cos\theta\}g$,
$F_T = [m_1 m_2 (\mu_2 - \mu_1)/ (m_1 + m_2)]g\cos\theta$.

25. (a) $\mu_k = (v_0^2/2gd\cos\theta) - \tan\theta$; (b) $\mu_s \geq \tan\theta$.

27. (a) 2.0 m/s^2 up the plane; (b) 5.4 m/s^2 up the plane.

29. $\mu_k = 0.40$.

31. (a) $c = 14$ kg/m; (b) 5.7×10^2 N.

33. $F_{\min} = (m + M)g(\sin\theta + \mu\cos\theta)/ (\cos\theta - \mu_s\sin\theta)$.

35. $v_{\max} = 21$ m/s, independent of the mass.

37. (a) 0.25 m/s^2 toward the center; (b) 6.3 N toward the center.

39. Yes, $v_{\text{top, min}} = (gR)^{1/2}$.

41. 0.34.

43. 2.1×10^2 N.

45. 5.91°, 14.3 N.

47. (a) 5.8×10^3 N; (b) 4.1×10^2 N; (c) 31 m/s.

49. $F_T = 2\pi m R f^2$.

51. 66 km/h $< v <$ 123 km/h.

53. (a) $(1.6$ m/s$^2)\mathbf{i}$; (b) $(0.98$ m/s$^2)\mathbf{i} - (1.7$ m/s$^2)\mathbf{j}$; (c) $-(4.9$ m/s$^2)\mathbf{i} - (1.6$ m/s$^2)\mathbf{j}$.

55. (a) 9.0 m/s^2; (b) 15 m/s^2.

57. $\tau = m/b$.

59. (a) $v = (mg/b) + [v_0 - (mg/b)]e^{-bt/m}$; (b) $v = -(mg/b) + [v_0 + (mg/b)]e^{-bt/m}$, $v \geq 0$.

61. (b) 1.8°.

63. 10 m.

65. $\mu_s = 0.41$.

67. 2.3.

69. 101 N, $\mu_k = 0.719$.

71. (b) Will slide.

73. Emerges with a speed of 13 m/s.

75. 27.6 m/s, 0.439 rev/s.

77. $\Sigma F_{\tan} = 3.3 \times 10^3$ N, $\Sigma F_R = 2.0 \times 10^3$ N.

79. (a) $F_{NC} > F_{NB} > F_{NA}$; (b) heaviest at C, lightest at A; (c) $v_{A\max} = (gR)^{1/2}$.

81. (a) 1.23 m/s; (b) 3.01 m/s.

83. $\phi = 31°$.

85. (a) $r = v^2/g\cos\theta$; (b) 92 m.

87. (a) 59 s; (b) greater normal force.

89. 29.2 m/s.

91. 302 m, 735 m.

93. $g(1 - \mu_s\tan\phi)/4\pi^2 f^2 (\tan\phi + \mu_s)$ $< r < g(1 + \mu_s\tan\phi)/ 4\pi^2 f^2 (\tan\phi - \mu_s)$.

CHAPTER 6

1. 1.52×10^3 N.

3. 1.6 m/s^2.

5. $g_h = 0.91 g_{\text{surface}}$.

7. 1.9×10^{-8} N toward center of square.

9. $Gm^2\{(2/x_0^2) + [3x_0/(x_0^2 + y_0^2)^{3/2}]\}\mathbf{i}$ $+ Gm^2\{[3y_0/(x_0^2 + y_0^2)^{3/2}]$ $+ (4/y_0^2)\}\mathbf{j}$.

11. 1.26.

13. 3.46×10^8 m from Earth's center.

15. (b) g decreases with an increase in height; (c) 9.493 m/s^2.

19. 7.56×10^3 m/s.

21. 2.0 h.

23. (a) 56 kg; (b) 56 kg; (c) 75 kg; (d) 38 kg; (e) 0.

25. (a) 22 N (toward the Moon); (b) -1.7×10^2 N (away from the Moon).

27. (a) Gravitational force provides required centripetal acceleration; (b) 9.6×10^{26} kg.

29. 7.9×10^3 m/s.

31. $v = (Gm/L)^{1/2}$.

33. 0.0587 days (1.41 h).

35. 1.6×10^2 yr.

37. 2×10^8 yr.

39. $r_{\text{Europa}} = 6.71 \times 10^5$ km, $r_{\text{Ganymede}} = 1.07 \times 10^6$ km, $r_{\text{Callisto}} = 1.88 \times 10^6$ km.

41. 9.0 Earth-days.

43. (a) 2.1×10^2 A.U. $(3.1 \times 10^{13}$ m); (b) 4.2×10^2 A.U.; (c) 4.2×10^2.

45. (a) 5.9×10^{-3} N/kg; (b) not significant.

47. 2.7×10^3 km.

49. 6.7×10^{12} m/s^2.

51. 4.4×10^7 m/s^2.

53. $G' = 1 \times 10^{-4}$ N·m^2/kg$^2 \approx 10^6 G$.

55. 5 h 35 min, 19 h 50 min.

57. (a) 10 h; (b) 6.5 km; (c) 4.2×10^{-3} m/s^2.

59. 5.4×10^{12} m, in the Solar System, Pluto.

61. $2.3 g_{\text{Earth}}$.

63. $m_P = g_P r^2/G$.

67. 7.9×10^3 m/s.

CHAPTER 7

1. 6.86×10^3 J.

3. 1.27×10^4 J.

5. 8.1×10^3 J.

7. 1 J $= 1 \times 10^7$ erg $= 0.738$ ft·lb.

9. 1.0×10^4 J.

13. (a) 3.6×10^2 N; (b) -1.3×10^3 J; (c) -4.6×10^3 J; (d) 5.9×10^3 J; (e) 0.

15. $W_{FN} = W_{mg} = 0$, $W_{FP} = -W_{\text{fr}} = 2.0 \times 10^2$ J.

21. (a) -16.1; (b) -238; (c) -3.9.

23. $\mathbf{C} = -1.3\mathbf{i} + 1.8\mathbf{j}$.

25. $\theta_x = 42.7°, \theta_y = 63.8°, \theta_z = 121°$.

27. 95°, $-35°$ from x-axis.

31. 0.089 J.

33. 2.3×10^3 J.

35. 2.7×10^3 J.

37. $(kX^2/2) + (aX^4/4) + (bX^5/5)$.

39. (a) 5.0×10^{10} J.

41. (a) $\sqrt{3}$; (b) $\frac{1}{4}$.

43. -5.02×10^5 J.

45. 3.0×10^2 N in the direction of the motion of the ball.

47. 24 m/s (87 km/h or 54 mi/h), the mass cancels.

49. (a) 72 J; (b) -35 J; (c) 37 J.

51. 10.2 m/s.

53. $\mu_k = F/2mg$.

55. (a) 6.5×10^2 J; (b) -4.9×10^2 J; (c) 0; (d) 4.0 m/s.

57. (a) 1.66×10^5 J; (b) 21.0 m/s; (c) 2.13 m.

59. $v_p = 2.0 \times 10^7$ m/s, $v_{pc} = 2.0 \times 10^7$ m/s; $v_e = 2.9 \times 10^8$ m/s, $v_{ec} = 8.4 \times 10^8$ m/s.

61. 1.74×10^3 J.

63. (a) 15 J; (b) 4.2×10^2 J; (c) -1.8×10^2 J; (d) -2.5×10^2 J; (e) 0; (f) 10 J.

65. (a) 12 J; (b) 10 J; (c) -2.1 J.

67. 86 kJ, $\theta = 42°$.

69. $(A/k)e^{-(0.10\,\mathrm{m})k}$.

71. 1.5 N.

73. 5.0×10^3 N/m.

75. (a) 6.6°; (b) 10.3°.

CHAPTER 8

1. 0.924 m.

3. 2.2×10^3 J.

5. (a) 51.7 J; (b) 15.1 J; (c) 51.7 J.

7. (a) Conservative; (b) $\frac{1}{2}kx^2 - \frac{1}{4}ax^4 - \frac{1}{5}bx^5$ + constant.

9. (a) $\frac{1}{2}k(x^2 - x_0^2)$; (b) same.

11. 45.4 m/s.

13. 6.5 m/s.

15. (a) 1.0×10^2 N/m; (b) 22 m/s^2.

17. (a) 8.03 m/s; (b) 3.44 m.

19. (a) $v_{\max} = [v_0^2 + (kx_0^2/m)]^{1/2}$; (b) $x_{\max} = [x_0^2 + (mv_0^2/k)]^{1/2}$.

21. (a) 2.29 m/s; (b) 1.98 m/s; (c) 1.98 m/s; (d) $F_{Ta} = 0.87$ N, $F_{Tb} = 0.80$ N, $F_{Tc} = 0.80$ N; (e) $v_a = 2.59$ m/s, $v_b = 2.31$ m/s, $v_c = 2.31$ m/s.

23. $k = 12Mg/h$.

25. 4.5×10^6 J.

27. (a) 22 m/s; (b) 2.9×10^2 m.

29. 13 m/s.

31. 0.23.

33. 0.40.

35. (a) 0.13 m; (b) 0.77; (c) 0.46 m/s.

37. (a) $K = GM_E m_S/2r_S$; (b) $U = -GM_E m_S/r_S$; (c) $-\frac{1}{2}$.

39. (a) 6.2×10^5 m/s; (b) 4.2×10^4 m/s, $v_{esc}/v_{orbit} = \sqrt{2}$.

45. (a) 1.07×10^4 m/s; (b) 1.17×10^4 m/s; (c) 1.12×10^4 m/s.

47. (a) $dv_{esc}/dr = -\frac{1}{2}(2GM_E/r^3)^{1/2}$ $= -v_{esc}/2r$; (b) 1.09×10^4 m/s.

49. $GmM_E/12r_E$.

51. 1.1×10^4 m/s.

55. 5.4×10^2 N.

57. (a) 1.0×10^3 J; (b) 1.0×10^3 W.

59. 2.1×10^4 W, 28 hp.

61. 4.8×10^2 W.

63. 1.2×10^3 W.

65. 1.8×10^6 W.

67. (a) -25 W; (b) $+4.3 \times 10^3$ W; (c) $+1.5 \times 10^3$ W.

69. (a) 80 J; (b) 60 J; (c) 80 J; (d) 5.7 m/s at $x = 0$; (e) 32 m/s^2 at $x = \pm x_0$.

71. $a^2/4b$.

73. 8.0 m/s.

75. 32.5 hp.

77. (a) 28 m/s; (b) 1.2×10^2 m.

79. (a) $(2gL)^{1/2}$; (b) $(1.2gL)^{1/2}$.

81. (a) 1.1×10^6 J; (b) 60 W (0.081 hp); (c) 4.0×10^2 W (0.54 hp).

83. (a) 40 m/s; (b) 2.6×10^5 W.

87. (a) 29°; (b) 6.4×10^2 N; (c) 9.2×10^2 N.

89. (a) $-\dfrac{U_0}{r}\left(\dfrac{r_0}{r} + 1\right)e^{-r/r_0}$; (b) 0.030; (c) $F(r) = -C/r^2$, 0.11.

91. 6.7 hp.

93. (a) 2.8 m; (b) 1.5 m; (c) 1.5 m.

95. 76 hp.

97. (a) 5.00×10^3 m/s; (b) 2.89×10^3 m/s.

CHAPTER 9

1. 6.0×10^7 N, up.

3. (a) 0.36 kg·m/s; (b) 0.12 kg·m/s.

5. $(26\,\text{N·s})\mathbf{i} - (28\,\text{N·s})\mathbf{j}$.

7. (a) $(8h/g)^{1/2}$; (b) $(2gh)^{1/2}$; (c) $-(8m^2gh)^{1/2}$ (up); (d) mg (down), a surprising result.

9. 3.4×10^4 kg.

11. 4.4×10^3 m/s.

13. -0.667 m/s (opposite to the direction of the package).

15. 2, lesser kinetic energy has greater mass.

17. $\frac{3}{2}v_0\mathbf{i} - v_0\mathbf{j}$.

19. 1.1×10^{-22} kg·m/s, 36° from the direction opposite to the electron's.

21. (a) $(100\,\text{m/s})\mathbf{i} + (50\,\text{m/s})\mathbf{j}$; (b) 3.3×10^5 J.

23. 130 N, not large enough.

25. 1.1×10^3 N.

27. (a) $2mv/\Delta t$; (b) $2mv/t$.

29. (a)

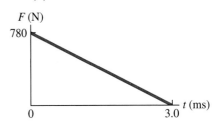

(b) 1.2 N·s; (c) 1.2 N·s; (d) 3.9 g.

31. (a) $(0.84\,\text{N}) + (1.2\,\text{N/s})t$; (b) 18.5 N; (c) $(0.12\,\text{kg/s})\{[(49\,\text{m}^2/\text{s}^2) - (1.18\,\text{m}^2/\text{s}^3)t]^{1/2} + (9.80\,\text{m/s}^2)t\}$, 18.3 N.

33. $v_1' = -1.40$ m/s (rebound), $v_2' = 2.80$ m/s.

35. (a) 2.7 m/s; (b) 0.84 kg.

37. 3.2×10^3 m/s.

39. (a) 1.00; (b) 0.89; (c) 0.29; (d) 0.019.

41. (a) 0.32 m; (b) -3.1 m/s (rebound), 4.9 m/s; (c) Yes.

43. (a) $+M/(m + M)$; (b) 0.964.

45. 141°.

47. (b) $e = (h'/h)^{1/2}$.

49. (a) $v_1' = v_2' = 1.9$ m/s; (b) $v_1' = -1.6$ m/s, $v_2' = 7.9$ m/s; (c) $v_1' = 0$, $v_2' = 5.2$ m/s; (d) $v_1' = 3.1$ m/s, $v_2' = 0$; (e) $v_1' = -4.0$ m/s, $v_2' = 12$ m/s; result for (c) is reasonable, result for (d) is not reasonable, result for (e) is not reasonable.

51. 61° from first eagles's direction, 6.8 m/s.

53. (a) 30°; (b) $v_2' = v/\sqrt{3}$, $v_1' = v/\sqrt{3}$; (c) $\frac{2}{3}$.

55. $\theta_1' = 76°$, $v_n' = 5.1 \times 10^5$ m/s,
$v_{He}' = 1.8 \times 10^5$ m/s.

59. 6.5×10^{-11} m from the carbon atom.

61. 0.030 nm above center of H triangle.

63. $x_{CM} = 1.10$ m (East),
$y_{CM} = -1.10$ m (South).

65. $x_{CM} = 0$, $y_{CM} = 2r/\pi$.

67. $x_{CM} = 0$, $y_{CM} = 0$,
$z_{CM} = 3h/4$ above the point.

69. (a) 4.66×10^6 m.

71. (a) $x_{CM} = 4.6$ m; (b) 4.3 m;
(c) 4.6 m.

73. $mv/(M + m)$ up, balloon will also stop.

75. 55 m.

77. 0.899 hp.

79. (a) 2.3×10^3 N; (b) 2.8×10^4 N;
(c) 1.1×10^4 hp.

81. A "scratch shot".

83. 1.4×10^4 N, 43.3°.

85. 5.1×10^2 m/s.

87. $m_2 = 4.00m$.

89. 50%.

91. (a) No; (b) $v_1/v_2 = -m_2/m_1$;
(c) m_2/m_1; (d) does not move;
(e) center of mass will move.

93. 8.29 m/s.

95. (a) 2.5×10^{-13} m/s; (b) 1.7×10^{-17};
(c) 0.19 J.

97. $m \le M/3$.

99. 29.6 km/s.

101. (a) 2.3 N·s; (b) 4.5×10^2 N.

103. (a) Inelastic collision; (b) 0.10 s;
(c) -1.4×10^5 N.

105. 0.28 m, 1.1 m.

CHAPTER 10

1. (a) $\pi/6$ rad $= 0.524$ rad;
(b) $19\pi/60 = 0.995$ rad;
(c) $\pi/2 = 1.571$ rad;
(d) $2\pi = 6.283$ rad;
(e) $7\pi/3 = 7.330$ rad.

3. 2.3×10^3 m.

5. (a) 0.105 rad/s;
(b) 1.75×10^{-3} rad/s;
(c) 1.45×10^{-4} rad/s; (d) zero.

7. (a) 464 m/s; (b) 185 m/s;
(c) 355 m/s.

9. (a) 262 rad/s;
(b) 46 m/s, 1.2×10^4 m/s² radial.

11. 7.4 cm.

13. (a) 1.75×10^{-4} rad/s²;
(b) $a_R = 1.17 \times 10^{-2}$ m/s²,
$a_{tan} = 7.44 \times 10^{-4}$ m/s².

15. (a) 0.58 rad/s2; (b) 12 s.

17. (a) $(1.67 \text{ rad/s}^4)t^3 - (1.75 \text{ rad/s}^3)t^2$;
(b) $(0.418 \text{ rad/s}^4)t^4 - (0.583 \text{ rad/s}^3)t^3$;
(c) 6.4 rad/s, 2.0 rad.

19. (a) ω_1 is in the $-x$-direction,
ω_2 is in the $+z$-direction;
(b) $\omega = 61.0$ rad/s, 35.0° above
$-x$-axis;
(c) $-(1.75 \times 10^3 \text{ rad/s}^2)\mathbf{j}$.

21. (a) 35 m·N; (b) 30 m·N.

23. 1.2 m·N (clockwise).

25. 3.5×10^2 N, 2.0×10^3 N.

27. 53 m·N.

29. (a) 3.5 kg·m²; (b) 0.024 m·N.

31. 2.25×10^3 kg·m², 8.8×10^3 m·N.

33. 9.5×10^4 m·N.

35. 10 m/s.

37. (a)

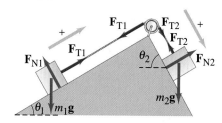

(b) $F_{T1} = 47$ N, $F_{T2} = 75$ N;
(c) 7.0 m·N, 1.7 kg·m².

39. Thin hoop (through center):
$k = R_0$;

Thin hoop (through diameter):
$k = [(R_0^2/2) + (w^2/12)]^{1/2}$;

Solid cylinder (through center):
$k = R/\sqrt{2}$;

Hollow cylinder (through center):
$k = [(R_1^2 + R_2^2)/2]^{1/2}$;

Uniform sphere (through center):
$k = (2r_0^2/5)^{1/2}$;

Rod (through center): $k = \ell/\sqrt{12}$;
Rod (through end): $k = \ell/\sqrt{3}$;

Plate (through center):
$k = [(\ell^2 + w^2)/12]^{1/2}$.

41. (a) 4.18 rad/s²; (b) 8.37 m/s²;
(c) 421 m/s²; (d) 3.07×10^3 N;
(e) 1.14°.

43. (a) $I_a = Ms^2/12$; (b) $I_b = Ms^2/12$.

45. (a) $5.30MR_0^2$; (b) -15%.

47. (a) $9MR_0^2/16$; (b) $MR_0^2/4$;
(c) $5MR_0^2/4$.

51. (b) $M\ell^2/12$, $Mw^2/12$.

53. 0.38 rev/s.

55. (a) As moment of inertia increases,
angular velocity must decrease;
(b) 1.6.

57. (a) 7.1×10^{33} kg·m²/s;
(b) 2.7×10^{40} kg·m²/s.

59. 0.45 rad/s, 0.80 rad/s.

61. 2.33×10^4 J.

63. 5×10^9, loss of gravitational
potential energy.

65. 1.4 m/s.

67. (a) 2.5 kg·m²; (b) 0.58 kg·m²;
(c) 0.35 s; (d) -72 J; (e) rotating.

69. 12.4 m/s.

71. 1.4×10^2 J.

73. (a) 4.48 m/s; (b) 1.21 J;
(c) $\mu_s \ge 0.197$.

75. $v = [10g(R_0 - r_0)/7]^{1/2}$.

77. (a) 4.5×10^5 J; (b) 0.18 (18%);
(c) 1.71 m/s²; (d) 6.4%.

79. (a) $12v_0^2/49\mu_k g$;
(b) $v = 5v_0/7$, $\omega = 5v_0/7R$.

81. (a) 4.5 m/s², 19 rad/s²; (b) 5.8 m/s;
(c) 15.3 J; (d) 1.4 J;
(e) $K = 16.7$ J, $\Delta E = 0$;
(f) $a = 4.5$ m/s², $v = 5.8$ m/s, 14.1 J.

83. $\theta_{Sun} = 9.30 \times 10^{-3}$ rad (0.53°),
$\theta_{Moon} = 9.06 \times 10^{-3}$ rad (0.52°).

85. $\omega_1/\omega_2 = R_2/R_1$.

87. $\ell/2$, $\ell/2$.

89. (a) $-(I_W/I_P)\omega_W$ (down);
(b) $-(I_W/2I_P)\omega_W$ (down);
(c) $(I_W/I_P)\omega_W$ (up); (d) 0.

91. (a) $\omega_R/\omega_F = N_F/N_R$; (b) 4.0;
(c) 1.5.

93. (a) 1.5×10^2 rad/s²;
(b) 1.2×10^3 N.

95. (a) 0.070 rad/s²; (b) 40 rpm.

97. 7.9 N.

99. (b) 2.2×10^3 rad/s; (c) 24 min.

101. (a) 2.9 m; (b) 3.6 m.

103. (a) 1.2 rad/s; (b) 2.0×10^3 J,
1.2×10^3 J, loss of 8.0×10^2 J,
decrease of 40%.

105. (a) 1.7 m/s; (b) 0.84 m/s.

107. (a) $h_{min} = 2.7R_0$;
(b) $h_{min} = 2.7R_0 - 1.7r_0$.

109. (a) 0.84 m/s; (b) 0.96.

CHAPTER 11

7. (a) $-7.0\mathbf{i} - 14.0\mathbf{j} + 19.3\mathbf{k}$; (b) 164°.

11. $-(30.3 \text{ m}\cdot\text{N})\mathbf{k}$ (in $-z$-direction).

13. $(18 \text{ m}\cdot\text{kN})\mathbf{i} \pm (14 \text{ m}\cdot\text{kN})\mathbf{j}$
$\mp (19 \text{ m}\cdot\text{kN})\mathbf{k}$.

19. $(55\mathbf{i} - 90\mathbf{j} + 42\mathbf{k})\,\mathrm{kg \cdot m^2/s}$.

21. (a) $\left[(7m/9) + (M/6)\right]\ell^2\omega^2$;
(b) $\left[(14m/9) + (M/3)\right]\ell^2\omega$.

23. $2.30\,\mathrm{m/s^2}$.

25. (a) $L = \left[R_0 M_1 + R_0 M_2 + (I/R_0)\right]v$;
(b) $a = M_2 g/\left[M_1 + M_2 + (I/R_0^2)\right]$.

27. Rod rotates at 7.8 rad/s about the center of mass, which moves with constant velocity of 0.21 m/s.

31. $F_1 = \left[(d + r\cos\phi)/2d\right]m_1 r\omega^2\sin\phi$,
$F_2 = \left[(d - r\cos\phi)/2d\right]m_1 r\omega^2\sin\phi$.

33. $16\,\mathrm{N}, -7.5\,\mathrm{N}$.

35. $3m^2v^2/g(3m + 4M)(m + M)$.

37. $(1 - 4.7 \times 10^{-13})\omega_E$.

39. (a) 14 m/s; (b) 6.8 rad/s.

41. $1.02 \times 10^{-3}\,\mathrm{kg \cdot m^2}$.

43. 2.2 rad/s (0.35 rev/s).

45. $\tan^{-1}\left(r\omega^2/g\right)$.

47. (a) g, along a radial line; (b) $0.998g$, $0.0988°$ south from a radial line;
(c) $0.997g$, along a radial line.

49. North or south direction.

51. (a) South; (b) $\omega D^2 \sin\lambda/v_0$;
(c) 0.46 m.

53. (a) $(-9.0\mathbf{i} + 12\mathbf{j} - 8.0\mathbf{k})\,\mathrm{kg \cdot m^2/s}$;
(b) $(9.0\mathbf{j} - 6.0\mathbf{k})\,\mathrm{m \cdot N}$.

55. (a) Turn in the direction of the lean;
(b) $\Delta L = 0.98\,\mathrm{kg \cdot m^2/s}$,
$\Delta L = 0.18 L_0$.

57. (a) $1.8 \times 10^3\,\mathrm{kg \cdot m^2/s^2}$;
(b) $1.8 \times 10^3\,\mathrm{m \cdot N}$; (c) $2.1 \times 10^3\,\mathrm{W}$.

59. $v_{\mathrm{CM}} = (3g\ell/4)^{1/2}$.

61. $(19\,\mathrm{m/s})(1 - \cos\theta)^{1/2}$.

63. (a) 2.3×10^4 rev/s;
(b) 5.7×10^3 rev/s.

CHAPTER 12

1. 379 N, 141°.

3. $1.6 \times 10^3\,\mathrm{m \cdot N}$.

5. 6.52 kg.

7. 2.84 m from the adult.

9. 0.32 m.

11. $F_{\mathrm{T1}} = 3.4 \times 10^3\,\mathrm{N}$,
$F_{\mathrm{T2}} = 3.9 \times 10^3\,\mathrm{N}$.

13. $F_1 = -2.94 \times 10^3\,\mathrm{N}$ (down),
$F_2 = 1.47 \times 10^4\,\mathrm{N}$.

15. Top hinge: $F_{Ax} = 55.2\,\mathrm{N}$,
$F_{Ay} = 63.7\,\mathrm{N}$; bottom hinge:
$F_{Bx} = -55.2\,\mathrm{N}, F_{By} = 63.7\,\mathrm{N}$.

17. (a)

(b) $1.5 \times 10^4\,\mathrm{N}$; (c) $6.7 \times 10^3\,\mathrm{N}$.

19. $F_{\mathrm{T}} = 1.4 \times 10^3\,\mathrm{N}$ (up),
$F_{\mathrm{bone}} = 2.1 \times 10^3\,\mathrm{N}$ (down).

21. $2.7 \times 10^3\,\mathrm{N}$.

23. 89.5 cm from the feet.

25. $F_1 = 5.8 \times 10^3\,\mathrm{N}, F_2 = 5.6 \times 10^3\,\mathrm{N}$.

27. (a) $2.1 \times 10^2\,\mathrm{N}$; (b) $2.0 \times 10^3\,\mathrm{N}$.

29. $7.1 \times 10^2\,\mathrm{N}$.

31. $F_{\mathrm{T}} = 2.5 \times 10^2\,\mathrm{N}$,
$F_{AH} = 2.5 \times 10^2\,\mathrm{N}$,
$F_{AV} = 2.0 \times 10^2\,\mathrm{N}$.

33. (a) 1.00 N; (b) 1.25 N.

35. $\theta_{\max} = 40°$, same.

37. (a) $F_{\mathrm{T}} = 182\,\mathrm{N}$; (b) $F_{\mathrm{N1}} = 352\,\mathrm{N}$,
$F_{\mathrm{N2}} = 236\,\mathrm{N}$; (c) $F_{\mathrm{B}} = 298\,\mathrm{N}, 52.4°$.

39. $1.0 \times 10^2\,\mathrm{N}$.

41. (a) $1.2 \times 10^5\,\mathrm{N/m^2}$; (b) 2.4×10^{-6}.

43. (a) $1.3 \times 10^5\,\mathrm{N/m^2}$; (b) 6.5×10^{-7};
(c) 0.0062 mm.

45. $9.6 \times 10^6\,\mathrm{N/m^2}$.

47. $9.0 \times 10^7\,\mathrm{N/m^2}, 9.0 \times 10^2$ atm.

49. $2.2 \times 10^7\,\mathrm{N}$.

51. (a) $1.1 \times 10^2\,\mathrm{m \cdot N}$; (b) wall;
(c) all three.

53. $3.9 \times 10^2\,\mathrm{N}$, thicker strings, maximum strength is exceeded.

55. (a) $4.4 \times 10^{-5}\,\mathrm{m^2}$; (b) 2.7 mm.

57. 1.2 cm.

61. (a) $F_{\mathrm{T}} = 129\,\mathrm{kN}$;
$F_{\mathrm{A}} = 141\,\mathrm{kN}, 23.5°$;
(b) $F_{DE} = 64.7\,\mathrm{kN}$ (tension),
$F_{CE} = 32.3\,\mathrm{kN}$ (compression),
$F_{CD} = 64.7\,\mathrm{kN}$ (compression),
$F_{BD} = 64.7\,\mathrm{kN}$ (tension),
$F_{BC} = 64.7\,\mathrm{kN}$ (tension),
$F_{AC} = 97.0\,\mathrm{kN}$ (compression),
$F_{AB} = 64.7\,\mathrm{kN}$ (compression).

63. (a) $4.8 \times 10^{-2}\,\mathrm{m^2}$; (b) $6.8 \times 10^{-2}\,\mathrm{m^2}$.

65. $F_{AB} = 5.44 \times 10^4\,\mathrm{N}$ (compression),
$F_{ACx} = 2.72 \times 10^4\,\mathrm{N}$ (tension),
$F_{BC} = 5.44 \times 10^4\,\mathrm{N}$ (tension),
$F_{BD} = 5.44 \times 10^4\,\mathrm{N}$ (compression),
$F_{CD} = 5.44 \times 10^4\,\mathrm{N}$ (tension),
$F_{CE} = 2.72 \times 10^4\,\mathrm{N}$ (tension),
$F_{DE} = 5.44 \times 10^4\,\mathrm{N}$ (compression).

67. 12 m.

69. $M_{\mathrm{C}} = 0.191\,\mathrm{kg}, M_{\mathrm{D}} = 0.0544\,\mathrm{kg}$,
$M_{\mathrm{A}} = 0.245\,\mathrm{kg}$.

71. (a) $Mg\left[h/(2R - h)\right]^{1/2}$;
(b) $Mg\left[h(2R - h)\right]^{1/2}/(R - h)$.

73. $\theta_{\max} = 29°$.

75. 6, 2.0 m apart.

77. 3.8.

79. $5.0 \times 10^5\,\mathrm{N}$, 3.2 m.

81. (a) 600 N; (b) $F_{\mathrm{A}} = 0, F_{\mathrm{B}} = 1200\,\mathrm{N}$;
(c) $F_{\mathrm{A}} = 150\,\mathrm{N}, F_{\mathrm{B}} = 1050\,\mathrm{N}$;
(d) $F_{\mathrm{A}} = 750\,\mathrm{N}, F_{\mathrm{B}} = 450\,\mathrm{N}$.

83. $6.5 \times 10^2\,\mathrm{N}$.

85. 0.67 m.

87. Right end is safe, left end is not safe, 0.10 m.

89. (a) $F_{\mathrm{L}} = 3.3 \times 10^2\,\mathrm{N}$ up,
$F_{\mathrm{R}} = 2.3 \times 10^2\,\mathrm{N}$ down;
(b) 65 cm from right hand;
(c) 123 cm from right hand.

91. $\theta \geq 40°$.

93. (b) beyond the table;
(c) $D = L \sum_{i=1}^{n} \dfrac{1}{2i}$; (d) 32 bricks.

95. $F_{\mathrm{TB}} = 134\,\mathrm{N}, F_{\mathrm{TA}} = 300\,\mathrm{N}$.

97. $2.6w, 31°$ above horizontal.

CHAPTER 13

1. $3 \times 10^{11}\,\mathrm{kg}$.

3. $4.3 \times 10^2\,\mathrm{kg}$.

5. 0.8477.

7. (a) $3 \times 10^7\,\mathrm{N/m^2}$;
(b) $2 \times 10^5\,\mathrm{N/m^2}$.

9. 1.1 m.

11. $8.28 \times 10^3\,\mathrm{kg}$.

13. $1.2 \times 10^5\,\mathrm{N/m^2}$,
$2.3 \times 10^7\,\mathrm{N}$ (down),
$1.2 \times 10^5\,\mathrm{N/m^2}$.

15. $6.54 \times 10^2\,\mathrm{kg/m^3}$.

17. $3.36 \times 10^4\,\mathrm{N/m^2}$ (0.331 atm).

19. (a) $1.41 \times 10^5\,\mathrm{Pa}$; (b) $9.8 \times 10^4\,\mathrm{Pa}$.

21. (a) 0.34 kg; (b) $1.5 \times 10^4\,\mathrm{N}$ (up).

23. (c) $\geq 0.38h$, no.

27. $4.70 \times 10^3\,\mathrm{kg/m^3}$.

29. $8.5 \times 10^2\,\mathrm{kg}$.

31. Copper.

33. (a) $1.14 \times 10^6\,\mathrm{N}$; (b) $4.0 \times 10^5\,\mathrm{N}$.

35. (b) Above the center of gravity.

37. 0.88.

39. $7.9 \times 10^2\,\mathrm{kg}$.

43. 4.1 m/s.

45. 9.5 m/s.

47. $1.5 \times 10^5\,\mathrm{N/m^2} = 1.5$ atm.

49. $4.11 \times 10^{-3}\,\mathrm{m^3/s}$.

51. $1.7 \times 10^6\,\mathrm{N}$.

59. (a) $2[h_1(h_2 - h_1)]^{1/2}$;
(b) $h_1' = h_2 - h_1$.

61. 0.072 Pa·s.

63. 4.0×10^3 Pa.

65. 11 cm.

67. (a) Laminar; (b) 3200, turbulent.

69. 1.9 m.

71. 9.1×10^{-3} N.

73. (a) $\gamma = F/4\pi r$; (b) 0.024 N/m.

75. (a) 0.88 m; (b) 0.55 m; (c) 0.24 m.

77. 1.5×10^2 N $\leq F \leq 2.2 \times 10^2$ N.

79. 0.051 atm.

81. 0.63 N.

83. 5 km.

85. 5.3×10^{18} kg.

87. 2.6 m.

89. 39 people.

91. 37 N, not float.

93. $d = D[v_0^2/(v_0^2 + 2gy)]^{1/4}$.

95. (a) 3.2 m/s; (b) 19 s.

97. 1.9×10^2 m/s.

CHAPTER 14

1. 0.60 m.

3. 1.15 Hz.

5. (a) 2.4 N/m; (b) 12 Hz.

7. (a) $0.866\, x_{max}$; (b) $0.500\, x_{max}$.

9. $0.866\, A$.

11. $[(k_1 + k_2)/m]^{1/2}/2\pi$.

13. (a) 8/7 s, 0.875 Hz; (b) 3.3 m, -10.4 m/s; (c) $+18$ m/s, -57 m/s^2.

15. 3.6 Hz.

19. (a) $y = -(0.220\text{ m})\sin[(37.1\text{ s}^{-1})t]$;
(b) maximum extensions at 0.0423 s, 0.211 s, 0.381 s, …; minimum extensions at 0.127 s, 0.296 s, 0.465 s, ….

21. $f = (3k/M)^{1/2}/2\pi$.

25. (a) $x = (12.0\text{ cm})\cos[(25.6\text{ s}^{-1})t + 1.89\text{ rad}]$;
(b) $t_{max} = 0.294$ s, 0.539 s, 0.784 s, …; $t_{min} = 0.171$ s, 0.416 s, 0.661 s, …;
(c) -3.77 cm; (d) $+13.1$ N (up); (e) 3.07 m/s, 0.110 s.

27. (a) 0.650 m; (b) 1.34 Hz; (c) 29.8 J; (d) $K = 25.0$ J, $U = 4.8$ J.

29. 9.37 m/s.

31. $A_1 = 2.24A_2$.

33. (a) 4.2×10^2 N/m; (b) 3.3 kg.

35. 352.6 m/s.

39. 0.9929 m.

41. (a)0.248 m; (b)2.01 s.

43. (a) $-12°$; (b) $+1.9°$; (c) $-13°$.

45. $\frac{1}{3}$.

47. 1.08 s.

49. 0.31 g.

51. (a) 1.6 s.

53. 3.5 s.

55. (a) 0.727 s; (b) 0.0755;
(c) $x = (0.189\text{ m})e^{-(0.108/\text{s})t}$
$\sin[(8.64\text{ s}^{-1})t]$.

57. (a) 8.3×10^{-4}%; (b) 39 periods.

59. (a) 5.03 Hz; (b) 0.0634 s^{-1};
(c) 110 oscillations.

61. $A_0 k/F_0$

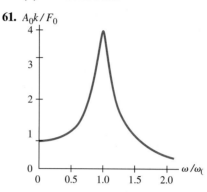

65. (a) 198 s; (b) 8.7×10^{-6} W;
(c) 8.8×10^{-4} Hz on either side of f_0.

69. (a) 0.63 Hz; (b) 0.65 m/s; (c) 0.077 J.

71. 151 N/m, 20.3 m.

73. 0.11 m.

75. 3.6 Hz.

77. (a) 1.1 Hz; (b) 13 J.

79. (a) 90 N/m; (b) 8.9 cm.

81. $k = \rho_{\text{water}}\, gA$.

83. Water will oscillate with SHM, $k = 2\rho gA$, the density and the cross section.

85. $T = 2\pi(ma/2k\,\Delta a)^{1/2}$.

87. (a) 1.64 s; (b) 0.67 m.

CHAPTER 15

1. 2.3 m/s.

3. 1.26 m.

5. 0.72 m.

7. 2.7 N.

9. (a) 75 m/s; (b) 7.8×10^3 N.

11. (a) 1.3×10^3 km;
(b) cannot be determined.

13. (a) 0.25; (b) 0.50.

17. (a) 0.30 W; (b) 0.25 cm.

19. $D = D_M \sin[2\pi(x/\lambda + t/T) + \phi]$.

21. (a) 41 m/s; (b) 6.4×10^4 m/s^2;
(c) 41 m/s, 8.2×10^3 m/s^2.

23. (a, c)

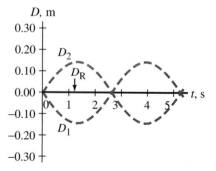

(b) $D = (0.45\text{ m})\cos[(3.0\text{ m}^{-1})x - (6.0\text{ s}^{-1})t + 1.2]$.
(d) $D = (0.45\text{ m})\cos[(3.0\text{ m}^{-1})x + (6.0\text{ s}^{-1})t + 1.2]$.

25. $D = -(0.020\text{ cm})\cos[(8.01\text{ m}^{-1})x - (2.76 \times 10^3\text{ s}^{-1})t]$.

27. The function is a solution.

31. (a) $v_2/v_1 = (\mu_1/\mu_2)^{1/2}$;
(b) $\lambda_2/\lambda_1 = v_2/v_1 = (\mu_1/\mu_2)^{1/2}$;
(c) lighter cord.

33. (c) $A_T = [2k_1/(k_2 + k_1)]A = [2v_2/(v_1 + v_2)]A$.

35. (b) $2D_M \cos(\frac{1}{2}\phi)$, purely sinusoidal;
(d) $D = \sqrt{2}\, D_M \sin(kx - \omega t + \pi/4)$.

37. 440 Hz, 880 Hz, 1320 Hz, 1760 Hz.

39. $f_n = n(0.50\text{ Hz}), n = 1, 2, 3, …$;
$T_n = (2.0\text{ s})/n, n = 1, 2, 3, …$.

41. 70 Hz.

45. 4.

47. (a) $D_2 = (4.2\text{ cm})\sin[(0.71\text{ cm}^{-1})x + (47\text{ s}^{-1})t + 2.1]$;
(b) $D_{\text{resultant}} = (8.4\text{ cm}) \sin[(0.71\text{ cm}^{-1})x + 2.1]\cos[(47\text{ s}^{-1})t]$.

49. 308 Hz.

51. (a)

(*b*)

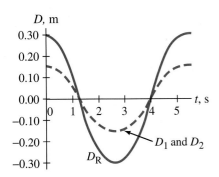

53. 5.4 km/s.

55. 29°.

57. 24°.

59. Speed will be greater in the less dense rod by a factor of $\sqrt{2}$.

61. (*a*) 0.050 m; (*b*) 2.3.

63. 0.99 m.

65. (*a*) solid curves,

(*c*) dashed curves;

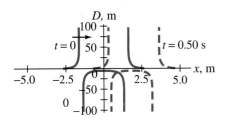

(*b*)
$$D = (4.0 \text{ m}^3)/\{[x - (3.0 \text{ m/s})t]^2 - 2.0 \text{ m}^2\};$$

(*d*)

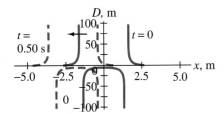

$$D = (4.0 \text{ m}^3)/\{[x + (3.0 \text{ m/s})t]^2 - 2.0 \text{ m}^2\}.$$

67. (*a*) 784 Hz, 1176 Hz, 880 Hz,

1320 Hz; (*b*) 1.26; (*c*) 1.12; (*d*) 0.794.

69. $\lambda_n = 4L/(2n - 1), n = 1, 2, 3, \ldots$.

71. $y = (3.5 \text{ cm}) \cos[(1.05 \text{ cm}^{-1})x$

$- (1.39 \text{ s}^{-1})t]$.

73.

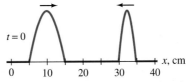

$t = 0$

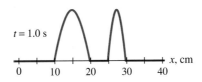

$t = 1.0$ s

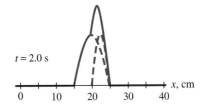

$t = 2.0$ s

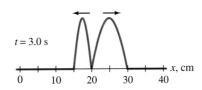

$t = 3.0$ s

CHAPTER 16

1. 2.6×10^2 m.

3. 5.4×10^2 m.

5. 1200 m, 300 m.

7. (*a*) 1.1×10^{-8} m; (*b*) 1.1×10^{-10} m.

9. (*a*) $\Delta P = (4 \times 10^{-5} \text{ Pa})$
$\sin[(0.949 \text{ m}^{-1})x$
$- (315 \text{ s}^{-1})t]$;

(*b*) $\Delta P = (4 \times 10^{-3} \text{ Pa})$
$\sin[(94.9 \text{ m}^{-1})x$
$- (3.15 \times 10^4 \text{ s}^{-1})t]$.

11. (*a*) 49 dB; (*b*) 3.2×10^{-10} W/m².

13. 150 Hz to 20,000 Hz.

15. (*a*) 9; (*b*) 9.5 dB.

17. (*a*) Higher frequency is greater by a factor of 2; (*b*) 4.

19. (*a*) 5.0×10^{-13} W; (*b*) 6.3×10^4 yr.

21. 87 dB.

23. (*a*) 5.10×10^{-5} m; (*b*) 29.8 Pa.

25. (*a*) 1.5×10^3 W; (*b*) 3.4×10^2 m.

27. (*b*) 190 dB.

29. (*a*) 570 Hz; (*b*) 860 Hz.

31. 8.6 mm < L < 8.6 m.

33. (*a*) 110 Hz, 330 Hz, 550 Hz, 770 Hz;

(*b*) 220 Hz, 440 Hz, 660 Hz, 880 Hz.

35. (*a*) 0.656 m; (*b*) 262 Hz, 1.31 m;

(*c*) 1.31 m, 262 Hz.

37. −2.6%.

39. (*a*) 0.578 m; (*b*) 869 Hz.

41. 215 m/s.

43. 0.64, 0.20, −2 dB, −7 dB.

45. 28.5 kHz.

47. (*a*) 130.5 Hz, or 133.5 Hz;

(*b*) ±2.3%.

49. (*a*) 343 Hz; (*b*) 1030 Hz, 1715 Hz.

53. 346 Hz.

57. (*a*) 1690 Hz; (*b*) 1410 Hz.

59. 30,890 Hz.

61. 120 Hz.

63. 91 Hz.

65. 90 beats/min.

67. (*a*) 570 Hz; (*b*) 570 Hz; (*c*) 570 Hz;

(*d*) 570 Hz; (*e*) 594 Hz; (*f*) 595 Hz.

71. (*a*) 120; (*b*) 0.96°.

73. (*a*) 37°; (*b*) 1.7.

75. 0.278 s.

77. 55 m.

79. 410 km/h (255 mi/h).

81. 1, 0.444, 0.198, 0.0878, 0.0389.

83. 18.1 W.

85. 15 W.

87. 2.3 Hz.

89. $\Delta P_M/\Delta P_{M0} = D_M/D_{M0} = 10^6$.

91. 50 dB.

93. 17.5 m/s.

95. 2.3 kHz.

97. 550 Hz.

99. (*a*) 2.8×10^3 Hz; (*b*) 1.80 m;

(*c*) 0.12 m.

101. (*a*) 2.2×10^{-7} m; (*b*) 5.4×10^{-5} m.

CHAPTER 17

1. 0.548.

3. (*a*) 20°C; (*b*) ≈3300°F.

5. 104.0°F.

7. −40°F = −40°C.

9. $\Delta L_{\text{Invar}} = 2.0 \times 10^{-6}$ m,

$\Delta L_{\text{steel}} = 1.2 \times 10^{-4}$ m,

$\Delta L_{\text{marble}} = 2.5 \times 10^{-5}$ m.

11. −69°C.

13. 5.1 mL.

15. 0.06 cm³.

19. −40 min.

21. -2.8×10^{-3} (0.28%).

23. 3.5×10^7 N/m².

25. (*a*) 27°C; (*b*) 4.3×10^3 N.

27. −459.7°F.

29. 1.07 m³.

31. 1.43 kg/m³.

33. (a) 14.8 m³; (b) 1.81 atm.

35. 1.80×10^3 atm.

37. 37°C.

39. 3.43 atm.

41. 0.588 kg/m³, water vapor is not an ideal gas.

43. 2.69×10^{25} molecules/m³.

45. 4.9×10^{22} molecules.

47. 7.7×10^3 N.

49. (a) 71.2 torr; (b) 157°C.

51. (a) 0.19 K; (b) 0.051%.

53. (a) Low; (b) 0.017%.

55. 1/6.

57. 5.1×10^{27} molecules, 8.4×10^3 mol.

59. 11 L, not advisable.

61. (a) 9.3×10^2 kg; (b) 1.0×10^2 kg.

63. 1.1×10^{44} molecules.

65. 3.3×10^{-7} cm.

67. 1.1×10^3 m.

69. 15 h.

71. 0.66×10^3 kg/m³.

73. ± 0.11 C°.

77. 3.6 m.

CHAPTER 18

1. (a) 5.65×10^{-21} J; (b) 7.3×10^3 J.

3. 1.17.

5. (a) 4.5; (b) 5.2.

7. $\sqrt{2}$.

11. (a) 461 m/s; (b) 19 s⁻¹.

13. 1.00429.

17. Vapor.

19. (a) Gas, liquid, vapor; (b) gas, liquid, solid, vapor.

21. 0.69 atm.

23. 11°C.

25. 1.96 atm.

27. 120°C.

29. (a) 5.3×10^6 Pa; (b) 5.7×10^6 Pa.

31. (b) $b = 4.28 \times 10^{-5}$ m³/mol, $a = 0.365$ N·m⁴/mol².

33. (a) 10^{-7} atm; (b) 300 atm.

35. (a) 6.3 cm; (b) 0.58 cm.

37. 2×10^{-7} m.

39. (b) 4.7×10^7 s⁻¹.

43. 7.8 h.

45. (b) 4×10^{-11} mol/s; (c) 0.7 s.

47. 2.6×10^2 m/s, 4×10^{-17} N/m² $\approx 4 \times 10^{-22}$ atm.

49. (a) 2.9×10^2 m/s; (b) 12 m/s.

51. Reasonable, 70 cm.

53. $mgh = 4.3 \times 10^{-5}(\frac{1}{2}mv_{rms}^2)$, reasonable.

55. $P_2/P_1 = 1.43$, $T_2/T_1 = 1.20$.

57. 1.4×10^5 K.

59. (a) 1.7×10^3 Pa; (b) 7.0×10^2 Pa.

61. 2×10^{13} m.

CHAPTER 19

1. 1.0×10^7 J.

3. 1.8×10^2 J.

5. 2.1×10^2 kg/h.

7. 83 kcal.

9. 4.7×10^6 J.

11. 40 C°.

13. 186°C.

15. 7.1 min.

17. (b) $mc_0\big[(T_2 - T_1) + a(T_2^2 - T_1^2)/2\big]$; (c) $c_{mean} = c_0\big[1 + \frac{1}{2}a(T_2 + T_1)\big]$.

19. 0.334 kg (0.334 L).

21. $\frac{2}{3}m$ steam and $\frac{4}{3}m$ water at 100°C.

23. 9.4 g.

25. 4.7×10^3 kcal.

27. 1.22×10^4 J/kg.

29. 360 m/s.

31. (a) 0; (b) 5.00×10^3 J.

33.
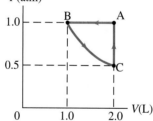

35. (a) 0; (b) −1300 kJ.

37. (a) 1.6×10^2 J; (b) $+1.6 \times 10^2$ J.

39. $W = 3.46 \times 10^3$ J, $\Delta U = 0$, $Q = +3.46 \times 10^3$ J (into the gas).

41. +129 J.

45. (a) +25 J; (b) +63 J; (c) −95 J; (d) −120 J; (e) −15 J.

47. $W = RT \ln\left(\dfrac{V_2 - b}{V_1 - b}\right) + a\left(\dfrac{1}{V_2} - \dfrac{1}{V_1}\right)$.

49. 22°C/h.

51. 4.98 cal/mol·K, 2.49 kcal/kg·K; 6.97 cal/mol·K, 3.48 kcal/kg·K.

53. 83.7 g/mol, krypton.

55. 46 C°.

57. (a) 2.08×10^3 J; (b) 8.32×10^2 J; (c) 2.91×10^3 J.

59. 0.379 atm, −51°C.

61. 1.33×10^3 J.

63. (a) $T_1 = 317$ K, $T_2 = 153$ K; (b) -1.59×10^4 J; (c) -1.59×10^4 J; (d) $Q = 0$.

65. (a)
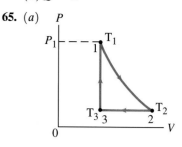

(b) 231 K;

(c) $Q_{1 \to 2} = 0$, $\Delta U_{1 \to 2} = -2.01 \times 10^3$ J, $W_{1 \to 2} = +2.01 \times 10^3$ J; $W_{2 \to 3} = -1.31 \times 10^3$ J, $\Delta U_{2 \to 3} = -1.97 \times 10^3$ J, $Q_{2 \to 3} = -3.28 \times 10^3$ J; $W_{3 \to 1} = 0$, $\Delta U_{3 \to 1} = +3.98 \times 10^3$ J, $Q_{3 \to 1} = +3.98 \times 10^3$ J;

(d) $W_{cycle} = +0.70 \times 10^3$ J, $Q_{cycle} = +0.70 \times 10^3$ J, $\Delta U_{cycle} = 0$.

67. (a) 64 W; (b) 22 W.

69. 4.8×10^2 W.

71. 31 h.

73. (a) 1.7×10^{17} W; (b) 278K (5°C).

75. (b) $\Delta Q/\Delta t = A(T_2 - T_1)/\Sigma(\ell_i/k_i)$.

77. 22%.

79. 4×10^{15} J.

81. 2.8 kcal/kg.

83. 30 C°.

85. 682 J.

87. 2.8 C°.

89. 2.58 cm, rod vaporizes.

91. 4.3 kg.

95. (a) 2.3 C°/s; (b) 84°C; (c) convection, conduction, evaporation.

97. (a) $\rho = m/V = (mP/nR)/T$; (b) $\rho = (m/nRT)P$.

99. (a) 1.9×10^5 J; (b) -1.4×10^5 J;

(c)
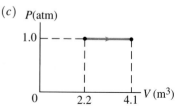

101. 3.2×10^5 s = 3.7 d.

103. 10 C°.

CHAPTER 20

1. 24%.

3. 816 MW.

5. 18%.

7. 13 km³/day, 63 km².

9. 28.0%.

13. 1.2×10^{13} J/h.

15. 1.4×10^3 m/day.

17. 660°C.

19. 3.7×10^8 kg/h.

21. (a) $P_a = 5.15 \times 10^5$ Pa,
    $P_b = 2.06 \times 10^5$ Pa;
    (b) $V_c = 30.0$ L, $V_d = 12.0$ L;
    (c) 2.83×10^3 J; (d) -2.14×10^3 J;
    (e) 0.69×10^3 J; (f) 24%.

23. 5.7.

25. −21°C.

27. 2.9.

29. (a) 3.9×10^4 J; (b) 3.0 min.

31. 76 L.

33. 0.15 J/K.

35. +11 kcal/K.

37. +0.0104 cal/K·s.

39. 1.7×10^2 J/K.

43. (a) 0.312 kcal/K;
    (b) > -0.312 kcal/K.

45. (a) Adiabatic process;
    (b) $\Delta S_i = -nR \ln 2$, $\Delta S_a = 0$;
    (c) $\Delta S_{surr,i} = nR \ln 2$, $\Delta S_{surr,a} = 0$.

47. (a) Entropy is a state function.

51. (a) 5/16; (b) 1/64.

53. (b), (a), (c), (d).

55. 69%.

57. 2.6×10^3 J/K.

59. (a) 17; (b) 5.9×10^7 J/h.

61. (a) 5.3 C°; (b) +77 J/kg·K.

63. $\Delta S = K/T$.

65. (a)

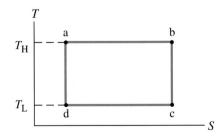

(b) area = $Q_{net} = W_{net}$.

67. $e_{Stirling} = (T_H - T_L) \ln(V_b/V_a) /$
    $[T_H \ln(V_b/V_a)$
    $+ \frac{3}{2}(T_H - T_L)]$,
    $e_{Stirling} < e_{Carnot}$.

69. 0.091 hp.

71. (a) 1/379; (b) $1/1.59 \times 10^{11}$.

CHAPTER 21

1. 6.3×10^9 N.

3. 2.7×10^{-3} N.

5. 5.5×10^3 N.

7. 8.66 cm.

9. -5.4×10^7 C.

11. 83.8 N away from the center of the triangle.

13. 2.96×10^7 N toward the center of the square.

15. $\mathbf{F}_1 = (kQ^2/\ell^2)[(-2 + 3\sqrt{2}/4)\mathbf{i}$
    $+ (4 - 3\sqrt{2}/4)\mathbf{j}]$, $\mathbf{F}_2 = (kQ^2/\ell^2)$
    $[(2 + 2\sqrt{2})\mathbf{i} + (-6 + 2\sqrt{2})\mathbf{j}]$,
    $\mathbf{F}_3 = (kQ^2/\ell^2)[(-12 - 3\sqrt{2}/4)\mathbf{i}$
    $+ (6 + 3\sqrt{2}/4)\mathbf{j}]$, $\mathbf{F}_4 = (kQ^2/\ell^2)$
    $[(12 - 2\sqrt{2})\mathbf{i} + (-4 - 2\sqrt{2})\mathbf{j}]$.

17. (a) $Q_1 = Q_2 = \frac{1}{2}Q_T$;
    (b) Q_1 (or Q_2) = 0.

19. $0.402Q_0$, 0.366ℓ from Q_0.

21. 60.2×10^{-6} C, 29.8×10^{-6} C;
    -16.8×10^{-6} C, 106.8×10^{-6} C.

23. $\mathbf{F} = -(1.90kQ^2/\ell^2)(\mathbf{i} + \mathbf{j} + \mathbf{k})$.

25. 2.18×10^{-16} N (west).

27. 7.43×10^6 N/C (up).

29. $(1.39 \times 10^2$ N/C)$\mathbf{j}$.

33.

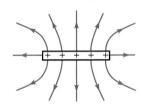

35. 8.26×10^{-10} N/C (south).

37. 4.5×10^6 N/C up, 1.2×10^7 N/C, 56° above the horizontal.

39. 5.61×10^4 N/C away from the opposite corner.

41. $Q_1/Q_2 = \frac{1}{4}$.

43. (a) $2Qy/4\pi\epsilon_0(y^2 + \ell^2)^{3/2}\,\mathbf{j}$.

45. $\dfrac{Q}{4\pi\epsilon_0}\left[\dfrac{x\mathbf{i} - (2a/\pi)\mathbf{j}}{(x^2 + a^2)^{3/2}}\right]$.

49. $\dfrac{-2\lambda \sin\theta_0}{4\pi\epsilon_0 R}\,\mathbf{i}$.

51. (a) $\dfrac{\lambda}{4\pi\epsilon_0 x(x^2 + L^2)^{1/2}}$
    $\{L\mathbf{i} + [x - (x^2 + L^2)^{1/2}]\mathbf{j}\}$.

53. $(\sigma/2\epsilon_0)\mathbf{k}$.

55. (a) $\mathbf{a} = -(3.5 \times 10^{15}$ m/s²$)\,\mathbf{i}$
    $-(1.41 \times 10^{16}$ m/s²$)\,\mathbf{j}$; (b) $\theta = -104°$.

57. $\theta = -28°$.

59. (b) $2\pi(4\pi\epsilon_0 mR^3/qQ)^{1/2}$.

61. (a) 3.4×10^{-20} C; (b) No;
    (c) 8.5×10^{-26} m·N;
    (d) 2.5×10^{-26} J.

63. (a) $\theta \ll 1$;
    (b) $(pE/I)^{1/2}/2\pi$.

65. (b) Direction of the dipole.

67. 6.8×10^3 C.

69. 5.7×10^{13} C.

71. $\mathbf{F}_1 = 0.30$ N, 265° from x-axis,
    $\mathbf{F}_2 = 0.26$ N, 139° from x-axis,
    $\mathbf{F}_3 = 0.26$ N, 30° from x-axis.

73. 4.2×10^5 N/C up.

75. $0.444Q_0$, 0.333ℓ from Q_0.

77. 5.60 m from the positive charge, and 3.60 m from the negative charge.

79. (a) In the direction of the velocity, to the right; (b) 2.1×10^2 N/C.

81. $\theta_0 = 18°$.

83. $(1.08 \times 10^7$ N/C$)/$
    $\{3.00 - \cos[(12.5$ s$^{-1})t]\}^2$, up.

85. $E_A = 4.2 \times 10^4$ N/C (right),
    $E_B = -1.4 \times 10^4$ N/C (left),
    $E_C = -2.8 \times 10^3$ N/C (left),
    $E_D = -4.2 \times 10^4$ N/C (left).

87. $d(1 + \sqrt{2})$ from the negative charge, and $d(2 + \sqrt{2})$ from the positive charge.

CHAPTER 22

1. (a) 41 N·m²/C; (b) 29 N·m²/C;
   (c) 0.

3. $\Phi_{net} = 0$,
   $\Phi_{x=0} = -(6.50 \times 10^3$ N/C$)\ell^2$,
   $\Phi_{x=\ell} = +(6.50 \times 10^3$ N/C$)\ell^2$,
   $\Phi_{all\,others} = 0$.

5. 12.8 nC.

7. -1.2 μC.

9. -3.75×10^{-11} C.

11. (a) -1.0×10^4 N/C (toward wire);
    (b) -2.5×10^4 N/C (toward wire).

13.

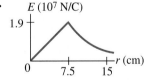

15. (a) 5.5×10^7 N/C (away from center);
    (b) 0.

17. (a) -8.00 μC;
    (b) $+1.00$ μC

19. (a) 0; (b) σ/ϵ_0;
    (c) unaffected.

21. (a) 0; (b) $\sigma_1 r_1^2/\epsilon_0 r^2$;
    (c) $(\sigma_1 r_1^2 + \sigma_2 r_2^2)/\epsilon_0 r^2$;
    (d) $\sigma_2/\sigma_1 = -(r_1/r_2)^2$; (e) $\sigma_1 = 0$.

23. (a) $q/4\pi\epsilon_0 r^2$;
    (b) $(1/4\pi\epsilon_0)[Q(r^3 - r_1^3)$
      $+ q(r_0^3 - r_1^3)]/(r_0^3 - r_1^3)r^2$;
    (c) $(q + Q)/4\pi\epsilon_0 r^2$.

25. (a) $q/4\pi\epsilon_0 r^2$; (b) $(q + Q)/4\pi\epsilon_0 r^2$;
    (c) $E(r < r_0) = Q/4\pi\epsilon_0 r^2$,
      $E(r > r_0) = 2Q/4\pi\epsilon_0 r^2$;
    (d) $E(r < r_0) = -Q/4\pi\epsilon_0 r^2$,
      $E(r > r_0) = 0$.

27. (a) $\sigma R_0/\epsilon_0 r$; (b) 0; (c) same.

29. (a) 0; (b) $Q/2\pi\epsilon_0 Lr$;
    (c) 0; (d) $eQ/4\pi\epsilon_0 L$.

31. (a) 0;
    (b) -2.3×10^5 N/C (toward the axis);
    (c) -1.8×10^4 N/C (toward the axis).

33. (a) $\rho_E r/2\epsilon_0$; (b) $\rho_E R_1^2/2\epsilon_0 r$;
    (c) $\rho_E(r^2 + R_1^2 - R_2^2)/2\epsilon_0 r$;
    (d) $\rho_E(R_3^2 + R_1^2 - R_2^2)/2\epsilon_0 r$;
    (e)

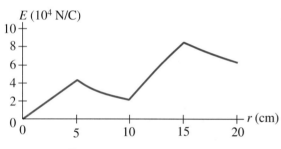

35. $Q/\epsilon_0\sqrt{2}$.

37. $\oint \mathbf{g}\cdot d\mathbf{A} = -4\pi GM$.

39. $Q_{enclosed} = \epsilon_0 b\ell^3$.

41. 3.95×10^2 N·m²/C,
    -1.69×10^2 N·m²/C.

43. (a) 0; (b) $Q/25\pi\epsilon_0 r_0^2 \le E \le Q/\pi\epsilon_0 r_0^2$;
    (c) not perpendicular;
    (d) not useful.

45. (a) $0.677e = +1.08 \times 10^{-19}$ C;
    (b) 3.5×10^{11} N/C.

47. (a) $\rho_E r_0/6\epsilon_0$ (right);
    (b) $-17\rho_E r_0/54\epsilon_0$ (left).

49. (a) 0; (b) 5.65×10^5 N/C (right);
    (c) 5.65×10^5 N/C (right);
    (d) -5.00×10^{-6} C/m²;
    (e) $+5.00 \times 10^{-6}$ C/m².

CHAPTER 23

1. -4.2×10^{-5} J (done by the field).

3. 3.4×10^{-15} J.

5. $V_a - V_b = +72.8$ V.

7. 7.04 V.

9. 0.8 μC.

11. (a) $V_{BA} = 0$; (b) $V_{CB} = -2100$ V;
    (c) $V_{CA} = -2100$ V.

13. (a) -9.6×10^8 V;
    (b) $V(\infty) = +9.6 \times 10^8$ V.

15. (a) The same;
    (b) $Q_2 = r_2 Q/(r_1 + r_2)$.

17. (a) $Q/4\pi\epsilon_0 r$;
    (b) $(Q/8\pi\epsilon_0 r_0)[3 - (r^2/r_0^2)]$;
    (c)

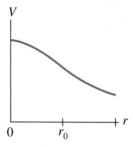

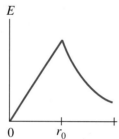

19. (a) $V_0 + (\sigma R_0/\epsilon_0) \ln(R_0/r)$;
    (b) $V = V_0$; (c) $V \ne 0$.

21. (a) 29 V;
    (b) -29 eV (-4.6×10^{-18} J).

23. $+0.19$ J.

25. 4.2 MV.

27. 2.33×10^7 m/s.

29. $V_{BA} = (1/2\pi\epsilon_0)q(2b - d)/b(d - b)$.

31. $\dfrac{\sigma}{2\epsilon_0}[(x^2 + R_2^2)^{1/2} - (x^2 + R_1^2)^{1/2}]$.

33. $\dfrac{Q}{8\pi\epsilon_0 L} \ln\left(\dfrac{x + L}{x - L}\right)$, $x > L$.

35. $\dfrac{a}{6\epsilon_0}[(x^2 + R^2)^{1/2}(R^2 - 2x^2) + 2x^3]$.

37. 3.2 mm.

39. (a) 8.5×10^{-30} C·m; (b) zero.

41. (a) -0.088 V; (b) 1%.

43. (a) 5.2×10^{-20} C.

47. $\mathbf{E} = 2y(2z - 1)\mathbf{i} - 2(y + x - 2xz)\mathbf{j}$
    $+ (4xy)\mathbf{k}$.

49. (a) 9.6×10^4 eV; (b) 1.9×10^5 eV.

51. -2.4×10^4 V.

53. (a) $U = (1/4\pi\epsilon_0)(Q_1 Q_2/r_{12}$
      $+ Q_1 Q_3/r_{13} + Q_1 Q_4/r_{14} + Q_2 Q_3/r_{23}$
      $+ Q_2 Q_4/r_{24} + Q_3 Q_4/r_{34})$.
    (b) $U = (1/4\pi\epsilon_0)(Q_1 Q_2/r_{12}$
      $+ Q_1 Q_3/r_{13} + Q_1 Q_4/r_{14} + Q_1 Q_5/r_{15}$
      $+ Q_2 Q_3/r_{23} + Q_2 Q_4/r_{24} + Q_2 Q_5/r_{25}$
      $+ Q_3 Q_4/r_{34} + Q_3 Q_5/r_{35} + Q_4 Q_5/r_{45})$.

55. (a) 2.0 keV; (b) 42.8.

57. (a) $(-4 + \sqrt{2})Q^2/4\pi\epsilon_0 b$; (b) 0.

59. $3Q^2/20\pi\epsilon_0 r$.

61. 5.4×10^5 V/m.

63. 9×10^2 V.

65. (a) 1.1 MV; (b) 13 kg.

67. 7.2 MV.

69. 1.58×10^{12} electrons.

71. 1.7×10^{-12} V.

73. 1.03×10^6 m/s.

75. $V_a = -3.5 \, Q/4\pi\epsilon_0 L$,
    $V_b = -5.2 \, Q/4\pi\epsilon_0 L$,
    $V_c = -6.8 \, Q/4\pi\epsilon_0 L$.

77. (a) 5.8×10^5 V; (b) 9.2×10^{-14} J.

79. $V_a - V_b = (\lambda/2\pi\epsilon_0) \ln(R_b/R_a)$.

83. (a) $\rho_E(r_2^3 - r_1^3)/3\epsilon_0 r$;
    (b) $(\rho_E/6\epsilon_0)[3r_2^2 - r^2 - (2r_1^3/r)]$;
    (c) $(\rho_E/2\epsilon_0)(r_2^2 - r_1^2)$, potential is
    continuous at r_1 and r_2.

CHAPTER 24

1. 2.6 μF.

3. 6.3 pF.

5. 0.80 μF.

7. 2.0 C.

9. 1.8×10^2 m².

11. 7.1×10^{-4} F.

13. 23 nC.

17. 4.5×10^4 V/m.

19. (a) $\epsilon_0 A/(d - \ell)$; (b) 3.

21. 2880 pF, yes.

23. (a) $(C_1 C_2 + C_1 C_3 + C_2 C_3)/$
      $(C_2 + C_3)$;
    (b) $Q_1 = 350$ μC, $Q_2 = 117$ μC.

25. (a) 3.71 μF; (b) $V_{ab} = V_1 = 26.0$ V,
    $V_2 = 14.9$ V, $V_3 = 11.1$ V.

27. 18 nF (parallel), 1.6 nF (series).

29. (a) $3C/5$; (b) $Q_1 = Q_2 = CV/5$,
    $Q_3 = 2CV/5$, $Q_4 = 3CV/5$,
    $V_1 = V_2 = V/5$, $V_3 = 2V/5$,
    $V_4 = 3V/5$.

31. $Q_1' = C_1 C_2 V_0/(C_1 + C_2)$,
    $Q_2' = C_2^2 V_0/(C_1 + C_2)$.

33. (a) $Q_1 = Q_3 = 30 \, \mu\text{C}$;
  $Q_2 = Q_4 = 60 \, \mu\text{C}$;
  (b) $V_1 = V_2 = V_3 = V_4 = 3.75 \, \text{V}$;
  (c) 7.5 V.

35. 3.0 μF.

37. $C \approx (\epsilon_0 A/d)[1 - \frac{1}{2}(\theta \sqrt{A}/d)]$.

39. 2.0×10^{-3} J.

41. 2.3×10^3 J.

43. 1.65×10^{-7} J.

45. (a) 2.5×10^{-5} J; (b) 6.2×10^{-6} J;
  (c) $Q_{\text{par}} = 4.2 \, \mu\text{C}$, $Q_{\text{ser}} = 1.0 \, \mu\text{C}$.

47. (a) 2.2×10^{-4} J; (b) 8.1×10^{-5} J;
  (c) -1.4×10^{-4} J; (d) stored
  potential energy is not conserved.

51. 1.5×10^{-10} F.

53. 0.46 μC.

55. 3.3×10^2 J.

57. $C = 2\epsilon_0 A K_1 K_2/d(K_1 + K_2)$.

59. (a) $0.40 Q_0$, $1.60 Q_0$; (b) $0.40 V_0$.

61. (a) 111 pF; (b) 1.66×10^{-8} C;
  (c) 1.84×10^{-8} C;
  (d) 1.17×10^5 V/m; (e) 3.34×10^4 V/m;
  (f) 150 V;
  (g) 172 pF; (h) 2.58×10^{-8} C.

63. 22%.

65. $Q = 4.41 \times 10^{-7}$ C,
  $Q_{\text{ind}} = 3.65 \times 10^{-7}$ C,
  $E_{\text{air}} = 2.69 \times 10^4$ V/m,
  $E_{\text{glass}} = 4.64 \times 10^3$ V/m;

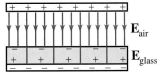

67. 11 μF.

69. (a) $4\times$; (b) $4\times$; (c) $\frac{1}{2}\times$.

71. 10.9 V.

73. (b) 1.5×10^{-10} F/m.

75. $U_2/U_1 = 1/2K$, $E_2/E_1 = 1/K$.

77. (a) 19 J; (b) 0.19 MW.

79. (a) 0.10 MV; (b) voltage will
  decrease exponentially.

81. 660 pF in parallel.

83. $Q_1 = 11 \, \mu\text{C}$, $Q_2 = Q_3 = 13 \, \mu\text{C}$,
  $V_1 = 11$ V, $V_2 = 6.5$ V, $V_3 = 4.4$ V.

85. $Q^2 x/2A\epsilon_0$.

87. (a) 7.4 nF, 0.33 μC, 1.5×10^4 V/m,
  7.5×10^{-6} J; (b) 27 nF, 1.2 μC,
  1.5×10^4 V/m, 2.7×10^{-5} J.

89. (a) 66 pF; (b) 30 μC; (c) 7.0 mm; (d)
  450 V.

CHAPTER 25

1. 9.38×10^{18} electron/s.

3. 2.1×10^{-11} A.

5. 7.5×10^2 V.

7. 2.1×10^{21} electron/min.

9. (a) 16 Ω; (b) 6.8×10^3 C.

11. 0.57 mm.

13. $R_{\text{Al}} = 1.2 R_{\text{Cu}}$.

15. 1/6 the length, 8.3 Ω, 1.7 Ω.

17. 58.3°C.

19. 1.8×10^3 °C.

21. $R_2 = \frac{1}{4}R_1$.

25. $R = (r_2 - r_1)/4\pi\sigma r_1 r_2$.

27. 3.2 W.

29. 37 V.

31. (a) 240 Ω, 0.50 A; (b) 96 Ω, 1.25 A.

33. 0.092 kWh, 22¢/month.

35. 1.1 kWh.

37. 3.

39. 5.3 kW.

41. 0.128 kg/s.

43. 0.094 A.

45. (a) Infinite; (b) 1.9×10^2 Ω.

47. 636 V, 5.66 A.

49. 1.5 kW, 3.0 kW, 0.

51. (a) 7.8×10^{-10} m/s; (b) 10.5 A/m^2
  along the wire; (c) 1.8×10^{-7} V/m.

53. 2.7 A/m^2 north.

55. 12 h.

57. 6.67×10^{-2} S.

59. (a) 8.6 Ω, 1.1 W; (b) $4\times$.

61. (a) $44; (b) 1.8×10^3 kg/yr.

63. (a) -19.5%; (b) percentage decrease
  in the power output would be less.

65. (a) 1.44×10^3 W; (b) 17 W; (c) 11 W;
  (d) 0.8¢/day.

67. (a) 1.5 kW; (b) 12.5 A.

69. 2.

71. 0.303 mm, 28.0 m.

73. (a) 1.2 kW; (b) 100 W.

75. 1.4×10^{12} protons.

77. $j_a = 2.8 \times 10^5$ A/m^2,
  $j_b = 1.6 \times 10^5$ A/m^2.

CHAPTER 26

1. (a) 8.39 V; (b) 8.49 V.

3. 0.060 Ω.

5. 360 Ω, 23 Ω.

7. 25 Ω, 70 Ω, 95 Ω, 18 Ω.

9. Series connection.

11. 4.6 kΩ.

13. 310 Ω, 3.7%.

15. 960 Ω in parallel.

17. 105 Ω.

19. (a) V_1 and V_2 increase; V_3 and V_4
  decrease; (b) I_1 ($= I$) and I_2
  increase; I_3 and I_4 decrease;
  (c) increases; (d) $I = I_1 = 0.300$ A,
  $I_2 = 0$, $I_3 = I_4 = 0.150$ A; $I = 0.338$ A,
  $I_2 = I_3 = I_4 = 0.113$ A.

21. 0.4 Ω.

23. 0.41 A.

25. $I_1 = 0.68$ A, $I_2 = -0.40$ A.

27. $I_1 = 0.18$ A right, $I_2 = 0.32$ A left,
  $I_3 = 0.14$ A up.

29. $I_1 = 0.274$ A, $I_2 = 0.222$ A,
  $I_3 = 0.266$ A, $I_4 = 0.229$ A,
  $I_5 = 0.007$ A, $I = 0.496$ A.

31. 52 V, –28 V.
  The negative value means the
  battery is facing the other direction.

33. 70 V.

35. $I_1 = 0.783$ A.

39. (a) $R(3R + 5R')/8(R + R')$; (b) $R/2$.

41. (a) 3.7 nF; (b) 22 μs.

43. $t = 1.23\tau$.

45. (a) $\tau = R_1 R_2 C/(R_1 + R_2)$;
  (b) $Q_{\text{max}} = \mathscr{E} R_2 C/(R_1 + R_2)$.

47. 2.1 μs.

49. 50 μA.

51. (a) 2.9×10^{-4} Ω in parallel;
  (b) 35 kΩ in series.

53. 22 V, 17 V, 14% low.

55. 0.85 mA, 4.3 V.

57. 9.6 V.

61. 3.6×10^{-2} C°.

63. Two resistors in series.

65. 2.2 V, 116 V.

67. 0.19 MΩ.

69. (a) 0.10 A; (b) 0.10 A; (c) 53 mA.

71. 2.5 V.

73. 46.1 V, 0.71 Ω.

75. (a) 72.0 W; (b) 14.2 W; (c) 3.76 W.

77. (a) 40 kΩ; (b) between b and c.

79. 375 cells, 3.8 m $\times$ 0.090 m.

81. (a) 0.50 A; (b) 0.17 A; (c) 3.3 μC;
  (d) 32 μs.

83. (a) + 6.8 V, 10.2 μC; (b) 28 μs.

CHAPTER 27

1. (a) 6.7 N/m; (b) 4.7 N/m.

3. 2.68 A.

5. 0.243 T.

7. (a) South pole; (b) 3.5×10^2 A;
  (c) 5.22 N.

9. 5.5×10^3 A.

13. 1.05×10^{-13} N north.

15. (a) Down; (b) in; (c) right.

21. 1.6 m.

23. (a) 2.7 cm; (b) 3.8×10^{-7} s.

25. 1.034×10^8 m/s (west), gravity can be ignored.

27. $(6.4\mathbf{i} - 10.3\mathbf{j} - 0.24\mathbf{k})] \times 10^{-16}$ N.

29. 5.3×10^{-5} m, 3.3×10^{-4} m.

31. (a) $45°$; (b) 3.5×10^{-3} m.

33. (a) $2\mu B$; (b) 0.

35. (a) 4.33×10^{-5} m $\cdot$ N;
(b) north edge.

39. $29 \ \mu$A.

41. 1.2×10^5 C/kg.

43. (a) 2.2×10^{-4} V/m;
(b) 2.7×10^{-4} m/s;
(c) 6.4×10^{28} electrons/m^3.

45. (a)Determine polarity of the emf;
(b) 0.43 m/s.

47. 1.53 mm, 0.76 mm.

51. 3.0 T up.

53. 1.1×10^{-6} m/s west.

55. 0.17 N, $68°$ above the horizontal toward the north.

57. 0.20 T, $26.6°$ from the vertical.

59. (c) 48 MeV.

61. Slower protons will deflect more, and faster protons will deflect less, $\theta = 12°$.

63. 2.0 A, down.

65. 7.3×10^{-3} T.

67. -2.1×10^{-20} J.

CHAPTER 28

1. 1.7×10^{-4} T, $3.1\times$.

3. 0.18 N attraction.

5.

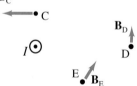

7. 8.9×10^{-5} T, $70°$ above horizontal.

9. 4.0×10^{-5} T, $15°$ below the horizontal.

11. (a) $(2.0 \times 10^{-5}$ T/A$)(15$ A $- I)$ up;
(b) $(2.0 \times 10^{-5}$ T/A$)(15$ A $+ I)$ down.

13. 21 A down.

15. $[(\mu_0/4\pi)2I(d - 2x)/x(d - x)]\mathbf{j}$.

17. 4.12×10^{-5} T.

19. (b) $(\mu_0/4\pi)(2I/y)$.

21. 0.123 A.

23. (a) 6.4×10^{-3} T; (b) 3.8×10^{-3} T;
(c) 2.1×10^{-3} T.

25. (a) 51 cm; (b) 1.3×10^{-2} T.

27. (a) $(\mu_0 I_0/2\pi R_1^2)r$ circular CCW;
(b) $\mu_0 I_0/2\pi r$ circular CCW;
(c) $(\mu_0 I_0/2\pi r)(R_3^2 - r^2)/(R_3^2 - R_2^2)$
circular CCW; (d) 0.

29. 3.6×10^{-6} T.

31. $\mu_0 I/8R$ out of the page.

33. (a) $\mu_0 I(R_1 + R_2)/4R_1 R_2$ into the page;
(b) $\frac{1}{2}\pi I(R_1^2 + R_2^2)$ into the page.

35. (a) $\dfrac{Q\omega R^2}{4}\mathbf{i}$;

(b)

$\dfrac{\mu_0 Q\omega}{2\pi R^2}\left[\dfrac{R^2 + 2x^2 - 2x\sqrt{R^2+x^2}}{\sqrt{R^2 + x^2}}\right]\mathbf{i}$;

(c) yes.

37. (b) $B = \mu_0 IL/4\pi y(L^2 + y^2)^{1/2}$ circular.

39. (a) $(\mu_0 I_0/2\pi R)n \tan{(\pi/n)}$ into the page.

41. $B = \dfrac{\mu_0 I}{4\pi}\Bigg\{\dfrac{(x^2 + y^2)^{1/2}}{xy}$

$+ \dfrac{[(b - x)^2 + y^2]^{1/2}}{y(b - x)}$

$+ \dfrac{[(a - y)^2 + (b - x)^2]^{1/2}}{(a - y)(b - x)}$

$+ \dfrac{[x^2 + (a - y)^2]^{1/2}}{x(a - y)}\Bigg\}$,

out of the page.

43. (a) 26 A $\cdot$ m^2; (b) 31 m $\cdot$ N.

45. 30 T.

47. $F_M/L = 5.84 \times 10^{-5}$ up,
$F_N/L = 3.37 \times 10^{-5}$ N/m $60°$ below the line toward P,
$F_P/L = 3.37 \times 10^{-5}$ N/m $60°$ below the line toward N.

49. 0.27 mm, 1.4 cm.

51. $B = \mu_0 jt/2$ parallel to the sheet, perpendicular to the current (opposite directions on the two sides).

53. Between long, thin and short, fat.

55. 3×10^9 A.

57. B will decrease.

59. 2.1×10^{-6} g.

61. $2\mu_0 I/L\pi$ (left).

63. 4×10^{-6} T, about 10% of the Earth's field.

CHAPTER 29

1. -3.8×10^2 V.

3. Counterclockwise.

5. 0.026 V.

7. (a) Counterclockwise; (b) clockwise;
(c) zero; (d) counterclockwise.

9. Counterclockwise.

11. (a) Clockwise; (b) 43 mV; (c) 17 mA.

13. 1.1×10^{-5} J.

15. 4.21 C.

17. (a) 5.2×10^{-2} A; (b) 0.32 mW.

19. 1.7×10^{-2} V.

21. $(\mu_0 Ia/2\pi) \ln 2$.

23. (a) 0.15 V; (b) 5.4×10^{-3} A;
(c) 4.5×10^{-4} N.

25. (a) Will move at constant speed;
(b) $v = v_0 e^{-B^2\ell^2 t/mR}$.

27. (a) $\dfrac{\mu_0 Iv}{2\pi}\ln\left(\dfrac{a + b}{b}\right)$ toward long wire;

(b) $\dfrac{\mu_0 Iv}{2\pi}\ln\left(\dfrac{a + b}{b}\right)$ away from long wire.

31. 0.33 kV, 120 rev/s.

33. 100 V.

35. 13 A.

37. 3.54×10^4 turns.

39. 0.18.

41. (a) 5.2 V;
(b) step-down transformer.

43. 549 V, 68.6 A.

45. 56.8 kW.

47. 0.188 V/m.

49. (b) Clockwise; (c) $dB/dt > 0$.

51. (a) IR/ℓ (constant); (b) $\dfrac{\mathcal{E}_0}{\ell}e^{-B^2\ell^2 t/mR}$.

53. 31 turns.

55. $v = 0.76$ m/s.

57. 184 kV.

59. 1.5×10^{17}.

61. (a) 23 A; (b) 90 V; (c) 6.9×10^2 W;
(d) 75%.

63. (a) 0.85 A; (b) 8.2.

65. $\frac{1}{2}B\omega L^2$ toward the center.

71. $B\omega r$, radially out from the axis.

CHAPTER 30

1. $M = \mu_0 N_1 N_2 A/\ell$.

3. $M/\ell = \mu_0 n_1 n_2 \pi r_2^2$.

5. $M = (\mu_0 w/2\pi) \ln(\ell_2/\ell_1)$.

7. 1.2 H.

9. 2.5×10^{-6} H.

11. $r_1 \geq 2.5$ mm.

13. 3.

15. (a) $L_1 + L_2$; (b) $= L_1L_2/(L_1 + L_2)$.

17. 15.9 J.

19. (a) $u_E = 4.4 \times 10^{-4}$ J/m^3,
$u_B = 1.6 \times 10^6$ J/m^3, $u_B \gg u_E$;
(b) $E = 6.0 \times 10^8$ V/m.

21. 4.4 J/m^3, 1.6×10^{-14} J/m^3.

23. $(\mu_0 I^2/4\pi) \ln(r_2/r_1)$.

25. $t/\tau = 4.6$.

27. $(dI/dt)_0 = V_0/L$.

29. (a) $(LV^2/2R^2)(1 - 2e^{-t/\tau} + e^{-2t/\tau})$;
(b) $t/\tau = 5.3$.

31. (a) 213 pF; (b) 46.5 μH.

35. (a) $Q = Q_0/\sqrt{2}$; (b) $T/8$.

37. $R = 2.30$ Ω.

41. Decrease, 1.15 kΩ.

43. 20 mH, 95 turns.

45. (a) 21 mH; (b) 45 mA;
(c) 2.2×10^{-5} J.

47. 3.0×10^3 turns, 95 turns.

51. (b) Positioning one coil
perpendicular to the other;
(c) $L_1L_2/(L_1 + L_2)$;
$(L_1L_2 - M^2)/(L_1 + L_2 \mp 2M)$.

55. (a) $\frac{1}{2}(Q_0^2/C)e^{-Rt/L}$.

57.

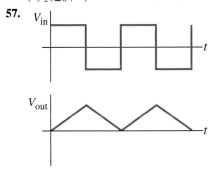

CHAPTER 31

1. (a) 3.7×10^2 Ω; (b) 2.2×10^{-2} Ω.

3. 9.90 Hz.

5.

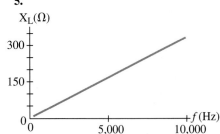

7. 0.13 H.

11. (a) 5.0%; (b) 98%.

13. (a) 9.0 kΩ; (b) 10.2 kΩ.

15. (a) 18 mA; (b) $-29°$; (c) 1.8 W;
(d) $V_R = 105$ V, $V_C = 58$ V.

17. (a) 0.38 A; (b) $-89°$; (c) 0.29 W.

19. 332 Ω.

21. (a) 0; (b) $\dfrac{2}{\pi}V_0$, $\overline{V}_{1/2} = \dfrac{2\sqrt{2}}{\pi}V_{rms}$.

23. 8.78 kΩ, $-7.66°$, 91.1 mA.

25. 265 Hz, 324 W.

27. 52.5 mA.

29. (b) $\omega'^2 = [(1/LC) - (R^2/2L^2)]$;
(c) $k \leftrightarrow 1/C, m \leftrightarrow L, b \leftrightarrow R$.

31. (a) $V_0^2 R/[2R^2 + 2(\omega L - 1/\omega C)^2]$;
(b) $\omega'^2 = 1/LC$; (c) $\Delta\omega = R/L$.

33. 4 Ω.

35. 9.76 nF.

37. 27.9 mH.

39. 1.6 kHz.

41. 14 Ω, 75 mH.

43. 2.2×10^3 Hz, 1.1×10^4 Hz.

45. (a) 23.6 kΩ, 10.8°; (b) 1.88×10^{-5} W;
(c) 2.8×10^{-5} A, 0.66 V, 4.7×10^{-4} V,
0.126 V.

49. $\{(R_1 + R_2)^2$
$+ [\omega(L_1 + L_2) -$
$(C_1 + C_2)/\omega C_1 C_2]^2\}^{1/2}$.

51. 19 Ω, 62 mH.

53. $I = \left(\dfrac{V_0}{Z}\right) \sin(\omega t + \phi)$,

$I_C = \left(\dfrac{V_0}{X_C}\right)\left[-\left(\dfrac{R}{Z}\right)\cos(\omega t + \phi) + \cos(\omega t)\right]$,

$I_L = \left(\dfrac{V_0}{X_L}\right)\left[\left(\dfrac{R}{Z}\right)\cos(\omega t + \phi) - \cos(\omega t)\right]$,

$Z = \sqrt{R^2 + \left(\dfrac{X_C X_L}{X_L - X_C}\right)^2}$,

$\tan\phi = \dfrac{X_C X_L}{(X_L - X_C)R}$.

CHAPTER 32

1. 9.2×10^4 V/m · s.

3. 7.9×10^{14} V/m · s.

7. $\oint \mathbf{B} \cdot d\mathbf{A} = \mu_0 Q_m$,
$\oint \mathbf{E} \cdot d\boldsymbol{\ell} = \mu_0\, dQ_m/dt - d\Phi_B/dt$.

9. 1.4×10^{-13} T.

11. (a) $B_0 = E_0/c$, $-y$-direction;
(b) $-z$-direction.

13. (a) 1.08 cm; (b) 3.0×10^{18} Hz.

15. 314 nm, ultraviolet.

17. (a) 4.3 min; (b) 71 min.

19. 1.77×10^{-6} W/m^2.

21. 7.82×10^{-7} J/h.

23. 4.50×10^{-6} J.

25. 3.8×10^{26} W.

29. $r < 3 \times 10^{-7}$ m.

31. 302 pF.

33. 2.59 nH $\leq L \leq$ 3.89 nH.

35. (a) 441 m; (b) 2.979 m.

37. 5.56 m, 0.372 m.

39. (a) 1.28 s; (b) 4.3 min.

43. 469 V/m.

45. Person at the radio hears the voice
0.14 s sooner.

47. (a) 0.40 W; (b) 12 V/m; (c) 12 V.

49. 1.5×10^{11} W.

51. (a) Parallel;
(b) 8.9 pF $\leq C \leq$ 80 pF;
(c) 1.05 mH $\leq L \leq$ 1.12 mH.

CHAPTER 33

1. (a) 2.21×10^8 m/s;
(b) 1.99×10^8 m/s.

3. 8.33 min.

5. 3 m.

7. 3.4×10^3 rad/s.

9. I_3 is the desired image:

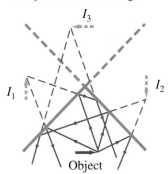

Depending on where you put your eye,
two other images may also be visible.

11. 5°.

15. 36.4 cm.

19. 4.5 m.

21. Convex, -20 m.

23. (a) Center of curvature; (b) real;
(c) inverted; (d) -1.

29. (a) Convex mirror; (b) 22 cm behind
surface; (c) -98 cm; (d) -196 cm.

33. 45.6°.

35. 24.9°.

37. 4.6 m.

43. 3.0%.

45. 0.22°.

47. 61.7°, lucite.

49. 93.5 cm.

51. $n_{\text{liquid}} \geq 1.5$.

55. 17.0 cm below the surface of the glass.

59. (a) 3.0 m, 4.0 m, 7.0 m; (b) toward you, away from you, toward you.

61. −3.80 m.

63. Chose different signs for the magnification; 13.3 cm, 26.7 cm.

65. $\geq 56.1°$.

69. 81 cm (inside the glass).

CHAPTER 34

1. (a)

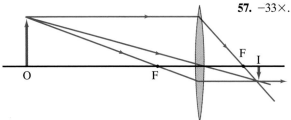

(b) 24.9 cm.

3. (a) 3.64 D, converging;
(b) −16.0 cm, diverging.

5. (a) −0.26 mm; (b) −0.47 mm;
(c) −1.9 mm.

7. (a) 81 mm; (b) 82 mm; (c) 87 mm;
(d) 24 cm.

9. (a) Virtual; (b) converging lens;
(c) 7.5 D.

11. (a) −10.5 cm (diverging), virtual;
(b) +203 cm (converging).

13. 22.9 cm, 53.1 cm.

15. Real and upright.

17. Real, 21.3 cm beyond second lens,
−0.708 (inverted).

19. (a) +7.14 cm; (b) −0.357 (inverted);
(c)

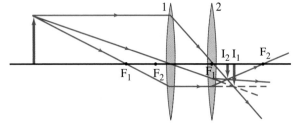

23. 1.54.

25. 8.1 cm.

27. −1.87 m (concave).

29. +1.15 D.

31. $f/2.3$.

35. 41 mm.

37. +2.3 D.

39. Glasses would be better.

41. (a) −1.33 D; (b) 38 cm.

43. −26.8 cm.

45. 17 cm, 100 cm.

47. 8.3 cm.

49. 6.3 cm from the lens, 3.9×.

51. (a) −49.4 cm; (b) 4.7×.

53. (a) 7.2×; (b) 2.2×.

55. 3.2 cm, 83 cm.

57. −33×.

59. 12×.

61. $f_o = 4.0$ m, $r = 8.0$ m.

63. 7.5×.

65. 1.7 cm.

67. (a) 0.85 cm; (b) 230×.

69. (a) 14.4 cm; (b) 137×.

71. (a) 15.9 cm; (b) 14.3 cm; (c) 1.6 cm;
(d) $r = 0.46$ cm.

73. 6.87 m $\leq d_o \leq \infty$.

75. 100 mm, 200 mm.

77. 79.4 cm, 75.5 cm.

79. 0.101 m, −2.7 m.

81. $0 < -d_o < -f$.

83. (c) $\Delta d = \sqrt{d_T^2 - 4d_T f}$,
$$m = \left(\frac{d_T + \sqrt{d_T^2 - 4d_T f}}{d_T - \sqrt{d_T^2 - 4d_T f}} \right)^2.$$

85. $1/f' = [(n/n') - 1][(1/R_1) + (1/R_2)]$;
$(1/d_o) + (1/d_i) = 1/f'$, where
$1/f' = [(n/n') - 1]/f(n - 1)$;
$m = -d_i/d_o$.

87. +3.6 D.

89. 2.9×, 4.1×.

91. (a) −2.5×; (b) 5.0 D.

93. −20×.

CHAPTER 35

3. 5.9 μm.

5. 3.9 cm.

7. 0.21 mm.

9. 613 nm.

11. 533 nm.

15. (a) $I_\theta/I_0 = (1 + 4\cos\delta + 4\cos^2\delta)/9$;
(b) $\sin\theta_{\text{max}} = m\lambda/d$, $m = 0, \pm1, \pm2, \dots$;
$\sin\theta_{\text{min}} = (m + \frac{1}{3}k)\lambda/d; k = 1, 2$;
$m = 0, \pm1, \pm2, \dots$.

17. Orange-red.

19. 179 nm.

21. 9.1 μm.

23. 120 nm, 240 nm.

25. 1.26.

29. 0.47, 0.23.

31. 0.221 mm.

33. 0.289 mm.

35. (a) 17 lm/W; (b) 156.

37. (a) Constructive; (b) destructive.

39. 464 nm.

41. 646 nm.

43. (a) 81.5 nm; (b) 127 nm.

45. 0.5 cm.

47. $\theta = 63.3°$.

49. $\sin\theta_{\text{max}} = (m + \frac{1}{2})\lambda/2S, m = 0,1,2,\dots$,
$\sin\theta_{\text{min}} = m\lambda/2S, m = 0,1,2,\dots$.

51. $I/I_0 = \cos^2(2\pi x/\lambda)$.

CHAPTER 36

1. 2.26°.

3. 2.4 m.

5. 5.8 cm.

7. 4.28 cm.

9. (b) Average intensity is 2×.

11. 10.7°.

13. $d = 4a$.

15. (a) 1.8 cm; (b) 11.0 cm.

17.

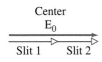

19. 2.4×10^{-7} rad $= (1.4 \times 10^{-5})°$
   $= 0.050''$.

21. 820 lines/mm, 102 lines/mm.

23. $5.61°$.

25. Two full orders.

27. 497 nm, 612 nm, 637 nm, 754 nm.

29. 600 nm to 750 nm of second order overlaps with 400 nm to 500 nm of third order.

31. 621 nm, 909 nm.

35. (*a*) Two orders; (*b*) 6.44×10^{-5} rad $= 13.3''$, 7.36×10^{-5} rad $= 15.2''$, 2.52×10^{-4} rad $= 52.0''$.

37. (*a*) 1.60×10^{4}, 3.20×10^{4}; (*b*) 0.026 nm, 0.013 nm.

39. $\Delta f = f/mN$.

41. (*a*) $62.0°$; (*b*) 0.21 nm.

43. 0.033.

45. $57.3°$.

47. (*a*) $35°$; (*b*) $63°$.

49. $36.9°$, $53.1°$.

51. $I_0/32$.

55. $31°$ on either side of the normal.

57. 12,500 lines/cm.

59. $\sin \theta = \sin 20° - (m\lambda/a)$, $m = \pm 1, \pm 2, \ldots$.

61. Two orders.

63. $11.7°$.

65. (*a*) 16 km; (*b*) $0.42'$.

67. (*a*) 0; (*b*) $0.094 I_0$; (*c*) no light gets transmitted.

69. (*a*) $30°$; (*b*) $18°$; (*c*) $5.7°$.

73. 0.245 nm.

CHAPTER 37

1. (*a*) 1.00; (*b*) 0.99995; (*c*) 0.995; (*d*) 0.436; (*e*) 0.141; (*f*) 0.0447.

3. 2.07×10^{-6} s.

5. $0.773c$.

7. $0.141c$.

9. (*a*) 99.0 yr; (*b*) 27.7 yr; (*c*) 26.6 ly; (*d*) $0.960c$.

11. $0.89c = 2.7 \times 10^{8}$ m/s.

13. (*a*) (470 m, 20 m, 0); (*b*) (1820 m, 20 m, 0).

15. (*a*) $0.80c$; (*b*) $-0.80c$.

17. 60 m/s, $24°$.

19. 2.7×10^{8} m/s, $43°$.

21. (*a*) $L_0\sqrt{1 - (v/c)^2 \cos^2\theta}$; (*b*) $\tan \theta' = \tan \theta / \sqrt{1 - (v/c)^2}$.

23. Not possible in the boy's frame.

25. $0.866c$.

27. (*a*) 0.5%; (*b*) 13%.

29. 5.36×10^{-13} kg.

31. 8.20×10^{-14} J, 0.511 MeV.

33. 9×10^{2} kg.

37. (*a*) 11.2 GeV (1.79×10^{-9} J); (*b*) 6.45×10^{-18} kg·m/s.

39. 7.49×10^{-19} kg·m/s.

41. $0.941c$.

43. $M = 2m/\sqrt{1 - (v^2/c^2)}$, $K_{loss} = 2mc^2\{[1/\sqrt{1 - (v^2/c^2)}] - 1\}$.

45. $0.866c$, 4.73×10^{-22} kg·m/s.

47. 39 MeV (6.3×10^{-12} J), 1.5×10^{-19} kg·m/s, -6%, -4%.

49. $0.804c$.

51. 3.0 T.

57. (*a*) $0.80c$; (*b*) $2.0c$.

61. 3.8×10^{-5} s.

63. $\rho = \rho_0/[1 - (v^2/c^2)]$.

65. (*a*) 1.5 m/s less than c; (*b*) 30 cm.

67. 1.02 MeV (1.64×10^{-13} J).

69. 2.2 mm.

71. 0.78 MeV.

73. Electron.

75. $5.19 \times 10^{-13}\%$.

81. (*a*) $\alpha = \tan^{-1}[(c^2/v^2) - 1]^{1/2}$; (*c*) $\tan \theta = c/v$, $u = \sqrt{v^2 + c^2}$.

CHAPTER 38

1. 6.59×10^{3} K.

3. 5.4×10^{-20} J, 0.34 eV.

5. (*a*) 114 J; (*b*) 228 J; (*c*) 342 J; (*d*) $114n$ J; (*e*) -456 J.

7. (*b*) $h = 6.63 \times 10^{-34}$ J·s.

9. 3.67×10^{-7} eV.

11. 2.4×10^{13} Hz, 1.2×10^{-5} m.

13. 400 nm.

15. (*a*) 2.18 eV; (*b*) 0.92 V.

17. 3.46 eV.

19. 1.88 eV, 43.3 kcal/mol.

21. (*a*) 2.43×10^{-12} m; (*b*) 1.32×10^{-15} m.

23. (*a*) 5.90×10^{-3}, 1.98×10^{-2}, 3.89×10^{-2}; (*b*) 60.8 eV, 204 eV, 401 eV.

27. 1.82 MeV.

29. 212 MeV, 5.85×10^{-15} m.

31. 3.2×10^{-32} m.

33. 19 V.

35. (*a*) 0.39 nm; (*b*) 0.12 nm; (*c*) 0.039 nm.

37. 1.84×10^{3}.

39. 3.3×10^{-38} m/s, no diffraction.

41. 0.026 nm.

43. 3.4 eV.

45. 122 eV.

49. 52.5 nm.

51.

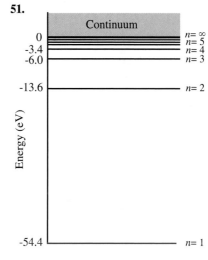

53. $U = -27.2$ eV, $K = +13.6$ eV.

55. Justified.

61. 3.28×10^{15} Hz.

65. 2.78×10^{21} photons/s·m².

67. 8.3×10^{6} photons/s.

69. $\theta = 89.4°$.

71. 4.7×10^{-14} m.

73. 10.2 eV.

75. 4.4×10^{-40}, yes.

77. 653 nm, 102 nm, 122 nm.

79. 0.64 V.

83. 5×10^{-12} m.

85. 3.

CHAPTER 39

1. 1.8×10^{-7} m.

3. $\pm 1.3 \times 10^{-11}$ m.

5. 7.2×10^{3} m/s.

7. 2.4×10^{-3} m, 1.4×10^{-32} m.

9. (a) 6.6×10^{-8} eV; (b) 6.5×10^{-9};
(c) 7.9×10^{-7} nm.

13. (a) $\psi = A \sin (3.5 \times 10^{9} \text{ m}^{-1})x +$
$B \cos (3.5 \times 10^{9} \text{ m}^{-1})x$;
(b) $\psi = A \sin (6.3 \times 10^{12} \text{ m}^{-1})x +$
$B \cos (6.3 \times 10^{12} \text{ m}^{-1})x$.

17. 3.6×10^{6} m/s.

19. (a) 52 nm; (b) 0.22 nm.

23. $E_1 = 0.094$ eV,
$\psi_1 = (1.00 \text{ nm}^{-1/2}) \sin (1.57 \text{ nm}^{-1} x)$;
$E_2 = 0.38$ eV,
$\psi_2 = (1.00 \text{ nm}^{-1/2}) \sin (3.14 \text{ nm}^{-1} x)$;
$E_3 = 0.85$ eV,
$\psi_3 = (1.00 \text{ nm}^{-1/2}) \sin (4.71 \text{ nm}^{-1} x)$;
$E_4 = 1.51$ eV,
$\psi_4 = (1.00 \text{ nm}^{-1/2}) \sin (6.28 \text{ nm}^{-1} x)$.

25. (a) 4 GeV; (b) 2 MeV; (c) 2 MeV.

27. (a) 0.18; (b) 0.50; (c) 0.50.

29.
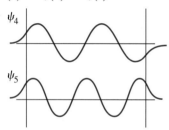
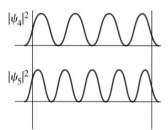

31. 0.03 nm.

33. 9.2 eV.

35. (a) Decreases by 8%;
(b) decreases by 5%.

37. (a) 32 MeV; (b) 56 fm;
(c) 8.8×10^{20} s^{-1}, 10^{10} yr.

39. 21 MeV.

41. 3.00×10^{-10} eV/c^2.

43. r_1, the Bohr radius.

45. 0.5 MeV, 5×10^{6} m/s.

47. (b) 4 s.

49. 14% decrease.

CHAPTER 40

1. $\ell = 0, 1, 2, 3, 4, 5$.

3. 32 states, $(4, 0, 0, -\frac{1}{2})$, $(4, 0, 0, +\frac{1}{2})$,
$(4, 1, -1, -\frac{1}{2})$, $(4, 1, -1, +\frac{1}{2})$, $(4, 1, 0, -\frac{1}{2})$,
$(4, 1, 0, +\frac{1}{2})$, $(4, 1, 1, -\frac{1}{2})$, $(4, 1, 1, +\frac{1}{2})$,
$(4, 2, -2, -\frac{1}{2})$, $(4, 2, -2, +\frac{1}{2})$,
$(4, 2, -1, -\frac{1}{2})$, $(4, 2, -1, +\frac{1}{2})$,
$(4, 2, 0, -\frac{1}{2})$, $(4, 2, 0, +\frac{1}{2})$, $(4, 2, 1, -\frac{1}{2})$,
$(4, 2, 1, +\frac{1}{2})$, $(4, 2, 2, -\frac{1}{2})$, $(4, 2, 2, +\frac{1}{2})$,
$(4, 3, -3, -\frac{1}{2})$, $(4, 3, -3, +\frac{1}{2})$,
$(4, 3, -2, -\frac{1}{2})$, $(4, 3, -2, +\frac{1}{2})$,
$(4, 3, -1, -\frac{1}{2})$, $(4, 3, -1, +\frac{1}{2})$, $(4, 3, 0, -\frac{1}{2})$,
$(4, 3, 0, +\frac{1}{2})$, $(4, 3, 1, -\frac{1}{2})$, $(4, 3, 1, +\frac{1}{2})$,
$(4, 3, 2, -\frac{1}{2})$, $(4, 3, 2, +\frac{1}{2})$, $(4, 3, 3, -\frac{1}{2})$,
$(4, 3, 3, +\frac{1}{2})$.

5. $n \geq 5$, $m_\ell = -4, -3, -2, -1, 0, 1, 2, 3, 4$,
$m_s = -\frac{1}{2}, +\frac{1}{2}$.

7. (a) 6; (b) -0.378 eV; (c) $\ell = 4$,
$\sqrt{20}\hbar = 4{,}72 \times 10^{-34}$ kg·m^2/s;
(d) $m_\ell = -4, -3, -2, -1, 0, 1, 2, 3, 4$.

13. (a) $-[3/(32\pi r_0^3)^{1/2}] \, e^{-5/2}$;
(b) $(9/32\pi r_0^3) \, e^{-5}$; (c) $(225/8r_0) \, e^{-5}$.

15. 1.85.

17. 17.3%.

19. (a) $1.34 r_0$; (b) $2.7 r_0$; (c) $4.2 r_0$.

21. $\dfrac{r^4}{24 r_0^5} e^{-r/r_0}$.

27. (a) $\dfrac{4r^2}{27 r_0^3}\left(1 - \dfrac{2r}{3r_0} + \dfrac{2r^2}{27 r_0^2}\right)^2 e^{-2r/3r_0}$;

(b)

P_r

0.10 ┤

0.05 ┤

0 ┤──┬────┬────┬────┬──
  0   10   20   30 r/r_0

(c) $r = 13 r_0$.

29. (a) $(1, 0, 0, -\frac{1}{2})$, $(1, 0, 0, +\frac{1}{2})$,
$(2, 0, 0, -\frac{1}{2})$, $(2, 0, 0, +\frac{1}{2})$, $(2, 1, -1, -\frac{1}{2})$,
$(2, 1, -1, +\frac{1}{2})$; (b) $(1, 0, 0, -\frac{1}{2})$,
$(1, 0, 0, +\frac{1}{2})$, $(2, 0, 0, -\frac{1}{2})$, $(2, 0, 0, +\frac{1}{2})$,
$(2, 1, -1, -\frac{1}{2})$, $(2, 1, -1, +\frac{1}{2})$, $(2, 1, 0, -\frac{1}{2})$,
$(2, 1, 0, +\frac{1}{2})$, $(2, 1, -1, -\frac{1}{2})$, $(2, 1, -1, +\frac{1}{2})$,
$(3, 0, 0, -\frac{1}{2})$, $(3, 0, 0, +\frac{1}{2})$.

31. (a) $1s^2 2s^2 2p^6 3s^2 3p^6 3d^{10} 4s^2 4p^4$; (b) $1s^2$
$2s^2 2p^6 3s^2 3p^6 3d^{10} 4s^2 4p^6 4d^{10} 4f^{14} 5s^2 5p^6$
$5d^{10} 6s^1$; (c) $1s^2 2s^2 2p^6 3s^2 3p^6 3d^{10} 4s^2 4p^6$
$4d^{10} 4f^{14} 5s^2 5p^6 5d^{10} 6s^2 6p^6 5f^3 6d^1 7s^2$.

33. 5.8×10^{-13} m, 0.115 MeV.

37. 0.041 nm, 1 nm.

41. 0.19 nm.

43. Chromium.

47. (a) 0.25 mm; (b) 0.13 mm.

49. (a) $\frac{1}{2}, \frac{3}{2}, \frac{1}{2}\sqrt{3}\,\hbar, \frac{1}{2}\sqrt{15}\hbar$;
(b) $\frac{5}{2}, \frac{7}{2}, \frac{1}{2}\sqrt{35}\hbar, \frac{1}{2}\sqrt{63}\hbar$;
(c) $\frac{3}{2}, \frac{5}{2}, \frac{1}{2}\sqrt{15}\hbar, \frac{1}{2}\sqrt{35}\hbar$.

51. (a) 0.4 T; (b) 0.4 T.

53. 5.6×10^{-4} rad, 1.7×10^{2} m.

55. 3.7×10^{4} K.

57. (a) 1.56; (b) 1.4×10^{-10} m.

59. (a) $1s^2 2s^2 2p^6 3s^2 3p^6 3d^7 4s^2$;
(b) $1s^2 2s^2 2p^6 3s^2 3p^6 3d^{10} 4s^2 4p^6$;
(c) $1s^2 2s^2 2p^6 3s^2 3p^6 3d^{10} 4s^2 4p^6 5s^2$.

61. (a) 2.5×10^{74}; (b) 5.0×10^{74}.

63. $r = 5.24 r_0$.

65. (a) ϕ is unknown;
(b) L_x and L_y are unknown.

67. (a) 1.2×10^{-4} eV;
(b) 1.1 cm; (c) no difference.

69. (a) 3×10^{-171}, 7×10^{-203};
(b) 1.1×10^{-8}, 6.3×10^{-10};
(c) 6.6×10^{15}, 3.8×10^{14};
(d) 7×10^{23} photons/s,
4×10^{22} photons/s.

71. 182.

73. 2.25.

CHAPTER 41

1. 5.1 eV.

3. 0.7 eV.

7. (a) 13.941 u; (b) 7.0034 u;
(c) 0.9801 u.

9. (a) 6.86 u; (b) 1.85×10^{3} N/m.

11. (a) 1.79×10^{-4} eV;
(b) 7.16×10^{-4} eV, 1.73 mm.

13. 2.36×10^{-10} m.

15. -7.9 eV.

17. 0.283 nm.

19. (b) -6.9 eV; (c) -10.8 eV; (d) 3%.

21. 1.8×10^{21}.

23. (a) 6.9 eV; (b) 6.8 eV.

25. 6.3%.

27. 3.2 eV, 1.05×10^{6} m/s.

29. (a) 1.79×10^{29} m^{-3}; (b) 3.

33. (a) 0.021, reasonable; (b) 0.979;
(c) 0.021.

37. A large energy is required to create
a conduction electron by raising an
electron from the valence band to
the conduction band.

39. 1.1 μm.

41. 5×10^{6}.

43. 1.91 eV.

45. 13 mA.

47. (*a*) 2.1 mA; (*b*) 4.3 mA.

49. (*a*) 9.4 mA (smooth);
(*b*) 6.7 mA (rippled).

51. 4.0 kΩ.

53. 0.43 mA.

55. (*a*) 3.5×10^4 K; (*b*) 1.2×10^3 K.

57. (*a*) 0.9801 u; (*b*) 482 N/m, 88% of
the constant for H_2.

59. States with higher values of *L* are
less likely to be occupied, so less
likely to absorb a photon; *I* will
depend on *L*.

61. 2.8×10^4 J/kg.

63. 1.24 eV.

65. 1.09 μm, could be used.

67. (*a*) 0.094 eV; (*b*) 0.63 nm.

69. (*a*) 145 V $\leq$ *V* $\leq$ 343 V;
(*b*) 3.34 kΩ $\leq$ R_{load} $<$ ∞.

CHAPTER 42

1. 3729 MeV/c^2.

3. 1.9×10^{-15} m.

5. (*a*) 2.3×10^{17} kg/m^3; (*b*) 184 m;
(*c*) 2.6×10^{-10} m.

7. 28 MeV.

9. $^{31}_{15}$P.

11. (*a*) 1.8×10^3 MeV;
(*b*) 7.3×10^2 MeV.

13. 7.48 MeV.

15. (*a*) 32.0 MeV, 5.33 MeV;
(*b*) 1636 MeV, 7.87 MeV.

17. 12.4 MeV, 7.0 MeV, neutron is more
closely bound in ^{23}Na.

19. (*b*) Stable.

21. 0.782 MeV.

23. β^+ emitter, 1.82 MeV.

25. $^{228}_{90}$Th, 228.02883 u.

27. (*a*) $^{32}_{16}$S; (*b*) 31.97207 u.

29. 0.862 MeV.

31. 5.31 MeV.

33. (*b*) 0.961 MeV, 0.961 MeV to 0.

35. 3.0 h.

37. 1.2×10^9 decays/s.

39. 1.78×10^{20} nuclei.

41. 7 α particles, 4 β^- particles.

43. (*a*) 4.8×10^{16} nuclei;
(*b*) 3.2×10^{15} nuclei;
(*c*) 7.2×10^{13} decays/s; (*d*) 26 min.

45. 1.68×10^{-10} g.

47. 2.6 min.

49. 3.4 mg.

51. 8.6×10^{-7}.

53. $^{211}_{82}$Pb.

55. $N_D = N_0(1 - e^{-\lambda t})$.

57. (*a*) 0.99946; (*b*) 1.2×10^{-14};
(*c*) 2.3×10^{17} kg/m^3, $10^{14}\times$.

59. 28.6 eV.

61. (*a*) 7.2×10^{55}; (*b*) 1.2×10^{29} kg;
(*c*) 3.2×10^{11} m/s^2.

63. 6×10^3 yr.

65. 6.64 $T_{1/2}$.

69. Calcium, stored by body in bones,
193 yr, $^{90}_{38}$SR $\rightarrow$ $^{90}_{39}$Y $+$ $^{0}_{-1}$e $+$ $\bar{\nu}$,
$^{90}_{39}$Y is radioactive,
$^{90}_{39}$Y $\rightarrow$ $^{90}_{40}$Zr $+$ $^{0}_{-1}$e $+$ $\bar{\nu}$,
$^{90}_{40}$Zr is stable.

73. (*a*) 0.002603 u, 2.425 MeV/c^2; (*b*) 0;
(*c*) -0.094909 u, -88.41 MeV/c^2;
(*d*) 0.043924 u, 40.92 MeV/c^2;
(*e*) $\Delta \geq 0$ for $0 \leq$ Z ≤ 8
and Z ≥ 85, $\Delta < 0$ for $9 \leq$ Z ≤ 84.

75. 0.083%.

77. $^{228}_{88}$RA, $^{228}_{89}$Ac, $^{228}_{90}$Th, $^{224}_{88}$Ra, $^{220}_{87}$Rn,
$^{231}_{90}$Th, $^{231}_{91}$Pa, $^{227}_{89}$Ac, $^{227}_{90}$Th, $^{223}_{88}$Ra.

79. (*b*) $\approx 10^{17}$ yr; (*c*) ≈ 60 yr; (*d*) 0.4.

CHAPTER 43

1. $^{28}_{13}$Al, β^- emitter, $^{28}_{14}$Si.

3. Possible.

5. 5.701 MeV is released.

7. (*a*) Can occur; (*b*) 19.85 MeV.

9. $+4.730$ MeV.

11. (*a*) $^{7}_{3}$Li; (*b*) neutron is stripped from
the deuteron; (*c*) $+5.025$ MeV,
exothermic.

13. (*a*) $^{31}_{15}$P(p, γ)$^{32}_{16}$S; (*b*) $+8.864$ MeV.

15. $\sigma = \pi(R_1 + R_2)^2$.

17. Rate at which incident particles pass
through target without scattering.

19. (*a*) 0.7 μm; (*b*) 1 mm.

21. 173.2 MeV.

23. 0.116 g.

25. 25.

27. 0.11.

29. 1.3 keV.

33. 6.1×10^{23} MeV/g, 4.9×10^{23} MeV/g,
2.1×10^{24} MeV/g,
5.1×10^{23} MeV/g.

35. Not independent.

37. 3.23×10^9 J, $65\times$.

39. (*a*) $4.9\times$; (*b*) 1.5×10^9 K.

41. 400 rad.

43. 167 rad.

45. 1.7×10^2 counts/s.

49. 0.225 μg.

51. (*a*) 0.03 mrem $\approx$ 0.006% of allowed
dose; (*b*) 0.3 mrem $\approx$ 0.06% of
allowed dose.

53. (*a*) 1; (*b*) 1 $\leq$ *m* $\leq$ 2.7.

55. (*a*) $^{12}_{6}$C; (*b*) $+5.70$ MeV.

57. 1.004.

59. 51 mrem/yr.

61. 4.6 m.

63. 18.000953 u.

65. 6.31×10^{14} J/kg, $\approx 10^7\times$ the heat of
combustion of coal.

67. 1×10^{24} neutrinos/yr.

69. (*a*) 6.8 bn; (*b*) 3×10^{-14} m.

71. (*a*) 3.7×10^3 decays/s;
(*b*) 5.2×10^{-4} Sv/yr $\approx$ 0.15
background.

CHAPTER 44

1. 7.29 GeV.

3. 1.8 T.

5. 13 MHz.

7. $\lambda_\alpha = 2.6 \times 10^{-15}$ m $\approx$ size of
nucleon, $\lambda_p = 5.2 \times 10^{-15}$ m $\approx$
2(size of nucleon), α particle is
better.

9. 1.4×10^{-18} m.

11. 2.2×10^6 km, 7.5 s.

15. 33.9 MeV.

17. 1.879 GeV.

19. 67.5 MeV.

21. 2.3×10^{-18} m.

23. (*b*) Uncertainty principle allows
energy to not be conserved.

25. 69.3 MeV.

27. 8.6 MeV, 57.4 MeV.

29. 52.3 MeV.

31. 7.5×10^{-21} s.

33. (*a*) 1.3 keV; (*b*) 8.9 keV.

35. (*a*) n = d d u; (*b*) $\bar{n}$ = $\bar{d}$ $\bar{d}$ $\bar{u}$;
(*c*) Λ^0 = u d s; (*d*) Σ^0 = $\bar{u}$ $\bar{d}$ $\bar{s}$.

37. D^0 = c$\bar{u}$.

39. (*a*)

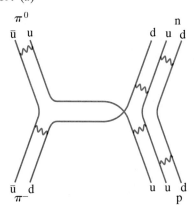

(*b*)

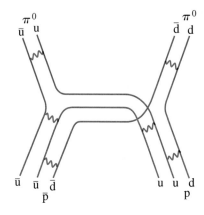

41. 26 GeV, 4.8×10^{-17} m.

43. 5.5 T.

45. (*a*) Possible, through the strong interaction; (*b*) possible, through the strong interaction; (*c*) possible, through the strong interaction; (*d*) forbidden, charge is not conserved; (*e*) possible, through the weak interaction; (*f*) forbidden, charge is not conserved; (*g*) possible, through the strong interaction; (*h*) forbidden, strangeness is not conserved; (*i*) possible, through the weak interaction; (*j*) possible, through the weak interaction.

49. -135.0 MeV, -140.9 MeV.

51. 64.

53. (*b*) 10^{27} K.

55. 6.58×10^{-5} m.

57. $\bar{u}\,\bar{u}\,\bar{d} + u\,d\,d \rightarrow \bar{u}\,d + d\,\bar{d}$.

CHAPTER 45

3. 4.8 ly.

5. $0.059''$, 17 pc.

7. Less, $\phi_1/\phi_2 = \frac{1}{2}$.

9. 48 W/m².

11. 1.4×10^{-4} kg/m³.

13. 1.8×10^{9} kg/m³, 3.3×10^{5}.

15. -92.2 keV, 7.366 MeV.

17. (*a*) 9.594 MeV is released; (*b*) 7.6 MeV; (*c*) 6×10^{10} K.

19. $d_2/d_1 = 6.5$.

23. 540°.

25. 1.4×10^{8} ly.

27. $0.86c$.

29. $0.328c$, 4.6×10^{9} ly.

31. 1.1 mm.

33. (*a*) 10^{-3}; (*b*) 10^{-10}; (*c*) 10^{-13}; (*d*) 10^{-27}.

35. (*a*) Temperature increases, luminosity is constant, size decreases; (*b*) temperature is constant, luminosity decreases, size decreases; (*c*) temperature decreases, luminosity increases, size increases.

37. 8×10^{3} rev/s.

39. 7×10^{24} W.

41. (*a*) 46 eV; (*b*) 18 eV.

43. $\approx (2 \times 10^{-5})°$.

45. 1.4×10^{16} K, hadron era.

47. Venus is brighter, $\ell_V/\ell_S = 16$.

49. $R \geq GM/c^2$.

51. $R = \dfrac{h^2}{16 m_n^{8/3} G M^{1/3}} \left(\dfrac{18}{\pi^2} \right)^{2/3}$.

Index

Copier, electrostatic, 555
Core, of reactor, 1093
Coriolis force, 292–94
Cornea, 850
Corrective lenses, 850–52
Correspondence principle, 943, 971, 976 *pr*, 978
Cosmic microwave background radiation, 1159–61, 1164
Cosmic rays, 1102, 1114
Cosmological principle, 1157 *ff*
  "perfect," 1159 *fn*
Cosmology, 1136, 1140–69
Coulomb, Charles, 549–50
Coulomb (unit), 550, 712
  operational definition of, 712
Coulomb barrier, 998, 1098, 1148
Coulomb's law, 549–53, 575–86, 966
  vector form of, 553
Coulomb potential, 1034
Counter emf, 742–43
Counter torque, 743
Covalent bond, 1030–33, 1044
Crane, 304
Creativity in science, 3
Crick, F.H.C., 906
Critical damping, 376
Critical density, of universe, 1165–68
Critical mass, 1092–95
Critical point, 473
Critical temperature, 473
Crossed Polaroids, 908
Cross product, 279–80
Cross section, 1088–90
CRT, 605–6, 700, 805
Crystal lattice, 446–47, 1044
Crystallography, 906
CT scan, 1105–6
Curie, M. and P., 726, 1067
Curie (unit), 1101
Curie temperature, 722
Curie's law, 726
Current (*see* Electric current)
Current density, 647–49
Current gain, 1055
Current sensitivity, 674
Curvature of field, 859
Curvature of universe (space-time), 149, 1151–55, 1165
Cycle (*defn*), 363
Cyclic universe, 1168
Cyclotron, 707 *pr*, 1116–17
Cyclotron frequency, 694, 1116

D

Damped harmonic motion, 374–77
Dark matter, 1166
Dating, radioactive, 1078–79
Daughter nucleus (*defn*), 1068
Davission, C. J., 960
dB (unit), 421
DC circuits, 658–77
DC generator, 741
DC motor, 698
de Broglie, Louis, 959–61, 971, 977–78
de Broglie wavelength, 959, 971–72
Debye's law, 514 *pr*
Decay:
  of elementary particles, 1124–34
  radioactive (*see* Radioactivity)
Decay constant, 1073–76
Decay series, 1077–78
Deceleration (*defn*), 24
Deceleration parameter, 1168
Decibel (db), 421
Declination, magnetic, 688
Decommissioning nuclear plant, 1094
Dee, 1116
Defects of the eye, 851
Defibrillator, 632 *pr*, 651 *fn*
Degrees of freedom, 500–501
Del, 603 *fn*
Delayed neutron, 1093
Delta particle, 1128
Demagnetization, 725
Density, 332–33
  and floating, 342
Density of occupied states, 1047
Density of states, 1045–48
Depth of field, 849
Derivatives, 21–22, A-4–A-5
  partial, 399
Derived quantities, 8
Descartes, R., 146
Destructive interference, 404, 430, 871, 1031
Detection of radiation, 1080
Determinism, 147, 984–85
Deuterium, 1086, 1092, 1095–1100
Deuteron, 1086
Dew point, 476
Diamagnetism, 725–26
Diamond, 826–27
Dielectric constant, 621–22
Dielectrics, 621–26
  molecular description of, 624–26

Dielectric strength, 622
Diesel engine, 540 *pr*
Differential cross section, 1089
Diffraction, 867, 887–911, 1020
  by circular opening, 896–97
  as distinguished from interference, 896
  in double slit experiment, 896
  of electrons, 960
  Fraunhofer, 888 *fn*
  Fresnel, 888 *fn*
  of light, 867, 887–911
  as limit to resolution, 896–97
  by single slit, 888–93
  of water waves, 410
  X-ray, 905–6
Diffraction grating, 900–905
  resolving power of, 903–5
Diffraction limit of lens resolution, 896–97
Diffraction patterns, 888 *ff*
  of circular opening, 896–97
  of single slit, 888–93
  X-ray, 905–6
Diffraction spot or disc, 896
Diffuse reflection, 813
Diffusion, 479–80
  Fick's law of, 480
Diffusion constant, 480
Diffusion equation, 480
Digital information, 749
Digital voltmeter, 675
Dimensional analysis, 12–13
Dimensions, 12–13
Diodes, 1052–54
Diopter, 838
Dip, angle of, 688
Dipole antenna, 793
Dipole bonds, 1036–37
Dipoles and dipole moments:
  of atoms, 1015–18
  electric, 565–67, 578, 601–3
  magnetic, 695–97, 721
Dirac, P. A. M., 1005, 1046 *fn*
Direct current (dc) (*see* Electric current)
Discharge tube, 963
"Discovery" in science, 700
Disintegration energy (*defn*), 1068–69
Disorder and order, 533–34
Dispersion, 402, 824–25
Displacement, 17–18, 46–47
  angular, 246–47

Electric power, 642–44
  in ac circuits, 778
    generation, 741
    in household circuits, 644–45
    and impedance matching, 781–82
    transmission of, 746–47
Electric shocks, 651–52
Electrocardiogram (ECG, EKG), 606
Electrode, 635
Electrolyte, 635
Electromagnet, 723–24
Electromagnetic force, 1121–22,
  1126–27, 1134–36
Electromagnetic induction, 734 *ff*
Electromagnetic oscillations, 764–65
Electromagnetic pumping, 703 *pr*
Electromagnetic spectrum, 798–800,
  874–75
Electromagnetic (EM) waves,
  792–806, 866 (*see also* Light)
Electrometer, 548
Electromotive force (emf), 659–60,
  734–35, 739–40, 742–43
  back, 742–43
  counter, 742–43
  of generator, 740–42
  Hall, 701
Electromotive force (emf) (*cont.*)
  induced, 734–35, 739–40
  motional, 739
  series and parallel, 668
  sources of, 659–60, 734–35
Electron:
  as beta particle, 1067, 1070
  charge on, 547, 550, 722–23
  discovery of, 699–700
  in double-slit experiment, 978–79
  as elementary particle, 1126
  free, 548
  mass of, 1063
  in pair production, 957–58
  wavelength of, 959–60
Electron band theory, 1048–50
Electron capture, 1072
Electron cloud, 1004, 1030–31
Electron configuration, 925–26
Electron gun, 606
Electronically steered phase array, 877
Electron microscope, 899, 961, 999
Electron sharing, 1030
Electron spin, 723, 1005–6, 1017–18
Electron spin resonance, 1029 *pr*
Electron volt (unit), 603–5, 1063
Electroscope, 548–49, 635 *fn*

Electrostatic copier, 555
Electrostatic potential energy, 603–5
Electrostatics, 551 *ff*
Electrostatic unit, 550 *fn*
Electroweak force, 148, 545 *fn*
Electroweak theory, 1132–34
Elementary charge, 550
Elementary particle physics, 1114–36
Elements:
  origin of in universe, 1149–50
  periodic table of, 1011–13, 1062 *fn*,
    inside rear cover
  transmutation of, 1068, 1070,
    1085–88
Elevator and counterweight, 92
Ellipse, 143
EMF (*see* Electromotive force)
Emission spectra, 902, 949–51,
  963–65, 968–69
Emission tomography, 1006–7
Emissivity, 506
Emitter, 1054
EM waves (*see* Electromagnetic
  waves)
Endoergic reaction (*defn*), 1087
Endoscope, 827
Endothermic reaction, 1086–87
Energy, 155, 165–69, 176–97, 260–65,
  487–88, 493–94
  activation, 471, 1034
  binding, 968, 1031, 1033, 1036,
    1064–66
  chemical, 190
  conservation of, 182–92, 493–95,
    1086
  degradation of, 535
  disintegration, 1068
  distinguished from heat and
    temperature, 487
  electric (*see* Electric energy)
  in EM waves, 800–802
  equipartition of, 500–501
  equivalence to mass, 938–42
  and first law of thermodynamics,
    493–94
  internal, 189–90, 487–88
  kinetic, 164–69, 182 *ff*, 260–65
  mechanical, 182–89
  molecular rotational and
    vibrational, 500–501
  nuclear, 990–1016
  potential (*see* Potential energy)
  quantization of, 951, 965–71,
    991–92

reaction, 1086
  relation to work, 164–69, 178–82,
    190–91, 195, 260–61
  rest, 939, 1063
  rotational, 260–65
  in simple harmonic motion, 369–70
  thermal, 189–90, 487–88
  threshold, 1088, 1113 *pr*
  total mechanical energy (*defn*), 183
  transformation of, 189–90
  unavailability of, 534–35
  and uncertainty principle, 982–83,
    996
  units of, 167
  vibrational, 369–70
  of waves, 395–96
  zero point, 992, 1041
Energy bands, 1048–50
Energy density:
  in electric field, 621
  in magnetic field, 761
Energy gap, 1049–50
Energy levels:
  in atoms, 967
  for fluorescence, 1019
  for lasers, 1019–22
Energy states, in atoms, 967
Energy transfer, heat as, 189–90,
  485 *ff*
Engine, heat (*see* Heat engine)
Enriched uranium, 1092
Entropy, 528–37
  in life processes, 534
  statistics and, 535–37
Environmental pollution, 525
Enzyme, 1037
Equally-tempered chromatic scale,
  424
Equation of continuity, 344
Equation of motion, oscillations, 364,
  375, 377
Equation of state, 454, 457, 477–78
  Clausius, 477
  ideal gas, 457
  van der Waals, 477–78
Equilibrium, 197–98, 300–303, 308
  conditions for, 301–3
  stable, unstable, neutral, 197–98,
    308
  thermal, 449–50
Equilibrium position (vibrational
  motion), 363
Equilibrium state (*defn*), 454
Equipartition of energy, 500–501

Equipotential lines, 600–601
Equipotential surface, 600–601
Equivalence, principle of, 148, 1151–52
Erg (unit), 156
Escape velocity, 192–94
Estimated uncertainty, 4
Estimating, 9–12
Ether, 919–22
Euclidean space, 1153
Evaporation, 474–75, 493
Event horizon, 1155
Everest, Mt., 138
Evolution and entropy, 534
Exact differential, 494 *fn*
Exchange coupling, 723
Excited state, of atom, 957 *ff*
Exclusion principle, 1010–12, 1046, 1132–33, 1150
Exoergic reaction (*defn*), 1086
Exothermic reaction, 1086
Expansion, thermal (*see* Thermal expansion)
Expansion of universe, 1156–59
Expansions, in waves, 391
Expansions, mathematical, A-1
Exponential curves, 1074–75
Exponential decay, 1074–75
Extragalactic (*defn*), 1143
Eye:
    accommodation, 851
    defects of, 851
    far and near points of, 851
    normal (*defn*), 851
    resolution of, 899
    structure and function of, 850–52
Eyeglass lenses, 851–52
Eyepiece, 854

F

Fahrenheit temperature scale, 448–49
Falling bodies, 31–36
Fallout, radioactive, 1095
Farad (unit of capacitance), 614
Faraday, Michael, 147, 554, 622, 734, 736, 787
Faraday's law, 736–38, 747–48, 756, 792
Far point of eye, 851
Farsighted eye, 851–52
Fermat's principle, 835 *pr*

Fermi, E., 11, 958, 978, 1011 *fn*, 1046 *fn*, 1071, 1088, 1090–95, 1128
Fermi-Dirac statistics, 1046–48
Fermi energy, 1046–48
Fermi factor, 1046–47
Fermi gas, 1046
Fermilab, 1114, 1117
Fermi level, 1046–48
Fermions, 1011 *fn*, 1046, 1132–33
Fermi speed, 1047
Fermi temperature, 1060 *pr*
Ferris wheel, 124 *pr*
Ferromagnetism and ferromagnetic materials, 687, 722–23
Feynman, R., 1121
Feynman diagram, 1121–22
Fiber optics, 826–27
Fick's law of diffusion, 480
Field, 147 (*see also* Electric field, Gravitational field, Magnetic field)
    conservative and nonconservative, 748–49
    in elementary particles, 1121
Film badge, 1103
Filter circuit, 776, 785 *pr*
Fine structure, 977, 1003, 1006, 1018
Fine structure constant, 1018
First law of motion, 78 *ff*
First law of thermodynamics, 493–98
Fission, nuclear, 1090–95
Fission bomb, 1095
Fission fragments, 1090–95
Fitzgerald, G. F., 922
Flashlight battery, 636
Flashlight bulb, 639
Flavor (of elementary particles), 1132–33
Floating, 342
Flow of fluids, 343–45
    laminar, 343–45
    streamline, 343–45
    in tubes, 351
    turbulent, 343
Flow rate, 343–45
Fluids, 332–61 (*see also* Gases, Liquids)
Fluorescence, 1019
Fluorescent lightbulbs, 1019
Flux:
    electric, 575–78, 789
    magnetic, 736, 791
Flying buttress, 319
FM radio, 804

Focal length:
    of lens, 838, 840–41
    of spherical mirror, 817, 821
Focal plane, 838
Focal point, 817, 821, 837–38
Focus, 817
Focusing, of camera, 849
Foot-candle (*defn*), 882 *fn*
Foot-pound (unit), 156, 167
Forbidden transition, 1007, 1020 *fn*, 1041 *fn*
Force, in general, 77–85, 147–48, 177–78 (*see also* Electric force, Gravitational force, Magnetic force)
    buoyant, 340–43
    centrifugal, 115
    centripetal, 114
    color, 1133–34
    conservative and nonconservative, 177–78
    contact, 86, 148
    Coriolis, 292–94
    damping, 374–77 (*see also* Drag force)
    dissipative, 189–92
    drag, 122–23
    elastic, 162
    electromagnetic, 1121–22, 1126–27, 1134–36
    electroweak, 148
    in equilibrium, 300–303
    exerted by inanimate object, 83–84
    fictitious, 291
    of friction, 78, 106–14, 118–20, 178, 190–92
    of gravity, 85–86, 133 *ff*, 1123, 1150–55
    impulsive, 212
    inertial, 291–92
    long and short range, 1066
    measurement of, 78
    in muscles and joints, 306
    net, 80, 88
    in Newton's laws, 78–85, 207–9
    nonconservative, 178
    normal, 85–87, 107
    nuclear, 1064–66, 1072, 1121–36
    pseudoforce, 291
    relation of momentum to, 206–8
    resistive, 122–23
    restoring, 162, 363
    strong, 1064–66, 1121–36

Induction, electromagnetic, 734 *ff*
Induction, Faraday's law, 736–38, 792
Inductive reactance, 774
Inductor, 758–59
  in circuits, 758–63, 772–83
  energy stored in, 760–61
  reactance of, 774
Inelastic collision, 214, 217–18
Inelastic scattering, 1089
Inertia, 79
  law of, 78–79
  moment of, 249–56, 374
  rotational (*defn*), 249–50
Inertial confinement, 1098–99
Inertial forces, 291–92
Inertial mass, 148, 1152
Inertial reference frame, 79, 291, 917 *ff*
  equivalence of all, 918–19, 922
  transformations, 932–36
Inflationary universe, 1164
Infrared radiation (IR), 506, 799, 825
Infrasonic waves, 419
Initial conditions, 365
Instantaneous axis, 246
Instruments, wind and stringed, 424–29
Insulators:
  electrical, 547–48, 640, 1049–50
  thermal, 504, 1049–50
Integrals and integration, 36–37, 161 *ff*, A-4–A-5, B-5
Integrated circuits, 1054–55
Intensity:
  of coherent and incoherent sources, 877
  in interference and diffraction patterns, 874–77, 890–93
  of light, 877, 882
  of sound, 420–23
  of waves, 395–96, 420–23
Intensity level (*see* Sound level)
Interference, 404–5
  constructive, 404, 430, 871, 1031
  destructive, 404, 430, 871, 1031
  as distinguished from diffraction, 895
  of light waves, 870–81, 894–96, 900
  of sound waves, 429–32
  by thin films, 877–81
  of water waves, 404–5
  of waves on a string, 404
Interference fringes, 871

Interference pattern:
  double slit, 870–73, 978–79
  double slit, including diffraction, 893–95
  multiple slit, 900–905
Interferometers, 881
Internal combustion engine, 518–19
Internal conversion, 1073
Internal energy, 189–90, 487–88
Internal reflection, total, 415 *pr*, 826–27
Internal resistance, 659
Intrinsic semiconductor (*defn*), 1049
Inverted population, 1020
Ion, 547
Ionic bonds, 1031–33, 1044
Ionic cohesive energy, 1044
Ionization energy, 968, 970
Ionizing radiation (*defn*), 1100
IR radiation, 506, 799, 825
Irreversible process, 520–25
Isobaric process (*defn*), 496–97
Isochoric process (*defn*), 496–97
Isolated system, 209, 493
Isomer, 1073
Isotherm, 495
Isothermal process (*defn*), 495
Isotopes, 702, 1062
  table of, A-9–A-13
Isotropic material, 867

J

J (total angular momentum), 1017–18
J/ψ particle, 1131
Jeweler's loupe, 854
Joint, 315
Joule, James Prescott, 486
Joule (unit), 156, 167, 248 *fn*
  relation to calorie, 486
Jump start, 668–69
Junction diode, 1054–55
Junction transistor, 1054–55
Jupiter, moons of, 152, 854

K

Kant, Immanuel, 1143
Kaon, 1127
K-capture, 1072
K lines, 1013

Kelvin (unit), 455
Kelvin-Planck statement of second law of thermodynamics, 520, 523, 533
Kelvin temperature scale, 455, 460, 538
Kepler, Johannes, 143, 854 *fn*, 1150
Keplerian telescope, 854
Kepler's laws, 143–46, 288–89
Keyboard, computer, 616
Kilocalorie (unit), 486–87
Kilogram (unit), 7, 79
Kilowatt-hour (unit), 643
Kinematics, 16–67, 240–44 (*see also* Motion)
  for rotational motion, 240–44
  for translational motion, 16–67
  for uniform circular motion, 63–65
  vector kinematics, 53–55
Kinetic energy, 164–69, 182 *ff*, 260–65
  in collisions, 214–17, 219
  molecular, relation to temperature, 469
  relativistic, 938–42
  rotational, 260–65
  total, 262–65
  translational, 165–69
Kinetic friction, 106 *ff*
Kinetic theory, 445, 466 *ff*
Kirchhoff, G. R., 665
Kirchhoff's rules, 664–69

L

Ladder, 306–7
Laminar flow, 343–45
Land, Edwin, 907
Large Magellanic Cloud, 1150
Laser light, 873
Lasers, 873, 887, 1019–22, 1100
Latent heats, 490–93
Lateral magnification, 819, 841
Lattice structure, 446–47, 1044, 1051–52
Laue, Max von, 905
Lawrence, E. O., 1116
Laws, 3–4 (*see also* specific name of law)
Lawson, J. D., 1099
Lawson criterion, 1099–1100
LC circuit, 764–65
LCD, 909

produces electric field and current, 747–49
of solenoid, 716–18
sources of, 709–26
of straight wire, 710, 720
of toroid, 708
Magnetic field lines, 687
Magnetic flux, 736, 791
Magnetic flux density (*see* Magnetic field)
Magnetic force, 686, 689–95
on electric current, 689–92
on moving electric charges, 692–95
Magnetic induction (*see* Magnetic field)
Magnetic lens, 961
Magnetic moment, 695–97, 1015–18, 1064
Magnetic permeability, 710, 724–25
Magnetic poles, 686–87
of Earth, 688
Magnetic quantum number, 1005–6, 1016
Magnetic resonance imaging, 1064, 1108–9
Magnetic tape and discs, 749
Magnetism, 686–726
Magnetization (vector), 726
Magnification:
angular, 853
of electron microscope, 961
lateral, 819, 841
of lens, 841–43
of lens combination, 843–45
longitudinal, 833 *pr*
of magnifying glass, 853–54
of microscope, 856–58, 898–99
of mirror, 819
sign conventions for, 819, 841
of telescope, 855, 898–99
useful, 899, 961
Magnifier, simple, 853–54
Magnifying glass, 836, 853–54
Magnifying power, 853, 855 (*see also* Magnification)
Magnitude (stars and galaxies), 1172 *pr*
Main-sequence, stars, 1146–49
Manhattan Project, 1095
Marconi, Guglielmo, 803
Manometer, 338
Mass, 7, 79, 82
atomic, 446
center of, 221–24

critical, 1092–95
gravitational vs. inertial, 148, 1152
inertial, 148, 1152
molecular, 446
in relativity theory, 936–38
rest, 938–40, 1063
standard of, 7
units of, 7, 79
variable, 227–29
Mass-energy transformation, 938–42
Mass excess, 1084 *pr*
Mass increase in relativity, 936–38
Mass spectrometer (spectrograph), 702
Matter, states of, 332, 446–47
Matter-dominated universe, 1161, 1164
Maxwell, James Clerk, 470, 787, 792, 797–98, 918
Maxwell distribution of molecular speeds, 470–72
Maxwell's equations, 787, 792–806, 918–19, 934
Mean free path, 478–79
Mean life, 1075, 1084 *pr*
Mean speed, 469
Measurement, 4–5, 981
Mechanical advantage, 93, 303
Mechanical energy, 182–89
Mechanical equivalent of heat, 486
Mechanical waves, 388–411
Mechanics (*defn*), 16
Medical imaging, 437, 1105–9
Meitner, L., 978, 1090
Melting point, 491 (*see also* Phase, changes of)
Mendeleev, D., 1012
Meson exchange, 1122
Mesons, 1122–36
Metallic bond, 1044
Metals, free electron theory of, 1045–48
Metastable state, 1019–20, 1073
Meter (unit), 6, 812 *fn*, 881
Meters, 674–76, 697
correction for resistance of, 676
Metric system, 6–8
MeV (million electron volts) (*see* Electron volt)
Mho (unit), 656 *pr*
Michelson, A., 811–12, 881, 919–22
Michelson interferometer, 881, 919–22

Michelson-Morley experiment, 919–23, 923 *fn*
Micrometer, 10
Microphones:
capacitor, 677
magnetic, 749
ribbon, 749
Microscope:
compound, 856–58
electron, 899, 961, 999
magnification of, 856–58, 898–99
resolving power of, 898–99
useful magnification, 899
Microscopic description of a system, 445
Microstate, 535–36
Microwaves, 798
Milky Way, 854, 1141
Millikan, R. A., 700, 953
Millikan oil-drop experiment, 700
Mirage, 869
Mirror equation, 819–22
Mirrors:
concave and convex, 819 *ff*
focal length of, 817, 821
plane, 812–15
spherical, 816–22
used in telescope, 856
Missing mass, 1166–68
MKS units (*see* SI units)
mm-Hg (unit), 339
Models, 3
Moderator, 1092–94
Modern physics (*defn*), 1, 916
Modulation, 804
Moduli of elasticity, 309–12
Molar specific heat, 498–99
Mole (*defn*), 456
volume of, for ideal gas, 457
Molecular mass, 446
Molecular rotation, 1038–40
Molecular spectra, 1037–43
Molecular speeds, 466–72
Molecular vibrations, 501, 1040–42
Molecules, 1030–43
bonding in, 1030–33
polar, 1032–33
P E diagrams for, 1033–36
spectra of, 1037–43
Moment arm, 247
Moment of a force, 247
Moment of inertia, 249–56, 374
Momentum, 206 *ff*, 936–38, 941
angular, 256–60, 281–89, 965

Total mechanical energy, 183
Total reaction cross section, 1089
Townsend, J. S., 700
Tracers, 1104
Transducers, 676–77
Transformations:
  Galilean, 932–36
  Lorentz, 932–36
Transformer, 744–47, 758
Transformer equation, 745
Transient ischemic attack (TIA), 349
Transistors, 1054–55
Transition elements, 1013
Transition temperature, 650
Translational motion, 16–238, 262–65
Transmission coefficient, 997
Transmission electron microscope, 961
Transmission grating, 900 *ff*
Transmission lines, 746–47, 782
Transmission of electricity, 746–47, 782
Transmutation of elements, 1068, 1070, 1085–88
Transverse waves, 391 *ff*, 794, 907
Trap, sink, 349
Triangulation, 11, 1144 *fn*, 1153
Trigonometric functions and identities, 49, A-2–A-3, inside back cover
Triple point, 460, 474
Tritium, 1084 *pr*, 1097–99
Trusses, 315–18
Tube, vibrating column of air in, 425–28
Tubes, flow in, 351
Tunnel diode, 998–99
Tunneling through a barrier, 996–99, 1069–70, 1155 *fn*, 1168
Turbine, 741
Turbulent flow, 343
Turning points, 197
Turn signal, automobile, 673
Tweeter, 776
Twin paradox, 926–29
Tycho Brahe, 143

U

UHF, 804–5
Ultimate speed, 938
Ultimate strength, 309, 312–13
Ultrasonic waves, 418, 437

Ultrasound, 437
Ultrasound imaging, 437
Ultraviolet (UV) light, 799, 825
Unavailability of energy, 534–35
Uncertainty (in measurements), 4, 981
Uncertainty principle, 981–84, 996, 1031
Underdamping, 376
Underexposure, 848
Unification scale, 1134
Unified atomic mass unit, 7, 446, 1063
Uniform circular motion, 63–65, 114 *ff*, 371
  dynamics of, 114–17
  kinematics of, 63–65
Uniformly accelerated motion, 26 *ff*, 54–55
Uniformly accelerated rotational motion, 243–44
Unit conversion, 8–9, inside front cover
Units of measurement, 6–8
Unit vectors, 52–53
Universal gas constant, 457
Universal law of gravitation, 133–37 and *ff*
Universe (*see also* Cosmology):
  age of, 1158–59
  Big Bang theory of, 1136, 1159–64
  curvature of, 149, 1151–55, 1165
  expanding, 1156–59
  future of, 1165–69
  inflation scenario of, 1164
  matter-dominater, 1161, 1164
  open or closed, 1151, 1165–69
  pulsating, 1168
  radiation-dominated, 1161, 1163
  standard model of, 1161–64
  steady state model of, 1159
Unpolarized light (*defn*), 907
Unstable equilibrium, 198, 308
Uranium, 1077, 1079, 1090–95
Uranus, 146
Useful magnification, 899, 961
UV light, 799, 825

V

Vacuum pump, 353
Vacuum tube, 803
Valence, 1013
Valence band, 1049–50

Van de Graaff generator and accelerator, 612 *pr*, 1115
van der Waals, J. D., 477
van der Waals bonds and forces, 1036–37
van der Waals equation of state, 477–78
Vapor, 473 (*see also* Gases)
Vaporization, latent heat of, 491
Vapor pressure, 474–76
Variable-mass systems, 227–29
Vector displacement, 46–47, 53
Vector model (atoms) 1028 *pr*
Vector product, 279–80
Vectors, 17, 46–53, 159–61, 279–80, A-3
  addition of, 46–52
  components of, 48–52
  cross product, 279–80
  multiplication of, 48, 159–61, 279–80
  multiplication by a scalar, 48
  position, 53, 55, 240, 243–44
  resolution of, 48–52
  resultant, 47, 50
  scalar (dot) product, 159–61
  subtraction of, 48
  sum, 46–52
  unit, 52–53
  vector (cross) product, 279–80
Velocity, 18–22 and *ff*, 36–37, 45, 54–55
  addition of, 66–68
  angular, 241–44
  average, 18–19, 21, 54
  drift, 647–49, 701
  of EM waves, 794–97
  escape, 192–94
  gradient, 350
  instantaneous, 20–22, 26–27, 54–55
  of light, 6, 797, 811–12, 867–68, 919–23, 938
  molecular, and relation to temperature, 466–72
  phase, 397
  relative, 66–68
  rms, 469
  of sound, 418
  supersonic, 418, 435–36
  terminal, 32 *fn*, 122–23
  of waves, 390–94, 397, 418
Velocity selector, 695
Venturi meter, 348
Venturi tube, 348

Photo Credits

CO-1 NOAA/Phil Degginger/Color-Pic, Inc. 1-1a Philip H. Coblentz/Tony Stone Images 1-1b Richard Berenholtz/The Stock Market 1-1c Antranig M. Ouzoonian, P.E./Weidlinger Associates, Inc. 1-2 Mary Teresa Giancoli 1-3a Oliver Meckes/E.O.S./MPI-Tubingen/Photo Researchers, Inc. 1-3b Douglas C. Giancoli 1-4 International Bureau of Weights and Measures, Sevres, France 1-5 Douglas C. Giancoli 1-6 Doug Martin/Photo Researchers, Inc. 1-9 David Parker/Science Photo Library/Photo Researchers, Inc. 1-10 The Image Works CO-2 Joe Brake Photography/Crest Communications, Inc. 2-22 Justus Sustermans, painting of Galileo Galilei/The Granger Collection 2-23 Photograph by Dr. Harold E. Edgerton, © The Harold E. Edgerton 1992 Trust. Courtesy Palm Press, Inc. CO-3 Michel Hans/Vandystadt Allsport Photography (USA), Inc 3-19 Berenice Abbott/Commerce Graphics Ltd., Inc. 3-21 Richard Megna/Fundamental Photographs 3-30a Don Farrall/PhotoDisc, Inc. 3-30b Robert Frerck/Tony Stone Images 3-30c Richard Megna/Fundamental Photographs CO-4 Mark Wagner/Tony Stone Images 4-1 AP/Wide World Photos 4-3 Central Scientific Company 4-5 Sir Godfrey Kneller, Sir Isaac Newton, 1702. Oil on canvass. The Granger Collection 4-6 Gerard Vandystadt/Agence Vandystadt/Photo Researchers, Inc. 4-8 David Jones/Photo Researchers, Inc. 4-11 Tsado/NASA/Tom Stack & Associates 4-31 Lars Ternblad/The Image Bank 4-34 Kathleen Schiaparelli 4-36 Jeff Greenberg/Photo Researchers, Inc. CO-5a Jess Stock/Tony Stone Images CO-5b Werner H. Muller/Peter Arnold, Inc. 5-12 Jay Brousseau/The Image Bank 5-17 Guido Alberto Rossi/The Image Bank 5-33 C. Grzimek/Okapia/Photo Researchers, Inc. 5-34 Photofest 5-38 Cedar Point Photo by Dan Feicht CO-6 Earth Imaging/Tony Stone Images 6-9 NASA/Johnson Space Center 6-13 I. M. House/Tony Stone Images 6-14L Jon Feingersh/The Stock Market 6-14M © Johan Elbers 1995 6-14R Peter Grumann/The Image Bank CO-7 Eric Miller/Reuters/Corbis 7-20a Stanford Linear Accelerator Center/Science Photo Library/Photo Researchers, Inc. 7-20b Account Phototake/Phototake NYC 7-24 Official U.S. Navy Photo CO-8 and 8-10 Photograph by Dr. Harold E. Edgerton, © The Harold E. Edgerton 1992 Trust. Courtesy Palm Press 8-11 David Madison/David Madison Photography 8-16 AP/Wide World Photos 8-23 M. C. Escher's "Waterfall" © 1996 Cordon Art-Baarn-Holland. All rights reserved. CO-9 Richard Megna/Fundamental Photographs 9-9 Photograph by Dr. Harold E. Edgerton, © The Harold E. Edgerton 1992 Trust. Courtesy Palm Press 9-16 D.J. Johnson 9-20 Courtesy Brookhaven National Laboratory 9-22 Berenice Abbott/Photo Researchers, Inc. CO-10 Ch. Russeil/Kipa/Sygma Photo News 10-10b Mary Teresa Giancoli 10-14a Richard Megna/Fundamental Photographs 10-14b Photoquest, Inc. 10-47 Jens Hartmann/AP/Wide World Photos 10-48 Focus on Sports, Inc. 10-49 Karl Weatherly/PhotoDisc, Inc CO-11 AP/Wide World Photos 11-19c NOAA/Phil Degginger/Color-Pic, Inc. 11-19d NASA/TSADO/Tom Stack & Associates 11-34 Michael Kevin Daly/The Stock Market CO-12 Steve Vidler/Leo de Wys, Inc. 12-2 AP/Wide World Photos 12-7 T. Kitchin/Tom Stack & Associates 12-10 The Stock Market 12-21 Douglas C. Giancoli 12-23 Mary Teresa Giancoli 12-27 Fabricius & Taylor/Liaison Agency, Inc. 12-36 Henryk T. Kaiser/Leo de Wys, Inc. 12-38 Douglas C. Giancoli 12-39 Galen Rowell/Mountain Light Photography, Inc. 12-41 Douglas C. Giancoli 12-43 Giovanni Paolo Panini (Roman, 1691-1765), Interior of the Pantheon, Rome c. 1734. Oil on canvas. 1.283 × .991 (50 1/2 × 39); framed: 1.441 × 1.143 (56 3/4 × 45). © 1995 Board of Trustees, National Gallery of Art, Washington. Samuel H. Kress Collection. Photo by Richard Carafelli. 12-44 Robert Holmes/Corbis 12-45 Italian Government Tourist Board CO-13 Steven Frink/Tony Stone Images 13-12 Corbis 13-20 Department of Mechanical and Aerospace Engineering, Princeton University 13-31 Rod Planck/Thomas Stack & Associates 13-33 Alan Blank/Bruce Coleman Inc. 13-42 Douglas C. Giancoli 13-45 Galen Rowell/Mountain Light Photography, Inc. 13-51 NASA/Goddard Space Flight Center/Science Source/Photo Researchers, Inc. CO-14 Richard Megna/Fundamental Photographs 14-4 Mark E. Gibson/Visuals Unlimited 14-14 Fundamental Photographs 14-15 Douglas C. Giancoli 14-22 Taylor Devices, Inc. 14-25 Martin Bough/Fundamental Photographs 14-26a AP/Wide World Photos 14-26b Paul X. Scott/Sygma Photo News CO-15 Richard Megna/Fundamental Photographs 15-1 Douglas C. Giancoli 15-24 Douglas C. Giancoli 15-30 Martin G. Miller/Visuals Unlimited 15-32 Richard Megna/Fundamental Photographs CO-16 Photographic Archives, Teatro alla Scala, Milan, Italy 16-6 Yoav Levy/Phototake NYC 16-9a David Pollack/The Stock Market 16-9b Andrea Brizzi/The Stock Market 16-10 Bob Daemmrich/The Image Works 16-13 Bildarchiv Foto Marburg/Art Resource, N.Y. 16-24a Norman Owen Tomalin/Bruce Coleman Inc. 16-25b Sandia National Laboratories, New Mexico 16-28a P. Saada/Eurelios/Science Photo Library/Photo Researchers, Inc. 16-28b Howard Sochurek/Medical Images Inc. CO-17 Le Matin de Lausanne/Sygma Photo News 17-3 Bob Daemmrich/Stock Boston 17-5 Leonard Lessin/Peter Arnold, Inc. 17-14 Leonard Lessin/Peter Arnold, Inc. 17-17 Brian Yarvin/Photo Researchers, Inc. CO-18 Tom Till/DRK Photo 18-9 Paul Silverman/Fundamental Photographs 18-14 Mary Teresa Giancoli CO-19 Tom Bean/DRK Photo 19-25 Science Photo Library/Photo Researchers, Inc. CO-20a David Woodfall/Tony Stone Images CO-20b AP/Wide World Photos 20-7a Sandia National Laboratories, New Mexico 20-7b Martin Bond/Science Photo Library/Photo Researchers, Inc 20-7c Lionel Delevingne/Stock Boston 20-13 Leonard Lessin/Peter Arnold, Inc. CO-21 Fundamental Photographs 21-38 Michael J. Lutch/Boston Museum of Science CO-23 Gene Moore/Phototake NYC 23-21 Jon Feingersh/The Stock Market CO-24 Paul Silverman/Fundamental Photographs CO-25 Mahaux Photography/The Image Bank 25-1 J.-L. Charmet/Science Photo Library/Photo Researchers, Inc. 25-10 T.J. Florian/Rainbow 25-13 Richard Megna/Fundamental Photographs 25-16 Barbara Filet/Tony Stone Images 25-26 Takeshi Takahara/Photo Researchers, Inc. 25-33 Liaison Agency, Inc. CO-26a Steve Weinrebe/Stock Boston CO-26b Sony Electronics, Inc. 26-22 Paul Silverman/Fundamental Photographs 26-26 Paul Silverman/Fundamental Photographs CO-27 Richard Megna/Fundamental Photographs 27-4 Richard Megna/Fundamental Photographs 27-6 Mary Teresa Giancoli 27-8 Richard Megna/Fundamental Photographs 27-19 Pekka Parviainen/Science Photo Library/Photo Researchers, Inc. CO-28 Manfred Kage/Peter Arnold, Inc. 28-21 Richard Megna/Fundamental Photographs CO-29 Richard Megna/Fundamental Photographs 29-10 Werner H. Muller/Peter Arnold, Inc.

29-15 Tomas D.W. Friedmann/Photo Researchers, Inc. **29-20** Jon Feingersh/Comstock **29-28b** National Earthquake Information Center, U.S.G.S. **CO-30** Adam Hart-Davis/Science Photo Library/Photo Researchers, Inc. **CO-31v1** Albert J. Copley/Visuals Unlimited **CO-31v2** Mary Teresa Giancoli **CO-32** Richard Megna/Fundamental Photographs **32-1** AIP Emilio Segre' Visual Archives **32-14** The Image Works **CO-33** Douglas C. Giancoli **33-6** Douglas C. Giancoli **33-11a** Mary Teresa Giancoli and Suzanne Saylor **33-11b** Mary Teresa Giancoli **33-21** Mary Teresa Giancoli **33-25** David Parker/Science Photo Library/Photo Researchers, Inc. **33-28b** Michael Giannechini/Photo Researchers, Inc. **33-34b** S. Elleringmann/Bilderberg/Aurora & Quanta Productions **33-40** Douglas C. Giancoli **33-42** Mary Teresa Giancoli **CO-34** Mary Teresa Giancoli **34-1c** Douglas C. Giancoli **34-1d** Douglas C. Giancoli **34-2** Douglas C. Giancoli and Howard Shugat **34-4** Douglas C. Giancoli **34-7a** Douglas C. Giancoli **34-7b** Douglas C. Giancoli **34-18** Mary Teresa Giancoli **34-19a** Mary Teresa Giancoli **34-19b** Mary Teresa Giancoli **34-26** Mary Teresa Giancoli **34-30a** Franca Principe/Istituto e Museo di Storia della Scienza, Florence, Italy **34-32** Yerkes Observatory, University of Chicago **34-33c** Palomar Observatory/California Institute of Technology **34-33d** Joe McNally/Joe McNally Photography **34-35b** Olympus America Inc. **CO-35** Larry Mulvehill/Photo Researchers, Inc. **35-4a** John M. Dunay IV/Fundamental Photographs **35-9a** Bausch & Lomb Incorporated **35-16a** Paul Silverman/Fundamental Photographs **35-16b** Richard Megna/Fundamental Photographs **35-16c** Yoav Levy/Phototake NYC **35-18b** Ken Kay/Fundamental Photographs **35-20b** Bausch & Lomb Incorporated **35-20c** Bausch & Lomb Incorporated **35-22** Kristen Brochmann/Fundamental Photographs **CO-36** Richard Megna/Fundamental Photographs **36-2a** Reprinted with permission from P.M. Rinard, American Journal of Physics, Vol. 44, #1, 1976, p. 70. Copyright 1976 American Association of Physics Teachers. **36-2b** Ken Kay/Fundamental Photographs **36-2c** Ken Kay/Fundamental Photographs **36-11a** Richard Megna/Fundamental Photographs **36-11b** Richard Megna/Fundamental Photographs **36-12a and b** Reproduced by permission from M. Cagnet, M. Francon, and J. Thrier, The Atlas of Optical Phenomena. Berlin: Springer-Verlag, 1962. **36-15** Space Telescope Science Institute **36-16** The Arecibo Observatory is part of the National Astronomy and Ionosphere Center which is operated by Cornell University under a cooperative agreement with the National Science Foundation. **36-21** Wabash Instrument Corp./Fundamental Photographs **36-26** Photo by W. Friedrich/Max von Laue. Burndy Library, Dibner Institute for the History of Science and Technology, Cambridge, Massachusetts. **36-29b** Bausch & Lomb Incorporated **36-36** Diane Schiumo/Fundamental Photographs **36-39a** Douglas C. Giancoli **36-39b** Douglas C. Giancoli **CO-37** Image of Albert Einstein licensed by Einstein Archives, Hebrew University, Jerusalem, represented by Roger Richman Agency, Beverly Hills, California **37-1** AIP Emilio Segre Visual Archives **CO-38** Wabash Instrument Corp./Fundamental Photographs **38-10** Photo by S.A. Goudsmit, AIP Emilio Segre' Visual Archives **38-11** Education Development Center, Inc. **38-20b** Richard Megna/Fundamental Photographs **38-21abc** Wabash Instrument Corp./Fundamental Photographs **CO-39** Institut International de Physique/American Institute of Physics/Emilio Segre Visual Archives **39-01** American Institute of Physics/Emilio Segre Visual Archives **39-02** F.D. Rosetti/American Institute of Physics/Emilio Segre Visual Archives **39-04** Advanced Research Laboratory/Hitachi Metals America, Ltd. **39-18** Driscoll, Youngquist, and Baldeschwieler, Caltech/Science Photo Library/Photo Researchers, Inc. **CO-40** Patricia Peticolas/Fundamental Photographs **40-16** Paul Silverman/Fundamental Photographs **40-22** Yoav Levy/Phototake NYC **40-24b** Paul Silverman/Fundamental Photographs **CO-41** Charles O'Rear/Corbis **CO-42** Reuters Newmedia Inc/Corbis **42-03** Chemical Heritage Foundation **42-07** University of Chicago, Courtesy of AIP Emilio Segre Visual Archives **CO-43** AP/Wide World Photos **43-07** Gary Sheahan "Birth of the Atomic Age" Chicago (Illinois); 1957. Chicago Historical Society. **43-10** Liaison Agency, Inc. **43-11** LeRoy N. Sanchez/Los Alamos National Laboratory **43-12** Corbis **43-16a** Lawrence Livermore National Laboratory/Science Source/Photo Researchers, Inc. **43-16b** Gary Stone/Lawrence Livermore National Laboratory **43-22a** Martin M. Rotker/Martin M. Rotker **43-22b** Simon Fraser/Science Photo Library/Photo Researchers, Inc. **43-26b** Southern Illinois University/Peter Arnold, Inc. **43-28** Mehau Kulyk/Science Photo Library/Photo Researchers, Inc. **CO-44** Fermilab Visual Media Services **44-06** CERN/Science Photo Library/Photo Researchers, Inc. **44-08** Fermilab Visual Media Services **CO-45** Jeff Hester and Paul Scowen, Arizona State University, and NASA **45-01** NASA Headquarters **45-02c** NASA/Johnson Space Center **45-03** U.S. Naval Observatory Photo/NASA Headquarters **45-04** National Optical Astronomy Observatories **45-05a** R.J. Dufour, Rice University **45-05b** U.S. Naval Observatory **45-05c** National Optical Astronomy Observatories **45-10** Space Telescope Science Institute/NASA Headquarters **45-11** National Optical Astronomy Observatories **45-15** European Space Agency/NASA Headquarters **45-22** Courtesy of Lucent Technologies/Bell Laboratories

Table of Contents Photos p. v (left) NOAA/Phil Degginger/Color-Pic, Inc. **p. v** (right) Mark Wagner/Tony Stone Images **p. vi** (left) Jess Stock/Tony Stone Images **p. vi** (right) Photograph by Dr. Harold E. Edgerton, © The Harold E. Edgerton 1992 Trust. Courtesy Palm Press **p. vii** (left) Ch. Russeil/Kipa/Sygma Photo News **p.vii** (right) Tibor Bognar/The Stock Market **p. viii** (left) Richard Megna/Fundamental Photographs **p. viii** (right) Douglas C. Giancoli **p. ix** (top) S. Feval/Le Matin de Lausanne/Sygma Photo News **p. ix** (bottom) Richard A. Cooke III/Tony Stone Images (right) David Woodfall/Tony Stone Images and AP/ Wide World Photos **p. x** (left) Fundamental Photographs **p. x** (right) Richard Kaylin/Tony Stone Images **p. xi** (left) Mahaux Photography/The Image Bank **p. xi** (right) Manfred Cage/Peter Arnold, Inc. **p. xii** Werner H. Muller/Peter Arnold, Inc. **p. xiii** (left) Mary Teresa Giancoli. **p. xiii** (right) Larry Mulvehill/Photo Researchers, Inc. **p. xiv** Image of Albert Einstein licensed by Einstein Archives, Hebrew University, Jerusalem, represented by Roger Richman Agency, Beverly Hills, California **p. xiv** (right) Donna McWilliam/AP/Wide World Photos **p. xv** (left) Charles O'Rear/Corbis (right) Fermilab **p. xvi** (left) Jeff Hester and Paul Scowen, Arizona State University, and NASA (right) R. J. Dufour, Rice University.